Current Topics in Behavioral Neurosciences

Volume 71

Current Topics in Behavioral Neurosciences provides critical and comprehensive discussions of the most significant areas of behavioral neuroscience research, written by leading international authorities. Each volume in the series represents the most informative and contemporary account of its subject available, making it an unrivalled reference source. Each volume will be made available in both print and electronic form.

With the development of new methodologies for brain imaging, genetic and genomic analyses, molecular engineering of mutant animals, novel routes for drug delivery, and sophisticated cross-species behavioral assessments, it is now possible to study behavior relevant to psychiatric and neurological diseases and disorders on the physiological level. The Behavioral Neurosciences series focuses on translational medicine and cutting-edge technologies. Preclinical and clinical trials for the development of new diagnostics and therapeutics as well as prevention efforts are covered whenever possible. Special attention is also drawn on epigenetical aspects, especially in psychiatric disorders.

CTBN series is indexed in PubMed and Scopus.

Wolfgang H. Sommer • Rainer Spanagel
Editors

Behavioral Neuroscience of Alcohol Addiction

Basic Mechanisms and Animal Studies

Springer

Editors
Wolfgang H. Sommer
Institute of Psychopharmacology
Central Institute of Mental Health
German Center for Mental Health (DZPG)
University of Heidelberg
Mannheim, Germany

Rainer Spanagel
Institute of Psychopharmacology
Central Institute of Mental Health
German Center for Mental Health (DZPG)
University of Heidelberg
Mannheim, Germany

ISSN 1866-3370 ISSN 1866-3389 (electronic)
Current Topics in Behavioral Neurosciences
ISBN 978-3-032-01580-8 ISBN 978-3-032-01581-5 (eBook)
https://doi.org/10.1007/978-3-032-01581-5

This Springer imprint is published by the registered company Springer Nature Switzerland AG
The registered company address is: Gewerbestrasse 11, 6330 Cham, Switzerland

Preface

A dozen years has passed since the first edition of *Behavioral Neurobiology of Alcohol Addiction* was published. At the time, our primary goal was to provide a comprehensive resource that served as a touchstone for researchers and a catalyst for reflection in the field of alcohol addiction. The perspectives of eminent scholars, coupled with a broad synthesis of knowledge, created a volume that was widely embraced—not only for its scholarly impact, as evidenced by numerous citations and downloads, but also for its utility in teaching and training future scientists. Yet, the field has since undergone major shifts that demand a fresh appraisal of where we stand and where we are headed.

We have sought to capture the evolving landscape of alcohol addiction research by curating a collection of Current Topics that exemplify recent methodological and conceptual advancements. These topics have not only substantially contributed to the field but have also profoundly influenced the way we think about alcohol addiction and its underlying mechanisms.

With this guiding principle, we have selected four Current Topics in Behavioral Neurobiology of Alcohol Addiction, which form the thematic backbone of this new edition. The current edition of *Behavioral Neurobiology of Alcohol Addiction* is organized into two volumes and four parts. The first volume, titled *Basic Mechanisms and Animal Studies*, focuses on foundational insights into the conceptualization of addiction and the neurobiological mechanisms that sustain it. The second volume, *Translational Studies and Human Phenotypes*, shifts focus to the interface of human and translational research.

In Volume 1, Part I begins with a chapter where we briefly sketch the current landscape of alcohol research (Sommer and Spanagel 2025). Then, Koob and Vendruscolo (2023) provide a comprehensive synthesis of the predominant theoretical framework. They delve into how reward deficits, the overactivation of stress systems, and chronic pain contribute to the persistence of alcohol addiction.

This theoretical overview is complemented by a series of chapters that explore various animal models of AUD, highlighting their strengths and limitations for uncovering mechanisms and guiding medication development. Becker and Lopez

(2024) provide a comprehensive review of five rodent models that are commonly used in alcohol research. Hitzemann et al. (2023) expand on this by showcasing the application of two such models to investigate brain gene expression, demonstrating how these findings contribute to the interpretation of human AUD-related genome-wide and transcriptome-wide association studies. Additionally, Scholz (2023) highlights the potential of invertebrate models in AUD research, underscoring their unique advantages, particularly for the efficient screening of novel therapeutic compounds.

The section concludes with a critical contribution from Meinhardt et al. (2024), who address the replication crisis that has challenged the reliability and validity of preclinical research. Their chapter not only scrutinizes the underlying causes of this crisis but extents our previous viewpoints by offering a set of practical guidelines aimed at improving research practices, ensuring greater reproducibility, and fostering confidence in preclinical findings.

Part II delves deeper into the specific mechanisms underpinning the actions of alcohol, providing a comprehensive exploration of its effects at multiple levels of biological organization. This section starts with Quintanilla and Israel (2023) focusing on the role metabolic pathways in alcohol preference, addiction, and treatment. The cellular and synaptic consequences of alcohol exposure are broadly discussed by Lovinger and Roberto (2023). This is followed by the chapter of Barbier et al. deliberating alcohol's effect on epigenetic mechanisms (Domi et al. 2023). A highlight of this part is the focus on the rise of neuronal population and circuit-based approaches, exemplified by chapters on manipulating the oxytocin system and dissecting the reward circuitry by Schimmer et al. (2023) and Doyle et al. (2023), respectively. These studies demonstrate the power of contemporary tools to refine our understanding of addiction's neural substrates. This part concludes with a forward-looking chapter by Lapish (2024) on computational methods for decoding the encoding of alcohol-related behaviors, illustrating the growing integration of data-driven approaches in addiction research.

Volume 2 starts with Part III that addresses the perennial challenge of modeling relapse in controlled laboratory settings. In their chapter, Milivojevic and Sinha discuss innovative strategies to replicate the conditions leading to relapse in human laboratory experiments (Milivojevic and Sinha 2023). Contrasting these approaches, Reichert et al. (2024) present the development and application of ecological momentary assessment (EMA), a methodology that provides real-time data on alcohol use and relapse risks in everyday life.

Bach et al. (2023) offer a compelling review of advances in molecular probes for positron emission tomography (PET), enabling precise imaging of alcohol-related phenotypes. Following this, Beck et al. (2023) synthesize findings from human PET and other neuroimaging studies, particularly focusing on dopamine's role in alcohol use and addiction. Their analysis reflects on both the coherence and the notable disparities between findings from animal and human studies, emphasizing the challenges of translating preclinical insights into human contexts.

This critical theme is further expanded by Crombag et al. (2024), who scrutinize the traditional reliance on "face validity" in translational research. They advocate for a paradigm shift, emphasizing the importance of identifying intermediate human behavioral phenotypes that capture discrete aspects of AUD. Such phenotypes, they argue, should serve as anchors for developing homologous animal models grounded in shared psychological and neurobiological processes. This part concludes with Sommer and Canals (2025) presenting examples of objectively translatable neuroimaging phenotypes based on diffusion tensor imaging (DTI). They discuss the implications of these phenotypes for studying AUD populations and highlight their potential for bridging the gap between animal and human research.

Part IV shifts the focus to novel concepts and advancements in the treatment of AUD. This section highlights innovative approaches aimed at reshaping therapeutic strategies. Three chapters delve into distinct learning mechanisms that hold promise for improving treatment outcomes. These include techniques such as memory retrieval and reconsolidation employed by the Kiefer and Barak Labs (Bach and Kiefer 2023; Barak and Goltseker 2023), and approach bias retraining discussed by Wiers et al. (2023), all of which leverage psychological constructs to enhance the efficacy of AUD psychotherapy. These contributions underscore the potential of targeting specific cognitive and behavioral processes to support long-term recovery.

An expert review by Ygael and Zangen (2024) provides a comprehensive examination of transcranial magnetic stimulation (TMS), an emerging neuromodulatory technique with growing evidence for its effectiveness in treating AUD and other addictions. The review offers insights into the mechanisms underlying TMS, its current clinical applications, and future directions for research in this area.

Leclercq and de Timary (2024) conclude the volume by exploring the critical role of inflammatory mechanisms in AUD. This includes a detailed discussion of the microbiome, the gut-brain axis, and the systemic inflammatory response as contributors to alcohol use disorder. These mechanisms are not only central to understanding the physiological consequences of chronic alcohol consumption but also represent promising targets for novel treatment interventions.

With this new edition of *Behavioral Neurobiology of Alcohol Addiction*, we aim to share our enthusiasm for the field of alcohol addiction research with a diverse and broad readership. Despite the challenges the field has faced, today's alcohol research generates more excitement and possibilities than ever before, offering innovative insights that continue to reshape our understanding of this complex disorder. We hope that this work not only serves as a comprehensive reference for seasoned scholars but also inspires a new generation of students and researchers to engage with the fascinating questions surrounding addiction.

Mannheim, Germany

Wolfgang H. Sommer
Rainer Spanagel

References

Bach P, Kiefer F (2023) New Approaches to addiction treatment based on learning and memory. Curr Top Behav Neurosci. https://doi.org/10.1007/7854_2023_457. Epub ahead of print. PMID: 38082110

Bach P, de Timary P, Gründer G, Cumming P (2023) Molecular imaging studies of alcohol use disorder. Curr Top Behav Neurosci. https://doi.org/10.1007/7854_2022_414. Epub ahead of print. PMID: 36639552

Barak S, Goltseker K (2023) New approaches for alcohol use disorder treatment via memory retrieval and reconsolidation manipulations. Curr Top Behav Neurosci. https://doi.org/10.1007/7854_2022_411. Epub ahead of print. PMID: 36627475

Beck A, Ebrahimi C, Rosenthal A, Charlet K, Heinz A (2023) The dopamine system in mediating alcohol effects in humans. Curr Top Behav Neurosci. https://doi.org/10.1007/7854_2022_415. Epub ahead of print. PMID: 36705911

Becker HC, Lopez MF (2024) Animal models of excessive alcohol consumption in rodents. Curr Top Behav Neurosci. https://doi.org/10.1007/7854_2024_461. Epub ahead of print. PMID: 38340255

Crombag HS, Duka T, Stephens DN (2024) The continuing challenges of studying parallel behaviours in humans and animal models. Curr Top Behav Neurosci. https://doi.org/10.1007/7854_2024_485. Epub ahead of print. PMID: 38976140

Domi E, Barchiesi R, Barbier E (2023) Epigenetic dysregulation in alcohol-associated behaviors: preclinical and clinical evidence. Curr Top Behav Neurosci. https://doi.org/10.1007/7854_2022_410. Epub ahead of print. PMID: 36717533

Doyle MA, Taylor A, Winder DG (2023) Neural circuitries and alcohol use disorder: cutting corners in the cycle. Curr Top Behav Neurosci. https://doi.org/10.1007/7854_2023_454. Epub ahead of print. PMID: 38082108

Hitzemann R, Ozburn AR, Lockwood D, Phillips TJ (2023) Modeling brain gene expression in alcohol use disorder with genetic animal models. Curr Top Behav Neurosci. https://doi.org/10.1007/7854_2023_455. Epub ahead of print. PMID: 37982929; PMCID: PMC11566292

Koob GF, Vendruscolo L (2023) Theoretical frameworks and mechanistic aspects of alcohol addiction: alcohol addiction as a reward deficit/stress surfeit disorder. Curr Top Behav Neurosci. https://doi.org/10.1007/7854_2023_424. Epub ahead of print. PMID: 37421551

Lapish CC (2024) Understanding how acute alcohol impacts neural encoding in the rodent brain. Curr Top Behav Neurosci. https://doi.org/10.1007/7854_2024_479. Epub ahead of print. PMID: 38858298

Leclercq S, de Timary P (2024) Role of the microbiome and the gut-brain axis in alcohol use disorder: potential implication for treatment development. Curr Top Behav Neurosci. https://doi.org/10.1007/7854_2024_478. Epub ahead of print. PMID: 38914878

Lovinger DM, Roberto M (2023) Synaptic effects induced by alcohol. Curr Top Behav Neurosci. https://doi.org/10.1007/7854_2022_412. Epub ahead of print. PMID: 36765015; PMCID: PMC11104446

Meinhardt MW, Gerlach B, Spanagel R (2024) Good practice guideline for preclinical alcohol research: the stringency framework. Curr Top Behav Neurosci. https://doi.org/10.1007/7854_2024_484. Epub ahead of print. PMID: 39117860

Milivojevic V, Sinha R (2023) Laboratory and real-world experimental approaches to understanding alcohol relapse. Curr Top Behav Neurosci. https://doi.org/10.1007/7854_2023_456. Epub ahead of print. PMID: 37985542

Quintanilla ME, Israel Y (2023) Role of metabolism on alcohol preference, addiction, and treatment. Curr Top Behav Neurosci. https://doi.org/10.1007/7854_2023_422. Epub ahead of print. PMID: 37221350

Reichert D, Brüßler S, Reichert M, Ebner-Priemer U (2024) Understanding alcohol consumption and its antecedents and consequences in daily life: the why and the how. Curr Top Behav Neurosci. https://doi.org/10.1007/7854_2024_486. Epub ahead of print. PMID: 39080235

Schimmer J, Patwell R, Küppers S, Grinevich V (2023) The relationship between oxytocin and alcohol dependence. Curr Top Behav Neurosci. https://doi.org/10.1007/7854_2023_444. Epub ahead of print. PMID: 37697074

Scholz H (2023) From natural behavior to drug screening: invertebrates as models to study mechanisms associated with alcohol use disorders. Curr Top Behav Neurosci. https://doi.org/10.1007/7854_2022_413. Epub ahead of print. PMID: 36598738

Sommer WH, Canals S (2025) Alcohol-induced changes of brain microstructure: uncovering novel pathophysiological mechanisms of AUD using translational DTI in humans and rodents. Curr Top Behav Neurosci. https://doi.org/10.1007/7854_2025_585. Epub ahead of print

Sommer WH, Spanagel R (2025) Behavioral neurobiology of alcohol addiction – a decade of great challenges, new hypes and hopes. Curr Top Behav Neurosci. https://doi.org/10.1007/7854_2025_585. Epub ahead of print

Wiers RW, Pan T, van Dessel P, Rinck M, Lindenmeyer J (2023) Approach-bias retraining and other training interventions as add-on in the treatment of AUD patients. Curr Top Behav Neurosci. https://doi.org/10.1007/7854_2023_421. Epub ahead of print. PMID: 37221351

Ygael N, Zangen A (2024) Modulation of alcohol use disorder by brain stimulation. Curr Top Behav Neurosci. https://doi.org/10.1007/7854_2024_487. Epub ahead of print. PMID: 39039357

Contents of Volume 71

Contents of Volume 72

Part I
Conceptualizing and Modeling Alcohol Addiction

Behavioral Neurobiology of Alcohol Addiction: A Decade of Great Challenges, New Hopes, and Hypes

Wolfgang H. Sommer and Rainer Spanagel

Contents

Abstract Over a decade after the first edition of "Behavioral Neurobiology of Alcohol Addiction," this chapter revisits the field at a critical juncture, marked by both persistent challenges and emerging opportunities. We reflect on the translational gap that has stalled the development of new treatments for alcohol use disorder (AUD), despite decades of promising preclinical findings. Particular attention is given to the replicability crisis in animal research, publication biases, and the limited predictive validity of existing models. At the same time, we highlight advances that offer renewed hope, including molecular and circuit-level technologies, AI-driven data analysis, real-world assessments, and new pharmacological candidates, such as GLP-1 agonists and psychedelics. These breakthroughs are considered alongside the increasing recognition of inflammation, pain, and neuroimmune factors as integral to AUD. However, we caution against exaggerated claims and urge the field to avoid oversimplified models, especially those that conflate habits and compulsions. Finally, we argue that neurobiological progress must be complemented by public health strategies aimed at reducing stigma and improving access to care. By fostering empirical rigor, embracing complexity, and maintaining critical self-reflection, addiction science can better align its innovations with real-world clinical and societal needs.

W. H. Sommer (✉) and R. Spanagel
Institute of Psychopharmacology, Central Institute of Mental Health, Medical Faculty Mannheim, Heidelberg University, Mannheim, Germany

German Center for Mental Health (DZPG), partner site Mannheim/Heidelberg/Ulm, Mannheim, Germany
e-mail: wolfgang.sommer@zi-mannheim.de; rainer.spanagel@zi-mannheim.de

Curr Topics Behav Neurosci (2025) 71: 3–16
https://doi.org/10.1007/7854_2025_586
Published Online: 12 July 2025

Keywords Addiction theory · Alcohol use disorder · Animal models · DELPHI method · GLP-1 agonists · Just-in-time-Adaptive-Interventions (JITAIs) · Psilocybin · Translation · Valley of death

A dozen years have passed since the first edition of "Behavioral Neurobiology of Alcohol Addiction" was published. The field has since undergone major shifts that demand a fresh appraisal of where we stand and where we are headed. The optimism that accompanied the initial publication was palpable. With the wealth of insights generated by decades of research, many believed that transformative new therapies for alcohol use disorder (AUD) were on the horizon. Yet, in the years since, those hopes have been tempered by sobering realities. Despite extensive efforts and promising preclinical findings, no novel AUD medications have gained approval from the FDA or EMA. This "valley of death" in translational science—where preclinical discoveries fail to materialize into clinical breakthroughs—has cast a shadow not only over addiction research but also over psychiatry as a whole.

In the following section, we examine the complex reasons why many promising drug candidates have failed to make the leap from bench to bedside. Despite substantial progress in human genetics and the rise of omics technologies and sophisticated data analysis tools, personalized treatments for AUD remain elusive. The gap between discovery and application has proven wider than anticipated.

Yet, there are reasons for renewed optimism. Innovations such as GLP-1 receptor agonists (Jerlhag 2025), the FDA's recent qualification of a drug development tool targeting reductions in risk drinking levels,[1] and the emergence of real-time digital interventions, (Liu et al. 2025) all suggest that a new era in treatment development may be within reach. These and other breakthroughs will be discussed in the second section of this chapter. As scientists, we are often called upon to project confidence in future therapies—sometimes prematurely. Still, we believe it is equally important to reflect a realistic but hopeful vision, grounded in the science and driven by the potential for real progress.

1 Reasons for Translational Failures

In our field, translational failures are perhaps best exemplified by the case of CRHR1 antagonists (Shaham and de Wit 2016; Murrough and Charney 2017). CRHR1 antagonists were once considered a promising therapeutic avenue based on robust preclinical data in animal models of stress and addiction (Heilig and Koob 2007; Sommer et al. 2008). Despite rigorous clinical trials with sophisticated methodologies, these efforts failed to yield even a glimmer of efficacy (Schwandt et al. 2016;

[1] https://www.fda.gov/drugs/drug-safety-and-availability/fda-qualifies-drug-development-tool-facilitate-clinical-trial-research-alcohol-use-disorder

Kwako et al. 2015). This disillusionment underscores deeper systemic issues. Beyond specific compounds or drug classes, broader methodological concerns hinder progress—ranging from the replicability crisis and neglect of relevant literature (e.g., (Sillaber et al. 2002; Refojo et al. 2011)) to the heterogeneity of biological systems—all of which have eroded confidence in the predictive power of animal models. Simultaneously, critics of the "brain disease model of addiction" have intensified their objections (Levy 2013), questioning the value of brain research in addressing addiction. These debates have not only shaped the trajectory of our field but also fueled an urgency to reassess our scientific paradigms and strategies for future progress (Heilig et al. 2019; Heilig et al. 2021).

A recent opinion paper outlined a number of factors contributing to replication failures in animal experiments (Spanagel 2022). These issues frequently stem from significant statistical challenges, inadequate methodologies—concerning the precision with which animal research is conducted and documented—and difficulties in generalization. Collectively, these challenges contribute to the replication crisis in animal studies. Several best practices have been proposed to enhance reproducibility: the use of alternatives to p-values (such as effect size estimation and Bayes factors), preregistration to prevent p-hacking and HARKing, and rigorous adherence to the ARRIVE guidelines to ensure methodological transparency and robustness.

Beyond replicability, other critical issues impede successful translation from animal models to human applications. Chief among these is publication bias. A preferential focus on positive results—and the corresponding underreporting of negative or contradictory data—skews the scientific literature. For instance, the observation that CRHR1 receptor knockouts exhibit increased alcohol consumption after stress (Sillaber et al. 2002) contradicts the expected model outcomes, yet such findings (even if prominently published) have often been overlooked. The publication of negative results remains difficult, as these studies usually demand greater methodological rigor and are often met with resistance from peer reviewers and journal editors, who may be difficult to persuade that negative results reflect genuine findings.

Systematic reviews and meta-analyses are increasingly recognized as essential tools in preclinical research. Though historically underutilized, they are now gaining traction, particularly in alcohol research (e.g., (Noori et al. 2014; Friske et al. 2025; Giannone et al. 2024)), where they serve to synthesize findings across studies and mitigate the effects of bias and replication failures.

Another underappreciated obstacle in translational research is tolerance development—the reduction of drug efficacy with repeated use. While prevalent in clinical contexts, tolerance is seldom evaluated in preclinical studies, which typically rely on acute dosing paradigms (Bespalov et al. 2016). Alarmingly, even pharmaceutical companies have often overlooked this factor. Investigational New Drug (IND) programs typically report sub-chronic treatment regimens merely to secure approval for initiating human clinical trials. Consequently, tolerance effects may not become apparent in initial clinical trials, potentially leading to failures in larger randomized controlled trials (RCTs). Notably, this issue has gained greater attention in recent years. The industry is increasingly incorporating chronic oral treatment studies in rodent models during early drug development to better account for tolerance effects.

In addition, the absence of a placebo effect in animal studies introduces another layer of complexity. In human clinical trials, placebo responses can significantly influence outcomes. A meta-regression involving 2000 individuals with AUD revealed placebo abstinence rates between 17% and 37%, depending on severity (Scherrer et al. 2021). These effects are absent in animal studies, potentially leading to an overestimation of treatment efficacy when translating to humans. Lower heterogeneity in animal models further compounds this issue, warranting caution when designing clinical trials based on preclinical data.

In summary, the challenges described above—along with several others detailed in (Levy 2013; Meinhardt et al. 2024)—underscore the structural and methodological hurdles that complicate replication and translation in behavioral neurobiology. However, ongoing initiatives in the scientific community and the open science movement are laying the groundwork for more reliable translational pipelines. One such initiative is the STRINGENCY framework presented in this book (Meinhardt et al. 2024), which provides comprehensive guidelines for preclinical alcohol research. It advocates for systematic reviews and meta-analyses prior to study initiation, careful sample size calculations, preregistration, multisite experimentation, adherence to FAIR data principles, and use of the ARRIVE guidelines for transparent reporting (including null results). Although these measures may not fully eliminate replication issues—given the inherent heterogeneity of biological systems—they offer a more rigorous foundation for translational success.

2 Breakthroughs Inspiring Confidence in Therapeutic Progress

While challenges persist, there are definite signs for optimism. One such area of progress lies in the growing body of convergent evidence from cross-species studies. These studies provide a robust foundation for replicability, as many key behavioral traits and molecular pathways have been preserved through evolution. For instance, transcriptomic and genetic alterations—as well as targeted genetic manipulations—in model organisms ranging from flies to rodents and monkeys have consistently yielded findings that effectively translate to human conditions (Friske et al. 2025; Stacey et al. 2012; Juraeva et al. 2015; Hirth et al. 2016).

Crucially, the past decade has witnessed transformative methodological advancements that are reshaping our understanding of alcohol addiction and informing the development of novel interventions. These innovations offer unprecedented opportunities to explore the complexity of addiction mechanisms at greater depth and resolution. High-throughput sequencing technologies, in particular, have revolutionized our capacity to investigate the genetic, epigenomic, and functional genomic dimensions of addiction. These technologies enable stratification of large datasets for meta-analyses, even across species, to reveal physiological and cellular processes that are altered in AUD (Friske et al. 2025).

Complementing these advances are new biostatistical methods, such as the generation of polygenic risk scores and causal inference through Mendelian randomization. These tools allow researchers to explore the intricate interplay between biological and environmental influences on addiction. For example, they make it possible to investigate the heritable relationships between intelligence, cognitive traits, and educational outcomes in relation to AUD liability (Rosoff et al. 2021; Kendler et al. 2021; Okbay et al. 2022). These investigations may yield complex and potentially uncomfortable insights, prompting important scientific and ethical debates around personalized prevention and treatment approaches, stigma, and social exclusion (Heilig et al. 2016).

This line of inquiry is further enabled by the emergence of deep-phenotyped study cohorts such as the IMAGEN and subsequent environMENTAL cohort, and the ReCoDe cohort (Mascarell Maričić et al. 2020; Spanagel et al. 2024; Desrivières et al. 2014). When combined with daily-life research methodologies like ecological momentary assessment (EMA) tools (Reichert et al. 2024), these cohorts provide a rich resource for understanding the dynamic and context-sensitive nature of addiction behaviors in real-time and in real-world settings.

Complementary advancements in proteomics, metabolomics, and large-scale compound screening are equally transformative (Noori et al. 2017; Noori et al. 2018). The application of artificial intelligence (AI) to big data derived from omics technologies has already provided new insights in the context of antidepressants (Corrivetti et al. 2024) and is expected to reshape our working hypotheses and theoretical frameworks when applied to human and animal models of addiction. Most importantly, in 2020, AlphaFold2, an advanced AI model, was introduced (Jumper et al. 2021). This breakthrough enables the prediction of the three-dimensional structures of nearly all the 200 million known proteins. AlphaFold2 represents a significant leap forward in understanding the atomic-level structures of complex molecules and paves the way for designing more effective small molecules for potential new drug targets. These cutting-edge analytical approaches allow for a more integrated systems-level understanding of addiction, facilitating the identification of novel therapeutic targets and biomarker profiles across species.

Beyond molecular and genetic insights, the field has embraced a suite of powerful tools for manipulating and monitoring specific neuronal populations and circuits. Genetically engineered tools now enable precise interventions, allowing researchers to dissect the intricate neural networks underlying addictive behaviors. As in all other areas of brain research, mapping neural circuits of a specific addiction-relevant behavior using optical tools, in vivo calcium imaging, and chemogenetics is now the standard approach to understanding brain function, sometimes without questioning whether dissection of a small circuit can explain complex behaviors at a systemic level. For bridging the gap between experimental neurobiology and real-world phenomena, systems biology/medicine approaches are emerging that use computational methods, coupled with high-density ecological assessments, for data-driven inferences of functional links and principles (Sommer et al. 2022). These innovations widen our understanding by uncovering an emerging complexity of addiction mechanisms, which may necessitate a shift away from targeting isolated molecular pathways to embracing broader, systems-level interventions.

New conceptual frameworks are also emerging, informed by the evolving understanding of the biological drivers of addiction. For example, chronic pain is increasingly recognized as a significant contributor to addictive behaviors, challenging the traditional narrative of alcohol use as merely a coping mechanism (Bilbao et al. 2018, 2019; Koob and Vendruscolo 2023). Likewise, neuroinflammation—both as a consequence of alcohol consumption and as a potential driver of continued use—has gained attention as a critical area of research (Friske et al. 2025; De Santis et al. 2020; Crews et al. 2024; Sommer and Canals 2025). Given that inflammatory and pain processes are closely interconnected, they jointly influence alcohol-related behaviors (Vozella et al. 2025). Immunomodulatory substances, as well as dietary interventions targeting the gut–brain axis via the microbiome, are increasingly being studied for their effects on alcohol consumption and craving (Meredith et al. 2021; Leclercq and de Timary 2024). Interestingly, the therapeutic potential of neuromodulation techniques, such as transcranial magnetic stimulation (TMS), may also be linked to these processes, as suggested by recent findings showing local recovery of alcohol-induced white matter damage under the area of stimulation (Selim et al. 2024). The integration of these innovative approaches—including AI-supported neuroimaging and behavioral data processing (Joutsa et al. 2022; Rawls et al. 2021)—into conceptual frameworks and clinical practice offers new hope for developing treatments that effectively address the complex interplay between brain, behavior, and environment in addiction.

As already mentioned at the beginning, the potential of incretins represents a significant breakthrough in the future of addiction medicine (Jerlhag 2024, 2025). Among the most promising candidates are the gut-derived peptides glucagon-like peptide-1 (GLP-1) and gastric inhibitory peptide (GIP). Ongoing phase II randomized controlled trials for the GLP-1 agonist semaglutide and the dual GLP-1/GIP receptor agonist tirzepatide are already yielding promising results. Case reports and emerging clinical evidence suggest that these drugs effectively suppress alcohol cravings, reducing alcohol consumption and relapse rates (Lähteenvuo et al. 2025; Hendershot et al. 2025; Klausen et al. 2022). Also, the renaissance of psychedelic research and the use of classic psychedelics such as psilocybin, but also new short-acting psychedelics with and without hallucinogenic properties (Tap 2024; Domanegg et al. 2023; Meinhardt and Sommer 2023), offer a completely new treatment approach with long-lasting efficacy, possibly after just a single application ((Sicignano et al. 2024; Bogenschutz et al. 2022) but see also (Rieser et al. 2025)). While there is optimism that fast-track FDA approval for these novel drugs—including both incretins and psychedelics—could be realized within this decade, the challenge of addressing the massive treatment gap for AUD remains daunting.

3 Separating Innovation from Illusion

The recent developments outlined above represent an exciting frontier in addiction research; however, it is crucial to temper expectations. The history of psychedelics offers a prime example, highlighting the need to avoid succumbing to the hype surrounding them as a "magic bullet." The inherent heterogeneity among patients with AUD will undoubtedly pose challenges, as only a subset of individuals may benefit from these innovative treatments, while others may not respond. Moreover, as psychedelics advance toward medical application and broader availability, the establishment of carefully considered regulatory frameworks will be essential to manage both the medical and nonmedical needs associated with medications that possess broad popular appeal (Andrews et al. 2025).

Another area where exaggeration is common concerns theory building in addiction research. While theoretical frameworks are important for guiding research efforts, the explanatory power of theories concerning biological phenomena is inherently low—far lower, for example, than the explanatory power of theories in physics—which calls for greater modesty and consilience (Epstein 2020). We should therefore exercise caution when promoting theories of addiction as dominant or overarching explanations.

We remain critical of the prominent idea that addiction develops along a continuum from controlled, goal-directed behavior to dysfunctional habits and ultimately to uncontrollable compulsions. Although this concept has inspired substantial research activity, we have repeatedly outlined our reservations (Giannone et al. 2024; Heinz et al. 2024; Ersche 2024). While habits play a role in the lay understanding of addictive behaviors, evidence that they account for a substantial proportion of the variance in the development or maintenance of AUD and other addictions remains limited. From a neurobiological perspective, most decisions are influenced or made by habitual or automatic processes to optimize resource allocation, with the degree of goal-directed control depending on situational complexity. There is thus no singular "bad habit"; rather, substance use may compromise the decision-making system, biasing it toward simpler, resource-efficient processing modes.

Compulsivity, by contrast, is typically defined as persistent behavior despite adverse consequences—clearly distinct from moment-to-moment decision-making processes. Both constructs, while influential, largely stem from animal models and capture distinct behavioral expressions. Crucially, they fall short of adequately representing the clinical phenomena and their underlying neurobiology. The habit theory of addiction represents one of numerous conceptual frameworks within the field (e.g., incentive salience theory, behavioral and cognitive models, etc.). While these theories collectively offer partial insights into the development and persistence of addictive behaviors, the field currently lacks a comprehensive, integrative model that synthesizes social, cognitive, emotional, and motivational dimensions. Moreover, there is a significant gap in incorporating evolutionary and genetic perspectives, including multilevel biological processes such as epigenetic modifications, persistent molecular and synaptic alterations, and the dynamics of cell assemblies and large-scale neural networks. Developing a unified theory of addiction—encompassing all system-level information—may be achievable through a Delphi process, wherein expert opinions are systematically collected and iteratively refined until consensus is reached. Although this endeavor presents a formidable challenge, establishing such an integrative model could fundamentally steer future research trajectories and accelerate advancements in understanding addiction over the coming decade.

Looking ahead, with these reflections we want to encourage the field to integrate emerging innovations with careful empirical validation and clinical reality. By embracing complexity and remaining open to revising dominant models, addiction research can avoid the pitfalls of overpromising and make meaningful progress toward meeting the diverse needs of patients.

4 Summary and Conclusions

Taken together, the reflections presented in this chapter highlight the complex and evolving landscape of alcohol addiction research. Drawing on lessons from past disappointments, emerging breakthroughs, and a healthy skepticism toward oversimplified narratives, we propose a forward-looking framework. The thematic contributions in our new edition of "Behavioral Neurobiology of Alcohol Addiction" reflect significant methodological and conceptual advances that are shaping a new phase of discovery in this area. Yet, while neuroscience provides crucial insights, it captures only one dimension of a deeply multifaceted disorder (Heilig et al. 2016, 2021). Addressing the crisis resulting from untreated AUD requires more than translational research breakthroughs but, more importantly, societal efforts and effective public health measures to reduce stigma, overcome isolation, and improve access to care (Kummetat et al. 2022; Morris et al. 2023). Promising tools—from telehealth to language-based AI systems and just-in-time-adaptive interventions (JITAIs)—may help bridge this gap but must be deployed thoughtfully (Uscher-Pines et al. 2024). As the field moves forward, embracing complexity, rejecting simplistic solutions, and aligning scientific innovation with clinical and societal needs will be essential. We look ahead to the coming decade with hope and critical resolve.

Funding Financial support for this work was provided by the Deutsche Forschungsgemeinschaft (DFG, German Research Foundation) for the TRR 265 (A05; Spanagel et al. 2024) to RS and B08 to WHS. Support was also provided by the German Federal Ministry of Education and Research (BMBF) for "A systems-medicine approach towards district and shared resilience and pathological mechanisms of substance use disorders" (01ZX01909) to RS and WHS. Further support by BMBF (01EW2402) through the ERA-NET Neuron framework project "Inflamed Brain Response in Alcohol Addiction (IBRAA): Understanding the role of neuroinflammation, immunotypes, and gut-brain reactivity for heterogeneity in disease course and therapy response" to WHS and the Ministry for Science, Research and Art of Baden-Wuerttemberg (MWK) for the 3R-Center Rhein-Neckar to RS.

References

Andrews CM, Hall W, Humphreys K, Marsden J (2025) Crafting effective regulatory policies for psychedelics: what can be learned from the case of cannabis? Addiction 120(2):201–206. https://doi.org/10.1111/add.16575. Epub 2024 Jun 6. PMID: 38845381; PMCID: PMC11707317

Bespalov A, Müller R, Relo AL, Hudzik T (2016) Drug tolerance: a known unknown in translational neuroscience. Trends Pharmacol Sci 37(5):364–378. https://doi.org/10.1016/j.tips.2016.01.008. PMID: 26935643

Bilbao A, Falfán-Melgoza C, Leixner S, Becker R, Singaravelu SK, Sack M, Sartorius A, Spanagel R, Weber-Fahr W (2018) Longitudinal structural and functional brain network alterations in a mouse model of neuropathic pain. Neuroscience 387:104–115. https://doi.org/10.1016/j.neuroscience.2018.04.020. Epub 2018 Apr 22. PMID: 29694917

Bilbao A, Leixner S, Wei S, Cantacorps L, Valverde O, Spanagel R (2019) Reduced sensitivity to ethanol and excessive drinking in a mouse model of neuropathic pain. Addict Biol 24(5): 1008–1018. https://doi.org/10.1111/adb.12784. Epub 2019 Jun 25. PMID: 31237390

Bogenschutz MP, Ross S, Bhatt S, Baron T, Forcehimes AA, Laska E, Mennenga SE, O'Donnell K, Owens LT, Podrebarac S, Rotrosen J, Tonigan JS, Worth L (2022) Percentage of heavy drinking days following psilocybin-assisted psychotherapy vs placebo in the treatment of adult patients with alcohol use disorder: a randomized clinical trial. JAMA Psychiatry 79(10):953–962. https://doi.org/10.1001/jamapsychiatry.2022.2982. Erratum in: JAMA Psychiatry. 2022. PMID: 36001306; PMCID: PMC9403854

Corrivetti G, Monaco F, Vignapiano A, Marenna A, Palm K, Fernández-Arroyo S, Frigola-Capell E, Leen V, Ibarrola O, Amil B, Caruson MM, Chiariotti L, Palacios-Ariza MA, Hoekstra PJ, Chiang HY, Floareş A, Fagiolini A, Fasano A (2024) Optimizing and predicting antidepressant efficacy in patients with major depressive disorder using multi-omics analysis and the Opade AI prediction tools. Brain Sci 14(7):658. https://doi.org/10.3390/brainsci14070658. PMID: 39061399; PMCID: PMC11275115

Crews FT, Coleman LG Jr, Macht VA, Vetreno RP (2024) Alcohol, HMGB1, and innate immune signaling in the brain. Alcohol Res 44(1):04. https://doi.org/10.35946/arcr.v44.1.04. PMID: 39135668; PMCID: PMC11318841

De Santis S, Cosa-Linan A, Garcia-Hernandez R, Dmytrenko L, Vargova L, Vorisek I, Stopponi S, Bach P, Kirsch P, Kiefer F, Ciccocioppo R, Sykova E, Moratal D, Sommer WH, Canals S (2020) Chronic alcohol consumption alters extracellular space geometry and transmitter diffusion in the brain. Sci Adv 6:eaba0154. https://doi.org/10.1126/sciadv.aba0154

Desrivières S, Miller A, Mathey CM, Yu X, Chen D, Agunbiade K, Heilmann-Heimbach S, Forstner AJ, Schumann G, Hoffmann P, Nöthen MM on behalf of the environMENTAL consortium (2014) Multi-omics analyses of the environMENTAL project provide insights into mental health and disease. Nat Ment Health 2:1131–1133. https://doi.org/10.1038/s44220-024-00317-y

Domanegg K, Sommer WH, Meinhardt MW (2023) Psychedelic targeting of metabotropic glutamate receptor 2 and its implications for the treatment of alcoholism. Cells 12(6):963. https://doi.org/10.3390/cells12060963. PMID: 36980303; PMCID: PMC10047550

Epstein DH (2020) Let's agree to agree: a comment on Hogarth (2020), with a plea for not-so-competing theories of addiction. Neuropsychopharmacology 45(5):715–716. https://doi.org/10.1038/s41386-020-0618-y. Epub 2020 Jan 22. PMID: 31969695; PMCID: PMC7265294

Ersche KD (2024) Defending and defining compulsive behaviour in addiction. Addict Biol 29(8): e13427. https://doi.org/10.1111/adb.13427. PMID: 39132705

Friske MM, Torrico EC, Haas MJW, Borruto AM, Giannone F, Hade AC, Yu Y, Gao L, Sutherland GT, Hitzemann R, Philips MA, Fei SS, Sommer WH, Mayfield RD, Spanagel R (2025) A systematic review and meta-analysis on the transcriptomic signatures in alcohol use disorder. Mol Psychiatry 30(1):310–326. https://doi.org/10.1038/s41380-024-02719-x. Epub 2024 Sep 6. PMID: 39242950; PMCID: PMC11649567

Giannone F, Ebrahimi C, Endrass T, Hansson AC, Schlagenhauf F, Sommer WH (2024) Bad habits-good goals? Meta-analysis and translation of the habit construct to alcoholism. Transl Psychiatry 14(1):298. https://doi.org/10.1038/s41398-024-02965-1. PMID: 39030169; PMCID: PMC11271507

Heilig M, Koob GF (2007) A key role for corticotropin-releasing factor in alcohol dependence. Trends Neurosci 30(8):399–406. https://doi.org/10.1016/j.tins.2007.06.006

Heilig M, Epstein DH, Nader MA, Shaham Y (2016) Time to connect: bringing social context into addiction neuroscience. Nat Rev Neurosci 17(9):592–599. https://doi.org/10.1038/nrn.2016.67. Epub 2016 Jun 9. PMID: 27277868; PMCID: PMC5523661

Heilig M, Augier E, Pfarr S, Sommer WH (2019) Developing neuroscience-based treatments for alcohol addiction: a matter of choice? Transl Psychiatry 9(1):255. https://doi.org/10.1038/s41398-019-0591-6. PMID: 31594920; PMCID: PMC6783461

Heilig M, MacKillop J, Martinez D, Rehm J, Leggio L, Vanderschuren LJMJ (2021) Addiction as a brain disease revised: why it still matters, and the need for consilience. Neuropsychopharmacology 46(10):1715–1723. https://doi.org/10.1038/s41386-020-00950-y. Epub 2021 Feb 22. PMID: 33619327; PMCID: PMC8357831

Heinz A, Gutwinski S, Bahr NS, Spanagel R, Di Chiara G (2024) Does compulsion explain addiction? Addict Biol 29(4):e13379. https://doi.org/10.1111/adb.13379. PMID: 38588458; PMCID: PMC11001268

Hendershot CS, Bremmer MP, Paladino MB, Kostantinis G, Gilmore TA, Sullivan NR, Tow AC, Dermody SS, Prince MA, Jordan R, McKee SA, Fletcher PJ, Claus ED, Klein KR (2025) Once-weekly Semaglutide in adults with alcohol use disorder: a randomized clinical trial. JAMA Psychiatry 82(4):395–405. https://doi.org/10.1001/jamapsychiatry.2024.4789. PMID: 39937469; PMCID: PMC11822619

Hirth N, Meinhardt MW, Noori HR, Salgado H, Torres-Ramirez O, Uhrig S, Broccoli L, Vengeliene V, Roßmanith M, Perreau-Lenz S, Köhr G, Sommer WH, Spanagel R, Hansson AC (2016) Convergent evidence from alcohol-dependent humans and rats for a hyperdopaminergic state in protracted abstinence. Proc Natl Acad Sci USA 113(11): 3024–3029. https://doi.org/10.1073/pnas.1506012113. Epub 2016 Feb 22. PMID: 26903621; PMCID: PMC4801269

Jerlhag E (2024) Ghrelin system and GLP-1 as potential treatment targets for alcohol use disorder. Int Rev Neurobiol 178:401–432. https://doi.org/10.1016/bs.irn.2024.07.006. Epub 2024 Aug 3. PMID: 39523062.

Jerlhag E (2025) GLP-1 receptor agonists: promising therapeutic targets for alcohol use disorder. Endocrinology 166(4):bqaf028. https://doi.org/10.1210/endocr/bqaf028

Joutsa J, Moussawi K, Siddiqi SH, Abdolahi A, Drew W, Cohen AL, Ross TJ, Deshpande HU, Wang HZ, Bruss J, Stein EA, Volkow ND, Grafman JH, van Wijngaarden E, Boes AD, Fox MD (2022) Brain lesions disrupting addiction map to a common human brain circuit. Nat Med 28(6): 1249–1255. https://doi.org/10.1038/s41591-022-01834-y. Epub 2022 Jun 13. PMID: 35697842; PMCID: PMC9205767

Jumper J, Evans R, Pritzel A, Green T, Figurnov M, Ronneberger O, Tunyasuvunakool K, Bates R, Žídek A, Potapenko A, Bridgland A, Meyer C, Kohl SAA, Ballard AJ, Cowie A, Romera-Paredes B, Nikolov S, Jain R, Adler J, Back T, Petersen S, Reiman D, Clancy E, Zielinski M, Steinegger M, Pacholska M, Berghammer T, Bodenstein S, Silver D, Vinyals O, Senior AW, Kavukcuoglu K, Kohli P, Hassabis D (2021) Highly accurate protein structure prediction with AlphaFold. Nature 596(7873):583–589. https://doi.org/10.1038/s41586-021-03819-2

Juraeva D, Treutlein J, Scholz H, Frank J, Degenhardt F, Cichon S, Ridinger M, Mattheisen M, Witt SH, Lang M, Sommer WH, Hoffmann P, Herms S, Wodarz N, Soyka M, Zill P, Maier W, Jünger E, Gaebel W, Dahmen N, Scherbaum N, Schmäl C, Steffens M, Lucae S, Ising M, Smolka MN, Zimmermann US, Müller-Myhsok B, Nöthen MM, Mann K, Kiefer F, Spanagel R, Brors B, Rietschel M (2015) XRCC5 as a risk gene for alcohol dependence: evidence from a genome-wide gene-set-based analysis and follow-up studies in Drosophila and humans.

Neuropsychopharmacology 40(2):361–371. https://doi.org/10.1038/npp.2014.178. PMID: 25035082

Kendler KS, Ohlsson H, Sundquist J, Sundquist K (2021) The moderation of the genetic risk for alcohol and drug use disorders in a Swedish national sample by the genetic aptitude for educational attainment. Psychol Med 53:3077. https://doi.org/10.1017/S0033291

Klausen MK, Jensen ME, Møller M, Le Dous N, Jensen AØ, Zeeman VA, Johannsen CF, Lee A, Thomsen GK, Macoveanu J, Fisher PM, Gillum MP, Jørgensen NR, Bergmann ML, Enghusen Poulsen H, Becker U, Holst JJ, Benveniste H, Volkow ND, Vollstädt-Klein S, Miskowiak KW, Ekstrøm CT, Knudsen GM, Vilsbøll T, Fink-Jensen A (2022) Exenatide once weekly for alcohol use disorder investigated in a randomized, placebo-controlled clinical trial. JCI Insight 7(19):e159863. https://doi.org/10.1172/jci.insight.159863. PMID: 36066977; PMCID: PMC9675448

Koob GF, Vendruscolo L (2023) Theoretical frameworks and mechanistic aspects of alcohol addiction: alcohol addiction as a reward deficit/stress surfeit disorder. Curr Top Behav Neurosci. https://doi.org/10.1007/7854_2023_424. Epub ahead of print. PMID: 37421551

Kummetat JL, Leonhard A, Manthey J, Speerforck S, Schomerus G (2022) Understanding the association between alcohol stigma and alcohol consumption within Europe: a cross-sectional exploratory study. Eur Addict Res 28(6):446–454. https://doi.org/10.1159/000526200. Epub 2022 Sep 9. PMID: 36088900

Kwako LE, Spagnolo PA, Schwandt ML, Thorsell A, George DT, Momenan R, Rio DE, Huestis M, Anizan S, Concheiro M, Sinha R, Heilig M (2015) The corticotropin releasing hormone-1 (CRH1) receptor antagonist pexacerfont in alcohol dependence: a randomized controlled experimental medicine study. Neuropsychopharmacology 40(5):1053–1063. https://doi.org/10.1038/npp.2014.306. PMID: 25409596; PMCID: PMC4367465

Lähteenvuo M, Tiihonen J, Solismaa A, Tanskanen A, Mittendorfer-Rutz E, Taipale H (2025) Repurposing semaglutide and liraglutide for alcohol use disorder. JAMA Psychiatry 82(1): 94–98. https://doi.org/10.1001/jamapsychiatry.2024.3599. PMID: 39535805; PMCID: PMC11561716

Leclercq S, de Timary P (2024) Role of the microbiome and the gut-brain Axis in alcohol use disorder: potential implication for treatment development. Curr Top Behav Neurosci. https://doi.org/10.1007/7854_2024_478. Epub ahead of print. PMID: 38914878

Levy N (2013) Addiction is not a brain disease (and it matters). Front Psych 4:24. https://doi.org/10.3389/fpsyt.2013.00024. PMID: 23596425; PMCID: PMC3622902

Liu S, Haucke M, Groß R, Schneider K, Shin J, Arntz F, Bach P, Banaschewski T, Beste C, Deserno L, Ebner-Priemer U, Endrass T, Ganz M, Ghadami A, Giurgiu M, Heinz A, Kiefer F, Kliegl R, Lenz B, Marciniak MA, Meyer-Lindenberg A, Neubauer AB, Rapp M, Smolka MN, Strehle J, Spanagel R, Spitta G, Tost H, Walter H, Zech H, Reichert D, Reichert M (2025) Real-time mechanism-based interventions for daily alcohol challenges: protocol for ecological momentary assessment and intervention. Digit Health 11:20552076241311731. https://doi.org/10.1177/20552076241311731

Mascarell Maričić L, Walter H, Rosenthal A, Ripke S, Quinlan EB, Banaschewski T, Barker GJ, Bokde ALW, Bromberg U, Büchel C, Desrivières S, Flor H, Frouin V, Garavan H, Itterman B, Martinot JL, Martinot MP, Nees F, Orfanos DP, Paus T, Poustka L, Hohmann S, Smolka MN, Fröhner JH, Whelan R, Kaminski J, Schumann G, Heinz A, IMAGEN consortium. (2020) The IMAGEN study: a decade of imaging genetics in adolescents. Mol Psychiatry 25(11): 2648–2671. https://doi.org/10.1038/s41380-020-0822-5. Epub 2020 Jun 29. PMID: 32601453; PMCID: PMC7577859

Meinhardt MW, Sommer WH (2023) Schrooms against booze: potential of mycotherapy for the treatment of AUD. Neuropsychopharmacology 48(1):211–212. https://doi.org/10.1038/s41386-022-01446-7. PMID: 36076019; PMCID: PMC9700780

Meinhardt MW, Gerlach B, Spanagel R (2024) Good practice guideline for preclinical alcohol research: the STRINGENCY framework. Curr top. Behav Neurosci. https://doi.org/10.1007/7854_2024_484. Online ahead of print. PMID: 39117860

Meredith LR, Burnette EM, Grodin EN, Irwin MR, Ray LA (2021) Immune treatments for alcohol use disorder: A translational framework. Brain Behav Immun. 97:349–364. https://doi.org/10.1016/j.bbi.2021.07.023. Epub 2021 Jul 31. PMID: 34343618; PMCID: PMC9044974.

Morris J, Boness CL, Burton R (2023) (Mis)understanding alcohol use disorder: Making the case for a public health first approach. Drug Alcohol Depend 253:111019. https://doi.org/10.1016/j.drugalcdep.2023.111019. Epub 2023 Nov 4. PMID: 37952353; PMCID: PMC11061885

Murrough JW, Charney DS (2017) Corticotropin-releasing factor type 1 receptor antagonists for stress-related disorders: time to call it quits? Biol Psychiatry 82(12):858–860. https://doi.org/10.1016/j.biopsych.2017.10.012

Noori HR, Helinski S, Spanagel R (2014) Cluster and meta-analyses on factors influencing stress-induced alcohol drinking and relapse in rodents. Addict Biol 19(2):225–232. https://doi.org/10.1111/adb.12125

Noori HR, Schöttler J, Ercsey-Ravasz M, Cosa-Linan A, Varga M, Toroczkai Z, Spanagel R (2017) A multiscale cerebral neurochemical connectome of the rat brain. PLoS Biol 15(7):e2002612. https://doi.org/10.1371/journal.pbio.2002612. PMID: 28671956; PMCID: PMC5507471

Noori HR, Mervin LH, Bokharaie V, Durmus Ö, Egenrieder L, Fritze S, Gruhlke B, Reinhardt G, Schabel HH, Staudenmaier S, Logothetis NK, Bender A, Spanagel R (2018) Systemic neurotransmitter responses to clinically approved and experimental neuropsychiatric drugs. Nat Commun 9(1):4699. https://doi.org/10.1038/s41467-018-07239-1. PMID: 30410047; PMCID: PMC6224407

Okbay A, Wu Y, Wang N, Jayashankar H, Bennett M, Nehzati SM, Sidorenko J, Kweon H, Goldman G, Gjorgjieva T, Jiang Y, Hicks B, Tian C, Hinds DA, Ahlskog R, PKE M, Oskarsson S, Hayward C, Campbell A, Porteous DJ, Freese J, Herd P, 23andMe Research Team; Social Science Genetic Association Consortium, Watson C, Jala J, Conley D, Koellinger PD, Johannesson M, Laibson D, Meyer MN, Lee JJ, Kong A, Yengo L, Cesarini D, Turley P, Visscher PM, Beauchamp JP, Benjamin DJ, Young AI (2022) Polygenic prediction of educational attainment within and between families from genome-wide association analyses in 3 million individuals. Nat Genet 54(4):437–449. https://doi.org/10.1038/s41588-022-01016-z. Epub 2022 Mar 31. PMID: 35361970; PMCID: PMC9005349

Rawls E, Kummerfeld E, Zilverstand A (2021) An integrated multimodal model of alcohol use disorder generated by data-driven causal discovery analysis. Commun Biol 4(1):435. https://doi.org/10.1038/s42003-021-01955-z. PMID: 33790384; PMCID: PMC8012376.

Refojo D, Schweizer M, Kuehne C, Ehrenberg S, Thoeringer C, Vogl AM, Dedic N, Schumacher M, von Wolff G, Avrabos C, Touma C, Engblom D, Schütz G, Nave KA, Eder M, Wotjak CT, Sillaber I, Holsboer F, Wurst W, Deussing JM (2011) Glutamatergic and dopaminergic neurons mediate anxiogenic and anxiolytic effects of CRHR1. Science 333(6051):1903–1907. https://doi.org/10.1126/science.1202107

Reichert D, Brüßler S, Reichert M, Ebner-Priemer U (2024) Understanding alcohol consumption and its antecedents and consequences in daily life: the why and the how. Curr Top Behav Neurosci. https://doi.org/10.1007/7854_2024_486. Epub ahead of print. PMID: 39080235

Rieser NM, Bitar R, Halm S, Rossgoderer C, Gubser LP, Thévenaz M, Kreis Y, von Rotz R, Nordt C, Visentini M, Moujaes F, Engeli EJE, Ort A, Seifritz E, Vollenweider FX, Herdener M, Preller KH (2025) Psilocybin-assisted therapy for relapse prevention in alcohol use disorder: a phase 2 randomized clinical trial. EClinicalMedicine 82:103149. https://doi.org/10.1016/j.eclinm.2025.103149

Rosoff DB, Clarke TK, Adams MJ, McIntosh AM, Davey Smith G, Jung J, Lohoff FW (2021) Educational attainment impacts drinking behaviors and risk for alcohol dependence: results from a two-sample Mendelian randomization study with ~780,000 participants. Mol Psychiatry 26(4):1119–1132. https://doi.org/10.1038/s41380-019-0535-9. Epub 2019 Oct 25. PMID: 31649322; PMCID: PMC7182503. 721005134

Scherrer B, Guiraud J, Addolorato G, Aubin HJ, de Bejczy A, Benyamina A, van den Brink W, Caputo F, Dematteis M, Goudriaan AE, Gual A, Kiefer F, Leggio L, Lesch OM, Maremmani I, Nutt DJ, Paille F, Perney P, Poulnais R, Raffaillac Q, Rehm J, Rolland B, Simon N,

Söderpalm B, Sommer WH, Walter H, Spanagel R (2021) Baseline severity and the prediction of placebo response in clinical trials for alcohol dependence: a meta-regression analysis to develop an enrichment strategy. Alcohol Clin Exp Res 45(9):1722–1734. https://doi.org/10.1111/acer.14670. PMID: 34418121

Schwandt ML, Cortes CR, Kwako LE, George DT, Momenan R, Sinha R, Grigoriadis DE, Pich EM, Leggio L, Heilig M (2016) The CRF1 antagonist Verucerfont in anxious alcohol-dependent women: translation of neuroendocrine, but not of anti-craving effects. Neuropsychopharmacology 41(12):2818–2829. https://doi.org/10.1038/npp.2016.61. Epub 2016 Apr 25. PMID: 27109623; PMCID: PMC5061889

Selim MK, Harel M, De Santis S, Perini I, Sommer WH, Heilig M, Zangen A, Canals S (2024) Repetitive deep TMS in alcohol dependent patients halts progression of white matter changes in early abstinence. Psychiatry Clin Neurosci 78(3):176–185. https://doi.org/10.1111/pcn.13624

Shaham Y, de Wit H (2016) Lost in translation: CRF1 receptor antagonists and addiction treatment. Neuropsychopharmacology 41(12):2795–2797. https://doi.org/10.1038/npp.2016.94

Sicignano D, Hernandez AV, Schiff B, Elmahy N, White CM (2024) The impact of psychedelics on patients with alcohol use disorder: a systematic review with meta-analysis. Curr Med Res Opin 40(2):293–302. https://doi.org/10.1080/03007995.2023.2296968. Epub 2024 Jan 24. PMID: 38111216

Sillaber I, Rammes G, Zimmermann S, Mahal B, Zieglgänsberger W, Wurst W, Holsboer F, Spanagel R (2002) Enhanced and delayed stress-induced alcohol drinking in mice lacking functional CRH1 receptors. Science 296(5569):931–933. https://doi.org/10.1126/science.1069836

Sommer WH, Rimondini R, Hansson AC, Hipskind PA, Gehlert DR, Barr CS, Heilig MA (2008) Upregulation of voluntary alcohol intake, behavioral sensitivity to stress, and amygdala crhr1 expression following a history of dependence. Biol Psychiatry 63(2):139–145. https://doi.org/10.1016/j.biopsych.2007.01.010. Epub 2007 Jun 21. PMID: 17585886.

Sommer WH, Canals S, Bifone A, Heilig M, Hyytiä P (2022) From a systems view to spotting a hidden Island: a narrative review implicating insula function in alcoholism. Neuropharmacology 209:108989. https://doi.org/10.1016/j.neuropharm.2022.108989. Epub 2022 Feb 22. PMID: 35217032

Sommer WH, Canals S (2025) Alcohol-Induced Changes in Brain Microstructure: Uncovering Novel Pathophysiological Mechanisms of AUD Using Translational DTI in Humans and Rodents. Curr Top Behav Neurosci 2025 May 14. https://doi.org/10.1007/7854_2025_585. Epub ahead of print. PMID: 40360929.

Spanagel R (2022) Ten points to improve reproducibility and translation of animal research. Front Behav Neurosci 16:869511. https://doi.org/10.3389/fnbeh.2022.869511. PMID: 35530730

Spanagel R, Bach P, Banaschewski T, Beck A, Bermpohl F, Bernardi RE, Beste C, Deserno L, Durstewitz D, Ebner-Priemer U, Endrass T, Ersche KD, Feld G, Gerchen MF, Gerlach B, Goschke T, Hansson AC, Heim C, Kiebel S, Kiefer F, Kirsch P, Kirschbaum C, Koppe G, Lenz B, Liu S, Marxen M, Meinhardt MW, Meyer-Lindenberg A, Montag C, Müller CP, Nagel WE, Oliveria AMM, Owald D, Pilhatsch M, Priller J, Rapp MA, Reichert M, Ripke S, Ritter K, Romanczuk-Seiferth N, Schlagenhauf F, Schwarz E, Schwöbel S, Smolka MN, Soekadar SR, Sommer WH, Stock AK, Ströhle A, Tost H, Vollstädt-Klein S, Walter H, Waschke T, Witt SH, Heinz A (2024) Other members of the ReCoDe consortium. The ReCoDe addiction research consortium: losing and regaining control over drug intake-findings and future perspectives. Addict Biol 29(7):e13419. https://doi.org/10.1111/adb.13419. PMID: 38949209; PMCID: PMC11215792

Stacey D, Bilbao A, Maroteaux M, Jia T, Easton AC, Longueville S, Nymberg C, Banaschewski T, Barker GJ, Büchel C, Carvalho F, Conrod PJ, Desrivières S, Fauth-Bühler M, Fernandez-Medarde A, Flor H, Gallinat J, Garavan H, Bokde AL, Heinz A, Ittermann B, Lathrop M, Lawrence C, Loth E, Lourdusamy A, Mann KF, Martinot JL, Nees F, Palkovits M, Paus T, Pausova Z, Rietschel M, Ruggeri B, Santos E, Smolka MN, Staehlin O, Jarvelin MR, Elliott P, Sommer WH, Mameli M, Müller CP, Spanagel R, Girault JA, Schumann G, IMAGEN

Consortium (2012) RASGRF2 regulates alcohol-induced reinforcement by influencing mesolimbic dopamine neuron activity and dopamine release. Proc Natl Acad Sci USA 109(51):21128–21133. https://doi.org/10.1073/pnas.1211844110. PMID: 23223532

Tap SC (2024) The potential of 5-methoxy-N,N-dimethyltryptamine in the treatment of alcohol use disorder: a first look at therapeutic mechanisms of action. Addict Biol 29(4):e13386. https://doi.org/10.1111/adb.13386. PMID: 38600715; PMCID: PMC11007263

Uscher-Pines L, Lin K, Busch AB (2024) Expanding treatment options for alcohol use disorder. JAMA Health Forum 5(8):e242184. https://doi.org/10.1001/jamahealthforum.2024.2184. PMID: 39150729.

Vozella V, Borgonetti V, Cruz B, Onge CMS, Bullard R, Vlkolinsky R, Ceballos DG, Ozburn AR, Roberts AJ, Ciccocioppo R, Bajo M, Roberto M (2025) Apremilast reduces co-occurring alcohol drinking and mechanical allodynia and regulates central amygdala GABAergic transmission. JCI Insight 10(8):e189732. https://doi.org/10.1172/jci.insight.189732. PMID: 40260918; PMCID: PMC12016922

Theoretical Frameworks and Mechanistic Aspects of Alcohol Addiction: Alcohol Addiction as a Reward Deficit/Stress Surfeit Disorder

George F. Koob and Leandro Vendruscolo

Contents

Abstract Alcohol use disorder (AUD) can be defined by a compulsion to seek and take alcohol, the loss of control in limiting intake, and the emergence of a negative emotional state when access to alcohol is prevented. Alcohol use disorder impacts multiple motivational mechanisms and can be conceptualized as a disorder that includes a progression from impulsivity (positive reinforcement) to compulsivity (negative reinforcement). Compulsive drug seeking that is associated with AUD can be derived from multiple neuroadaptations, but the thesis argued herein is that a key

G. F. Koob (✉) and L. Vendruscolo
Integrative Neuroscience Research Branch, National Institute on Drug Abuse, Intramural Research Program, National Institutes of Health, Baltimore, MD, USA
e-mail: george.koob@nih.gov

Curr Topics Behav Neurosci (2025) 71: 17–77
https://doi.org/10.1007/7854_2023_424
Published Online: 09 July 2023

component involves the construct of negative reinforcement. Negative reinforcement is defined as drug taking that alleviates a negative emotional state. The negative emotional state that drives such negative reinforcement is hypothesized to derive from the dysregulation of specific neurochemical elements that are involved in reward and stress within basal forebrain structures that involve the ventral striatum and extended amygdala, respectively. Specific neurochemical elements in these structures include decreases in reward neurotransmission (e.g., decreases in dopamine and opioid peptide function in the ventral striatum) and the recruitment of brain stress systems (e.g., corticotropin-releasing factor [CRF]) in the extended amygdala, which contributes to hyperkatifeia and greater alcohol intake that is associated with dependence. Glucocorticoids and mineralocorticoids may play a role in sensitizing the extended amygdala CRF system. Other components of brain stress systems in the extended amygdala that may contribute to the negative motivational state of withdrawal include norepinephrine in the bed nucleus of the stria terminalis, dynorphin in the nucleus accumbens, hypocretin and vasopressin in the central nucleus of the amygdala, and neuroimmune modulation. Decreases in the activity of neuropeptide Y, nociception, endocannabinoids, and oxytocin in the extended amygdala may also contribute to hyperkatifeia that is associated with alcohol withdrawal. Such dysregulation of emotional processing may also significantly contribute to pain that is associated with alcohol withdrawal and negative urgency (i.e., impulsivity that is associated with hyperkatifeia during hyperkatifeia). Thus, an overactive brain stress response system is hypothesized to be activated by acute excessive drug intake, to be sensitized during repeated withdrawal, to persist into protracted abstinence, and to contribute to the compulsivity of AUD. The combination of the loss of reward function and recruitment of brain stress systems provides a powerful neurochemical basis for a negative emotional state that is responsible for the negative reinforcement that at least partially drives the compulsivity of AUD.

Keywords Alcohol use disorder · Opponent process · Stress · Extended amygdala · Corticotropin-releasing factor · Hypothalamic-pituitary-adrenal axis

1 Definitions and Conceptual Framework for Hyperkatifeia in Alcohol Use Disorder

Alcohol use disorder (AUD) can be defined as a chronically relapsing disorder, characterized by (1) compulsion to seek and take alcohol, (2) the loss of control in limiting alcohol intake, and (3) the emergence of a negative emotional state, termed hyperkatifeia (e.g., dysphoria, anxiety, and irritability), when access to alcohol is prevented (Koob and Le Moal 1997). Hyperkatifeia can be defined as the greater intensity of negative emotional/motivational signs and symptoms during withdrawal from drugs of addiction, including alcohol, in the withdrawal/negative affect stage of the drug addiction cycle (Shurman et al. 2010). The 5th edition of the *Diagnostic and*

Statistical Manual of Mental Disorders (DSM-5; American Psychiatric Association 2013) combined what was previously conceptualized as two separate disorders (alcohol abuse and alcohol dependence from the DSM-IV; American Psychiatric Association 1994) into one construct, defining AUD on a continuum from mild to moderate to severe, with severity of the disorder depending on how many of the diagnostic criteria apply. The presence of two to three criteria indicates a mild disorder, four to five criteria indicate a moderate disorder, and six or more criteria indicate a severe disorder (American Psychiatric Association 2013). This nomenclature better captures dimensionality of the disease and the complex progression of neural and behavioral impairments that afflict individuals with addiction. Clinically, the occasional but limited use of alcohol is distinct from compulsive-like alcohol seeking that characterizes moderate to severe AUD. Similarly in animal models, the escalation of alcohol intake and emergence of a chronic alcohol-dependent state are distinct from limited access to alcohol. The thesis argued herein is that AUD, similar to other substance use disorders (drug addiction), is a reward deficit/stress surfeit disorder, and the emergence of a negative emotional state (hyperkatifeia) plays an important role in defining and perpetuating AUD. Alcohol use disorder also involves substantial neuroadaptations that persist beyond acute withdrawal and trigger relapse and deficits in cognitive function that can also fuel compulsive-like drinking. However, the argument is that a core deficit that sets up vulnerability to relapse in AUD and possibly even deficits in cognitive function is in fact hyperkatifeia.

Supporting this hypothesis, a holistic view of AUD is presented with the following arguments. A negative emotional state (i.e., hyperkatifeia) is a common presentation in individuals with moderate to severe AUD during withdrawal and protracted abstinence. Compulsivity in AUD has an important negative reinforcement component that perpetuates AUD. Such negative emotional states become sensitized over time and set up an allostatic state that perpetuates addiction. Hyperkatifeia sets up a powerful motivational state for relapse. The neurobiological substrates that underlie the motivation to seek alcohol are reviewed herein, and an argument is presented that the loss of reward function and gain of brain stress function mediate hyperkatifeia as a key component of AUD. The overall hypothesis is that hyperkatifeia is a major driving force of AUD that is mediated by multiple components of motivational neurocircuits (Koob 2021).

1.1 Cycle of Addiction, Impulsivity, Compulsivity

A heuristic framework for addiction generally includes a three-stage cycle, represented by binge/intoxication, withdrawal/negative affect, and preoccupation/anticipation (craving) stages (Koob and Le Moal 1997; Koob et al. 2019). Under this framework, stage-related dysregulations occur in three functional domains (incentive salience/habits, negative emotional states, and executive function) that are mediated by three major neurocircuitry elements (basal ganglia, extended amygdala, and prefrontal cortex, respectively; Koob and Le Moal 1997). These three stages

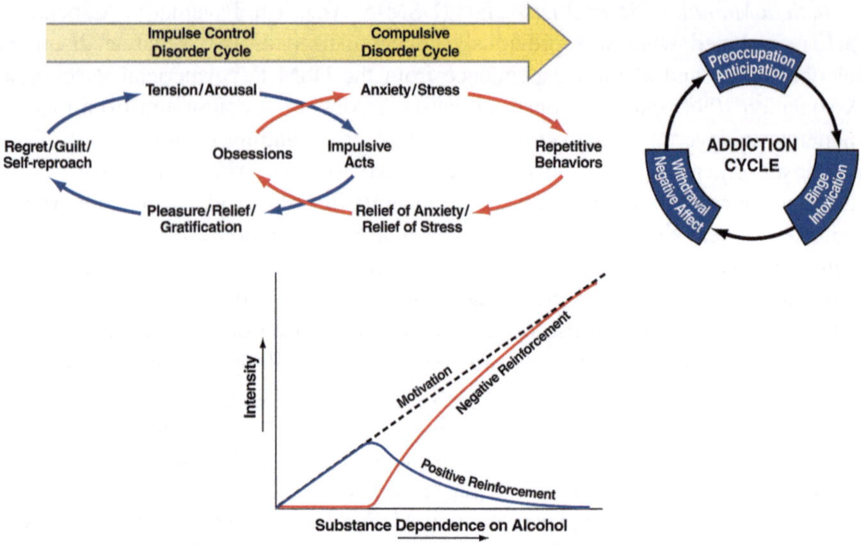

Fig. 1 (*Top left*) Diagram showing the stages of impulse control disorder and compulsive disorder cycles related to the sources of reinforcement. In impulse control disorders, such as AUD, an increasing tension and arousal occurs before the impulsive act, with pleasure, gratification, or relief during the act. Following the act, there may or may not be regret or guilt. In compulsive disorders, there are recurrent and persistent thoughts (obsessions) that cause marked anxiety and stress followed by repetitive behaviors (compulsions) that are aimed at preventing or reducing distress (American Psychiatric Association 1994). Positive reinforcement (pleasure/gratification) is more closely associated with impulse control disorders. Negative reinforcement (relief of anxiety or relief of stress or relief of hyperkatifeia) is more closely associated with compulsive disorders. (*Top right*) Collapsing the cycles of impulsivity and compulsivity results in the addiction cycle, conceptualized as three major components: preoccupation/anticipation, binge/intoxication, and withdrawal/negative affect. [Taken with permission from (Koob 2008b).] (*Bottom*) Change in the relative contribution of positive and negative reinforcement constructs during the development of substance dependence on alcohol (i.e., AUD)

feed into each other, become more intense, and ultimately lead to the pathological state of AUD (Koob and Le Moal 1997; Fig. 1).

Drug addiction has historically been conceptualized as a disorder that involves elements of both impulsivity and compulsivity, in which *impulsivity* can be defined behaviorally as "a predisposition toward rapid, unplanned reactions to internal and external stimuli without regard for the negative consequences of these reactions to themselves or others" (Moeller et al. 2001). Impulsivity is measured in two domains: the choice of a smaller, immediate reward over a larger, delayed reward (Rachlin and Green 1972) and the inability to inhibit behavior by changing the course of action or to stop a response once it is initiated (Logan et al. 1997). Impulsivity is a core deficit in substance use disorders (Allen et al. 1998) and such neuropsychiatric disorders as attention-deficit/hyperactivity disorder. Operationally, delay-to-gratification tasks (e.g., delayed discounting tasks and impulsive choice) and the stop-signal or Go/

No-Go tasks (behavioral impulsivity) have been used as measures of impulsivity (Fillmore and Rush 2002; Green et al. 1994). Compulsivity can be defined as elements of behavior that result in the perseveration of responding in the face of adverse consequences or perseveration in the face of incorrect responses in choice situations (e.g., operationally, responding for a drug or alcohol in the face of adverse consequences; Wolffgramm and Heyne 1995) or responding for a drug or alcohol in the face of punishment (Vendruscolo et al. 2012) or in tasks with increasing demands for the reinforcer (e.g., behavioral economic paradigms [Kim and Kearns 2019] and progressive-ratio schedules of reinforcement [Walker et al. 2008]). Compulsivity is analogous to symptoms of AUD that are outlined by the American Psychiatric Association: continued alcohol use despite having persistent or recurrent social or interpersonal problems that are caused or exacerbated by the effects of alcohol and a great deal of time spent in activities that are necessary to obtain alcohol (American Psychiatric Association 2013). However, impulsivity persists through the addiction cycle and the shift to an allostatic state and can be driven from the hyperkatifeia perspective as illustrated by negative urgency (Zorrilla and Koob 2019).

Collapsing the cycles of impulsivity and compulsivity yields a composite addiction cycle that comprises three stages (binge/intoxication, withdrawal/negative affect, and preoccupation/anticipation), in which impulsivity often dominates at the early stages, and compulsivity dominates at later stages (Fig. 1). As an individual moves from impulsivity to compulsivity, a shift occurs from positive reinforcement that drives the motivated behavior to negative reinforcement that drives the motivated behavior (Koob 2004). Negative reinforcement can be defined as the process by which the removal of an aversive stimulus (e.g., negative emotional state of drug withdrawal) increases the probability of a response (e.g., dependence-induced drug intake to relieve the negative emotional state). Notably, negative reinforcement is *not* punishment, although both involve an aversive stimulus. In punishment, the aversive stimulus suppresses behavior, including drug taking (e.g., disulfiram [Antabuse]). Negative reinforcement can be perhaps described in lay terms as pleasure via relief (i.e., relief reward), such as the removal of pain or, in the case of AUD, removal of the negative emotional state of acute withdrawal or protracted abstinence.

The three stages are conceptualized as interacting with each other, becoming more intense, and ultimately leading to the pathological state known as addiction (Koob and Le Moal 1997; Fig. 1). From the perspective of AUD development, entrance into the three-stage cycle at any stage can engage neuroadaptations that lead to hyperkatifeia. The hypothesis that is elaborated herein is that such engagement ultimately triggers a break from hedonic homeostasis and subsequent compensatory responses in brain reward and stress systems that generate hyperkatifeia in the withdrawal/negative affect stage (Koob and Le Moal 1997). Although entrance into the addiction cycle often occurs via the positive reinforcing or rewarding properties of drugs, neuroadaptations ultimately occur that limit the rewarding properties of alcohol (i.e., tolerance; Elvig et al. 2021). The termination of alcohol taking inevitably during acute and protracted withdrawal ultimately leads to hyperkatifeia, which generates an additional motivational drive that derives from negative reinforcement. Here, negative reinforcement becomes a major source of motivation for drug seeking as the individual works to reduce, terminate, or prevent the negative emotional state or hyperkatifeia of drug withdrawal. However, one can enter the addiction cycle at any stage of the addiction cycle. For example, one can enter through the withdrawal/negative affect stage via negative reinforcement in individuals who attempt to self-medicate pain, either physical pain or emotional pain or both (Khantzian 1997; Boissoneault et al. 2019; Ferguson et al. 2021). An individual can also enter through the preoccupation/anticipation stage, such as when an individual with attention-deficit/hyperactivity disorder enters the cycle via the impulsivity/executive function deficits that one associates with the preoccupation/anticipation stage.

In AUD, a broad spectrum of drug misuse evolves that can range from engagement in binge drinking in daily episodes to continuous intoxication. A binge is defined as consuming five standard drinks for males and four standard drinks for females within 2 h or reaching blood alcohol levels (BALs) of 0.08 g% (National Institute on Alcohol Abuse and Alcoholism 2004). Many individuals with AUD continue with a binge/withdrawal pattern for extended periods of time. For others, the pattern evolves into prolonged days of heavy drinking to continual drinking for fear of withdrawal, such as in opioid use disorder, in which an individual must always have alcohol available to avoid the consequences of abstinence (Koob et al. 2019, 2021). Withdrawal from an alcohol binge or chronic high alcohol intake is characterized by a severe emotional and somatic withdrawal syndrome.

Intense preoccupation with obtaining alcohol (i.e., craving) develops that is linked to stimuli that are associated with obtaining the drug and stimuli that are associated with withdrawal and the aversive motivational state. A pattern develops in which the drug must be taken to avoid the severe dysphoria and discomfort of abstinence.

The pattern of alcohol addiction can be amply illustrated by excerpts from two case histories from Knapp (1996) and Goodwin (1981). In the first representative case history, an individual progresses from:

> "I drank when I was happy and I drank when I was anxious and I drank when I was bored and I drank when I was depressed, which was often," to:
>
> "I loved the way drink made me feel, and I loved its special power of deflection, its ability to shift my focus away from my own awareness of self and onto something else, something less painful than my own feelings," and finally to:
>
> "There's a sense of deep need, and the response is a grabbiness, a compulsion to latch on to something outside yourself in order to assuage some deep discomfort" (Knapp 1996).

Similarly, in a second representative case history:

> "Alcohol seemed to satisfy some specific need I had, which I can't describe,"
>
> "There were always reasons to drink. I was low, tense, tired, mad, happy,"
>
> "The goal, always, was to maintain a glow, not enough, I hoped, that people would notice, but a glow,"
>
> "By now I was hooked and knew it, but desperately did not want others to know it. I had been sneaking drinks for years – slipping out to the kitchen during parties and such – but now I began hiding alcohol, in my desk, bedroom, car glove compartment, so it would never be far away, ever. I grew panicky even thinking I might not have alcohol when I needed it, which was just about always," and

> "I loathed myself. I was waking early and thinking what a mess I was, how I had hurt so many others and myself. The words 'guilty' and 'depression' sound superficial in trying to describe how I felt. The loathing was almost physical – a dead weight that could be lifted in only one way, and that was by having a drink" (Goodwin 1981; see Appendix in Koob and Le Moal 2006, for full quotations).

These case histories illustrate numerous key points of the present treatise, but the main point is the transition from drinking to feel good to drinking to avoid feeling bad. To some extent, this transition is facilitated by personality differences that are presumably shaped by genetics and developmental and even social factors. As Khantzian cogently argued, addiction can be considered a type of chronic emotional distress syndrome that varies with the individual from physical and emotional pain to chronic dysphoria to stress and anxiety to interpersonal difficulties for which drugs can be argued to be sources of self-medication for such negative emotional states (Khantzian 1997). Additionally, Khantzian argued that self-medication may be drug-specific. Patients may engage in the preferential use of a drug that fits with the nature of the painful feelings that they are self-medicating (e.g., opioids to counter intense anger and rage, stimulants as augmenting agents for high-energy individuals, energizing agents for low-energy individuals, and depressants [e.g., alcohol] for individuals who are tense and anxious). The common element that is argued by Khantzian is that each class of drugs serves as antidotes or correctives to dysphoric states and acts as a "replacement for a defect in the psychological structure" (Kohut 1971, p. 46) of such individuals (Khantzian 2003).

1.2 Theoretical Framework: Motivation, Withdrawal, and Opponent Process

Motivation is a state that can be defined as a "tendency of the whole animal to produce organized activity" (Hebb 1972), and such motivational states are not constant but rather vary over time. Early work by Wikler stressed the role of changes in drive states that are associated with dependence. People described changes during withdrawal as a "hunger" or primary need, and the effects of morphine on such a state were described as "satiation" or gratification of the primary need (Wikler 1952). Although Wikler argued that positive reinforcement was retained even in heavily dependent individuals (e.g., thrill of the intravenous opioid injection), dependence produced a new source of gratification, that of negative reinforcement (Pantazis et al. 2021).

The concept of motivation in addiction was inextricably linked with hedonic, affective, or emotional states in the context of temporal dynamics by Solomon's opponent process theory of motivation. Solomon and Corbit (1974) postulated that hedonic, affective, or emotional states, once initiated by drugs, are automatically modulated by the central nervous system with mechanisms that reduce the intensity of hedonic feelings. In their postulation, the *a-process* includes affective or hedonic habituation (or tolerance), and the *b-process* includes affective or hedonic withdrawal (abstinence). The *a-process* in drug use consists of positive hedonic responses, occurs shortly after the presentation of a stimulus, correlates closely with the intensity, quality, and duration of the reinforcer, and shows tolerance. In contrast, the *b-process* in drug use appears after the *a-process* has terminated, consists of negative hedonic responses, and is sluggish in onset, is slow to build up to an asymptote, is slow to decay, and gets larger with repeated exposure. The thesis herein is that opponent processes begin early in drug taking, reflect changes in brain reward and stress systems, and later form one of the major motivations for compulsivity in drug taking.

Thus, dependence or the manifestation of a withdrawal syndrome after the removal of chronic drug administration is defined in terms of *motivational* aspects of dependence, such as the emergence of hyperkatifeia when access to the drug is prevented (Koob and Le Moal 2001), rather than on the *physical* (somatic) signs of dependence.

1.3 Alcohol Withdrawal, Dependence, and Hyperkatifeia

The constructs of *withdrawal* and *dependence* have a complex history, which stimulated a change in the nosology of addiction and the conceptual framework shift that is outlined in this chapter. Withdrawal can be defined simply as abstinence from or the removal of chronic drug use, usually characterized by signs and symptoms that are opposite to acute effects of the drug (Koob et al. 2019). Withdrawal from drugs with addiction potential is one symptom of what is defined symptomatically as substance use disorder in the DSM-5 and *International Statistical Classification of Diseases and Related Health Problems*, 10th revision (ICD-10; World Health Organization 1992). The word *dependence*, though, has multiple meanings. Dependence can be defined as the manifestation of a withdrawal syndrome upon the cessation of drug use. Under this definition, any drug, even drugs without addiction potential, can produce dependence. Historically, dependence was initially defined as:

> "...an arbitrary term used to denote the presence of an acquired abnormal state wherein the regular administration of adequate amount of a drug has, through previous prolonged use, become requisite to physiologic equilibrium. Since it is not yet possible to diagnose physical dependence objectively without withholding drugs, the sine qua non of physical dependence remains the demonstration of a physical abstinence syndrome" (Himmelsbach 1943).

In contrast, psychological dependence was later defined as:

> "A condition in which a drug produces 'a feeling of satisfaction and a psychic drive that require periodic or continuous administration of the drug to produce pleasure or to avoid discomfort" (Eddy et al. 1965).

Somatic symptoms of withdrawal are reflected by signs and symptoms of a physical nature that are usually opposite to acute effects of the drug itself. For example, for alcohol, sympathetic-like responses, such as hyperthermia, indicate withdrawal, whereas hypothermia characterizes acute intoxication. However, the argument here in is that the symptoms that are associated with hyperkatifeia have significantly more motivational significance than somatic signs of withdrawal (Koob et al. 2019). From the perspective of negative reinforcement, drug seeking and craving that are associated with acute and protracted abstinence are key to the thesis herein, in which somatic measures of withdrawal can be viewed as an index of dependence but do not always reflect the more motivational measures of withdrawal or hyperkatifeia. Notably, however, hyperkatifeia-associated symptoms of withdrawal generally have an earlier onset and are manifested at lower doses of chronic drug intake.

From the motivational perspective, the emergence of hyperkatifeia when access to the drug is prevented, rather than somatic signs, drives AUD. This is not a new hypothesis. Some theoreticians argued that the development of such a negative affective state can define dependence as it relates to addiction:

> "The notion of dependence on a drug, object, role, activity, or any other stimulus-source requires the crucial feature of negative affect experienced in its absence. The degree of dependence can be equated with the amount of this negative affect, which may range from mild discomfort to extreme distress, or it may be equated with the amount of difficulty or effort required to do without the drug, object, etc." (Russell 1976).

Ample evidence indicates that individuals with AUD exhibit dramatic evidence of dysphoric states during acute withdrawal that persist into protracted abstinence. Alcohol withdrawal in humans produces well-documented somatic symptoms, such as tremor, autonomic hyperactivity, nausea, vomiting, and seizures. More importantly, however, it produces significant affective symptoms of anxiety, dysphoria, and depression-like symptoms. Acute withdrawal (i.e., the first week post-alcohol) is characterized by Beck Depression Inventory score of approximately 20, which is within the range of moderate depression (15–30; Potokar et al. 1997), and Hamilton Depression Score of 18, which is close to 20 (the cutoff for antidepressant medication in affective disorder; Brown and Schuckit 1988). Depression scores decline during subsequent weeks of treatment but remain close to 10 for Hamilton Depression Scores for up to 4 weeks in an inpatient treatment program (Brown and Schuckit 1988). In a study of inpatient individuals with AUD during withdrawal, the Beck Depression Inventory score was 15 at withdrawal and remained at 12.8 two days into withdrawal and 9.4 two weeks post-withdrawal (de Timary et al. 2008). Similar results were reported for anxiety measures (Potokar et al. 1997; de Timary et al. 2008). In another study with a long-term follow-up of 6 months after 4-week inpatient detoxification, Beck Depression Inventory scores remained at approximately 6, and trait anxiety scores on the State-Trait Anxiety Inventory remained above 33 even in individuals without comorbid anxiety or depression (Driessen et al. 2001). Independent of comorbidity status, individuals who relapsed had higher trait anxiety scores than those who abstained (Driessen et al. 2001). Thus, although individuals with AUD exhibit significant decreases in measures of depression and anxiety during withdrawal, there are measurable levels of depression-like symptoms that persist long after acute withdrawal into protracted abstinence that may be clinically relevant, especially for treatment.

More compelling for the present thesis, during a 2-week inpatient withdrawal study, alexithymia (i.e., a state of deficiency in understanding, processing, or describing emotions, from the Greek *a* for "lack," *lexis* for "word," and *thymos* for "emotion"; Sifneos 1973; Taylor and Bagby 2000), which results in poor emotional regulation and stress management abilities, remained high and stable during the 2-week period (de Timary et al. 2008). Alexithymia scores did not decline between day 0 and day 2 but remained high at a score of 57 and declined only to 53 at 3 weeks (de Timary et al. 2008). These authors argued that alexithymia is a stable personality trait in people with AUD rather than a state-dependent phenomenon, providing support for the self-medication hypothesis.

Animal models can also be used to test the hypothesis that there are opponent process-like motivational changes that are associated with the development of alcohol dependence. Electrical brain stimulation reward or intracranial self-stimulation has a long history as a measure of activity of the brain reward system and acute reinforcing effects of drugs of addiction. All drugs of addiction, when administered acutely, lower brain stimulation reward thresholds (Kornetsky and Esposito 1979). When administered chronically, they elevate reward thresholds during withdrawal. Brain stimulation reward involves widespread neurocircuitry in the brain. The most sensitive sites, defined by the lowest reward thresholds, involve the trajectory of the medial forebrain bundle that connects the ventral tegmental area (VTA) with the basal forebrain (Olds and Milner 1954; Koob et al. 1977). Although much emphasis was placed initially on the role of ascending monoamine systems in the medial forebrain bundle in brain stimulation reward, other nondopaminergic systems in the medial forebrain bundle clearly play a key role (Hernandez et al. 2006).

Rats that were made dependent using chronic, intermittent, alcohol vapor exposure at BALs that were sufficient to drive excessive drinking exhibited an elevation of brain reward thresholds during withdrawal that lasted up to 3 days post-withdrawal (Schulteis et al. 1995). However, data suggest that, similar to other drugs of addiction, such opponent-like processes can begin with a single dose (Fig. 2).

An acute elevation of brain reward thresholds was observed during repeated acute withdrawal from alcohol, bearing a striking resemblance to human subjective reports (Schulteis and Liu 2006; Fig. 2). These results demonstrate that the elevation of brain reward thresholds following prolonged access to alcohol may fail to return to baseline levels between repeated, prolonged exposure to alcohol self-administration (i.e., a residual reward deficit), thus creating greater elevations of reward thresholds during withdrawal from chronic alcohol. Rapid acute tolerance and opponent process-like effects in response to hedonic effects of alcohol have been reported in human studies using the alcohol clamp procedure (Morzorati et al. 2002). These data provide compelling evidence of brain reward dysfunction with chronic alcohol exposure, providing strong support for a hedonic allostasis model of AUD (Koob 2003).

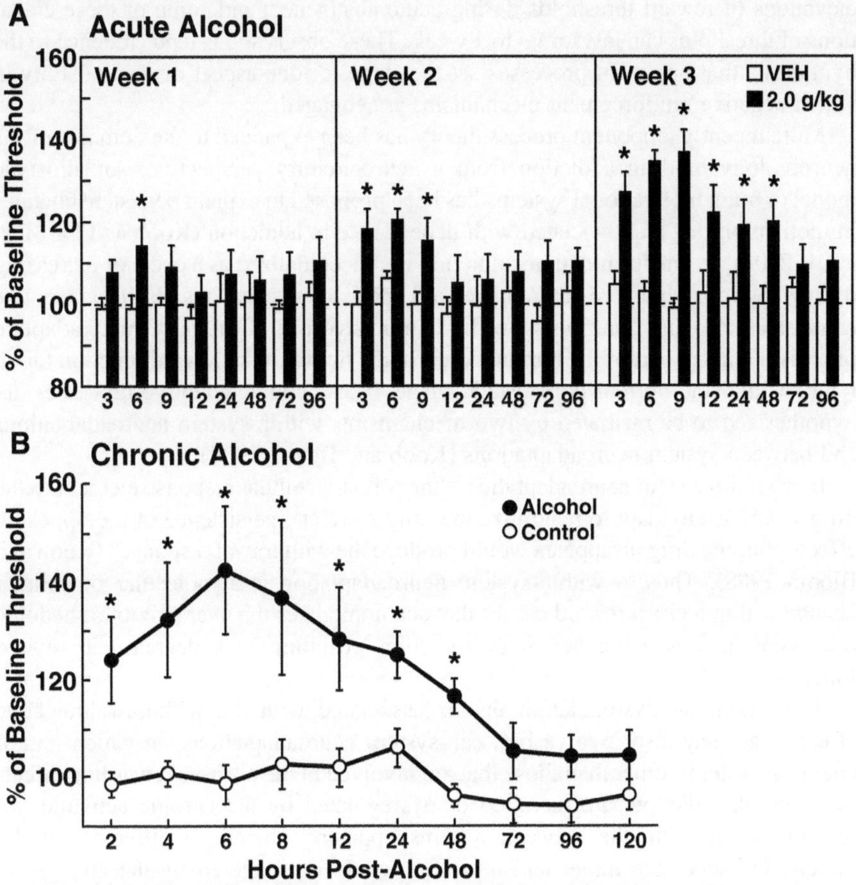

Fig. 2 (**a**) Withdrawal from a single bout of acute alcohol intoxication (week 1) resulted in a transient but significant elevation of brain reward thresholds ($*p < 0.05$, compared with vehicle controls at given time-point post-injection). The effect was significant at 6 h, a time when BALs had declined to virtually undetectable levels following this dose of alcohol. Repeated treatment with this dose for two additional weeks resulted in a progressive broadening of the duration of significant threshold elevations. Data are expressed as the mean ± SEM percentage of baseline threshold. $n = 8$–10 per dose group. [Taken with permission from (Schulteis and Liu 2006).] (**b**) Time-dependent elevation of intracranial self-stimulation thresholds during alcohol withdrawal. Mean BALs were around 0.2 g%. Data are expressed as the mean ± SEM percentage of baseline threshold. $*p < 0.05$, thresholds that were significantly elevated above control levels at 2–48 h post-alcohol. Open circles indicate the control condition. Closed circles indicate the alcohol withdrawal condition. [Taken with permission from (Schulteis et al. 1995)]

The dysregulation of brain reward function that is associated with withdrawal from chronic drug administration is a common element of all drugs of addiction. Withdrawal from chronic cocaine (Markou and Koob 1991), amphetamine (Paterson et al. 2000), opioids (Schulteis et al. 1994), cannabinoids (Gardner and Vorel 1998), nicotine (Epping-Jordan et al. 1998), and alcohol (Schulteis et al. 1995) leads to elevations of reward thresholds during acute abstinence, and some of these elevations of thresholds can last for up to 1 week. These observations lend credence to the hypothesis that opponent processes set the stage for one aspect of compulsivity in which negative reinforcement mechanisms are engaged.

More recently, opponent process theory has been expanded to the domains of the neurobiology of drug addiction from a neurocircuitry perspective. An allostatic model of brain motivational systems has been proposed to explain persistent changes in motivation that are associated with dependence in addiction (Koob and Le Moal 2001, 2008). In this formulation, addiction is conceptualized as a cycle of increasing dysregulation of brain reward/anti-reward mechanisms that results in a negative emotional state that contributes to the compulsive use of drugs. Counteradaptive processes that are part of the normal homeostatic limitation of reward function fail to return within the normal homeostatic range. These counteradaptive processes are hypothesized to be mediated by two mechanisms: within-system neuroadaptations and between-system neuroadaptations (Koob and Bloom 1988).

In a within-system neuroadaptation, "the primary cellular response element to the drug would itself adapt to neutralize the drug's effects; persistence of the opposing effects after the drug disappears would produce the withdrawal response" (Koob and Bloom 1988). Thus, a within-system neuroadaptation is a molecular or cellular change within a given reward circuit that accommodates the overactivity of hedonic processing that is associated with addiction, resulting in a decrease in reward function.

The emotional dysregulation that is associated with the withdrawal/negative affect stage may also involve between-system neuroadaptations, in which neurochemical systems, other than those that are involved in the positive rewarding effects of drugs of addiction, are recruited or dysregulated by the chronic activation of reward systems. "In the between-systems opposing process, a different cellular system and separable molecular apparatus would be triggered by the changes in the primary drug response neurons and would produce the adaptation and tolerance" (Koob and Bloom 1988). Thus, a between-system neuroadaptation is a circuitry change in which another, different circuit (anti-reward circuit) is activated by the reward circuit and has opposing actions, again limiting reward function. The remainder of this chapter explores neuroadaptations that occur in brain emotional systems to account for neurocircuitry changes that produce opponent processes and are hypothesized to play a key role in the compulsivity of addiction. Thus, the conceptual framework herein focuses on hyperkatifeia in AUD (Koob 2021). Indeed, the hyperkatifeia construct in addiction has gained traction in the context of the exacerbation of alcohol-related pathology that is associated with the coronavirus 2019 (COVID-19) pandemic (Koob et al. 2020; White et al. 2022).

2 Animal Models for Compulsive Alcohol Seeking

Methods of inducing binge-like alcohol drinking range from having animals drink alcohol solutions that are made more palatable with the addition of a sweetener (Ji et al. 2008) to restricting intake to specific periods of the dark cycle (drinking-in-the-dark procedure; Rhodes et al. 2005) to models that involve alcohol dependence in animals, such as alcohol vapor inhalation (Vendruscolo and Roberts 2014), intragastric alcohol infusion, and an alcohol liquid diet (Tunstall et al. 2020). The compulsive use of alcohol derives from multiple sources of reinforcement. Animal models have been developed for the acute positive reinforcing effects of alcohol and negative reinforcing effects that are associated with removal of the aversive effects of alcohol withdrawal or an existing aversive state (i.e., self-medication of aversive effects of abstinence from chronic alcohol or self-medication of a preexisting negative affective state; Koob and Le Moal 1997). A major early breakthrough was the development of a training procedure that involved access to a sweetened solution and the subsequent fading in of alcohol to avoid aversiveness of the taste of alcohol (for a review, see Samson 1987). Subsequent work extended these procedures to measures of self-administration in dependent rats and post-dependent rats (Roberts et al. 1996; O'Dell et al. 2004).

High doses of alcohol solutions are self-administered intragastrically after animals are made dependent via passive intragastric infusion, and rats self-infuse 4–7 g/kg per day of alcohol (Fidler et al. 2006). In this paradigm, BALs average 0.12 g%, measured 30 min after the start of a bout in which rats infuse 1.5 g/kg alcohol per 30 min.

In an alcohol liquid diet procedure, the diet is typically the sole source of calories that is available to rats (Moy et al. 1997), thereby forcing rats to consume alcohol. Typically, rats are provided a palatable liquid diet that contains 5–8.7% (v/v) alcohol as their sole source of calories, which is sufficient to produce dependence and maintain BALs of 0.1–0.13 g% during the dark (active drinking) cycle (Schulteis et al. 1996; Brown et al. 1998; Valdez et al. 2004). High responders during withdrawal from the liquid diet reach BALs of approximately 0.08–0.1 g% (Schulteis et al. 1996; Gilpin et al. 2009).

Reliable alcohol self-administration in dependent animals using alcohol vapor exposure has been extensively characterized in rats, in which animals achieve BALs in the 0.1–0.15 g% range (Roberts et al. 1999, 2000). Similarly, rats with a history of alcohol dependence exhibit an increase in alcohol self-administration, even weeks after acute withdrawal (Roberts et al. 2000). In a variant of alcohol vapor exposure with additional face validity, intermittent exposure to chronic alcohol using alcohol vapor chambers (14 h on/10 h off) produces the more rapid escalation of alcohol intake (O'Dell et al. 2004; Rimondini et al. 2002), and BALs are reliably above 0.14 g% after a 30 min session of self-administration in dependent animals (Richardson et al. 2008). In both the liquid diet and alcohol vapor procedures, alcohol intake is directly related to the blood alcohol range and pattern of intermittent high-dose alcohol exposure (Gilpin et al. 2009). Although the alcohol vapor model may have limited face validity, considering that alcohol is passively administered to animals, numerous studies demonstrated that it also has robust predictive validity for alcohol addiction (Heilig and Koob 2007; Koob et al. 2009). Moreover, studies indicated that rats are motivated to self-administer alcohol vapor to the point of alcohol dependence (de Guglielmo et al. 2017).

A similar procedure has been developed for mice and produces reliable increases in alcohol self-administration during withdrawal. Termed withdrawal-induced drinking, C57BL/6 mice are exposed to intermittent alcohol vapor (three cycles of 16 h of vapor and 8 h of air) and then tested in a 2 h limited access alcohol preference drinking test during the circadian dark period (Becker and Lopez 2004; Lopez and Becker 2005; Finn et al. 2007). Intermittent alcohol vapor exposure significantly increased 15% (v/v) alcohol intake by 30–50% in the post-vapor period, usually after multiple cycles and usually after 24 h of withdrawal (Finn et al. 2007). Similar results were reported using an operant response in mice in 60 min test sessions for 10% (w/v) alcohol with intermittent vapor exposure of 14 h on/10 h off (Chu et al. 2007).

3 Neural Substrates for the Negative Emotional State Associated with Alcohol Use Disorder

3.1 Within-System Neuroadaptations that Contribute to the Compulsivity Associated with the "Negative Emotional Side" of Alcohol Use Disorder

3.1.1 Within-System Neuroadaptations: Dopamine

Within-system neuroadaptations to chronic drug exposure include decreases in function of the same neurotransmitter systems in the same neurocircuits that are implicated in the acute reinforcing effects of drugs of addiction. One prominent hypothesis is that dopamine systems are compromised in crucial phases of the addiction cycle, such as withdrawal and protracted abstinence. This decrease in dopamine function is hypothesized to lead to lower motivation for non-drug-related stimuli and greater sensitivity to drugs of addiction (Melis et al. 2005). Activation of the mesolimbic dopamine system has long been known to be critical for the acute rewarding properties of psychostimulant drugs and associated with the acute reinforcing effects of alcohol (Koob 1992; McBride and Li 1998; Nestler 2005; Lewis 1996). In humans, using positron emission tomography, oral doses of alcohol increased extracellular dopamine concentrations during intoxication in healthy volunteers, consistent with animal findings (Boileau et al. 2003; Urban et al. 2010). However, in animal studies, the magnitude of the increase in dopaminergic activity that is produced by alcohol pales in comparison to psychostimulant intoxication. For example, intravenous cocaine self-administration produces a 200% increase in

Table 1 Effects of intravenous self-administration of D-amphetamine, cocaine, and heroin and oral self-administration of alcohol on extracellular dopamine levels in the nucleus accumbens using *in vivo* microdialysis

Drug	Increase in dopamine over baseline	Reference
D-Amphetamine	700%	Di Ciano et al. (1995)
Cocaine	200–500%	Di Ciano et al. (1995), Weiss et al. (1992a)
Alcohol	25–50%	Weiss et al. (1992b, 1996)
Heroin	<20%	Hemby et al. (1995)

extracellular dopamine (Weiss et al. 1992b) compared with alcohol, which produces a 20% increase in extracellular dopamine in the nucleus accumbens (NAc; Doyon et al. 2003), and heroin, which, in some studies, does not increase extracellular dopamine in the NAc (Table 1). Such a relationship changes with the development of dependence.

Acute withdrawal in animals is characterized by decreases in activity of the mesolimbic dopamine system and decreases in serotonergic neurotransmission in the NAc (Rossetti et al. 1992; Weiss et al. 1992a, 1996; Diana 2011). In dependent male Wistar rats that were trained to self-administer alcohol during withdrawal, the release of dopamine and serotonin was monitored by microdialysis in the NAc at the end of a 3–5-week alcohol (8.7%, w/v) liquid diet regimen, during 8 h of withdrawal, and during the renewed availability of alcohol that involved the opportunity to operantly self-administer alcohol (10% w/v) for 60 min, followed by unlimited access to the alcohol liquid diet. In nondependent rats, operant alcohol self-administration increased both dopamine and serotonin release in the NAc. Withdrawal from the chronic alcohol diet produced progressive suppression of the release of these neurotransmitters over the 8 h withdrawal period. Alcohol self-administration reinstated and maintained dopamine release at pre-withdrawal levels but failed to completely restore serotonin efflux. These findings suggest that deficits in NAc monoamine release may contribute to negative affective consequences of alcohol withdrawal and thus motivate alcohol-seeking behavior in dependent individuals (Weiss et al. 1996). Similar dramatic decreases in extracellular dopamine in the NAc, measured by microdialysis, were found in a study in which animals were tested 8 h into alcohol withdrawal from chronic repeated alcohol injections of up to 5 g/kg every 6 h for 6 consecutive days (Majchrowicz 1975), which correlated with withdrawal scores (Rossetti et al. 1999). Thus, alcohol-dependent animals may exhibit a much higher percentage increase in dopamine release in the NAc during alcohol self-administration during withdrawal because baseline levels of dopamine are low during withdrawal (Weiss et al. 1996). Additionally, substantial electrophysiological data show profound decreases in both the origins and terminals of the mesocorticolimbic dopamine system during withdrawal from chronic alcohol that drive these changes (Diana et al. 1993, 2003; Bailey et al. 2001).

Imaging studies in humans with drug addiction have consistently shown long-lasting decreases in the number of dopamine D_2 receptors compared with controls

(Volkow et al. 1996, 2002; Kuikka et al. 2000). Additionally, alcohol-dependent individuals exhibit dramatic reductions of dopamine release in the striatum in response to a pharmacological challenge with the psychostimulant drug methylphenidate (Volkow et al. 2007). Decreases in the number of D_2 receptors, coupled with the decrease in dopaminergic activity, in chronic users of cocaine, nicotine, and alcohol are hypothesized to produce a decrease in the sensitivity of reward circuits to stimulation by natural reinforcers (Martin-Solch et al. 2001; Volkow and Fowler 2000). These findings suggest an overall reduction of the sensitivity of the dopamine component of reward circuitry to natural reinforcers and other drugs in individuals with addiction.

Another mechanism that may explain the hypodopaminergic state is that alcohol activates dynorphin, particularly in the shell of the NAc, triggered by a cascade of molecular events that involve cyclic adenosine monophosphate (cAMP) and dynorphin activation (Carlezon et al. 2000; Chavkin and Koob 2016). In protracted abstinence from alcohol, there may be a rebound in dopamine activity that contributes to greater vulnerability to the reinstatement of alcohol seeking. For example, at 3 weeks of withdrawal in rats, there was evidence of elevations of extracellular dopamine levels in the NAc, a lack of the synaptic response to D_1 receptor stimulation in the NAc, and an increase in motor activity that paralleled what was seen in post-mortem analyses of humans that exhibited the downregulation of D_1 receptor- and dopamine transporter-binding sites in the ventral striatum (Hirth et al. 2016).

3.1.2 Within-System Neuroadaptations: Enkephalin/Endorphin Opioid Peptides

Endogenous enkephalin/endorphin opioid peptide systems have long been hypothesized to play a role in the reinforcing effects of alcohol, largely based on pharmacological studies with opioid receptor antagonists in animal models and humans. The hypothesis is that alcohol promotes the release of opioid peptides to act on μ-opioid receptors (MORs; Nutt 2014). However, evidence of opponent process-like neuroadaptations is limited. A post-mortem study revealed a reduction of MOR-binding sites in post-mortem striatal tissue in individuals with AUD (Hermann et al. 2016). Acute alcohol potentiates receptor-activated cAMP production, but chronic exposure to alcohol decreases cAMP-protein kinase A (PKA) activity and decreases cAMP response element binding protein (CREB) phosphorylation (Gordon et al. 1986). In the central nucleus of the amygdala (CeA) and medial nucleus of the amygdala, a decrease in phosphorylated CREB is associated with anxiety-like behavior during withdrawal and an increase in alcohol self-administration (Pandey 2004; Pandey et al. 2003). Decreases in CREB activity have been associated with a decrease in neuropeptide Y (NPY) activity, linking within-system molecular changes with between-system neurotransmitter circuit changes.

3.1.3 Within-System Neuroadaptations: γ-Aminobutyric Acid

Acute alcohol enhances γ-aminobutyric acid (GABA) function in the CeA (Roberto and Varodayan 2017). GABA has long been associated with neuroadaptations that are associated with the chronic use of alcohol (Roberto and Varodayan 2017). For example, chronic alcohol decreases GABA receptor function, possibly through downregulation of the α1 subunit (Mhatre et al. 1993; Devaud et al. 1997). However, in microcircuits, chronic alcohol can also increase presynaptic GABA release, with no evidence of tolerance (Roberto et al. 2004). Thus, the action of alcohol on GABA does not readily demonstrate within-system evidence of obvious opponent processes. In local microcircuits, even in dependent animals, alcohol continues to presynaptically activate GABA (Roberto and Varodayan 2017). In the extended amygdala, chronic alcohol exposure increased GABAergic tone in dependent animals, which was hypothesized to drive compulsive-like drinking (Roberto et al. 2008). In the VTA, alcohol-exposed neurons showed electrophysiological evidence of a higher probability of action potential-independent GABA release after alcohol exposure, consistent with actions that would drive a hypodopaminergic state, such as with opioids (Melis et al. 2002).

3.1.4 Within-System Neuroadaptations: Glutamate

In contrast, alcohol withdrawal is characterized by a hyperglutamatergic state that is opposite to the acute effects of alcohol (Hwa et al. 2017). Chronic alcohol exposure upregulates *N*-methyl-D-aspartate (NMDA) receptor function in the brain (Chandler et al. 1993), including the upregulation of different NMDA receptor subunits with chronic alcohol and withdrawal in some brain regions (Trevisan et al. 1994; Follesa and Ticku 1995). One hypothesis is that glutamate plays a key role in sensitization or hyperexcitability that is elicited by repeated alcohol withdrawal (McCown and Breese 1990). From the perspective of hyperkatifeia, competitive glutamate receptor antagonists can partially reverse the anxiogenic-like effects of alcohol withdrawal (Gatch et al. 1999).

In people who are dependent on alcohol, alcohol detoxification produced significant increases in brain glutamate levels, measured by magnetic resonance spectroscopy, in prefrontocortical regions during acute alcohol withdrawal (Hermann et al. 2012). Similar results were reported in alcohol-dependent rats during withdrawal using magnetic resonance spectroscopy (Hermann et al. 2012). Chronic acamprosate, which is a glutamate modulator and Food and Drug Administration (FDA)-approved treatment for AUD, blocked the alcohol deprivation effect-induced increase in drinking in rodents (Heyser et al. 1998) and blocked increases in glutamate in the brain in rats (Dahchour et al. 1998) and humans (Hermann et al. 2012; Umhau et al. 2010).

3.2 Between-System Neuroadaptations that Contribute to Compulsivity Associated with Hyperkatifeia of Alcohol Use Disorder

Brain neurochemical systems that are involved in arousal-stress modulation may also be engaged within the neurocircuitry of brain stress systems to overcome the chronic presence of the perturbing drug (alcohol) and restore normal function despite the presence of drug (Fig. 3). The neuroanatomical entity of the extended amygdala (Heimer and Alheid 1991) may represent a common anatomical substrate that integrates brain arousal-stress systems with hedonic processing systems to produce a between-system opponent process. The extended amygdala is composed of the CeA, bed nucleus of the stria terminalis (BNST), and a transition zone in the medial

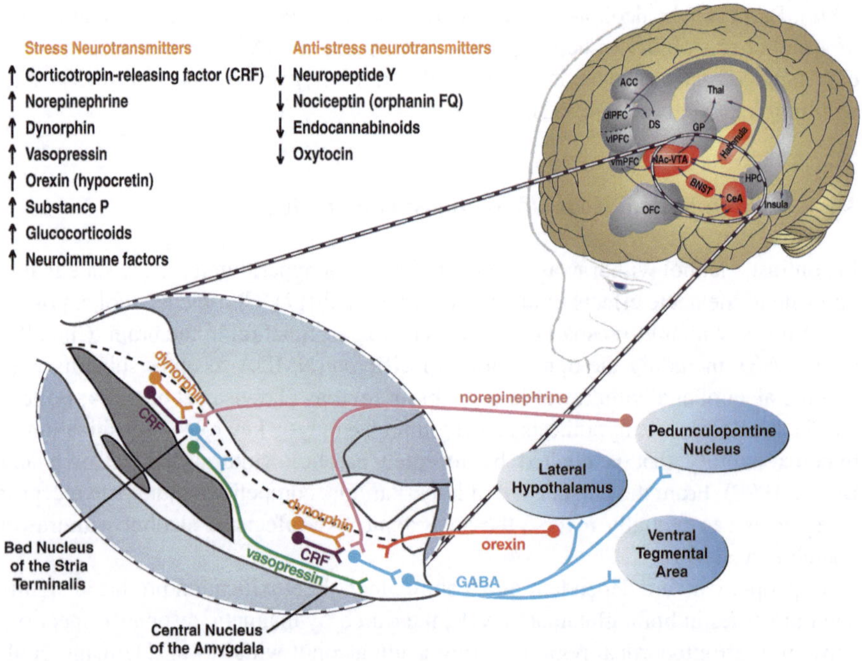

Fig. 3 Between-system extended amygdala circuitry associated with hyperkatifeia in the withdrawal/negative affect stage. Note the gain of stress neurotransmitter and neuromodulator function and the loss of anti-stress neurotransmitter and neuromodulator function throughout the neurocircuitry of the extended amygdala. The extended amygdala is composed of several basal forebrain structures, including the BNST, the CeA, and possibly a transition area in the medial portion (shell) of the NAc. ACC, anterior cingulate cortex; BNST, bed nucleus of the stria terminalis; CeA, central nucleus of the amygdala; CRF, corticotropin-releasing factor; dlPFC, dorsolateral prefrontal cortex; DS, dorsal striatum; GABA, γ-aminobutyric acid; GP, globus pallidus; HPC, hippocampus; OFC, orbitofrontal cortex; Thal, thalamus; vlPFC, ventrolateral prefrontal cortex; vmPFC, ventromedial prefrontal cortex; VTA, ventral tegmental area. [Taken with permission from (George and Koob 2013)]

(shell) subregion of the NAc. Each of these regions has cytoarchitectural and circuitry similarities (Heimer and Alheid 1991). The extended amygdala receives numerous afferents from limbic structures, such as the basolateral amygdala (BLA) and hippocampus, and sends efferents to the medial part of the ventral pallidum and a large projection to the lateral hypothalamus, thus further defining specific brain areas that interface classic limbic (emotional) structures with the extrapyramidal motor system (Alheid et al. 1995). The extended amygdala has long been hypothesized to play a key role in fear conditioning (Le Doux 2000) and the emotional component of pain processing (Neugebauer et al. 2004). For a detailed review of the neurochemistry of between-system neuroadaptations, see Koob (2021).

3.2.1 Between-System Neuroadaptations: Corticotropin-Releasing Factor

The brain stress system that is mediated by corticotropin-releasing factor (CRF) systems in both the extended amygdala and hypothalamic-pituitary-adrenal (HPA) axis is dysregulated by the chronic administration of all major drugs with addiction potential, with a common response of elevations of adrenocorticotropic hormone (ACTH), corticosterone, and amygdala CRF during acute withdrawal from chronic drug administration (Rivier et al. 1984; Merlo-Pich et al. 1995; Koob et al. 1994; Rasmussen et al. 2000; Olive et al. 2002; Delfs et al. 2000; Koob 2008a).

Alcohol withdrawal produces anxiety-like behavior in animal models that can be reversed by CRF receptor antagonists (Koob 2008a). Alcohol withdrawal produces anxiety-like behavior that is reversed by the intracerebroventricular administration of CRF_1/CRF_2 peptidergic receptor antagonists (Baldwin et al. 1991), small-molecule CRF_1 receptor antagonists (Knapp et al. 2004; Overstreet et al. 2004; Funk et al. 2007), and the intracerebral administration of a peptidergic CRF_1/CRF_2 receptor antagonist in the amygdala (Rassnick et al. 1993). Corticotropin-releasing factor receptor antagonists that are injected intracerebroventricularly or systemically also block the potentiation of anxiety-like behavior in response to stressors during protracted abstinence from chronic alcohol (Breese et al. 2005; Valdez et al. 2003; Sommer et al. 2008). The effects of CRF receptor antagonists on anxiety-like responses have been localized to the CeA (Rassnick et al. 1993). Additionally, systemic CRF_1 receptor antagonist administration also blunted the hyperalgesic response to alcohol withdrawal (Edwards et al. 2012b).

Perhaps more relevant to the present thesis are studies that showed that intermittent alcohol exposure facilitated withdrawal-induced anxiety-like behavior, and the administration of pharmacological treatments during withdrawal from the first and second alcohol cycles blocked this sensitization of withdrawal (Knapp et al. 2004). Diazepam, flumazenil (a $GABA_A$ receptor partial agonist), and baclofen (a $GABA_B$ receptor agonist) blocked the sensitization of withdrawal, consistent with a within-system neuroadaptation (Knapp et al. 2004, 2005, 2007). A CRF_1 receptor antagonist also prevented the sensitization of withdrawal-induced anxiety-like behavior (Overstreet et al. 2004, 2005). These results are consistent with a prolonged history of alcohol exposure that produces the persistent upregulation of both CRF and CRF_1 receptors in the brain (Roberto et al. 2010; Sommer et al. 2008; Zorrilla et al. 2001).

The ability of CRF receptor antagonists to block the anxiogenic-like and aversive-like motivational effects of drug withdrawal would predict motivational effects of CRF receptor antagonists in animal models of extended access to drugs. A particularly dramatic example of the motivational effects of CRF in dependence can be observed in animal models of alcohol self-administration in dependent animals. During alcohol withdrawal, extrahypothalamic CRF systems become hyperactive, with an increase in extracellular CRF in the CeA and BNST in dependent rats (Funk et al. 2006; Merlo-Pich et al. 1995; Olive et al. 2002). In mice, CRF neurons in the CeA were more active than non-CRF cells in the CeA (Aroni et al. 2021). The inactivation of a neuronal ensemble in the CeA during abstinence decreased alcohol drinking in dependent and nondependent rats, with a prolonged effect in alcohol-dependent rats. These effects were accompanied by a significant reduction of somatic withdrawal signs (de Guglielmo et al. 2016).

The dysregulation of brain CRF systems is hypothesized to underlie the enhancement of anxiety-like behavior and enhancement of alcohol self-administration that is associated with alcohol withdrawal. Supporting this hypothesis, the pharmacological blockade of CRF_1 receptors or vasopressin (a co-regulator of the HPA axis that potentiates CRF's effects) V_{1b} receptors reduced alcohol drinking and seeking in rodents (Edwards et al. 2012b; Funk et al. 2006; Marinelli et al. 2007; Richardson et al. 2008; Roberto et al. 2010; Sommer et al. 2008; Valdez et al. 2002). When administered directly in the CeA, a CRF_1/CRF_2 receptor antagonist blocked alcohol self-administration in alcohol-dependent rats (Funk et al. 2006). Systemic injections of small-molecule CRF_1 receptor antagonists also blocked the increase in alcohol intake that was associated with acute withdrawal (Knapp et al. 2004; Overstreet et al. 2004; Funk et al. 2007). These data suggest an important role for CRF, primarily in the CeA, in mediating the increase in self-administration that is associated with dependence. In mice (Lowery et al. 2008; Lowery-Gionta et al. 2012) and in high-intake nondependent rats (Cippitelli et al. 2012; Simms et al. 2014), CRF receptor antagonists reduced binge-like, excessive drinking, and stress-induced increases in alcohol intake. CRF_1 receptor antagonism (systemic or intra-CeA) also decreased allodynia/hyperalgesia (Edwards et al. 2012b) and anxiety- and depression-like behavior (Baldwin et al. 1991; Breese et al. 2005; Knapp et al. 2004; Overstreet et al. 2004; Rassnick et al. 1993; Valdez et al. 2004) in alcohol dependence. Both anxiety and pain are hypothesized to contribute to compulsive alcohol drinking (Aoun et al. 2018; Egli et al. 2012) in the withdrawal/negative affect stage.

Cellular studies have shown that alcohol's effects on the presynaptic activation of GABAergic interneurons in the CeA were enhanced in dependent rats, and these effects were abolished by CRF_1 receptor antagonists. Intra-CeA CRF_1 receptor antagonist administration reversed alcohol dependence-induced elevations of GABA dialysate (Roberto et al. 2010). However, there are prominent CRF projections from the CeA to the BNST (de Guglielmo et al. 2019; Vranjkovic et al. 2017). Corticotropin-releasing factor microcircuits in the BNST are activated by alcohol (Pati et al. 2020), and the BNST projects to the midbrain and hypothalamus

(Pati et al. 2020; Vranjkovic et al. 2017). An optogenetic study showed that the activation of CRF neurons in the CeA during alcohol withdrawal that projected to the BNST may mediate dependence-induced excessive alcohol intake (de Guglielmo et al. 2016, 2019).

3.2.2 Between-System Neuroadaptations: Glucocorticoid Receptors

Hans Selye was an endocrinologist who first described a role for the HPA axis in stress responses. He conceptualized stress and HPA axis function as adaptive responses to environmental challenges, termed "general adaptive syndrome" (Selye 1950). These discoveries motivated the search for releasing factors in the hypothalamus (Guillemin 1978) and discovery of CRF (Vale et al. 1981). In response to environmental stimuli, the paraventricular nucleus of the hypothalamus (PVN) stimulates the anterior pituitary to release ACTH into the bloodstream via CRF that is released into the portal system. Adrenocorticotropic hormone activates melanocortin receptor 2 (an ACTH receptor) in the cortex of the adrenal glands, causing the production and release of glucocorticoids (cortisol, corticosterone) into the blood circulation. Circulating glucocorticoids act on the body to produce many physiological and behavioral effects (Garabedian et al. 2017; Kadmiel and Cidlowski 2013; McEwen 2007; McEwen et al. 2015; Myers et al. 2014; Packard et al. 2016). Negative feedback mechanisms along the HPA axis prevent further corticosterone release.

In the brain, glucocorticoids bind to type I mineralocorticoid receptors and type II glucocorticoid receptors. Mineralocorticoid receptors have high affinity for glucocorticoids, whereas glucocorticoid receptors have lower affinity for glucocorticoids. Thus, glucocorticoid receptors are activated at high circulating glucocorticoid levels (McEwen 2007). Although activation of the HPA axis is adaptive and critical for survival, intense and sustained HPA axis activation may cause long-lasting detrimental neuroadaptations that contribute to the development of mental disorders (Koob and Kreek 2007; Packard et al. 2016; Sinha et al. 2011; Stephens and Wand 2012), including AUD.

Like excessive stress exposure (e.g., life adversities), alcohol exposure acutely activates the HPA axis in humans (Mendelson and Stein 1966), nonhuman primates (Jimenez and Grant 2017), and rodents (Ellis 1966; Ogilvie et al. 1998; Richardson et al. 2008). In rats, removal of the adrenal glands reduced alcohol drinking, an effect that was restored by glucocorticoid replacement (Fahlke et al. 1995). Glucocorticoid receptor antagonism also blocked alcohol reward in mice (Rotter et al. 2012). These findings indicate that glucocorticoids may be necessary for alcohol's acute rewarding/reinforcing effects in a nondependent state (Piazza and Le Moal 1998). In rats that were subjected to adrenalectomy, extracellular dopamine levels decreased in the NAc shell (Barrot et al. 2000), and changes in dopamine levels may affect alcohol intake (Juarez et al. 2017). Thus, glucocorticoids, via interactions with the dopamine system, appear to be required for reward function in response to alcohol.

Repeated HPA axis activation is hypothesized to drive cumulative neuroadaptations in brain reward and stress systems that facilitate the transition to, and maintain, alcohol dependence (Edwards et al. 2015; Somkuwar et al. 2017; Vendruscolo et al. 2015, 2012; Vendruscolo and Koob 2018). Excessive activation of the HPA axis by repeated alcohol exposure and withdrawal may lead to HPA axis activity dysregulation (Adinoff et al. 2003, 2017; Rasmussen et al. 2000; Richardson et al. 2008; Sinha et al. 2011). Alcohol-dependent humans and rodents (Richardson et al. 2008; Stephens and Wand 2012) exhibit neuroendocrine tolerance, in which alcohol-induced activation of the HPA axis is blunted. This blunted HPA axis response to alcohol may contribute to the decrease in alcohol-induced rewarding effects in alcohol dependence (i.e., alcohol tolerance; Elvig et al. 2021). Importantly, alcohol craving and relapse have been associated with HPA axis dysregulation in alcohol dependence (Kiefer et al. 2003; O'Malley et al. 2002). The opioid receptor antagonist naltrexone, which is approved by the FDA for the treatment of AUD, has anti-craving effects and activates the HPA axis (O'Malley et al. 2002). This anti-craving effect is likely attributable to the blockade of MORs in limbic brain regions. Activation of the HPA axis may be a consequence of the blockade of endogenous opioids (e.g., endorphins; Stephens and Wand 2012) that have an inhibitory effect on the PVN and cause a direct stress response in rats and humans that consume high levels of alcohol (Mitchell et al. 2009). Moreover, alcohol-associated cues stimulate glucocorticoid release during abstinence in people with a history of AUD (Fox et al. 2007) and produce craving. These findings suggest that glucocorticoids may also contribute to conditioned responses to promote relapse (Pantazis et al. 2021).

The glucocorticoid receptor is a steroid hormone-activated transcription factor that is ubiquitously expressed throughout the brain and peripheral tissues. Glucocorticoid receptor-dependent CRF downregulation in the PVN may be a mechanism of blunted HPA axis function in alcohol dependence (Makino et al. 1994; Richardson et al. 2008; Swanson and Simmons 1989). However, the glucocorticoid receptor-dependent upregulation of CRF levels in the CeA and BNST has been observed following the onset of high circulating glucocorticoids (Makino et al. 1994; Roberto et al. 2010; Shepard et al. 2000; Sommer et al. 2008) and may be involved in dysphoria, hypohedonia, and stress sensitization in alcohol dependence (Vendruscolo and Koob 2018). This bidirectional regulation of CRF has been hypothesized to depend on interactions between glucocorticoid receptors and various steroid-related co-regulators (Carmack et al. 2022; Edwards et al. 2015; Zalachoras et al. 2016). Dysregulation of the HPA axis and extrahypothalamic stress systems is argued to contribute to negative emotional states that drive compulsive alcohol drinking and seeking via negative reinforcement. These same neuroadaptations may underlie the vulnerability to other reward- and stress-related mental conditions, such as anxiety, depression, and heightened pain sensitivity, which are highly comorbid with AUD.

There is an association between glucocorticoid receptor-dependent plasticity in brain reward and stress regions and alcohol dependence. The downregulation of glucocorticoid receptor mRNA in the prefrontal cortex (PFC), NAc, and BNST was observed during acute alcohol withdrawal. This adaptation was hypothesized to be attributable to a compensatory effect of receptor overactivation, an effect that was possibly attributable to an increase in glucocorticoid levels in the brain and at least partially independent from HPA axis activity (Little et al. 2008; Somkuwar et al. 2017; Vendruscolo et al. 2012; Vendruscolo and Koob 2018). In a comparative study of plasma and brain levels of corticosterone in mice, corticosterone levels increased in several brain regions in alcohol-dependent mice compared with nondependent mice, although both dependent and nondependent mice had similar levels of corticosterone in blood (Little et al. 2008). These results likely suggest greater glucocorticoid receptor activation during alcohol dependence, which may be attributable to alterations of the enzyme 11β-hydroxysteroid dehydrogenase 1 (11β-HSD1). This enzyme colocalizes with glucocorticoid receptors and converts inactive glucocorticoids (e.g., cortisone and 11β-dehydrocorticosterone) into active glucocorticoids (e.g., cortisol and corticosterone). The pharmacological inhibition of 11β-HSD with carbenoxolone reduced the escalation of alcohol drinking in both rats and mice (Sanna et al. 2016). Consistent with this overactivation hypothesis, glucocorticoid receptor signaling, measured by phosphorylated glucocorticoid receptors, increased in the CeA in alcohol-dependent rats (Vendruscolo et al. 2015).

The functional role of glucocorticoid receptors in alcohol dependence-induced drinking during acute withdrawal was tested using the glucocorticoid receptor antagonist mifepristone (also called RU-38486 or RU-486) and several other glucocorticoid receptor modulators. The infusion of mifepristone directly in the CeA decreased alcohol drinking specifically in dependent rats (Vendruscolo et al. 2015). Chronic, systemic mifepristone administration blocked the escalation of alcohol drinking in alcohol-dependent rats but not in nondependent rats (Vendruscolo et al. 2012). Systemic, acute treatment with mifepristone and the selective glucocorticoid receptor antagonist CORT113176 decreased alcohol drinking in dependent rats but not in nondependent rats (Vendruscolo et al. 2015) and binge-like alcohol drinking in high-drinking rats but not in their low-drinking founder line (Savarese et al. 2020). Additionally, relative to their founder line, high-drinking mice exhibited alterations of the expression of several genes that are related to the glucocorticoid system in the NAc (Savarese et al. 2022). The glucocorticoid receptor modulators CORT118335, CORT122928, and CORT125134 significantly reduced alcohol self-administration in alcohol-dependent and nondependent rats, whereas CORT108297 had no effect on alcohol drinking in either group (McGinn et al. 2021), suggesting that different glucocorticoid receptor modulators may have different behavioral effects. Acute, systemic treatment with mifepristone reduced heavy alcohol drinking in rhesus macaques but did not block alcohol-induced relapse-like behavior in early abstinence (Jimenez et al. 2020).

Mifepristone did not affect alcohol drinking in nondependent, unstressed male rodents (Fahlke et al. 1995, 1996; Logrip and Gainey 2020; Repunte-Canonigo et al. 2015; Simms et al. 2012; Vendruscolo et al. 2015) or baboons (Holtyn and Weerts 2019). However, mifepristone reduced alcohol consumption in female rats (Logrip and Gainey 2020) and reduced the stress-induced reinstatement of alcohol seeking (Simms et al. 2012). Mifepristone prevented the increase in preference for alcohol in low-drinking male and female mice (O'Callaghan et al. 2005). Mifepristone also decreased alcohol drinking and seeking in rodents under stressful experimental conditions (Koenig and Olive 2004; O'Callaghan et al. 2005). The systemic and intra-CeA (but not intra-BLA) administration of mifepristone suppressed the stress-induced reinstatement of alcohol seeking (Simms et al. 2012). These findings suggest that the reinstatement of alcohol seeking may involve a contribution from the dysregulation of glucocorticoid receptor activity in the CeA, which may have implications for the contribution of stress to relapse in the preoccupation/anticipation stage of AUD.

These findings suggest preferential effects of mifepristone and CORT113176 in reducing excessive alcohol drinking under multiple conditions, including binge-like drinking, heavy drinking, dependence, and stress, and that the recruitment of different cellular responses by different glucocorticoid receptor modulators may affect behavior differently in alcohol-dependent and nondependent states (McGinn et al. 2021).

After several weeks of abstinence from alcohol (i.e., protracted abstinence), the expression of glucocorticoid receptor mRNA increased in the NAc, BNST, and CeA in dependent rats compared with nondependent rats (Vendruscolo et al. 2012), an effect that was opposite to findings during acute alcohol withdrawal. These results indicate the dynamic nature of glucocorticoid receptor expression in an alcohol-dependent state relative to a post-dependent state and suggest involvement of the glucocorticoid receptor system in long-lasting symptoms that persist into protracted abstinence. Studies that employed a systems-biology approach (i.e., based on the assembly and interrogation of gene regulatory networks to investigate genome-wide gene activity in protracted alcohol abstinence) provided additional evidence of the role of glucocorticoid receptors in alcohol dependence during protracted abstinence. The gene that encodes the glucocorticoid receptor, *nr3c1*, was found to be a master regulator in multiple brain regions, including the PFC, NAc, CeA, and VTA, in alcohol dependence (Repunte-Canonigo et al. 2015). The chronic, systemic administration of mifepristone (Vendruscolo et al. 2012) and acute mifepristone infusions in the NAc and VTA (Repunte-Canonigo et al. 2015) reduced alcohol drinking in rats with a history of alcohol dependence. Furthermore, rats that were chronically exposed to alcohol drinking and dependence exhibited the robust cue/context-induced reinstatement of alcohol seeking during protracted abstinence, an effect that was accompanied by higher glucocorticoid receptor activity in the medial PFC (Somkuwar et al. 2017).

The electrophysiological properties of neurons in the CeA are altered in alcohol dependence. Mifepristone reduced the frequency of spontaneous inhibitory postsynaptic currents (sIPSCs) in the CeA, without affecting postsynaptic measures, suggesting a decrease in GABA release, with the largest effect in dependent rats compared with nondependent rats. CORT118335 did not significantly alter GABA transmission in naive rats but decreased sIPSC frequency in dependent rats. Similarly, mifepristone decreased amplitudes of evoked inhibitory postsynaptic potentials only in dependent rats and during protracted withdrawal. These findings suggest that alcohol dependence enhances the efficacy of mifepristone and CORT118335 (Khom et al. 2022).

Glucocorticoid receptors and glucocorticoids are implicated in other alcohol-related behaviors. Treatment with mifepristone during acute withdrawal attenuated memory deficits in mice during protracted alcohol abstinence (Jacquot et al. 2008). Neurogenesis in the dentate gyrus of the hippocampus decreased in alcohol dependence (Mandyam 2013). Mifepristone had a neuroprotective effect in this brain region in rats that were exposed to binge-like alcohol (Cippitelli et al. 2014). Mifepristone also attenuated motor cross-sensitization between stress and alcohol in mice (Roberts et al. 1995) and reduced somatic signs of alcohol withdrawal in mice and rats (Jacquot et al. 2008; Sharrett-Field et al. 2013). However, mifepristone did not change the increase in anxiety-like behaviors in male or female alcohol-preferring rats and and did not reverse the reduction of sleeping bout duration in male alcohol-preferring rats (Vozella et al. 2021). These findings suggest that decreasing glucocorticoid receptor activity normalizes several signs of heightened stress and alcohol dependence (for review, see Lu and Richardson 2014).

Encouraged by these positive findings, a double-blind, randomized, placebo-controlled, human laboratory study was conducted to test the effect of mifepristone in 56 non-treatment-seeking alcohol-dependent individuals (Vendruscolo et al. 2015). Subjects who received mifepristone (600 mg daily, orally, for 1 week) exhibited lower cue-induced alcohol craving, reported a reduction of alcohol drinking, and exhibited improvements in liver function compared with those who received placebo. A single-nucleotide polymorphism genetic alteration of the FK506-binding protein, a glucocorticoid receptor negative regulator, was associated with problematic drinking in humans (Nylander et al. 2017). In humans with AUD, *NR3C1* methylation was altered in the PFC, and glucocorticoid receptor mRNA and protein levels were lower in the PFC (Gatta et al. 2021). During alcohol withdrawal, glucocorticoid levels positively correlated with the severity of cognitive deficits in humans with alcohol dependence (Errico et al. 2002).

In summary, chronic stress, alcohol misuse, and alcohol withdrawal alter brain reward and stress systems, and glucocorticoid-dependent plasticity contributes to these changes. The findings that are discussed above provide compelling evidence of a role for glucocorticoids and glucocorticoid receptors in all three stages of the alcohol addiction cycle. Glucocorticoid receptor antagonism, combined with behavioral treatments to cope with stress (e.g., cognitive-behavioral therapy; Longabaugh and Morgenstern 1999), may be potentially effective in normalizing sensitized stress systems and decrease compulsive alcohol drinking and seeking.

3.2.3 Between-System Neuroadaptations: Mineralocorticoids

Compared with glucocorticoid receptors, much less is known about the role of mineralocorticoid receptors in alcohol dependence (Leko et al. 2022). Both glucocorticoids (centrally and peripherally) and aldosterone (mainly peripherally) bind to mineralocorticoid receptors. Aldosterone is a mineralocorticoid steroid hormone that is produced in the zona glomerulosa of the adrenal gland. Aldosterone controls blood pressure and electrolyte levels through mineralocorticoid receptors, which are encoded by the *NR3C2* gene. However, mineralocorticoid receptors are also expressed in the CeA, hippocampus, and PFC. In the brain, mineralocorticoid receptors are preferentially activated by glucocorticoids. The expression of mineralocorticoid receptor mRNA in the CeA but not PFC negatively correlated with anxiety-like behavior and compulsive-like alcohol drinking in dependent rats but not in nondependent rats (Aoun et al. 2018). Similarly, alcohol drinking negatively correlated with mineralocorticoid receptor expression in the CeA but not PFC in long-term drinking rhesus macaques (Aoun et al. 2018). Both alcohol-drinking rhesus macaques and humans with AUD exhibited higher plasma aldosterone levels compared with controls, and aldosterone levels correlated with craving, anxiety, and the number of drinks that were consumed by humans with AUD (Aoun et al. 2018).

Systemic or intracerebroventricular administration of the nonselective mineralocorticoid receptor antagonist spironolactone and mineralocorticoid receptor antagonist RU28318 did not reduce alcohol drinking in male rats (Fahlke et al. 1995, 1996) or mice (O'Callaghan et al. 2005) in a continuous (24 h) two-bottle (water *vs.* alcohol) choice model or in a limited (1 h) two-bottle choice model following fluid restriction (Koenig and Olive 2004). However, 7 days of oral spironolactone treatment decreased alcohol drinking and blood pressure in high-drinking but not low-drinking male rats that were given continuous two-bottle choice access (Kashkin et al. 2018). In nondependent male and female rats, the systemic administration of spironolactone reduced operant alcohol self-administration (Makhijani et al. 2018). In nondependent male and female mice, spironolactone dose-dependently reduced binge-like alcohol drinking (Farokhnia et al. 2022). In male and female alcohol-dependent and nondependent rats, spironolactone dose-dependently reduced operant alcohol self-administration (Farokhnia et al. 2022). The intra-CeA infusion of eplerenone, a selective mineralocorticoid receptor antagonist, and mineralocorticoid receptor knockdown in the CeA reduced alcohol consumption in nondependent rats for 1 day after injection (Makhijani et al. 2020). In humans, two pharmacoepidemiological studies that used high-dimensional propensity score matching found that spironolactone dispensation was associated with a decrease in alcohol drinking (Palzes et al. 2021; Farokhnia et al. 2022). These findings suggest that mineralocorticoid receptors may be implicated in alcohol reinforcement and that spironolactone may be further studied as a potential pharmacotherapy for AUD.

3.2.4 Between-System Neuroadaptations: Norepinephrine

Although less well developed, evidence supports a role for norepinephrine systems in the extended amygdala in the negative motivational state and increase in self-administration that are associated with dependence. Evidence suggests that in animals and humans, central noradrenergic systems are activated during acute withdrawal from alcohol. Alcohol withdrawal in humans is associated with the activation of noradrenergic function, and signs and symptoms of alcohol withdrawal in humans are blocked by a postsynaptic β-adrenergic receptor antagonist (Romach and Sellers 1991). Alcohol withdrawal signs are also blocked by the administration of α_1- and β-adrenergic antagonists and selective blockade of norepinephrine synthesis (Trzaskowska and Kostowski 1983) in rodents. In alcohol-dependent rats, the α_1-adrenergic antagonist prazosin selectively blocked the increase in drinking that was associated with acute withdrawal (Walker et al. 2008), and the β-adrenergic receptor antagonist propranolol blocked the increase in drinking that was associated with acute withdrawal in dependent rats at doses that did not block nondependent drinking (Gilpin and Koob 2010). Thus, converging data suggest that noradrenergic neurotransmission is enhanced during alcohol withdrawal and that noradrenergic functional receptor antagonists can block aspects of alcohol withdrawal.

Consistent with a role for norepinephrine in hyperkatifeia, α_2-adrenergic receptor agonists have also been shown to be effective in blocking somatic alcohol withdrawal signs, reducing anxiogenic-like withdrawal responses, and reducing alcohol intake in alcohol-preferring rodents (Trzaskowska and Kostowski 1983; Arora and Vohora 2016). α_2-adrenergic receptor agonists decreased alcohol intake in alcohol-preferring rats (Opitz 1990).

In an animal study, prolonged treatment with alcohol (5 g/kg/day, orally) increased the sensitivity of α_2-adrenergic receptors to clonidine's actions on protein kinase activity, measured by the inhibition of an endogenous inhibitor of protein kinase in the hippocampus, NAc, and hypothalamus in rats (Szmigielski et al. 1977; Szmigielski et al. 1989). Postsynaptic α_2-adrenergic receptor function was downregulated in patients during acute withdrawal but also during late withdrawal after heavy alcohol intake, measured by the growth hormone response to clonidine (Berggren et al. 2000). These results are consistent with the hypothesis that the greater sensitivity of α_2-adrenergic receptors may be a marker of alcohol dependence in humans (Berggren et al. 2000; Balldin et al. 1992).

The dynamic nature of the brain stress system response to challenge is illustrated by the pronounced interaction of central nervous system CRF systems and central nervous system norepinephrine systems. Conceptualized as a feed-forward system at multiple levels of the pons and basal forebrain, CRF activates norepinephrine, and norepinephrine in turn activates CRF (Koob 1999). Much pharmacological, physiological, and anatomical evidence supports an important role for a CRF–norepinephrine interaction in the locus coeruleus in response to stressors (Valentino et al. 1991, 1993; Van Bockstaele et al. 1998). However, norepinephrine also stimulates CRF release in the PVN (Alonso et al. 1986), BNST, and CeA. Such feed-forward systems were further hypothesized to have functional significance for mobilizing an organism's response to environmental challenge, but such a mechanism may be particularly vulnerable to pathology (Koob 1999).

3.2.5 Between-System Neuroadaptations: Dynorphin

Dynorphin, an opioid peptide that binds to κ-opioid receptors (KORs), has long been known to show activation with chronic administration of psychostimulants and opioids (Nestler 2004; Koob 2008a), and KOR agonists produce aversive effects in animals and humans (Mucha and Herz 1985; Pfeiffer et al. 1986) and decrease dopamine activity (Wee and Koob 2010). Although KOR agonists suppress nondependent drinking, possibly via aversive stimulus effects (Wee and Koob 2010), KOR antagonists block excessive drinking that is associated with alcohol withdrawal and dependence (Holter et al. 2000; Walker and Koob 2008). Land et al. argued that the effects of CRF in producing negative emotional states are mediated by the activation of KOR systems (Land et al. 2008). However, KOR activation can activate CRF systems in the spinal cord (Song and Takemori 1992), and pharmacological evidence indicates that dynorphin systems can also activate the CRF system. A CRF_1 receptor antagonist blocked the KOR agonist-induced reinstatement of cocaine seeking in squirrel monkeys (Valdez et al. 2007).

Increases in prodynorphin mRNA expression are observed in the CeA and NAc during acute withdrawal and in alcohol-preferring rats (Kissler et al. 2014; Przewlocka et al. 1997; Zhou et al. 2013). With chronic, intermittent exposure to alcohol in mice, there was an increase in anxiety-like responses during acute withdrawal and extended withdrawal (72 h), and these anxiety-like responses were reversed by a KOR antagonist (Rose et al. 2016). In this study, the KOR activation-induced inhibition of dopamine release, measured by fast-scan voltammetry, was produced by a KOR agonist, and these effects were exacerbated in mice that were exposed to chronic, intermittent alcohol (Rose et al. 2016). In mice that were exposed to chronic, intermittent alcohol, an acute alcohol challenge decreased extracellular dopamine levels in the NAc, measured by microdialysis, which was reversed by the blockade of KORs, confirming the hypothesis that an increase in dynorphin/KOR system activity drives the reduction of dopamine release in the NAc (Karkhanis et al. 2016; Karkhanis and Al-Hasani 2020).

Both systemic and intracerebral KOR antagonist administration blocked the increase in alcohol intake that was associated with extended access to, and dependence on, alcohol (Holter et al. 2000; Walker and Koob 2008). The stress-induced escalation of intake in mice that were exposed to chronic, intermittent alcohol was also blocked by a KOR antagonist (Anderson et al. 2016). Microinjection studies suggest that these effects may be mediated by the extended amygdala, including the CeA and shell of the NAc (Nealey et al. 2011; Anderson et al. 2019). In a human post-mortem study, expression of the prodynorphin (*PDYN*) and KOR (*OPRK1*) genes in NAc samples from humans with AUD did not differ, but downregulation of the expression of the dopamine D_1 (*DRD1*) but not D_2 (*DRD2*) receptor genes was seen in samples from humans with AUD (Bazov et al. 2018). The expression of *DRD1* and *DRD2* significantly correlated with *PDYN* and *OPRK1* expression, suggesting high levels of transcriptional coordination between these genes and supporting the hypothesis of a role for the dynorphin and dopamine systems in the negative affective state that is associated with AUD (Bazov et al. 2018).

3.2.6 Between-System Neuroadaptations: Vasopressin

Vasopressin is a neuropeptide that is synthesized in the hypothalamus and transported to the posterior pituitary where it is released into the bloodstream during dehydration to act as a hormone via vasopressin V_1 receptors to raise blood pressure and via vasopressin V_2 receptors to produce water retention (Hays 1980; Kamm et al. 1928). Vasopressin also has a neurotropic action by being centrally localized and originating in the PVN, BNST, medial amygdala, and suprachiasmatic nucleus and projecting extensively throughout the basal forebrain (De Vries and Buijs 1983).

Vasopressin was originally hypothesized to play a role in learning and memory and the maintenance of tolerance to hypothermic and sedative effects of alcohol via actions on central vasopressin V_1 receptors (de Wied and Versteeg 1979; Le Moal et al. 1984; Hoffman et al. 1978). Later, it became clear that vasopressin V_1 receptor antagonism could produce anxiolytic-like effects in rodent models, with a focus on sites of action in the extended amygdala and its inputs (Salome et al. 2006). A small-molecule V_{1b} receptor antagonist attenuated the increase in levels of alcohol self-administration in dependent animals, without affecting alcohol drinking in nondependent animals (Edwards et al. 2012a), and reduced alcohol intake in Sardinian alcohol-preferring rats (Zhou et al. 2011).

In humans, the selective V_{1b} receptor antagonist ABT-436 significantly increased the percentage of days abstinent compared with placebo in a 12-week multisite randomized clinical trial in 150 individuals with AUD (Ryan et al. 2017). Individuals who reported higher baseline levels of stress responded better to ABT-436 treatment than to placebo on drinking outcomes in an analysis of moderators (Ryan et al. 2017), suggesting that vasopressin may also contribute to hyperkatifeia that is associated with alcohol withdrawal.

3.2.7 Between-System Neuroadaptations: Hypocretin

The neuropeptide hypocretin (orexin) has been associated with sleep–wake regulation, arousal, stress, and drug-seeking behavior (Sutcliffe and de Lecea 2002; Johnson et al. 2012; Mahler et al. 2012). Two neuropeptides, hypocretin 1 (Hcrt-1; also called orexin A) and Hcrt-2 (also called orexin B) that target two G-protein-coupled receptors (Hcrt-1 and Hcrt-2), have been identified. Hypocretin-containing neurons are found in restricted regions of the dorsal hypothalamus (de Lecea et al. 1998; Sakurai et al. 1998) and project widely throughout the brain (Peyron et al. 1998).

Hypocretin neuron projections include reciprocal connections to the extended amygdala and other basal forebrain regions (Baldo et al. 2003; Peyron et al. 1998), thus providing a neuroanatomical basis for the hypothesis that hypocretin neurotransmission plays an important role in hyperkatifeia (for a review, see Koob 2008a; Koob et al. 2014). Consistent with this hypothesis, the stress-induced activation of Hcrt neurons through CRF_1 receptor activation has been observed (Winsky-Sommerer et al. 2004, 2005). Intraventricular Hcrt-1 elevates intracranial self-stimulation thresholds (Boutrel et al. 2005), and intra-VTA infusions of Hcrt-1 also elevated reward thresholds via the activation of CRF in the CeA (Hata et al. 2011). Altogether, these results support the hypothesis that Hcrt may have anti-reward/brain stress actions, possibly via the CRF system in the extended amygdala.

Chronic alcohol consumption increased prepro-Hcrt mRNA expression in the lateral hypothalamus in inbred alcohol-preferring rats (Lawrence et al. 2006), and both Hcrt-1 and Hcrt-2 receptor blockade decreased alcohol self-administration in alcohol-preferring rats (Jimenez et al. 2020). A Hcrt-1 receptor antagonist dose-dependently decreased alcohol intake in alcohol-dependent mice (Lopez et al. 2016). In a rat study, the Hcrt-2-selective antagonist NBI-80713 and dual Hcrt-1/2 antagonist NBI-87571 dose-dependently decreased overall alcohol drinking in alcohol-dependent rats, whereas the Hcrt-1-selective antagonist SB-408124 decreased alcohol drinking in both alcohol-dependent and nondependent rats at the highest dose. A significant decrease in Hcrt-1 mRNA expression was observed in the CeA in dependent rats under acute withdrawal conditions compared with nondependent rats (Aldridge et al. 2022). The knockdown of *Hcrt* in CeA-projecting neurons from the lateral hypothalamus with a short-hairpin RNA-encoding adeno-associated viral vector with retrograde function significantly attenuated alcohol self-administration in alcohol-dependent rats (Aldridge et al. 2022). A Hcrt-1 receptor antagonist also blocked the stress (yohimbine)-induced reinstatement of alcohol seeking (Richards et al. 2008). Altogether, Hcrt likely exerts modulatory actions on brain stress systems to contribute to alcohol withdrawal-induced hyperkatifeia.

3.2.8 Between-System Neuroadaptations: Neuroimmune Systems

Neuroimmune systems that interface with the central nervous system are also hypothesized to contribute to neuroadaptive processes that mediate the dysregulation of hyperkatifeia that is associated with the withdrawal/negative affect stage of the addiction cycle (Koob and Volkow 2010; Crews and Vetreno 2016). Both microglia and astrocytes mediate interactions between the immune system and stress system that are associated with hyperkatifeia (Crews et al. 2017). In response to insults, both microglia and astrocytes undergo a process of activation and can adopt proinflammatory and anti-inflammatory states (Jang et al. 2013; Kettenmann et al. 2013). Repeated cycles of stress and excessive drug use are hypothesized to result in increasingly sensitized/activated microglia, contributing to signs and symptoms that are associated with the withdrawal/negative affect stage of the addiction cycle (Crews et al. 2017). With repeated episodes of alcohol administration or stress, immune danger signals and glial cell activation are amplified within glia and across other brain cells, resulting in increases in Toll-like receptor 4 (TLR4) danger signals, the expression of nuclear factor-κB target genes, and tumor necrosis factor-α (TNFα; Watkins et al. 2007; Graeber 2010). For example, chronic alcohol treatment in mice induced proinflammatory gene expression in the brain that persisted for at least 1 week of abstinence (Qin et al. 2008). Mice that lacked TLR4 were protected from alcohol-induced glial cell activation, anxiety-like behavior, and cognitive impairments (Pascual et al. 2011). Mice that lacked CD14, a key TLR4 accessory signaling protein, drank significantly less alcohol than normal mice (Blednov et al. 2012). One likely site of action for neuroimmune cytokines is the extended amygdala. Injections of TNFα in the amygdala in rats increased anxiety-like behavior and mimicked the amplification of anxiety-like behavior that was observed with repeated cycles of alcohol drinking (Breese et al. 2008). Injections of cytokines in the amygdala also increased withdrawal-induced hyperkatifeia-like responses, like exposure to stress or alcohol (Knapp et al. 2011). Furthermore, TLR4 knockdown in the amygdala reduced responding for alcohol in alcohol-dependent rats (Liu et al. 2011), consistent with the hypothesis that the activation of innate immune danger signals in the amygdala may contribute to hyperkatifeia that drives dependence-induced alcohol drinking. However, the genetic and pharmacologic blockade of TLR4 has a minimal effect on alcohol drinking in dependent and nondependent rodents (Harris et al. 2017), suggesting that TLR4 blockade alone is insufficient to modulate alcohol drinking.

Additionally, CRF receptor antagonists that blocked alcohol dependence-induced excessive drinking also blocked the induction of negative affect-like and anxiety-like responses that were produced by cytokines (Knapp et al. 2011). Another link between alcohol misuse and innate immunity may involve inflammation in the gut and alcohol-induced increase in permeability of the gut to endotoxins, such as lipopolysaccharides, which cause a sickness-like response. Activated cytokines may be transported to the brain and result in neuroinflammation and microglial activation (Dantzer et al. 2008; Qin et al. 2008; Banks and Erickson 2010; Szabo et al. 2011). These studies support the hypothesis that the amplification of glial activation in the amygdala and other components of brain fear/anxiety/stress neurocircuitry may contribute to the progression and persistence of hyperkatifeia, possibly by interacting with brain stress systems.

4 Anti-Stress Neurotransmitters/Neuromodulators and Hyperkatifeia

Hyperkatifeia that is associated with AUD may not only derive from the activation of pro-stress neurotransmitter systems but also derive from anti-stress neurotransmitter systems (see Fig. 3 above). One view of anti-stress neurotransmitter systems has been that they serve as neuroadaptive buffers to pro-stress actions that are described above, in which such neurotransmitter systems are activated in response to the engagement of pro-stress neurotransmitters (Heilig 2004; Pleil et al. 2015) or in anticipation of the activation of stress (Heilig and Koob 2007). Neurotransmitter/neuromodulatory systems that are implicated in anti-stress actions include NPY, nociceptin, endocannabinoids, and oxytocin.

4.1 Neuropeptide Y

Neuropeptide Y has powerful orexigenic and anxiolytic-like effects in animal models and has been hypothesized to have actions that are opposite to CRF in animal models of AUD (Heilig et al. 1994). Numerous studies showed that the activation of NPY receptors in the brain blocked the increase in self-administration in alcohol-preferring rats and in alcohol dependence (Gilpin et al. 2003, 2008, 2011; Thorsell et al. 2005a, b, 2007), and these effects have been localized to the extended amygdala (Gilpin et al. 2008; Kelley et al. 2001; Pleil et al. 2015; for a review, see Koob 2021). Chronic NPY treatment blocked the escalation of operant alcohol drinking that was associated with alcohol dependence and gradual increases in alcohol drinking in intermittently tested nondependent rats. Neuropeptide Y decreased baseline GABAergic transmission and reversed the alcohol-induced enhancement of inhibitory transmission in the CeA by suppressing GABA release via actions at presynaptic NPY Y_2 receptors (Gilpin et al. 2011). However, neither systemic (JNJ-31020028) nor intra-CeA (BIIE0246) Y_2 receptor antagonism changed operant alcohol self-administration in alcohol-dependent or nondependent rats, but BIIE0246 in the CeA reduced anxiety-like behavior in both alcohol-dependent and alcohol-naive rats (Kallupi et al. 2014). These findings suggest that anxiety-like behavior and alcohol drinking may occur via different functions of Y_2 receptors (e.g., as an autoreceptor *vs.* heteroreceptor). Given that the activation of NPY in the CeA has opposite effects to CRF, one hypothesis is that NPY may act as a buffer to the stress-driving effects of CRF. By extrapolation, low functional NPY activity may contribute to the etiology and vulnerability to hyperkatifeia.

4.2 Nociceptin

Nociceptin/orphanin QF (N/OFQ) is an endogenous opioid-like peptide that is found in the brain that is the endogenous ligand for the nociceptin receptor, and numerous nociceptin neurons are found in the extended amygdala (Mollereau and Mouledous 2000). Nociceptin has anti-stress-like effects in animals (Ciccocioppo et al. 2003; Martin-Fardon et al. 2010) and can block both opioid and alcohol withdrawal and attenuate alcohol drinking in high alcohol-preferring rats and alcohol self-administration in dependent animals (Economidou et al. 2008, 2011). For example, synthetic nociceptin receptor agonists blocked high alcohol consumption in Marchigian Sardinian preferring (msP) rats, a genetically selected line of rats that is known to be hypersensitive to stressors (Economidou et al. 2008). However, small-molecule nociceptin receptor antagonists also decreased high levels of drinking in msP rats (Rorick-Kehn et al. 2016), leading to the hypothesis that agonist effects may reflect receptor desensitization, producing the antagonist-like functional blockade of subchronic agonist dosing (Rorick-Kehn et al. 2016; Toll et al. 2016).

4.3 *Endocannabinoids*

Endocannabinoids have also been hypothesized to have stress-buffering actions that may be involved in the response to repeated withdrawal from alcohol and the vulnerability to negative emotional dysregulation that drives excessive drinking (Serrano and Parsons 2011; Koob 2021). Increasing endogenous endocannabinoid levels through the inhibition of endocannabinoid clearance mechanisms produces anxiolytic-like effects in various animal models of anxiety, particularly under stressful or aversive conditions (Serrano and Parsons 2011). For example, both fatty acid amide hydrolase inhibitors and monoacylglycerol inhibitors, which functionally increase endocannabinoids, decreased withdrawal-induced anxiety-like behavior and decreased alcohol consumption in alcohol-dependent rats (Serrano et al. 2018). Some of these effects may be mediated by actions in the extended amygdala, in which microinjections of the selective fatty acid amide hydrolase inhibitor URB597 in the CeA in alcohol-preferring rats significantly reduced alcohol self-administration (Stopponi et al. 2018).

4.4 *Oxytocin*

Oxytocin, in addition to its hormonal role in stimulating uterine contractions in childbirth and stimulating bonding and milk production post-partum, has been hypothesized to play a role in such diverse functions as memory, learning, social behavior, fear, and anxiety (Lee et al. 2009; Stoop 2012). Oxytocin has also been hypothesized to normalize stress function and attenuate hyperkatifeia in the context of addiction (Lee and Weerts 2016; Koob 2021). Early work with alcohol showed that oxytocin and oxytocin analogs inhibited the development of tolerance and decreased withdrawal signs in alcohol-dependent mice (Kovacs et al. 1998; Szabo et al. 1987). Oxytocin was also shown to attenuate the increase in drinking that was associated with dependence in animal models after both systemic and central administration at doses that did not alter non-alcohol-related behaviors or alcohol drinking in nondependent rats (Tunstall et al. 2019). As with the other stress buffer neurotransmitter systems, *ex vivo* electrophysiological recordings in the CeA indicated that oxytocin blocked the facilitatory effects of alcohol on GABA release in the CeA in dependent rats but not in nondependent rats (Tunstall et al. 2019). Intraperitoneal oxytocin administration also blocked the stress-induced reinstatement of alcohol seeking in mice (King and Becker 2019).

In summary (Table 2), acute withdrawal from alcohol increases CRF in the extended amygdala, which contributes to hyperkatifeia, and the increase in drug intake that is associated with dependence. Glucocorticoids may play a role in sensitizing the extended amygdala CRF system. Acute withdrawal may also increase the release of norepinephrine in the BNST, dynorphin in the NAc, hypocretin and vasopressin in the CeA, and neuroimmune modulation, all of which may contribute

Table 2 Role of corticotropin-releasing factor in dependence

Drug	CRF antagonist effects on withdrawal-induced anxiety-like responses	Withdrawal-induced changes in extracellular CRF in CeA	CRF antagonist effects on dependence-induced increases in self-administration
Cocaine	↓	↑	↓
Opioids	↓[a]	↑	↓
Alcohol	↓	↑	↓
Nicotine	↓	↑	↓
Δ^9-tetrahydrocannabinol	↓	↑	nt

[a] Aversive effects with place conditioning. nt, not tested. CeA, central nucleus of the amygdala

to the negative emotional state that is associated with dependence. The lower activity of NPY, nociceptin, endocannabinoids, and oxytocin in the extended amygdala may contribute to hyperkatifeia that is associated with alcohol dependence. The activation of brain stress systems, combined with the inactivation of brain anti-stress systems, elicits powerful emotional dysregulation via the extended amygdala. Such dysregulation of emotional processing may be a significant contribution to between-system opponent processes that maintain dependence and set the stage for more prolonged state changes in emotionality, such as in protracted abstinence.

5 Hyperalgesia and Hyperkatifeia: Interface Between Pain and Alcohol Use Disorder

Acute alcohol administration in human studies has analgesic effects (Thompson et al. 2017) and anesthetic effects at high doses. However, alcohol withdrawal in human studies is associated with an increase in pain sensitivity. In a subgroup analysis, more severe withdrawal was associated with lower pain thresholds and lower pain tolerance (Jochum et al. 2010; Edwards et al. 2020). Lower thermal pain tolerance during withdrawal was exaggerated by negative emotional states (Jochum et al. 2010), and individuals with comorbid pain and AUD may drink to alleviate pain-related negative affect (Moskal et al. 2018; Witkiewitz et al. 2015). Chronic alcohol consumption in rodents produced hyperalgesia in multiple pain assays (Gatch and Lal 1999; Gatch 2009; Egli et al. 2012; Edwards et al. 2012b, 2015, 2020).

The neurobiological substrates for alcohol withdrawal-induced hyperalgesia range from molecular elements that are relevant to molecular signal transduction function, metabolism, and epigenetic factors to neurocircuits that are relevant to hyperkatifeia. For example, an antisense oligodeoxynucleotide to protein kinase Cε blocked alcohol withdrawal-induced hyperalgesia (Dina et al. 2006; Shumilla et al. 2005). Melanocortin-4 receptor antagonists, which are known to produce antidepressant-like and anxiolytic-like effects, reduced thermal hyperalgesia in alcohol-dependent rats during withdrawal (Roltsch Hellard et al. 2017; Avegno et al. 2018). A histone deacetylase inhibitor blunted alcohol-induced hyperalgesia, implicating epigenetic mechanisms (Pradhan et al. 2019).

Hyperalgesia that is associated with alcohol withdrawal can be blunted by the blockade of CRF_1 receptors (Edwards et al. 2012b). Corticotropin-releasing factor receptor antagonists also blunted nociceptive hypersensitivity in a wide variety of pain assays in animals (Hummel et al. 2010). More relevant to a hyperalgesia-hyperkatifeia connection, withdrawal-induced hyperalgesia and relapse-like alcohol consumption were reversed by the chemogenetic inhibition of lateral habenula neurons, the pharmacological activation of M-channels, and overexpression of the M-channel subunit KCNQ3 (Kang et al. 2019). An intra-lateral habenula infusion of serotonin 5-hydroxytryptamine-2 receptor antagonists or a serotonin reuptake blocker decreased nociceptive sensitivity and alcohol intake in rats during alcohol withdrawal (Zuo et al. 2019). Corticotropin-releasing factor in the CeA has been shown to play an important role in affect-like responses that are associated with pain and pain modulation (Ji et al. 2013).

6 Negative Urgency

Impulsivity in addiction, and AUD in particular, takes on multiple forms, such as attentional ("decreased ability to focus") and motor ("acting without thinking") impulsivity and negative urgency (Coskunpinar et al. 2013; VanderVeen et al. 2016; Stautz and Cooper 2013). Some have argued that the most important impulsivity-related trait for alcohol-related problems and alcohol dependence is negative urgency (VanderVeen et al. 2016). Negative urgency can be defined as a tendency to experience strong impulses, frequently under conditions of negative affect (Whiteside and Lynam 2003), or impulsive risk-taking behavior during extreme negative emotional states (VanderVeen et al. 2016). Negative urgency has been particularly associated with hyperkatifeia (VanderVeen et al. 2016) because such symptoms as irritability, anxiety, and dysphoria may elicit intense urges. A decrease in the capacity to resist urges to pursue substance use in the preoccupation/anticipation stage, in which behavior occurs rapidly without forethought of potential harm (impulsively) despite actual negative or incorrect consequences (compulsively), may be mediated by impairments in executive control. Indeed, some have argued that negative urgency may be an endophenotype of AUD (VanderVeen et al. 2016; Cyders and Smith 2008).

Negative urgency has been observed in alcohol-seeking individuals with and without AUD. In a study of alcohol in adults in a community setting, negative urgency increased alcohol self-administration via an increase in emotional reactivity to negative events and an increase in alcohol craving in response to an initial alcohol exposure (VanderVeen et al. 2016). In a study of undergraduate students who reported a history of self-harm, negative urgency was significantly associated with several self-harming measures, problematic alcohol use, and eating problems (Dir et al. 2013).

With regard to alcohol misuse, in a community sample of college and non-college young-adult Australians, negative urgency was directly related to greater alcohol intake, binge drinking, and alcohol-related problems (Tran et al. 2018). In a study of moderate to heavy drinking in college students, sadness and anxiety were each directly associated with AUD symptoms, and negative urgency moderated the within-person association between negative affect and risk behavior (Simons et al. 2010). Negative urgency was also linked to the higher prevalence of comorbid AUD in adults with attention-deficit/hyperactivity disorder (Daurio et al. 2018).

In a study that was designed to identify core functional domains that are associated with AUD in a large diverse sample of 454 individuals who represented the spectrum of alcohol use to AUD and who were screened in the National Institute on Alcohol Abuse and Alcoholism (NIAAA) outpatient clinic for participation in research studies between January 2015 and February 2017, factor analysis identified three intercorrelated functional dimensions: impairments in executive control, negative emotionality, and incentive salience (Kwako et al. 2019). Each domain discriminated participants with AUD *vs.* without AUD. Negative urgency was 50% higher in participants with AUD than those without AUD and loaded strongly on the impaired inhibitory control factor and less so on the negative emotionality factor (Kwako et al. 2019).

Neurobiological data on negative urgency are limited to date, and the data are almost completely derived from human imaging studies (Um et al. 2019). Negative urgency has been hypothesized to reflect impairments in the top-down cortical control over both basal ganglia and extended amygdala function (Smith and Cyders 2016; Zorrilla and Koob 2019). For example, compulsive alcohol seeking has been hypothesized to reflect the loss of control over pathological habits that involve the basal ganglia (Giuliano et al. 2019; Lovinger and Kash 2015; Belin et al. 2013). Preparing for action in the context of emotional experiences has long been hypothesized to involve interconnections between the amygdala and prefrontal cortex, particularly its more medial aspect, the ventromedial PFC (Bechara et al. 2000; Lewis and Todd 2007). Thus, in the presence of negative emotional states, and by extrapolation hyperkatifeia, one could hypothesize a reduction of inhibitory control over alcohol seeking and habits as reflected by alterations of the function and connectivity of OFC/ventromedial PFC projections to the basal ganglia and extended amygdala (Cyders and Smith 2008; Smith and Cyders 2016; Belin et al. 2013; Giuliano et al. 2019).

Supporting this hypothesis, trait negative urgency in social drinkers predicted an increase in activation of the ventromedial PFC in response to an alcohol odor cue in social drinkers and mediated the association between activation of the ventromedial PFC and alcohol craving and problematic drinking (Cyders et al. 2014). Similarly, negative urgency predicted greater OFC and amygdala activation in response to negative visual stimuli and mediated the relationship between activation and risky behavior (Cyders et al. 2015). Negative urgency predicted greater caudate responses to alcohol-related images in alcohol-dependent individuals (Chester et al. 2016).

From a neurotransmitter-specific neurocircuitry perspective, many neurochemical studies have linked deficiencies in brain serotonin function to a greater risk for alcohol use and/or dependence (Ballenger et al. 1979). Low serotonin function has been linked to impulsivity, violence, suicide, and AUD (Roy et al. 1990). Such impulsivity was shown to involve negative urgency in a study of participants from a larger longitudinal study of familial AUD that spanned three generations (Chassin et al. 1992). A composite polygenic 5-hydroxytryptamine score predicted alcohol problems via an increase in negative urgency and not via other measures of impulsivity (Wang and Chassin 2018).

Thus, from a negative emotional state and hyperkatifeia perspective, negative urgency forms an additional pathway for impulsivity deficits to continue throughout the addiction cycle, not being simply limited to positive urgency, reward, or basal ganglia function. Undergraduate subjects with high negative urgency exhibited greater recruitment of inhibitory brain regions than controls on an emotional, inhibitory Go/No-Go task (Chester et al. 2016). These findings suggest that negative urgency might manifest as an increase in attention to rewarding cues in the environment, perhaps attributable to the excessive activation of, and fatigue in, inhibitory brain regions (Chester et al. 2016).

7 Compulsivity in Alcohol Use Disorder: An Allostatic View

Compulsivity in AUD can derive from multiple sources, including the enhancement of incentive salience, the engagement of habit function, and impairments in executive function. However, underlying each of these sources is a negative emotional state that may strongly impact compulsivity. Development of the negative emotional state that drives the negative reinforcement of addiction has been defined as the negative emotional side of addiction (Koob and Le Moal 2005, 2008) and is hypothesized to be the *b-process* of the hedonic dynamic that is known as the opponent process when the *a-process* is euphoria. The negative emotional state (hyperkatifeia) that comprises the withdrawal/negative affect stage consists of key motivational elements, such as chronic irritability, emotional pain, malaise, dysphoria, alexithymia, and the loss of motivation for natural rewards, and is characterized in animals by elevations of reward thresholds during withdrawal from all major drugs of addiction. Two processes are hypothesized to form the neurobiological basis for the *b-process*: loss of function in reward systems (within-system neuroadaptation) and recruitment of brain stress or anti-reward systems (between-system neuroadaptation; Koob and Bloom 1988; Koob and Le Moal 1997). Anti-reward is a construct that is based on the hypothesis that brain systems are in place to limit reward (Koob and Le Moal 2008). As dependence and withdrawal develop, brain stress systems, such as CRF, norepinephrine, and dynorphin, are recruited, producing aversive or stress-like states (Koob 2003; Nestler 2001; Aston-Jones et al. 1999). Within motivational circuits of the ventral striatum-extended amygdala, reward function simultaneously decreases. The combination of decreases in reward

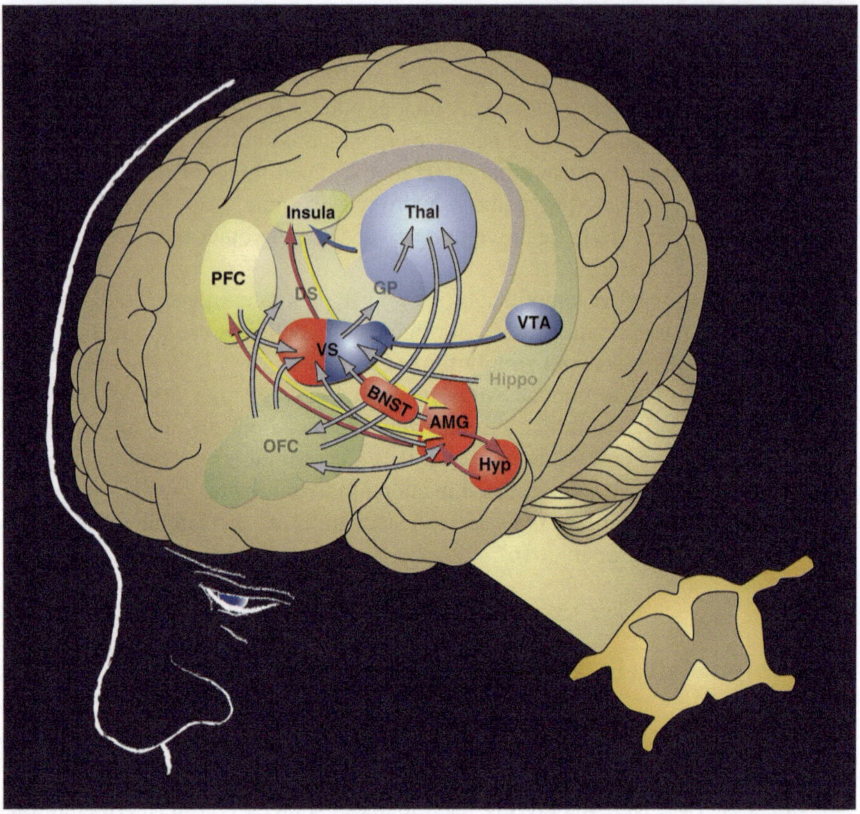

Fig. 4 Pathways for key elements of addiction circuitry (including AUD) implicated in negative emotional states. Addiction circuitry is composed of structures that are involved in the three stages of the addiction cycle: binge/intoxication (ventral striatum, dorsal striatum, thalamus), withdrawal/negative affect (ventral striatum, bed nucleus of the stria terminalis, central nucleus of the amygdala), and preoccupation/anticipation (prefrontal cortex, orbitofrontal cortex, hippocampus). Highlighted here for the withdrawal/negative affect stage is an increase in activity in the extended amygdala and decrease in activity in the reward system, illustrated with the use of imaging colors (i.e., red for high activity and blue for low activity). [Modified with permission from Blackburn-Munro and Blackburn-Munro (2003) and Koob et al. (2008).] AMG, amygdala; BNST, bed nucleus of the stria terminalis; DS, dorsal striatum; GP, globus pallidus; Hippo, hippocampus; Hyp, hypothalamus; Insula, insular cortex; OFC, orbitofrontal cortex; PFC, prefrontal cortex; Thal, thalamus; VS, ventral striatum; VTA, ventral tegmental area. [Modified with permission from (Zald and Kim 2001)]

neurotransmitter function and recruitment of anti-reward systems provides a powerful source of negative reinforcement that contributes to compulsive drug-seeking behavior and addiction (Fig. 4).

An overall conceptual theme that is argued herein is that drug addiction represents a break with homeostatic brain regulatory mechanisms that regulate the emotional state of the individual. The dysregulation of emotion begins with the binge and subsequent acute withdrawal but leaves a residual neuroadaptive trace that allows rapid "re-addiction" even months and years after detoxification and abstinence. Thus, the emotional dysregulation of alcohol addiction represents more than simply a homeostatic dysregulation of hedonic function—it also represents a dynamic break with homeostasis of this system that has been termed *allostasis* (Koob 2003).

Allostasis, originally conceptualized to explain persistent morbidity of arousal and autonomic function, can be defined simply as "stability through change" (Sterling and Eyer 1988). Allostasis is different from homeostasis. Allostasis involves a feed-forward mechanism rather than negative feedback mechanisms of homeostasis. Allostasis involves a change in set point with the continuous reevaluation of need and continuous readjustment of all parameters toward new set points. The set point here is an elevation of reward threshold. An *allostatic state* can be defined as a state of chronic deviation of the reward system from its normal (homeostatic) operating level. *Allostatic load* is defined as the "long-term cost of allostasis that accumulates over time and reflects the accumulation of damage that can lead to pathological states" (McEwen 2000). Although the concept of allostatic state has not received much attention, the argument here is that AUD reflects largely a movement to an allostatic state, often before sufficient pathology has ensued to produce allostatic load that is sufficient for pathology (Koob and Le Moal 2001).

Allostatic mechanisms have been hypothesized to be involved in maintaining a functioning brain reward system that has relevance to the pathology of addiction (Koob and Le Moal 2001). Two components are hypothesized to adjust to challenges to the brain that are produced by drugs of addiction: underactivation of brain reward transmitters and circuits and recruitment of the brain anti-reward or brain stress systems (Fig. 4). Thus, the physiological mechanism that allows rapid responses to environmental challenge becomes the source of pathology if adequate time or resources are not available to shut off the response. One example is the interaction between CRF and norepinephrine in the brainstem and basal forebrain that could lead to pathological anxiety (Koob 1999) (Fig. 5).

Repeated challenges, such as with repeated alcohol binge episodes, lead to attempts of the brain via molecular, cellular, and neurocircuitry changes to maintain stability but at a cost. For the AUD framework that is elaborated herein, residual deviation from normal brain reward thresholds is termed the *allostatic state*. This state represents a combination of chronic elevation of the reward set point that is fueled by a decrease in the function of reward circuits and recruitment of anti-reward systems, both of which lead to the compulsivity of alcohol seeking and alcohol taking. How these systems are modulated by other known brain emotional systems that are localized to the basal forebrain, where the ventral striatum and extended amygdala project to convey emotional valence, how the dysregulation of brain emotional systems impacts the cognitive domain that is linked to impairments in executive function, and how individuals differ at the molecular-genetic level of analysis to convey loading on these circuits remain challenges for future research (George and Koob 2010).

The present thesis does not exclude key roles for other systems that are associated with the addiction process, including the mesolimbic dopamine system that is

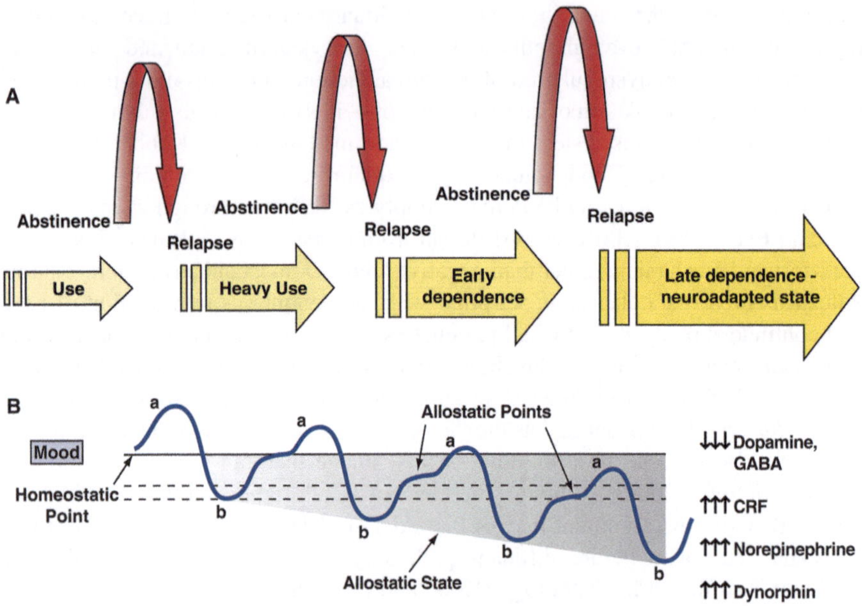

Fig. 5 (**a**) Schematic of the progression of alcohol dependence over time, illustrating the shift in underlying motivational mechanisms. From initial, positive reinforcing, pleasurable alcohol effects, the addictive process progresses over time to being maintained by negative reinforcing relief from a negative emotional state. Data suggest that neuroadaptations that encompass the recruitment of extrahypothalamic CRF systems are key to this shift. [Taken with permission from (Heilig and Koob 2007).] (**b**) The *a-process* represents a positive hedonic or positive mood state, and the *b-process* represents the negative hedonic or negative mood state. The affective stimulus (state) has been argued to be the sum of both the *a-process* and *b-process*. An individual who experiences a positive hedonic mood state from a drug of addiction, such as alcohol, with sufficient time between re-administering the drug is hypothesized to retain the *a-process*. An appropriate counteradaptive opponent process (*b-process*) that balances the activational process (*a-process*) does not lead to an allostatic state. The changes in the affective stimulus (state) in an individual with repeated frequent drug use may represent a transition to an allostatic state in the brain reward systems and, by extrapolation, a transition to addiction. Notice that the apparent *b-process* never returns to the original homeostatic level before drug taking begins again, thus creating a progressively greater allostatic state in the brain reward system. The counteradaptive opponent process (*b-process*) does not balance the activational process (*a-process*) but in fact shows residual hysteresis. Although these changes in the figure are exaggerated and condensed over time, the hypothesis is that even during post-detoxification (a period of "protracted hyperkatifeia"), the reward system still bears allostatic changes. The following definitions apply: *allostasis*, the process of achieving stability through change; *allostatic state*, a state of chronic deviation of the regulatory system from its normal (homeostatic) operating level; *allostatic load*, the cost to the brain and body of the deviation, accumulating over time, and reflecting in many cases pathological states and accumulation of damage. [Modified with permission from (Koob and Le Moal 2001)]

involved in incentive salience, the dorsal striatum that is involved in habit formation, the parabrachial-amygdala and spinothalamocortical systems that are involved in pain, and the PFC that is involved in decision-making (Koob and Volkow 2010; George and Koob 2010). Such modules are driven by bottom-up signals from both the external world and interoceptive signals and by top-down signals from higher-order systems that mediate cognitive control. Indeed, the failure of a specific module may differ from one individual to another and may represent a neuropsychobiological mechanism that underlies individual differences in the vulnerability to drug addiction. For example, we have hypothesized that individual differences in the function of the incentive salience mesolimbic dopamine system and the habit/striatum modules may be particularly important for craving-type 1 (or reward craving), defined as craving for the rewarding effects of alcohol and usually induced by stimuli that have been paired with alcohol self-administration, such as environmental cues. Additionally, hypoactivity of the decision-making/prefrontal cortex module may lead to a loss of control over drug intake despite negative consequence because of impairments in inhibitory control and decision-making, leading to choices of immediate rewards over delayed rewards (Goldstein and Volkow 2002).

Nevertheless, the hypothesis here is that a core component of AUD involves hyperactivity of the negative emotional state/extended amygdala system that is associated with increases in emotional pain and stress and might be a risk factor for drug use as self-medication for emotional pain, dysphoria, and stress (Khantzian 1997). A subhypothesis is that vulnerability in the emotional pain parabrachial-amygdala system (Besson 1999; Shurman et al. 2010) may lead to an increase in emotional pain during withdrawal and intense craving-type 2 (or withdrawal relief craving; Pantazis et al. 2021), which is conceptualized as an excessive motivation for the drug to obtain relief from a state change that is characterized by anxiety and dysphoria after protracted abstinence (Heinz et al. 2003), thus contributing to the preponderant role of the withdrawal/negative affect stage that characterizes AUD. Greater reactivity of the stress/HPA axis module may be critical in the initiation of alcohol intake and for the maintenance of drug intake, which have little initial rewarding value, such as with nicotine. Activation of the HPA axis can potentiate the reinforcing effects of drugs (Piazza and Le Moal 1998). However, this activation can drive amygdala CRF, further exacerbating the development of negative emotional states (Koob and Kreek 2007; Vendruscolo and Koob 2019). Although the initial deficit in a specific functional circuit that drives excessive drinking might be specific to one stage of the addiction cycle, as the transition to addiction progresses, an individual is ultimately likely to show a progressive and generalized loss of control over many, if not all, systems. However, the thesis here is that as excessive alcohol intake progresses to AUD, a common dysregulated functional element is a reward system deficit and stress surfeit.

Acknowledgments The author thanks Michael Arends for assistance with manuscript preparation and editing. Research was supported by the National Institute on Drug Abuse Intramural Research Program.

References

Adinoff B, Ruether K, Krebaum S, Iranmanesh A, Williams MJ (2003) Increased salivary cortisol concentrations during chronic alcohol intoxication in a naturalistic clinical sample of men. Alcohol Clin Exp Res 27(9):1420–1427

Adinoff B, Leonard D, Price J, Javors MA, Walker R, Brown ES, Xiao H, Rao U (2017) Adrenocortical sensitivity, moderated by ongoing stress, predicts drinking intensity in alcohol-dependent men. Psychoneuroendocrinology 76:67–76

Aldridge GM, Zarin TA, Brandner AJ, George O, Gilpin NW, Vendruscolo LF, Koob GF, Schmeichel BE (2022) Effects of single and dual hypocretin-receptor blockade or knockdown of hypocretin amygdalar projections on alcohol drinking in dependent male rats. Addict Neurosci. In press

Alheid GF, De Olmos JS, Beltramino CA (1995) Amygdala and extended amygdala. In: Paxinos G (ed) The rat nervous system. Academic Press, San Diego, pp 495–578

Allen TJ, Moeller FG, Rhoades HM, Cherek DR (1998) Impulsivity and history of drug dependence. Drug Alcohol Depend 50:137–145

Alonso G, Szafarczyk A, Balmefrezol M, Assenmacher I (1986) Immunocytochemical evidence for stimulatory control by the ventral noradrenergic bundle of parvocellular neurons of the paraventricular nucleus secreting corticotropin-releasing hormone and vasopressin in rats. Brain Res 397:297–307

American Psychiatric Association (1994) Diagnostic and statistical manual of mental disorders, 4th edn. American Psychiatric Press, Washington

American Psychiatric Association (2013) Diagnostic and statistical manual of mental disorders, 5th edn. American Psychiatric Publishing, Washington

Anderson RI, Lopez MF, Becker HC (2016) Stress-induced enhancement of ethanol intake in C57BL/6J mice with a history of chronic ethanol exposure: involvement of kappa opioid receptors. Front Cell Neurosci 10:45

Anderson RI, Lopez MF, Griffin WC, Haun HL, Bloodgood DW, Pati D, Boyt KM, Kash TL, Becker HC (2019) Dynorphin-kappa opioid receptor activity in the central amygdala modulates binge-like alcohol drinking in mice. Neuropsychopharmacology 44(6):1084–1092

Aoun EG, Jimenez VA, Vendruscolo LF, Walter NAR, Barbier E, Ferrulli A, Haass-Koffler CL, Darakjian P, Lee MR, Addolorato G, Heilig M, Hitzemann R, Koob GF, Grant KA, Leggio L (2018) A relationship between the aldosterone-mineralocorticoid receptor pathway and alcohol drinking: preliminary translational findings across rats, monkeys and humans. Mol Psychiatry 23:1466–1473

Aroni S, Marino RAM, Girven KS, Irving JM, Cheer JF, Sparta DR (2021) Repeated binge ethanol drinking enhances electrical activity of central amygdala corticotropin releasing factor neurons in vivo. Neuropharmacology 189:108527

Arora S, Vohora D (2016) Comparative evaluation of partial α2-adrenoceptor agonist and pure α2-adrenoceptor antagonist on the behavioural symptoms of withdrawal after chronic alcohol administration in mice. Basic Clin Pharmacol Toxicol 119:202–209

Aston-Jones G, Delfs JM, Druhan J, Zhu Y (1999) The bed nucleus of the stria terminalis: a target site for noradrenergic actions in opiate withdrawal. In: McGinty JF (ed) Advancing from the ventral striatum to the extended amygdala: implications for neuropsychiatry and drug abuse, Annals of the New York Academy of Sciences, vol 877. New York Academy of Sciences, New York, pp 486–498

Avegno EM, Lobell TD, Itoga CA, Baynes BB, Whitaker AM, Weera MM, Edwards S, Middleton JW, Gilpin NW (2018) Central amygdala circuits mediate hyperalgesia in alcohol-dependent rats. J Neurosci 38:7761–7773

Bailey CP, O'Callaghan MJ, Croft AP, Manley SJ, Little HJ (2001) Alterations in mesolimbic dopamine function during the abstinence period following chronic ethanol consumption. Neuropharmacology 41:989–999

Baldo BA, Daniel RA, Berridge CW, Kelley AE (2003) Overlapping distributions of orexin/hypocretin- and dopamine-β-hydroxylase immunoreactive fibers in rat brain regions mediating arousal, motivation, and stress. J Comp Neurol 464:220–237

Baldwin HA, Rassnick S, Rivier J, Koob GF, Britton KT (1991) CRF antagonist reverses the "anxiogenic" response to ethanol withdrawal in the rat. Psychopharmacology 103:227–232

Balldin J, Berggren U, Engel J, Lindstedt G, Sundkler A, Walinder J (1992) Alpha-2-adrenoceptor sensitivity in early alcohol withdrawal. Biol Psychiatry 31:712–719

Ballenger JC, Goodwin FK, Major LF, Brown GL (1979) Alcohol and central serotonin metabolism in man. Arch Gen Psychiatry 36:224–227

Banks WA, Erickson MA (2010) The blood-brain barrier and immune function and dysfunction. Neurobiol Dis 37:26–32

Barrot M, Marinelli M, Abrous DN, Rouge-Pont F, Le Moal M, Piazza PV (2000) The dopaminergic hyper-responsiveness of the shell of the nucleus accumbens is hormone-dependent. Eur J Neurosci 12:973–979

Bazov I, Sarkisyan D, Kononenko O, Watanabe H, Yakovleva T, Hansson AC, Sommer WH, Spanagel R, Bakalkin G (2018) Dynorphin and κ-opioid receptor dysregulation in the dopaminergic reward system of human alcoholics. Mol Neurobiol 55:7049–7061

Bechara A, Damasio H, Damasio AR (2000) Emotion, decision making and the orbitofrontal cortex. Cereb Cortex 10:295–307

Becker HC, Lopez MF (2004) Increased ethanol drinking after repeated chronic ethanol exposure and withdrawal experience in C57BL/6 mice. Alcohol Clin Exp Res 28:1829–1838

Belin D, Belin-Rauscent A, Murray JE, Everitt BJ (2013) Addiction: failure of control over maladaptive incentive habits. Curr Opin Neurobiol 23:564–572

Berggren U, Fahlke C, Norrby A, Zachrisson O, Balldin J (2000) Subsensitive alpha-2-adrenoceptor function in male alcohol-dependent individuals during 6 months of abstinence. Drug Alcohol Depend 57:255–260

Besson JM (1999) The neurobiology of pain. Lancet 353:1610–1615

Blackburn-Munro G, Blackburn-Munro R (2003) Pain in the brain: are hormones to blame? Trends Endocrinol Metab 14:20–27

Blednov YA, Ponomarev I, Geil C, Bergeson S, Koob GF, Harris RA (2012) Neuroimmune regulation of alcohol consumption: behavioral validation of genes obtained from genomic studies. Addict Biol 17:108–120

Boileau I, Assaad JM, Pihl RO, Benkelfat C, Leyton M, Diksic M, Tremblay RE, Dagher A (2003) Alcohol promotes dopamine release in the human nucleus accumbens. Synapse 49:226–231

Boissoneault J, Lewis B, Nixon SJ (2019) Characterizing chronic pain and alcohol use trajectory among treatment-seeking alcoholics. Alcohol 75:47–54

Boutrel B, Kenny PJ, Specio SE, Martin-Fardon R, Markou A, Koob GF, de Lecea L (2005) Role for hypocretin in mediating stress-induced reinstatement of cocaine-seeking behavior. Proc Natl Acad Sci U S A 102:19168–19173

Breese GR, Overstreet DH, Knapp DJ, Navarro M (2005) Prior multiple ethanol withdrawals enhance stress-induced anxiety-like behavior: inhibition by CRF_1- and benzodiazepine-receptor antagonists and a $5\text{-}HT_{1a}$-receptor agonist. Neuropsychopharmacology 30:1662–1669

Breese GR, Knapp DJ, Overstreet DH, Navarro M, Wills TA, Angel RA (2008) Repeated lipopolysaccharide (LPS) or cytokine treatments sensitize ethanol withdrawal-induced anxiety-like behavior. Neuropsychopharmacology 33:867–876. Erratum: 33(7):1777

Brown SA, Schuckit MA (1988) Changes in depression among abstinent alcoholics. J Stud Alcohol 49:412–417

Brown G, Jackson A, Stephens DN (1998) Effects of repeated withdrawal from chronic ethanol on oral self-administration of ethanol on a progressive ratio schedule. Behav Pharmacol 9:149–161

Carlezon WA Jr, Nestler EJ, Neve RL (2000) Herpes simplex virus-mediated gene transfer as a tool for neuropsychiatric research. Crit Rev Neurobiol 14:47–67

Carmack SA, Vendruscolo JCM, McGinn MA, Miranda-Barrientos J, Hinrich AJ, Bosse G, Repunte-CAnonigo V, Jodelka FM, Ling K, Messing RO, Peterson R, Rigo F, Edwards S,

Sanna PP, Morales M, Hastings ML, Koob GF, Vendruscolo LF (2022) Corticosteroid sensitization drives opioid addiction. Mol Psychiatry 27:2492–2501

Chandler LJ, Newsom H, Sumners C, Crews F (1993) Chronic ethanol exposure potentiates NMDA excitotoxicity in cerebral cortical neurons. J Neurochem 60:1578–1581

Chassin L, Barrera M Jr, Bech K, Kossak-Fuller J (1992) Recruiting a community sample of adolescent children of alcoholics: a comparison of three subject sources. J Stud Alcohol 53:316–319

Chavkin C, Koob GF (2016) Dynorphin, dysphoria and dependence: the stress of addiction. Neuropsychopharmacology 41:373–374

Chester DS, Lynam DR, Milich R, DeWall CN (2016) Craving versus control: negative urgency and neural correlates of alcohol cue reactivity. Drug Alcohol Depend 163(Suppl 1):S25–S28

Chu K, Koob GF, Cole M, Zorrilla EP, Roberts AJ (2007) Dependence-induced increases in ethanol self-administration in mice are blocked by the CRF_1 receptor antagonist antalarmin and by CRF_1 receptor knockout. Pharmacol Biochem Behav 86:813–821

Ciccocioppo R, Economidou D, Fedeli A, Massi M (2003) The nociceptin/orphanin FQ/NOP receptor system as a target for treatment of alcohol abuse: a review of recent work in alcohol-preferring rats. Physiol Behav 79:121–128

Cippitelli A, Damadzic R, Singley E, Thorsell A, Ciccocioppo R, Eskay RL, Heilig M (2012) Pharmacological blockade of corticotropin-releasing hormone receptor 1 (CRH1R) reduces voluntary consumption of high alcohol concentrations in non-dependent Wistar rats. Pharmacol Biochem Behav 100:522–529

Cippitelli A, Damadzic R, Hamelink C, Brunnquell M, Thorsell A, Heilig M, Eskay RL (2014) Binge-like ethanol consumption increases corticosterone levels and neurodegeneration whereas occupancy of type II glucocorticoid receptors with mifepristone is neuroprotective. Addict Biol 19:27–36

Coskunpinar A, Dir AL, Cyders MA (2013) Multidimensionality in impulsivity and alcohol use: a meta-analysis using the UPPS model of impulsivity. Alcohol Clin Exp Res 37:1441–1450

Crews FT, Vetreno RP (2016) Mechanisms of neuroimmune gene induction in alcoholism. Psychopharmacology 233:1543–1557

Crews FT, Lawrimore CJ, Walter TJ, Coleman LG Jr (2017) The role of neuroimmune signaling in alcoholism. Neuropharmacology 122:56–73

Cyders MA, Smith GT (2008) Emotion-based dispositions to rash action: positive and negative urgency. Psychol Bull 134:807–828

Cyders MA, Dzemidzic M, Eiler WJ, Coskunpinar A, Karyadi K, Kareken DA (2014) Negative urgency and ventromedial prefrontal cortex responses to alcohol cues: fMRI evidence of emotion-based impulsivity. Alcohol Clin Exp Res 38:409–417

Cyders MA, Dzemidzic M, Eiler WJ, Coskunpinar A, Karyadi KA, Kareken DA (2015) Negative urgency mediates the relationship between amygdala and orbitofrontal cortex activation to negative emotional stimuli and general risk-taking. Cereb Cortex 25:4094–4102

Dahchour A, de Witte P, Bolo N, Nedelec JF, Muzet M, Durbin P, Macher JP (1998) Central effects of acamprosate: Part 1. Acamprosate blocks the glutamate increase in the nucleus accumbens microdialysate in ethanol withdrawn rats. Psychiatry Res 82:107–114

Dantzer R, O'Connor JC, Freund GG, Johnson RW, Kelley KW (2008) From inflammation to sickness and depression: when the immune system subjugates the brain. Nat Rev Neurosci 9:46–56

Daurio AM, Aston SA, Schwandt ML, Bukhari MO, Bouhlal S, Farokhnia M, Lee MR, Leggio L (2018) Impulsive personality traits mediate the relationship between adult attention-deficit/hyperactivity symptoms and alcohol dependence severity. Alcohol Clin Exp Res 42:173–183

de Guglielmo G, Crawford E, Kim S, Vendruscolo LF, Hope BT, Brennan M, Cole M, Koob GF, George O (2016) Recruitment of a neuronal ensemble in the central nucleus of the amygdala is required for alcohol dependence. J Neurosci 36:9446–9453 de Guglielmo G, Kallupi M, Cole MD, George O (2017) Voluntary induction and maintenance of alcohol dependence in rats using alcohol vapor self-administration. Psychopharmacology 234: 2009–2018

de Guglielmo G, Kallupi M, Pomrenze M, Crawford E, Simpson S, Schweitzer P, Koob GF, Messing RO, George O (2019) Inactivation of a CRF-dependent amygdalofugal pathway reverses addiction-like behaviors in alcohol-dependent rats. Nat Commun 10:1238

de Lecea L, Kilduff TS, Peyron C, Gao X, Foye PE, Danielson PE, Fukuhara C, Battenberg EL, Gautvik VT, Bartlett FS 2nd, Frankel WN, van den Pol AN, Bloom FE, Gautvik KM, Sutcliffe JG (1998) The hypocretins: hypothalamus-specific peptides with neuroexcitatory activity. Proc Natl Acad Sci U S A 95:322–327

de Timary P, Luts A, Hers D, Luminet O (2008) Absolute and relative stability of alexithymia in alcoholic inpatients undergoing alcohol withdrawal: relationship to depression and anxiety. Psychiatry Res 157:105–113

De Vries GJ, Buijs RM (1983) The origin of the vasopressinergic and oxytocinergic innervation of the rat brain with special reference to the lateral septum. Brain Res 273:307–317

de Wied D, Versteeg DH (1979) Neurohypophyseal principles and memory. Fed Proc 38:2348–2354

Delfs JM, Zhu Y, Druhan JP, Aston-Jones G (2000) Noradrenaline in the ventral forebrain is critical for opiate withdrawal-induced aversion. Nature 403:430–434

Devaud LL, Fritschy JM, Sieghart W, Morrow AL (1997) Bidirectional alterations of $GABA_A$ receptor subunit peptide levels in rat cortex during chronic ethanol consumption and withdrawal. J Neurochem 69:126–130

Di Ciano P, Coury A, Depoortere RY, Egilmez Y, Lane JD, Emmett-Oglesby MW, Lepiane FG, Phillips AG, Blaha CD (1995) Comparison of changes in extracellular dopamine concentrations in the nucleus accumbens during intravenous self-administration of cocaine or d-amphetamine. Behav Pharmacol 6:311–322

Diana M (2011) The dopamine hypothesis of drug addiction and its potential therapeutic value. Front Psychiatry 2:64

Diana M, Pistis M, Carboni S, Gessa GL, Rossetti ZL (1993) Profound decrement of mesolimbic dopaminergic neuronal activity during ethanol withdrawal syndrome in rats: electrophysiological and biochemical evidence. Proc Natl Acad Sci U S A 90:7966–7969

Diana M, Brodie M, Muntoni A, Puddu MC, Pillolla G, Steffensen S, Spiga S, Little HJ (2003) Enduring effects of chronic ethanol in the CNS: basis for alcoholism. Alcohol Clin Exp Res 27: 354–361

Dina OA, Messing RO, Levine JD (2006) Ethanol withdrawal induces hyperalgesia mediated by PKCε. Eur J Neurosci 24:197–204

Dir AL, Karyadi K, Cyders MA (2013) The uniqueness of negative urgency as a common risk factor for self-harm behaviors, alcohol consumption, and eating problems. Addict Behav 38:2158–2162

Doyon WM, York JL, Diaz LM, Samson HH, Czachowski CL, Gonzales RA (2003) Dopamine activity in the nucleus accumbens during consummatory phases of oral ethanol self-administration. Alcohol Clin Exp Res 27:1573–1582

Driessen M, Meier S, Hill A, Wetterling T, Lange W, Junghanns K (2001) The course of anxiety, depression and drinking behaviours after completed detoxification in alcoholics with and without comorbid anxiety and depressive disorders. Alcohol Alcohol 36:249–255

Economidou D, Hansson AC, Weiss F, Terasmaa A, Sommer WH, Cippitelli A, Fedeli A, Martin-Fardon R, Massi M, Ciccocioppo R, Heilig M (2008) Dysregulation of nociceptin/orphanin FQ activity in the amygdala is linked to excessive alcohol drinking in the rat. Biol Psychiatry 64: 211–218

Economidou D, Cippitelli A, Stopponi S, Braconi S, Clementi S, Ubaldi M, Martin-Fardon R, Weiss F, Massi M, Ciccocioppo R (2011) Activation of brain NOP receptors attenuates acute and protracted alcohol withdrawal symptoms in the rat. Alcohol Clin Exp Res 35:747–755

Eddy NB, Halbach H, Isbell H, Seevers MH (1965) Drug dependence: its significance and characteristics. Bull World Health Organ 32:721–733

Edwards S, Guerrero M, Ghoneim OM, Roberts E, Koob GF (2012a) Evidence that vasopressin V1b receptors mediate the transition to excessive drinking in ethanol-dependent rats. Addict Biol 17:76–85

Edwards S, Vendruscolo LF, Schlosburg JE, Misra KK, Wee S, Park PE, Schulteis G, Koob GF (2012b) Development of mechanical hypersensitivity in rats during heroin and ethanol dependence: alleviation by CRF1 receptor antagonism. Neuropharmacology 62:1142–1151

Edwards S, Little HJ, Richardson HN, Vendruscolo LF (2015) Divergent regulation of distinct glucocorticoid systems in alcohol dependence. Alcohol 49:811–816

Edwards S, Vendruscolo LF, Gilpin NW, Wojnar M, Witkiewitz K (2020) Alcohol and pain: a translational review of preclinical and clinical findings to inform future treatment strategies. Alcohol Clin Exp Res 44:368–383

Egli M, Koob GF, Edwards S (2012) Alcohol dependence as a chronic pain disorder. Neurosci Biobehav Rev 36:2179–2192

Ellis FW (1966) Effect of ethanol on plasma corticosterone levels. J Pharmacol Exp Ther 153:121–127

Elvig SK, McGinn A, Smith C, Arends MA, Koob GF, Vendruscolo LF (2021) Tolerance to alcohol: a critical yet understudied factor in alcohol addiction. Pharmacol Biochem Behav 204: 173155

Epping-Jordan MP, Watkins SS, Koob GF, Markou A (1998) Dramatic decreases in brain reward function during nicotine withdrawal. Nature 393:76–79

Errico AL, King AC, Lovallo WR, Parsons OA (2002) Cortisol dysregulation and cognitive impairment in abstinent male alcoholics. Alcohol Clin Exp Res 26:1198–1204

Fahlke C, Hard E, Eriksson CPJ, Engel JA, Hansen S (1995) Consequence of long-term exposure to corticosterone or dexamethasone on ethanol consumption in the adrenalectomized rat, and the effect of type I and type II corticosteroid receptor antagonists. Psychopharmacology 117:216–224

Fahlke C, Hard E, Hansen S (1996) Facilitation of ethanol consumption by intracerebroventricular infusions of corticosterone. Psychopharmacology 127:133–139

Farokhnia M, Rentsch CT, Chuong V, McGinn A, Elvig SK, Douglass EA, Sanfilippo JE, Marchette RCN, Tunstall BJ, Fiellin DA, Koob GF, Justice AC, Leggio L, Vendruscolo LF (2022) Spironolactone as a potential new pharmacotherapy for alcohol use disorder: convergent evidence from rodent and human studies. J Clin Invest. In press

Ferguson E, Zale E, Ditre J, Wesolowicz D, Stennett B, Robinson M, Boissoneault J (2021) CANUE: a theoretical model of pain as an antecedent for substance use. Ann Behav Med 55: 489–502

Fidler TL, Clews TW, Cunningham CL (2006) Reestablishing an intragastric ethanol self-infusion model in rats. Alcohol Clin Exp Res 30:414–428

Fillmore MT, Rush CR (2002) Impaired inhibitory control of behavior in chronic cocaine users. Drug Alcohol Depend 66:265–273

Finn DA, Snelling C, Fretwell AM, Tanchuck MA, Underwood L, Cole M, Crabbe JC, Roberts AJ (2007) Increased drinking during withdrawal from intermittent ethanol exposure is blocked by the CRF receptor antagonist D-Phe-CRF$_{12\text{-}41}$. Alcohol Clin Exp Res 31:939–949

Follesa P, Ticku MK (1995) Chronic ethanol treatment differentially regulates NMDA receptor subunit mRNA expression in rat brain. Mol Brain Res 29:99–106

Fox HC, Bergquist KL, Hong KI, Sinha R (2007) Stress-induced and alcohol cue-induced craving in recently abstinent alcohol-dependent individuals. Alcohol Clin Exp Res 31:395–403

Funk CK, O'Dell LE, Crawford EF, Koob GF (2006) Corticotropin-releasing factor within the central nucleus of the amygdala mediates enhanced ethanol self-administration in withdrawn, ethanol-dependent rats. J Neurosci 26:11324–11332

Funk CK, Zorrilla EP, Lee MJ, Rice KC, Koob GF (2007) Corticotropin-releasing factor 1 antagonists selectively reduce ethanol self-administration in ethanol-dependent rats. Biol Psychiatry 61:78–86

Garabedian MJ, Harris CA, Jeanneteau F (2017) Glucocorticoid receptor action in metabolic and neuronal function. F1000Res 6:1208

Gardner EL, Vorel SR (1998) Cannabinoid transmission and reward-related events. Neurobiol Dis 5:502–533

Gatch MB (2009) Ethanol withdrawal and hyperalgesia. Curr Drug Abuse Rev 2:41–50

Gatch MB, Lal H (1999) Effects of ethanol and ethanol withdrawal on nociception in rats. Alcohol Clin Exp Res 23:328–333

Gatch MB, Wallis CJ, Lal H (1999) Effects of NMDA antagonists on ethanol-withdrawal induced "anxiety" in the elevated plus maze. Alcohol 19:207–211

Gatta E, Grayson DR, Auta J, Saudagar V, Dong E, Chen Y, Krishnan HR, Drnevich J, Pandey SC, Guidotti A (2021) Genome-wide methylation in alcohol use disorder subjects: implications for an epigenetic regulation of the cortico-limbic glucocorticoid receptors (NR3C1). Mol Psychiatry 26:1029–1041

George O, Koob GF (2010) Individual differences in prefrontal cortex function and the transition from drug use to drug dependence. Neurosci Biobehav Rev 35:232–247

George O, Koob GF (2013) Control of craving by the prefrontal cortex. Proc Natl Acad Sci U S A 110:4165–4166

Gilpin NW, Koob GF (2010) Effects of β-adrenoceptor antagonists on alcohol drinking by alcohol-dependent rats. Psychopharmacology 212:431–439

Gilpin NW, Stewart RB, Murphy JM, Li TK, Badia-Elder NE (2003) Neuropeptide Y reduces oral ethanol intake in alcohol-preferring (P) rats following a period of imposed ethanol abstinence. Alcohol Clin Exp Res 27:787–794

Gilpin NW, Misra K, Koob GF (2008) Neuropeptide Y in the central nucleus of the amygdala suppresses dependence-induced increases in alcohol drinking. Pharmacol Biochem Behav 90: 475–480

Gilpin NW, Smith AD, Cole M, Weiss F, Koob GF, Richardson HN (2009) Operant behavior and alcohol levels in blood and brain of alcohol-dependent rats. Alcohol Clin Exp Res 33:2113–2123. Erratum: 34:382

Gilpin NW, Misra K, Herman MA, Cruz MT, Koob GF, Roberto M (2011) Neuropeptide Y opposes alcohol effects on gamma-aminobutyric acid release in amygdala and blocks the transition to alcohol dependence. Biol Psychiatry 69:1091–1099

Giuliano C, Belin D, Everitt BJ (2019) Compulsive alcohol seeking results from a failure to disengage dorsolateral striatal control over behavior. J Neurosci 39:1744–1754

Goldstein RZ, Volkow ND (2002) Drug addiction and its underlying neurobiological basis: neuroimaging evidence for the involvement of the frontal cortex. Am J Psychiatry 159:1642–1652

Goodwin DW (1981) Alcohol use disorder: the facts. Oxford University Press, New York

Gordon AS, Collier K, Diamond I (1986) Ethanol regulation of adenosine receptor-stimulated cAMP levels in a clonal neural cell line: an in vitro model of cellular tolerance to ethanol. Proc Natl Acad Sci U S A 83:2105–2108

Graeber MB (2010) Changing face of microglia. Science 330:783–788

Green L, Fristoe N, Myerson J (1994) Temporal discounting and preference reversals in choice between delayed outcomes. Psychonom Bull Rev 1:383–389

Guillemin R (1978) Peptides in the brain: the new endocrinology of the neuron. Science 202:390–402

Harris RA, Bajo M, Bell RL, Blednov YA, Varodayan FP, Truitt JM, de Guglielmo G, Lasek AW, Logrip ML, Vendruscolo L, Roberts AJ, Roberts E, George O, Mayfield J, Billiar TR, Hackam DJ, Mayfield RD, Koob GF, Roberto M, Homanics GE (2017) Genetic and pharmacologic manipulation of TLR4 has minimal impact on ethanol consumption in rodents. J Neurosci 37: 1139–1155

Hata T, Chen J, Ebihara K, Date Y, Ishida Y, Nakahara D (2011) Intra-ventral tegmental area or intracerebroventricular orexin-A increases the intra-cranial self-stimulation threshold via activation of the corticotropin-releasing factor system in rats. Eur J Neurosci 34:816–826

Hays RM (1980) Agents affecting renal conservation of water. In: Gilman AG, Goodman LS (eds) The pharmacological basis of therapeutics. Macmillan, New York, pp 916–928

Hebb DO (1972) Textbook of psychology, 3rd edn. WB Saunders, Philadelphia

Heilig M (2004) The NPY system in stress, anxiety and depression. Neuropeptides 38:213–224

Heilig M, Koob GF (2007) A key role for corticotropin-releasing factor in alcohol dependence. Trends Neurosci 30:399–406

Heilig M, Koob GF, Ekman R, Britton KT (1994) Corticotropin-releasing factor and neuropeptide Y: role in emotional integration. Trends Neurosci 17:80–85

Heimer L, Alheid G (1991) Piecing together the puzzle of basal forebrain anatomy. In: Napier TC, Kalivas PW, Hanin I (eds) The Basal Forebrain: anatomy to function, Advances in experimental medicine and biology, vol 295. Plenum Press, New York, pp 1–42

Heinz A, Lober S, Georgi A, Wrase J, Hermann D, Rey ER, Wellek S, Mann K (2003) Reward craving and withdrawal relief craving: assessment of different motivational pathways to alcohol intake. Alcohol Alcohol 38:35–39

Hemby SE, Martin TJ, Co C, Dworkin SI, Smith JE (1995) The effects of intravenous heroin administration on extracellular nucleus accumbens dopamine concentrations as determined by *in vivo* microdialysis. J Pharmacol Exp Ther 273:591–598

Hermann D, Weber-Fahr W, Sartorius A, Hoerst M, Frischknecht U, Tunc-Skarka N, Perreau-Lenz S, Hansson AC, Krumm B, Kiefer F, Spanagel R, Mann K, Ende G, Sommer WH (2012) Translational magnetic resonance spectroscopy reveals excessive central glutamate levels during alcohol withdrawal in humans and rats. Biol Psychiatry 71:1015–1021

Hermann A, Stark R, Milad MR, Merz CJ (2016) Renewal of conditioned fear in a novel context is associated with hippocampal activation and connectivity. Soc Cogn Affect Neurosci 11:1411–1421

Hernandez G, Hamdani S, Rajabi H, Conover K, Stewart J, Arvanitogiannis A, Shizgal P (2006) Prolonged rewarding stimulation of the rat medial forebrain bundle: neurochemical and behavioral consequences. Behav Neurosci 120:888–904

Heyser CJ, Schulteis G, Durbin P, Koob GF (1998) Chronic acamprosate eliminates the alcohol deprivation effect while having limited effects on baseline responding for ethanol in rats. Neuropsychopharmacology 18:125–133

Himmelsbach CK (1943) Can the euphoric, analgetic, and physical dependence effects of drugs be separated? IV. With reference to physical dependence. Fed Proc 2:201–203

Hirth N, Meinhardt MW, Noori HR, Salgado H, Torres-Ramirez O, Uhrig S, Broccoli L, Vengeliene V, Roßmanith M, Perreau-Lenz S, Köhr G, Sommer WH, Spanagel R, Hansson AC (2016) Convergent evidence from alcohol-dependent humans and rats for a hyperdopaminergic state in protracted abstinence. Proc Natl Acad Sci U S A 113:3024–3029

Hoffman PL, Ritzmann RF, Walter R, Tabakoff B (1978) Arginine vasopressin maintains ethanol tolerance. Nature 276:614–616

Holter SM, Henniger MS, Lipkowski AW, Spanagel R (2000) Kappa-opioid receptors and relapse-like drinking in long-term ethanol-experienced rats. Psychopharmacology 153:93–102

Holtyn AF, Weerts EM (2019) Evaluation of mifepristone effects on alcohol-seeking and self-administration in baboons. Exp Clin Psychopharmacol 27:227–235

Hummel M, Cummons T, Lu P, Mark L, Harrison JE, Kennedy JD, Whiteside GT (2010) Pain is a salient "stressor" that is mediated by corticotropin-releasing factor-1 receptors. Neuropharmacology 59:160–166

Hwa L, Besheer J, Kash T (2017) Glutamate plasticity woven through the progression to alcohol use disorder: a multi-circuit perspective. F1000Res 6:298

Jacquot C, Croft AP, Prendergast MA, Mulholland P, Shaw SG, Little HJ (2008) Effects of the glucocorticoid antagonist, mifepristone, on the consequences of withdrawal from long term alcohol consumption. Alcohol Clin Exp Res 32:2107–2116

Jang E, Kim JH, Lee S, Kim JH, Seo JW, Jin M, Lee MG, Jang IS, Lee WH, Suk K (2013) Phenotypic polarization of activated astrocytes: the critical role of lipocalin-2 in the classical inflammatory activation of astrocytes. J Immunol 191:5204–5219

Ji D, Gilpin NW, Richardson HN, Rivier CL, Koob GF (2008) Effects of naltrexone, duloxetine, and a corticotropin-releasing factor type 1 receptor antagonist on binge-like alcohol drinking in rats. Behav Pharmacol 19:1–12

Ji G, Fu Y, Adwanikar H, Neugebauer V (2013) Non-pain-related CRF_1 activation in the amygdala facilitates synaptic transmission and pain Mol. Pain 9:2–16

Jimenez VA, Grant KA (2017) Studies using macaque monkeys to address excessive alcohol drinking and stress interactions. Neuropharmacology 122:127–135

Jimenez VA, Walter NAR, Shnitko TA, Newman N, Diem K, Vanderhooft L, Hunt H, Grant KA (2020) Mifepristone decreases chronic voluntary ethanol consumption in rhesus macaques. J Pharmacol Exp Ther 375:258–267

Jochum T, Boettger MK, Burkhardt C, Juckel G, Bär KJ (2010) Increased pain sensitivity in alcohol withdrawal syndrome. Eur J Pain 14:713–718

Johnson PL, Samuels BC, Fitz SD, Federici LM, Hammes N, Early MC, Truitt W, Lowry CA, Shekhar A (2012) Orexin 1 receptors are a novel target to modulate panic responses and the panic brain network. Physiol Behav 107:733–742

Juarez B, Morel C, Ku SM, Liu Y, Zhang H, Montgomery S, Gregoire H, Ribeiro E, Crumiller M, Roman-Ortiz C, Walsh JJ, Jackson K, Croote DE, Zhu Y, Zhang S, Vendruscolo LF, Edwards S, Roberts A, Hodes GE, Lu Y, Calipari ES, Chaudhury D, Friedman AK, Han MH (2017) Midbrain circuit regulation of individual alcohol drinking behaviors in mice. Nat Commun 8: 2220. Erratum: Nat Commun 9(1):653

Kadmiel M, Cidlowski JA (2013) Glucocorticoid receptor signaling in health and disease. Trends Pharmacol Sci 34:518–530

Kallupi M, Vendruscolo LF, Carmichael CY, George O, Koob GF, Gilpin NW (2014) Neuropeptide Y Y_2R blockade in the central amygdala reduces anxiety-like behavior but not alcohol drinking in alcohol-dependent rats. Addict Biol 19:755–757

Kamm O, Aldrich TB, Grote IW, Rowe LW, Bugbee EP (1928) The active principles of the posterior lobe of the pituitary gland: I. The demonstration of the presence of two active principles. II. The separation of the two principles and their concentration in the form of potent solid preparations. J Am Chem Soc 50:573–601

Kang S, Li J, Zuo W, Chen P, Gregor D, Fu R, Han X, Bekker A, Ye JH (2019) Downregulation of M-channels in lateral habenula mediates hyperalgesia during alcohol withdrawal in rats. Sci Rep 9:2714

Karkhanis AN, Al-Hasani R (2020) Dynorphin and its role in alcohol use disorder. Brain Res 1735: 146742

Karkhanis AN, Huggins KN, Rose JH, Jones SR (2016) Switch from excitatory to inhibitory actions of ethanol on dopamine levels after chronic exposure: role of kappa opioid receptors. Neuropharmacology 110:190–197

Kashkin VA, Shekunova EV, Egorov AY, Bagrov AY (2018) Marinobufagenin in urine: a potential marker of predisposition to ethanol and a target for spironolactone. Curr Hypertens Rev 14:35–38

Kelley SP, Nannini MA, Bratt AM, Hodge CW (2001) Neuropeptide-Y in the paraventricular nucleus increases ethanol self-administration. Peptides 22:515–522

Kettenmann H, Kirchhoff F, Verkhratsky A (2013) Microglia: new roles for the synaptic stripper. Neuron 77:10–18

Khantzian EJ (1997) The self-medication hypothesis of substance use disorders: a reconsideration and recent applications. Harvard Rev Psychiatry 4:231–244

Khantzian EJ (2003) Understanding addictive vulnerability: an evolving psychodynamic perspective. Neuro-Psychoanalysis 5:5–7

Khom S, Rodriguez L, Gandhi P, Kirson D, Bajo M, Oleata CS, Vendruscolo LF, Mason BJ, Roberto M (2022) Alcohol dependence and withdrawal increase sensitivity of central amygdalar

GABAergic synapses to the glucocorticoid receptor antagonist mifepristone in male rats. Neurobiol Dis 164:105610

Kiefer F, Jahn H, Tarnaske T, Helwig H, Briken P, Holzbach R, Kämpf P, Stracke R, Baehr M, Naber D, Wiedemann K (2003) Comparing and combining naltrexone and acamprosate in relapse prevention of alcoholism: a double-blind, placebo-controlled study. Arch Gen Psychiatry 60:92–99

Kim JS, Kearns DN (2019) Reduced ethanol self-administration in rats produced by the introduction of a high value non-drug alternative reinforcer. Pharmacol Biochem Behav 184:172744

King CE, Becker HC (2019) Oxytocin attenuates stress-induced reinstatement of alcohol seeking behavior in male and female mice. Psychopharmacology 236:2613–2622

Kissler JL, Sirohi S, Reis DJ, Jansen HT, Quock RM, Smith DG, Walker BM (2014) The one-two punch of alcoholism: role of central amygdala dynorphins/kappa-opioid receptors. Biol Psychiatry 75:774–782

Knapp C (1996) Drinking: a love story. Dial Press, New York

Knapp DJ, Overstreet DH, Moy SS, Breese GR (2004) SB242084, flumazenil, and CRA1000 block ethanol withdrawal-induced anxiety in rats. Alcohol 32:101–111

Knapp DJ, Overstreet DH, Breese GR (2005) Modulation of ethanol withdrawal-induced anxiety-like behavior during later withdrawals by treatment of early withdrawals with benzodiazepine/γ-aminobutyric acid ligands. Alcohol Clin Exp Res 29:553–563

Knapp DJ, Overstreet DH, Breese GR (2007) Baclofen blocks expression and sensitization of anxiety-like behavior in an animal model of repeated stress and ethanol withdrawal. Alcohol Clin Exp Res 31:582–595

Knapp DJ, Whitman BA, Wills TA, Angel RA, Overstreet DH, Criswell HE, Ming Z, Breese GR (2011) Cytokine involvement in stress may depend on corticotrophin releasing factor to sensitize ethanol withdrawal anxiety. Brain Behav Immun 25(Suppl 1):S146–S154

Koenig HN, Olive MF (2004) The glucocorticoid receptor antagonist mifepristone reduces ethanol intake in rats under limited access conditions. Psychoneuroendocrinology 29:999–1003

Kohut H (1971) The analysis of the self, The psychoanalytic study of the child, vol 4. International Universities Press, New York

Koob GF (1992) Drugs of abuse: anatomy, pharmacology, and function of reward pathways. Trends Pharmacol Sci 13:177–184

Koob GF (1999) Corticotropin-releasing factor, norepinephrine and stress. Biol Psychiatry 46: 1167–1180

Koob GF (2003) Alcohol use disorder: allostasis and beyond. Alcohol Clin Exp Res 27:232–243

Koob GF (2004) Allostatic view of motivation: implications for psychopathology. In: Bevins RA, Bardo MT (eds) Motivational factors in the etiology of drug abuse, Nebraska symposium on motivation, vol 50. University of Nebraska Press, Lincoln, pp 1–18

Koob GF (2008a) A role for brain stress systems in addiction. Neuron 59:11–34

Koob GF (2008b) Neurobiology of addiction. In: Galanter M, Kleber HD (eds) Textbook of substance abuse treatment, 4th edn. American Psychiatric Publishing, Washington, pp 3–16

Koob GF (2021) Drug addiction: hyperkatifeia/negative reinforcement as a framework for medications development. Pharmacol Rev 73:163–201

Koob GF, Bloom FE (1988) Cellular and molecular mechanisms of drug dependence. Science 242: 715–723

Koob GF, Kreek MJ (2007) Stress, dysregulation of drug reward pathways, and the transition to drug dependence. Am J Psychiatry 164:1149–1159

Koob GF, Le Moal M (1997) Drug abuse: hedonic homeostatic dysregulation. Science 278:52–58

Koob GF, Le Moal M (2001) Drug addiction, dysregulation of reward, and allostasis. Neuropsychopharmacology 24:97–129

Koob GF, Le Moal M (2005) Plasticity of reward neurocircuitry and the 'dark side' of drug addiction. Nat Neurosci 8:1442–1444

Koob GF, Le Moal M (2006) Neurobiology of addiction. Academic Press, London

Koob GF, Le Moal M (2008) Addiction and the brain antireward system. Annu Rev Psychol 59:29–53

Koob GF, Volkow ND (2010) Neurocircuitry of addiction. Neuropsychopharmacol Rev 35:217–238. Erratum: 35:1051

Koob GF, Winger GD, Meyerhoff JL, Annau Z (1977) Effects of D-amphetamine on concurrent self-stimulation of forebrain and brain stem loci. Brain Res 137:109–126

Koob GF, Heinrichs SC, Menzaghi F, Pich EM, Britton KT (1994) Corticotropin releasing factor, stress and behavior. Semin Neurosci 6:221–229

Koob GF, Everitt BJ, Robbins TW (2008) Reward, motivation, and addiction. In: Squire LG, Berg D, Bloom FE, Du Lac S, Ghosh A, Spitzer N (eds) Fundamental neuroscience, 3rd edn. Academic Press, Amsterdam, pp 987–1016

Koob GF, Lloyd GK, Mason BJ (2009) Development of pharmacotherapies for drug addiction: a Rosetta Stone approach. Nat Rev Drug Discov 8:500–515

Koob GF, Buck CL, Cohen A, Edwards S, Park PE, Schlosburg JE, Schmeichel B, Vendruscolo LF, Wade CL, Whitfield TW Jr, George O (2014) Addiction as a stress surfeit disorder. Neuropharmacology 76:370–382

Koob GF, Arends MA, McCracken M, Le Moal M (2019) Neurobiology of addiction: volume 1. Introduction to addiction. Elsevier, New York. ISBN: 978-0128168639

Koob GF, Powell P, White A (2020) Addiction as a coping response: hyperkatifeia, deaths of despair, and COVID-19. Am J Psychiatry 177:1031–1037

Koob GF, Arends MA, McCracken M, Le Moal M (2021) Neurobiology of addiction: volume 3. Alcohol. ISBN: 9780128167939. Elsevier, New York

Kornetsky C, Esposito RU (1979) Euphorigenic drugs: effects on the reward pathways of the brain. Fed Proc 38:2473–2476

Kovacs GL, Sarnyai Z, Szabo G (1998) Oxytocin and addiction: a review. Psychoneuroendocrinology 23:945–962

Kuikka JT, Repo E, Bergstrom KA, Tupala E, Tiihonen J (2000) Specific binding and laterality of human extrastriatal dopamine D2/D3 receptors in late onset type 1 alcoholic patients. Neurosci Lett 292:57–59

Kwako LE, Schwandt ML, Ramchandani VA, Diazgranados N, Koob GF, Volkow ND, Blanco C, Goldman D (2019) Neurofunctional domains derived from deep behavioral phenotyping in alcohol use disorder. Am J Psychiatry 176:744–753

Land BB, Bruchas MR, Lemos JC, Xu M, Melief EJ, Chavkin C (2008) The dysphoric component of stress is encoded by activation of the dynorphin κ-opioid system. J Neurosci 28:407–414

Lawrence AJ, Cowen MS, Yang HJ, Chen F, Oldfield B (2006) The orexin system regulates alcohol-seeking in rats. Br J Pharmacol 148:752–759

Le Doux JE (2000) Emotion circuits in the brain. Annu Rev Neurosci 23:155–184

Le Moal M, Dantzer R, Mormede P, Baduel A, Lebrun C, Ettenberg A, van der Kooy D, Wenger J, Deyo S, Koob GF, Bloom FE (1984) Behavioral effects of peripheral administration of arginine vasopressin: a review of our search for a mode of action and a hypothesis. Psychoneuroendocrinology 9:319–341

Lee MR, Weerts EM (2016) Oxytocin for the treatment of drug and alcohol use disorders. Behav Pharmacol 27:640–648

Lee HJ, Macbeth AH, Pagani JH, Young WS 3rd (2009) Oxytocin: the great facilitator of life. Prog Neurobiol 88:127–151

Leko AH, McGinn MA, Farokhnia M (2022) The mineralocorticoid receptor: an emerging pharmacotherapeutic target for alcohol use disorder? ACS Chem Neurosci 13:1832–1834

Lewis MJ (1996) Alcohol reinforcement and neuropharmacological therapeutics. Alcohol Alcohol 31(Suppl 1):17–25

Lewis MD, Todd RM (2007) The self-regulating brain: cortical-subcortical feedback and the development of intelligent action. Cogn Dev 22:406–430

Little HJ, Croft AP, O'Callaghan MJ, Brooks SP, Wang G, Shaw SG (2008) Selective increases in regional brain glucocorticoid: a novel effect of chronic alcohol. Neuroscience 156:1017–1027

Liu J, Yang AR, Kelly T, Puche A, Esoga C, June HL Jr, Elnabawi A, Merchenthaler I, Sieghart W, June HL Sr, Aurelian L (2011) Binge alcohol drinking is associated with GABAA alpha2-regulated Toll-like receptor 4 (TLR4) expression in the central amygdala. Proc Natl Acad Sci U S A 108:4465–4470

Logan GD, Schachar RJ, Tannock R (1997) Impulsivity and inhibitory control. Psychol Sci 8:60–64

Logrip ML, Gainey SC (2020) Sex differences in the long-term effects of past stress on alcohol self-administration, glucocorticoid sensitivity and phosphodiesterase 10A expression. Neuropharmacology 164:107857

Longabaugh R, Morgenstern J (1999) Cognitive-behavioral coping-skills therapy for alcohol dependence: current status and future directions. Alcohol Res Health 23:78–85

Lopez MF, Becker HC (2005) Effect of pattern and number of chronic ethanol exposures on subsequent voluntary ethanol intake in C57BL/6J mice. Psychopharmacology 181:688–696

Lopez MF, Moorman DE, Aston-Jones G, Becker HC (2016) The highly selective orexin/hypocretin 1 receptor antagonist GSK1059865 potently reduces ethanol drinking in ethanol dependent mice. Brain Res 1636:74–80

Lovinger DM, Kash TL (2015) Mechanisms of neuroplasticity and ethanol's effects on plasticity in the striatum and bed nucleus of the stria terminalis. Alcohol Res 37:109–124

Lowery EG, Sparrow AM, Breese GR, Knapp DJ, Thiele TE (2008) The CRF-1 receptor antagonist, CP-154,526, attenuates stress-induced increases in ethanol consumption by BALB/cJ mice. Alcohol Clin Exp Res 32:240–248

Lowery-Gionta EG, Navarro M, Li C, Pleil KE, Rinker JA, Cox BR, Sprow GM, Kash TL, Thiele TE (2012) Corticotropin releasing factor signaling in the central amygdala is recruited during binge-like ethanol consumption in C57BL/6J mice. J Neurosci 32:3405–3413

Lu YL, Richardson HN (2014) Alcohol, stress hormones, and the prefrontal cortex: a proposed pathway to the dark side of addiction. Neuroscience 277:139–151

Mahler SV, Smith RJ, Moorman DE, Sartor GC, Aston-Jones G (2012) Multiple roles for orexin/hypocretin in addiction. Prog Brain Res 198:79–121

Majchrowicz E (1975) Induction of physical dependence upon ethanol and the associated behavioral changes in rats. Psychopharmacologia 43:245–254

Makhijani VH, Van Voorhies K, Besheer J (2018) The mineralocorticoid receptor antagonist spironolactone reduces alcohol self-administration in female and male rats. Pharmacol Biochem Behav 175:10–18

Makhijani VH, Irukulapati P, Van Voorhies K, Fortino B, Besheer J (2020) Central amygdala mineralocorticoid receptors modulate alcohol self-administration. Neuropharmacology 181:108337

Makino S, Gold PW, Schulkin J (1994) Corticosterone effects on corticotropin-releasing hormone mRNA in the central nucleus of the amygdala and the parvocellular region of the paraventricular nucleus of the hypothalamus. Brain Res 640:105–112

Mandyam CD (2013) The interplay between the hippocampus and amygdala in regulating aberrant hippocampal neurogenesis during protracted abstinence from alcohol dependence. Front Psychiatry 4:61

Marinelli PW, Funk D, Juzytsch W, Harding S, Rice KC, Shaham Y, Le AD (2007) The CRF1 receptor antagonist antalarmin attenuates yohimbine-induced increases in operant alcohol self-administration and reinstatement of alcohol seeking in rats. Psychopharmacology 195:345–355

Markou A, Koob GF (1991) Post-cocaine anhedonia: an animal model of cocaine withdrawal. Neuropsychopharmacology 4:17–26

Martin-Fardon R, Zorrilla EP, Ciccocioppo R, Weiss F (2010) Role of innate and drug-induced dysregulation of brain stress and arousal systems in addiction: focus on corticotropin-releasing factor, nociceptin/orphanin FQ, and orexin/hypocretin. Brain Res 1314:145–161

Martin-Solch C, Magyar S, Kunig G, Missimer J, Schultz W, Leenders KL (2001) Changes in brain activation associated with reward processing in smokers and nonsmokers: a positron emission tomography study. Exp Brain Res 139:278–286

McBride WJ, Li TK (1998) Animal models of alcohol use disorder: neurobiology of high alcohol-drinking behavior in rodents. Crit Rev Neurobiol 12:339–369

McCown TJ, Breese GR (1990) Multiple withdrawals from chronic ethanol "kindles" inferior collicular seizure activity: evidence for kindling of seizures associated with alcoholism. Alcohol Clin Exp Res 14:394–399

McEwen BS (2000) Allostasis and allostatic load: implications for neuropsychopharmacology. Neuropsychopharmacology 22:108–124

McEwen BS (2007) Physiology and neurobiology of stress and adaptation: central role of the brain. Physiol Rev 87:873–904

McEwen BS, Gray JD, Nasca C (2015) 60 Years of neuroendocrinology: redefining neuroendocrinology: stress, sex and cognitive and emotional regulation. J Endocrinol 226:T67–T83

McGinn MA, Tunstall BJ, Schlosburg JE, Gregory-Flores A, George O, de Guglielmo G, Mason BJ, Hunt HJ, Koob GF, Vendruscolo LF (2021) Glucocorticoid receptor modulators decrease alcohol self-administration in male rats. Neuropharmacology 188:108510

Melis M, Camarini R, Ungless MA, Bonci A (2002) Long-lasting potentiation of GABAergic synapses in dopamine neurons after a single in vivo ethanol exposure. J Neurosci 22:2074–2082

Melis M, Spiga S, Diana M (2005) The dopamine hypothesis of drug addiction: hypodopaminergic state. Int Rev Neurobiol 63:101–154

Mendelson JH, Stein S (1966) Serum cortisol levels in alcoholic and nonalcoholic subjects during experimentally induced ethanol intoxication. Psychosom Med 28:616–626

Merlo-Pich E, Lorang M, Yeganeh M, Rodriguez de Fonseca F, Raber J, Koob GF, Weiss F (1995) Increase of extracellular corticotropin-releasing factor-like immunoreactivity levels in the amygdala of awake rats during restraint stress and ethanol withdrawal as measured by microdialysis. J Neurosci 15:5439–5447

Mhatre MC, Pena G, Sieghart W, Ticku MK (1993) Antibodies specific for $GABA_A$ receptor alpha subunits reveal that chronic alcohol treatment down-regulates α-subunit expression in rat brain regions. J Neurochem 61:1620–1625

Mitchell JM, Bergren LJ, Chen KS, Rowbotham MC, Fields HL (2009) Naltrexone aversion and treatment efficacy are greatest in humans and rats that actively consume high levels of alcohol. Neurobiol Dis 33:72–80

Moeller FG, Barratt ES, Dougherty DM, Schmitz JM, Swann AC (2001) Psychiatric aspects of impulsivity. Am J Psychiatry 158:1783–1793

Mollereau C, Mouledous L (2000) Tissue distribution of the opioid receptor-like (ORL1) receptor. Peptides 21:907–917

Morzorati SL, Ramchandani VA, Flury L, Li TK, O'Connor S (2002) Self-reported subjective perception of intoxication reflects family history of alcohol use disorder when breath alcohol levels are constant. Alcohol Clin Exp Res 26:1299–1306

Moskal D, Maisto SA, De Vita M, Ditre JW (2018) Effects of experimental pain induction on alcohol urge, intention to consume alcohol, and alcohol demand. Exp Clin Psychopharmacol 26: 65–76

Moy SS, Knapp DJ, Criswell HE, Breese GR (1997) Flumazenil blockade of anxiety following ethanol withdrawal in rats. Psychopharmacology 131:354–360

Mucha RF, Herz A (1985) Motivational properties of kappa and mu opioid receptor agonists studied with place and taste preference conditioning. Psychopharmacology 86:274–280

Myers B, McKlveen JM, Herman JP (2014) Glucocorticoid actions on synapses, circuits, and behavior: implications for the energetics of stress. Front Neuroendocrinol 35:180–196

National Institute on Alcohol Abuse and Alcoholism (2004) NIAAA Council approves definition of binge drinking. NIAAA Newsletter, No. 3 (NIH pub. no. 04-5346). National Institute on Alcohol Abuse and Alcoholism, Bethesda

Nealey KA, Smith AW, Davis SM, Smith DG, Walker BM (2011) κ-opioid receptors are implicated in the increased potency of intra-accumbens nalmefene in ethanol-dependent rats. Neuropharmacology 61:35–42

Nestler EJ (2001) Molecular basis of long-term plasticity underlying addiction. Nat Rev Neurosci 2: 119–128
Nestler EJ (2004) Historical review: molecular and cellular mechanisms of opiate and cocaine addiction. Trends Pharmacol Sci 25:210–218
Nestler EJ (2005) Is there a common molecular pathway for addiction? Nat Neurosci 8:1445–1449
Neugebauer V, Li W, Bird GC, Han JS (2004) The amygdala and persistent pain. Neuroscientist 10: 221–234
Nutt DJ (2014) The role of the opioid system in alcohol dependence. J Psychopharmacol 28:8–22
Nylander I, Todkar A, Granholm L, Vrettou M, Bendre M, Boon W, Andershed H, Tuvblad C, Nilsson KW, Comasco E (2017) Evidence for a link between Fkbp5/FKBP5, early life social relations and alcohol drinking in young adult rats and humans. Mol Neurobiol 54:6225–6234
O'Callaghan MJ, Croft AP, Jacquot C, Little HJ (2005) The hypothalamopituitary-adrenal axis and alcohol preference. Brain Res Bull 68:171–178
O'Dell LE, Roberts AJ, Smith RT, Koob GF (2004) Enhanced alcohol self-administration after intermittent versus continuous alcohol vapor exposure. Alcohol Clin Exp Res 28:1676–1682
O'Malley SS, Krishnan-Sarin S, Farren C, Sinha R, Kreek MJ (2002) Naltrexone decreases craving and alcohol self-administration in alcohol-dependent subjects and activates the hypothalamo-pituitary-adrenocortical axis. Psychopharmacology 160:19–29
Ogilvie K, Lee S, Weiss B, Rivier C (1998) Mechanisms mediating the influence of alcohol on the hypothalamic-pituitary-adrenal axis responses to immune and nonimmune signals. Alcohol Clin Exp Res 22(5 Suppl):243S–247S
Olds J, Milner P (1954) Positive reinforcement produced by electrical stimulation of septal area and other regions of rat brain. J Comp Physiol Psychol 47:419–427
Olive MF, Koenig HN, Nannini MA, Hodge CW (2002) Elevated extracellular CRF levels in the bed nucleus of the stria terminalis during ethanol withdrawal and reduction by subsequent ethanol intake. Pharmacol Biochem Behav 72:213–220
Opitz K (1990) The effect of clonidine and related substances on voluntary ethanol consumption in rats. Drug Alcohol Depend 25:43–48
Overstreet DH, Knapp DJ, Breese GR (2004) Modulation of multiple ethanol withdrawal-induced anxiety-like behavior by CRF and CRF_1 receptors. Pharmacol Biochem Behav 77:405–413
Overstreet DH, Knapp DJ, Breese GR (2005) Pharmacological modulation of repeated ethanol withdrawal-induced anxiety-like behavior differs in alcohol-preferring P and Sprague-Dawley rats. Pharmacol Biochem Behav 81:122–130
Packard AE, Egan AE, Ulrich-Lai YM (2016) HPA axis interactions with behavioral systems. Compr Physiol 6:1897–1934
Palzes VA, Farokhnia M, Kline-Simon AH, Elson J, Sterling S, Leggio L, Weisner C, Chi FW (2021) Effectiveness of spironolactone dispensation in reducing weekly alcohol use: a retrospective high-dimensional propensity score-matched cohort study. Neuropsychopharmacology 46:2140–2147
Pandey SC (2004) The gene transcription factor cyclic AMP-responsive element binding protein: role in positive and negative affective states of alcohol addiction. Pharmacol Ther 104:47–58
Pandey SC, Roy A, Zhang H (2003) The decreased phosphorylation of cyclic adenosine monophosphate (cAMP) response element binding (CREB) protein in the central amygdala acts as a molecular substrate for anxiety related to ethanol withdrawal in rats. Alcohol Clin Exp Res 27:396–409
Pantazis CB, Gonzalez LA, Tunstall BJ, Carmack SA, Koob GF, Vendruscolo LF (2021) Cues conditioned to withdrawal and negative reinforcement: neglected but key motivational elements driving opioid addiction. Sci Adv 7:eabf0364
Pascual M, Baliño P, Alfonso-Loeches S, Aragón CM, Guerri C (2011) Impact of TLR4 on behavioral and cognitive dysfunctions associated with alcohol-induced neuroinflammatory damage. Brain Behav Immun 25(Suppl 1):S80–S91
Paterson NE, Myers C, Markou A (2000) Effects of repeated withdrawal from continuous amphetamine administration on brain reward function in rats. Psychopharmacology 152:440–446

Pati D, Marcinkiewcz CA, DiBerto JF, Cogan ES, McElligott ZA, Kash TL (2020) Chronic intermittent ethanol exposure dysregulates a GABAergic microcircuit in the bed nucleus of the stria terminalis. Neuropharmacology 168:107759

Peyron C, Tighe DK, van den Pol AN, de Lecea L, Heller HC, Sutcliffe JG, Kilduff TS (1998) Neurons containing hypocretin (orexin) project to multiple neuronal systems. J Neurosci 18: 9996–10015

Pfeiffer A, Brantl V, Herz A, Emrich HM (1986) Psychotomimesis mediated by κ opiate receptors. Science 233:774–776

Piazza PV, Le Moal M (1998) The role of stress in drug self-administration. Trends Pharmacol Sci 19:67–74

Pleil KE, Rinker JA, Lowery-Gionta EG, Mazzone CM, McCall NM, Kendra AM, Olson DP, Lowell BB, Grant KA, Thiele TE, Kash TL (2015) NPY signaling inhibits extended amygdala CRF neurons to suppress binge alcohol drinking. Nat Neurosci 18:545–552

Potokar J, Coupland N, Glue P, Groves S, Malizia A, Bailey J, Wilson S, Nutt D (1997) Flumazenil in alcohol withdrawal: a double-blind placebo-controlled study. Alcohol Alcohol 32:605–611

Pradhan AA, Tipton AF, Zhang H, Akbari A, Pandey SC (2019) Effect of histone deacetylase inhibitor on ethanol withdrawal-induced hyperalgesia in rats. Int J Neuropsychopharmacol 22: 523–527

Przewlocka B, Turchan J, Lasoń W, Przewłocki R (1997) Ethanol withdrawal enhances the prodynorphin system activity in the rat nucleus accumbens. Neurosci Lett 238:13–16

Qin L, He J, Hanes RN, Pluzarev O, Hong JS, Crews FT (2008) Increased systemic and brain cytokine production and neuroinflammation by endotoxin following ethanol treatment. J Neuroinflammation 5:10

Rachlin H, Green L (1972) Commitment, choice and self-control. J Exp Anal Behav 17:15–22

Rasmussen DD, Boldt BM, Bryant CA, Mitton DR, Larsen SA, Wilkinson CW (2000) Chronic daily ethanol and withdrawal: 1. Long-term changes in the hypothalamo-pituitary-adrenal axis. Alcohol Clin Exp Res 24:1836–1849

Rassnick S, Heinrichs SC, Britton KT, Koob GF (1993) Microinjection of a corticotropin-releasing factor antagonist into the central nucleus of the amygdala reverses anxiogenic-like effects of ethanol withdrawal. Brain Res 605:25–32

Repunte-Canonigo V, Shin WK, Vendruscolo LF, Lefebvre C, van der Stap L, Kawamura T, Schlosburg JE, Alvarez M, Koob GF, Califano A, Sanna PP (2015) Identifying candidate drivers of alcohol dependence-induced excessive drinking by assembly and interrogation of brain-specific regulatory networks. Genome Biol 16:68

Rhodes JS, Best K, Belknap JK, Finn DA, Crabbe JC (2005) Evaluation of a simple model of ethanol drinking to intoxication in C57BL/6J mice. Physiol Behav 84:53–63

Richards JK, Simms JA, Steensland P, Taha SA, Borgland SL, Bonci A, Bartlett SE (2008) Inhibition of orexin-1/hypocretin-1 receptors inhibits yohimbine-induced reinstatement of ethanol and sucrose seeking in Long-Evans rats. Psychopharmacology 199:109–117

Richardson HN, Zhao Y, Fekete EM, Funk CK, Wirsching P, Janda KD, Zorrilla EP, Koob GF (2008) MPZP: a novel small molecule corticotropin-releasing factor type 1 receptor (CRF_1) antagonist. Pharmacol Biochem Behav 88:497–510

Rimondini R, Arlinde C, Sommer W, Heilig M (2002) Long-lasting increase in voluntary ethanol consumption and transcriptional regulation in the rat brain after intermittent exposure to alcohol. FASEB J 16:27–35

Rivier C, Bruhn T, Vale W (1984) Effect of ethanol on the hypothalamic-pituitary-adrenal axis in the rat: role of corticotropin-releasing factor (CRF). J Pharmacol Exp Ther 229:127–131

Roberto M, Varodayan FP (2017) Synaptic targets: chronic alcohol actions. Neuropharmacology 122:85–99

Roberto M, Madamba SG, Stouffer DG, Parsons LH, Siggins GR (2004) Increased GABA release in the central amygdala of ethanol-dependent rats. J Neurosci 24:10159–10166

Roberto M, Bajo M, Cruz MT, Messing RO, Siggins GR (2008) Ethanol and CRF: which is driving GABA release in the amygdala? Alcoholism and stress: a framework for future treatment strategies, Volterra, Italy, 6–8 May 2008.

Roberto M, Cruz MT, Gilpin NW, Sabino V, Schweitzer P, Bajo M, Cottone P, Madamba SG, Stouffer DG, Zorrilla EP, Koob GF, Siggins GR, Parsons LH (2010) Corticotropin releasing factor-induced amygdala gamma-aminobutyric acid release plays a key role in alcohol dependence. Biol Psychiatry 67:831–839

Roberts AJ, Lessov CN, Phillips TJ (1995) Critical role for glucocorticoid receptors in stress- and ethanol-induced locomotor sensitization. J Pharmacol Exp Ther 275:790–797

Roberts AJ, Cole M, Koob GF (1996) Intra-amygdala muscimol decreases operant ethanol self-administration in dependent rats. Alcohol Clin Exp Res 20:1289–1298

Roberts AJ, Heyser CJ, Koob GF (1999) Operant self-administration of sweetened versus unsweetened ethanol: effects on blood alcohol levels. Alcohol Clin Exp Res 23:1151–1157

Roberts AJ, Heyser CJ, Cole M, Griffin P, Koob GF (2000) Excessive ethanol drinking following a history of dependence: animal model of allostasis. Neuropsychopharmacology 22:581–594

Roltsch Hellard EA, Impastato RA, Gilpin NW (2017) Intra-cerebral and intra-nasal melanocortin-4 receptor antagonist blocks withdrawal hyperalgesia in alcohol-dependent rats. Addict Biol 22: 692–701

Romach MK, Sellers EM (1991) Management of the alcohol withdrawal syndrome. Annu Rev Med 42:323–340

Rorick-Kehn LM, Ciccocioppo R, Wong CJ, Witkin JM, Martinez-Grau MA, Stopponi S, Adams BL, Katner JS, Perry KW, Toledo MA, Diaz N, Lafuente C, Jiménez A, Benito A, Pedregal C, Weiss F, Statnick MA (2016) A novel, orally bioavailable nociceptin receptor antagonist, LY2940094, reduces ethanol self-administration and ethanol seeking in animal models. Alcohol Clin Exp Res 40:945–954

Rose JH, Karkhanis AN, Chen R, Gioia D, Lopez MF, Becker HC, McCool BA, Jones SR (2016) Supersensitive kappa opioid receptors promotes ethanol withdrawal-related behaviors and reduce dopamine signaling in the nucleus accumbens. Int J Neuropsychopharmacol 19:pyv127

Rossetti ZL, Hmaidan Y, Gessa GL (1992) Marked inhibition of mesolimbic dopamine release: a common feature of ethanol, morphine, cocaine and amphetamine abstinence in rats. Eur J Pharmacol 221:227–234

Rossetti ZL, Isola D, De Vry J, Fadda F (1999) Effects of nimodipine on extracellular dopamine levels in the rat nucleus accumbens in ethanol withdrawal. Neuropharmacology 38:1361–1369

Rotter A, Biermann T, Amato D, Schumann G, Desrivieres S, Kornhuber J, Müller CP (2012) Glucocorticoid receptor antagonism blocks ethanol-induced place preference learning in mice and attenuates dopamine D2 receptor adaptation in the frontal cortex. Brain Res Bull 88:519–524

Roy A, Lamparski D, De Jong J, Adinoff B, Ravitz B, George DT, Nutt D, Linnoila M (1990) Cerebrospinal fluid monoamine metabolites in alcoholic patients who attempt suicide. Acta Psychiatr Scand 81:58–61

Russell MAH (1976) What is dependence? In: Edwards G (ed) Drugs and drug dependence. Lexington Books, Lexington, pp 182–187

Ryan ML, Falk DE, Fertig JB, Rendenbach-Mueller B, Katz DA, Tracy KA, Strain EC, Dunn KE, Kampman K, Mahoney E, Ciraulo DA, Sickles-Colaneri L, Ait-Daoud N, Johnson BA, Ransom J, Scott C, Koob GF, Litten RZ (2017) A phase 2, double-blind, placebo-controlled randomized trial assessing the efficacy of ABT-436, a novel V1b antagonist, for alcohol dependence. Neuropsychopharmacology 42:1012–1023

Sakurai T, Amemiya A, Ishii M, Matsuzaki I, Chemelli RM, Tanaka H, Williams SC, Richardson JA, Kozlowski GP, Wilson S, Arch JR, Buckingham RE, Haynes AC, Carr SA, Annan RS, McNulty DE, Liu WS, Terrett JA, Elshourbagy NA, Bergsma DJ, Yanagisawa M (1998) Orexins and orexin receptors: a family of hypothalamic neuropeptides and G protein-coupled receptors that regulate feeding behavior. Cell 92:573–585

Salome N, Stemmelin J, Cohen C, Griebel G (2006) Differential roles of amygdaloid nuclei in the anxiolytic- and antidepressant-like effects of the V1b receptor antagonist, SSR149415, in rats. Psychopharmacology 187:237–244

Samson HH (1987) Initiation of ethanol-maintained behavior: a comparison of animal models and their implication to human drinking. In: Thompson T, Dews PB, Barrett JE (eds) Neurobehavioral pharmacology, Advances in behavioral pharmacology, vol 6. Lawrence Erlbaum, Hillsdale, pp 221–248

Sanna PP, Kawawmura T, Chen J, Koob G, Roberts A, Vendruscolo L, Repunte-Canonigo V (2016) 11β-hydroxysteroid dehydrogenase inhibition as a new potential therapeutic target for alcohol abuse. Transl Psychiatry 6:e760

Savarese AM, Ozburn AR, Metten P, Schlumbohm JP, Hack WR, LeMoine K, Hunt H, Hausch F, Bauder M, Crabbe JC (2020) Targeting the glucocorticoid receptor reduces binge-like drinking in high drinking in the dark (HDID-1) mice. Alcohol Clin Exp Res 44:1025–1036

Savarese AM, Grigsby KB, Jensen BE, Borrego MB, Finn DA, Crabbe JC, Ozburn AR (2022) Corticosterone levels and glucocorticoid receptor gene expression in high drinking in the dark mice and their heterogeneous stock (HS/NPT) founder line. Front Behav Neurosci 16:821859

Schulteis G, Liu J (2006) Brain reward deficits accompany withdrawal (hangover) from acute ethanol in rats. Alcohol 39:21–28

Schulteis G, Markou A, Gold LH, Stinus L, Koob GF (1994) Relative sensitivity to naloxone of multiple indices of opiate withdrawal: a quantitative dose-response analysis. J Pharmacol Exp Ther 271:1391–1398

Schulteis G, Markou A, Cole M, Koob G (1995) Decreased brain reward produced by ethanol withdrawal. Proc Natl Acad Sci U S A 92:5880–5884

Schulteis G, Hyytia P, Heinrichs SC, Koob GF (1996) Effects of chronic ethanol exposure on oral self-administration of ethanol or saccharin by Wistar rats. Alcohol Clin Exp Res 20:164–171

Selye H (1950) Stress and the general adaptation syndrome. Br Med J 1(4667):1383–1392

Serrano A, Parsons LH (2011) Endocannabinoid influence in drug reinforcement, dependence and addiction-related behaviors. Pharmacol Ther 132:215–241

Serrano A, Pavon FJ, Buczynski MW, Schlosburg J, Natividad LA, Polis IY, Stouffer DG, Zorrilla EP, Roberto M, Cravatt BF, Martin-Fardon R, Rodriguez de Fonseca F, Parsons LH (2018) Deficient endocannabinoid signaling in the central amygdala contributes to alcohol dependence-related anxiety-like behavior and excessive alcohol intake. Neuropsychopharmacology 43: 1840–1850

Sharrett-Field L, Butler TR, Berry JN, Reynolds AR, Prendergast MA (2013) Mifepristone pretreatment reduces ethanol withdrawal severity in vivo. Alcohol Clin Exp Res 37:1417–1423

Shepard JD, Barron KW, Myers DA (2000) Corticosterone delivery to the amygdala increases corticotropin-releasing factor mRNA in the central amygdaloid nucleus and anxiety-like behavior. Brain Res 861:288–295

Shumilla JA, Liron T, Mochly-Rosen D, Kendig JJ, Sweitzer SM (2005) Ethanol withdrawal-associated allodynia and hyperalgesia: age-dependent regulation by protein kinase C epsilon and gamma isoenzymes. J Pain 6:535–549

Shurman J, Koob GF, Gutstein HB (2010) Opioids, pain, the brain, and hyperkatifeia: a framework for the rational use of opioids for pain. Pain Med 11:1092–1098

Sifneos PE (1973) The prevalence of 'alexithymic' characteristics in psychosomatic patients. Psychother Psychosom 22:255–262

Simms JA, Haass-Koffler CL, Bito-Onon J, Li R, Bartlett SE (2012) Mifepristone in the central nucleus of the amygdala reduces yohimbine stress-induced reinstatement of ethanol-seeking. Neuropsychopharmacology 37:906–918

Simms JA, Nielsen CK, Li R, Bartlett SE (2014) Intermittent access ethanol consumption dysregulates CRF function in the hypothalamus and is attenuated by the CRF-R1 antagonist, CP-376395. Addict Biol 19:606–611

Simons JS, Dvorak RD, Batien BD, Wray TB (2010) Event-level associations between affect, alcohol intoxication, and acute dependence symptoms: effects of urgency, self-control, and drinking experience. Addict Behav 35:1045–1053

Sinha R, Fox HC, Hong KI, Hansen J, Tuit K, Kreek MJ (2011) Effects of adrenal sensitivity, stress- and cue-induced craving, and anxiety on subsequent alcohol relapse and treatment outcomes. Arch Gen Psychiatry 68:942–952

Smith GT, Cyders MA (2016) Integrating affect and impulsivity: the role of positive and negative urgency in substance use risk. Drug Alcohol Depend 163(Suppl 1):S3–S12

Solomon RL, Corbit JD (1974) An opponent-process theory of motivation: 1. Temporal dynamics of affect. Psychol Rev 81:119–145

Somkuwar SS, Vendruscolo LF, Fannon MJ, Schmeichel BE, Nguyen TB, Guevara J, Sidhu H, Contet C, Zorrilla EP, Mandyam CD (2017) Abstinence from prolonged ethanol exposure affects plasma corticosterone, glucocorticoid receptor signaling and stress-related behaviors. Psychoneuroendocrinology 84:17–31

Sommer WH, Rimondini R, Hansson AC, Hipskind PA, Gehlert DR, Barr CS, Heilig MA (2008) Upregulation of voluntary alcohol intake, behavioral sensitivity to stress, and amygdala *crhr1* expression following a history of dependence. Biol Psychiatry 63:139–145

Song ZH, Takemori AE (1992) Stimulation by corticotropin-releasing factor of the release of immunoreactive dynorphin A from mouse spinal cords in vitro. Eur J Pharmacol 222:27–32

Stautz K, Cooper A (2013) Impulsivity-related personality traits and adolescent alcohol use: a meta-analytic review. Clin Psychol Rev 33:574–592

Stephens MA, Wand G (2012) Stress and the HPA axis: role of glucocorticoids in alcohol dependence. Alcohol Res 34:468–483

Sterling P, Eyer J (1988) Allostasis: a new paradigm to explain arousal pathology. In: Fisher S, Reason J (eds) Handbook of life stress, cognition and health. John Wiley, Chichester, pp 629–649

Stoop R (2012) Neuromodulation by oxytocin and vasopressin. Neuron 76:142–159

Stopponi S, Fotio Y, Domi A, Borruto AM, Natividad L, Roberto M, Ciccocioppo R, Cannella N (2018) Inhibition of fatty acid amide hydrolase in the central amygdala alleviates co-morbid expression of innate anxiety and excessive alcohol intake. Addict Biol 23:1223–1232

Sutcliffe JG, de Lecea L (2002) The hypocretins: setting the arousal threshold. Nat Rev Neurosci 3: 339–349

Swanson LW, Simmons DM (1989) Differential steroid hormone and neural influences on peptide mRNA levels in CRH cells of the paraventricular nucleus: a hybridization histochemical study in the rat. J Comp Neurol 285:413–435

Szabo G, Kovács GL, Baláspiri L, Telegdy G (1987) D-pipecolyl-leucyl-glycinamide, a substituted tripeptide analogue of the C-terminal part of oxytocin, influences tolerance to and dependence on ethanol in mice. Alcohol Drug Res 7:99–105

Szabo G, Mandrekar P, Petrasek J, Catalano D (2011) The unfolding web of innate immune dysregulation in alcoholic liver injury. Alcohol Clin Exp Res 35:782–786

Szmigielski A, Guidotti A, Costa E (1977) Endogenous protein kinase inhibitors. Purification, characterization, and distribution in different tissues. J Biol Chem 252:3848–3853

Szmigielski A, Szmigielska H, Wejman I (1989) The effect of prolonged ethanol administration on central alpha2-adrenoceptors sensitivity. Pol J Pharmacol Pharm 41:263–272

Taylor GJ, Bagby RM (2000) An overview of the alexithymia construct. In: Bar-On R, Parker JDA (eds) The handbook of emotional intelligence: theory, development, assessment, and application at home, school, and in the workplace. Jossey-Bass, San Francisco, pp 40–67

Thompson T, Oram C, Correll CU, Tsermentseli S, Stubbs B (2017) Analgesic effects of alcohol: a systematic review and meta-analysis of controlled experimental studies in healthy participants. J Pain 18:499–510

Thorsell A, Slawecki CJ, Ehlers CL (2005a) Effects of neuropeptide Y and corticotropin-releasing factor on ethanol intake in Wistar rats: interaction with chronic ethanol exposure. Behav Brain Res 161:133–140

Thorsell A, Slawecki CJ, Ehlers CL (2005b) Effects of neuropeptide Y on appetitive and consummatory behaviors associated with alcohol drinking in Wistar rats with a history of ethanol exposure. Alcohol Clin Exp Res 29:584–590

Thorsell A, Rapunte-Canonigo V, O'Dell L, Chen SA, King A, Lekic D, Koob GF, Sanna PP (2007) Viral vector-induced amygdala NPY overexpression reverses increased alcohol intake caused by repeated deprivations in Wistar rats. Brain 130:1330–1337

Toll L, Bruchas MR, Calo G, Cox BM, Zaveri NT (2016) Nociceptin/orphanin FQ receptor structure, signaling, ligands, functions, and interactions with opioid systems. Pharmacol Rev 68:419–457

Tran J, Teese R, Gill PR (2018) UPPS-P facets of impulsivity and alcohol use patterns in college and noncollege emerging adults. Am J Drug Alcohol Abuse 44:695–704

Trevisan L, Fitzgerald LW, Brose N, Gasic GP, Heinemann SF, Duman RS, Nestler EJ (1994) Chronic ingestion of ethanol up-regulates NMDAR1 receptor subunit immunoreactivity in rat hippocampus. J Neurochemi 62:1635–1638

Trzaskowska E, Kostowski W (1983) Further studies on the role of noradrenergic mechanisms in ethanol withdrawal syndrome in rats. Pol J Pharmacol 35:351–358

Tunstall BJ, Kirson D, Zallar LJ, McConnell SA, Vendruscolo JCM, Ho CP, Oleata CS, Khom S, Manning M, Lee MR, Leggio L, Koob GF, Roberto M, Vendruscolo LF (2019) Oxytocin blocks enhanced motivation for alcohol in alcohol dependence and blocks alcohol effects on GABAergic transmission in the central amygdala. PLoS Biol 17:e20066421

Tunstall BJ, Vendruscolo LF, Allen-Worthington K (2020) Rat models of alcohol use disorder. In: Suckow MA, Hankenson FC, Wilson RP, Foley PL (eds) The Laboratory Rat, 3rd edn. Academic Press, New York, pp 967–986

Um M, Hershberger AR, Cyders MA (2019) The relationship among depressive symptoms, urgency, and problematic alcohol and cannabis use in community adults. Addict Behav 88: 36–42

Umhau JC, Momenan R, Schwandt ML, Singley E, Lifshitz M, Doty L, Adams LJ, Vengeliene V, Spanagel R, Zhang Y, Shen J, George DT, Hommer D, Heilig M (2010) Effect of acamprosate on magnetic resonance spectroscopy measures of central glutamate in detoxified alcohol-dependent individuals: a randomized controlled experimental medicine study. Arch Gen Psychiatry 67:1069–1077

Urban NB, Kegeles LS, Slifstein M, Xu X, Martinez D, Sakr E, Castillo F, Moadel T, O'Malley SS, Krystal JH, Abi-Dargham A (2010) Sex differences in striatal dopamine release in young adults after oral alcohol challenge: a positron emission tomography imaging study with [^{11}C] raclopride. Biol Psychiatry 68:689–696

Valdez GR, Roberts AJ, Chan K, Davis H, Brennan M, Zorrilla EP, Koob GF (2002) Increased ethanol self-administration and anxiety-like behavior during acute withdrawal and protracted abstinence: regulation by corticotropin-releasing factor. Alcohol Clin Exp Res 26:1494–1501

Valdez GR, Zorrilla EP, Roberts AJ, Koob GF (2003) Antagonism of corticotropin-releasing factor attenuates the enhanced responsiveness to stress observed during protracted ethanol abstinence. Alcohol 29:55–60

Valdez GR, Sabino V, Koob GF (2004) Increased anxiety-like behavior and ethanol self-administration in dependent rats: reversal via corticotropin-releasing factor-2 receptor activation. Alcohol Clin Exp Res 28:865–872

Valdez GR, Platt DM, Rowlett JK, Rüedi-Bettschen D, Spealman RD (2007) κ Agonist-induced reinstatement of cocaine seeking in squirrel monkeys: a role for opioid and stress-related mechanisms. J Pharmacol Exp Ther 323:525–533

Vale W, Spiess J, Rivier C, Rivier J (1981) Characterization of a 41-residue ovine hypothalamic peptide that stimulates the secretion of corticotropin and β-endorphin. Science 213:1394–1397

Valentino RJ, Page ME, Curtis AL (1991) Activation of noradrenergic locus coeruleus neurons by hemodynamic stress is due to local release of corticotropin-releasing factor. Brain Res 555:25–34

Valentino RJ, Foote SL, Page ME (1993) The locus coeruleus as a site for integrating corticotropin-releasing factor and noradrenergic mediation of stress responses. In: Tache Y, Rivier C (eds) Corticotropin-releasing factor and cytokines: role in the stress response, Annals of the New York Academy of Sciences, vol 697. New York Academy of Sciences, New York, pp 173–188

Van Bockstaele EJ, Colago EE, Valentino RJ (1998) Amygdaloid corticotropin-releasing factor targets locus coeruleus dendrites: substrate for the co-ordination of emotional and cognitive limbs of the stress response. J Neuroendocrinol 10:743–757

VanderVeen JD, Plawecki MH, Millward JB, Hays J, Kareken DA, O'Connor S, Cyders MA (2016) Negative urgency, mood induction, and alcohol seeking behaviors. Drug Alcohol Depend 165:151–158

Vendruscolo LF, Koob GF (2018) Alcohol dependence conceptualized as a stress disorder. In: Harkness KL, Hayden EP (eds) Oxford handbook of stress and mental health. Oxford University Press, New York, pp 1–37. Print Publication 2020, pp 199–219

Vendruscolo LF, Koob GF (2019) Alcohol dependence conceptualized as a stress disorder. In: Harkness KL, Hayden EP (eds) Oxford handbook of stress and mental health. Oxford University Press, New York, pp 1–37

Vendruscolo LF, Barbier E, Schlosburg JE, Misra KK, Whitfield T Jr, Logrip ML, Rivier CL, Repunte-Canonigo V, Zorrilla EP, Sanna PP, Heilig M, Koob GF (2012) Corticosteroid-dependent plasticity mediates compulsive alcohol drinking in rats. J Neurosci 32:7563–7571

Vendruscolo LF, Estey D, Goodell V, Macshane LG, Logrip ML, Schlosburg JE, McGinn MA, Zamora-Martinez ER, Belanoff JK, Hunt HJ, Sanna PP, George O, Koob GF, Edwards S, Mason BJ (2015) Glucocorticoid receptor antagonism decreases alcohol seeking in alcohol-dependent individuals. J Clin Invest 125:3193–3197

Vendruscolo LF, Roberts AJ (2014) Operant alcohol self-administration in dependent rats: focus on the vapor model. Alcohol 48:277–286

Volkow ND, Fowler JS (2000) Addiction, a disease of compulsion and drive: involvement of the orbitofrontal cortex. Cereb Cortex 10:318–325

Volkow ND, Wang GJ, Fowler JS, Logan J, Hitzemann R, Ding YS, Pappas N, Shea C, Piscani K (1996) Decreases in dopamine receptors but not in dopamine transporters in alcoholics. Alcohol Clin Exp Res 20:1594–1598

Volkow ND, Fowler JS, Wang GJ (2002) Role of dopamine in drug reinforcement and addiction in humans: results from imaging studies. Behav Pharmacol 13:355–366

Volkow ND, Wang GJ, Telang F, Fowler JS, Logan J, Jayne M, Ma Y, Pradhan K, Wong C (2007) Profound decreases in dopamine release in striatum in detoxified alcoholics: possible orbitofrontal involvement. J Neurosci 27:12700–12706

Vozella V, Cruz B, Natividad LA, Benvenuti F, Cannella N, Edwards S, Zorrilla EP, Ciccocioppo R, Roberto M (2021) Glucocorticoid receptor antagonist mifepristone does not alter innate anxiety-like behavior in genetically-selected Marchigian Sardinian (msP) rats. Int J Mol Sci 22:3095

Vranjkovic O, Pina M, Kash TL, Winder DG (2017) The bed nucleus of the stria terminalis in drug-associated behavior and affect: a circuit-based perspective. Neuropharmacology 122:100–106

Walker BM, Koob GF (2008) Pharmacological evidence for a motivational role of κ-opioid systems in ethanol dependence. Neuropsychopharmacology 33:643–652

Walker BM, Rasmussen DD, Raskind MA, Koob GF (2008) α_1-Noradrenergic receptor antagonism blocks dependence-induced increases in responding for ethanol. Alcohol 42:91–97

Wang FL, Chassin L (2018) Negative urgency mediates the relation between genetically-influenced serotonin functioning and alcohol problems. Clin Psychol Sci 6:106–122

Watkins LR, Hutchinson MR, Milligan ED, Maier SF (2007) "Listening" and "talking" to neurons: implications of immune activation for pain control and increasing the efficacy of opioids. Brain Res Rev 56:148–169

Wee S, Koob GF (2010) The role of the dynorphin-κ opioid system in the reinforcing effects of drugs of abuse. Psychopharmacology 210:121–135

Weiss F, Hurd YL, Ungerstedt U, Markou A, Plotsky PM, Koob GF (1992a) Neurochemical correlates of cocaine and ethanol self-administration. In: Kalivas PW, Samson HH (eds) The neurobiology of drug and alcohol addiction, Annals of the New York Academy of Sciences, vol 654. New York Academy of Sciences, New York, pp 220–241

Weiss F, Markou A, Lorang MT, Koob GF (1992b) Basal extracellular dopamine levels in the nucleus accumbens are decreased during cocaine withdrawal after unlimited-access self-administration. Brain Res 593:314–318

Weiss F, Parsons LH, Schulteis G, Hyytia P, Lorang MT, Bloom FE, Koob GF (1996) Ethanol self-administration restores withdrawal-associated deficiencies in accumbal dopamine and 5-hydroxytryptamine release in dependent rats. J Neurosci 16:3474–3485

White AM, Castle IP, Powell PA, Hingson RW, Koob GF (2022) Alcohol-related deaths during the COVID-19 pandemic. JAMA 327:1704–1706

Whiteside SP, Lynam DR (2003) Understanding the role of impulsivity and externalizing psychopathology in alcohol abuse: application of the UPPS impulsive behavior scale. Exp Clin Psychopharmacol 11:210–217

Wikler A (1952) A psychodynamic study of a patient during experimental self-regulated re-addiction to morphine. Psychiatr Q 26:270–293

Winsky-Sommerer R, Yamanaka A, Diano S, Borok E, Roberts AJ, Sakurai T, Kilduff TS, Horvath TL, de Lecea L (2004) Interaction between the corticotropin-releasing factor system and hypocretins (orexins): a novel circuit mediating stress response. J Neurosci 24:11439–11448

Winsky-Sommerer R, Boutrel B, de Lecea L (2005) Stress and arousal: the corticotrophin-releasing factor/hypocretin circuitry. Mol Neurobiol 32:285–294

Witkiewitz K, Vowles KE, McCallion E, Frohe T, Kirouac M, Maisto SA (2015) Pain as a predictor of heavy drinking and any drinking lapses in the COMBINE study and the UK alcohol treatment trial. Addiction 110:1262–1271

Wolffgramm J, Heyne A (1995) From controlled drug intake to loss of control: the irreversible development of drug addiction in the rat. Behav Brain Res 70:77–94

World Health Organization (1992) International statistical classification of diseases and related health problems, 10th revision. World Health Organization, Geneva

Zalachoras I, Verhoeve SL, Toonen LJ, van Weert LT, van Vlodrop AM, Mol IM, Meelis W, de Kloet ER, Meijer OC (2016) Isoform switching of steroid receptor co-activator-1 attenuates glucocorticoid-induced anxiogenic amygdala CRH expression. Mol Psychiatry 21:1733–1739

Zald DH, Kim SW (2001) The orbitofrontal cortex. In: Salloway SP, Malloy PF, Duffy JD (eds) The frontal lobes and neuropsychiatric illness. American Psychiatric Press, Washington, pp 33–70

Zhou Y, Colombo G, Carai MA, Ho A, Gessa GL, Kreek MJ (2011) Involvement of arginine vasopressin and V1b receptor in alcohol drinking in Sardinian alcohol-preferring rats. Alcohol Clin Exp Res 35:1876–1883

Zhou Y, Leri F, Grella SL, Aldrich JV, Kreek MJ (2013) Involvement of dynorphin and kappa opioid receptor in yohimbine-induced reinstatement of heroin seeking in rats. Synapse 67:358–361

Zorrilla EP, Koob GF (2019) Impulsivity derived from the dark side: neurocircuits that contribute to negative urgency. Front Behav Neurosci 13:136

Zorrilla EP, Valdez GR, Weiss F (2001) Changes in levels of regional CRF-like-immunoreactivity and plasma corticosterone during protracted drug withdrawal in dependent rats. Psychopharmacology 158:374–381

Zuo W, Wu L, Mei Q, Zuo Q, Zhou Z, Fu R, Li W, Wu W, Matthew L, Ye JH (2019) Adaptation in 5-HT_2 receptors-CaMKII signaling in lateral habenula underlies increased nociceptive-sensitivity in ethanol-withdrawn rats. Neuropharmacology 158:107747

Animal Models of Excessive Alcohol Consumption in Rodents

Howard C. Becker and Marcelo F. Lopez

Contents

Abstract The development of animal models that demonstrate excessive levels of alcohol consumption has played an important role in advancing our knowledge about neurobiological underpinnings and environmental circumstances that engender such maladaptive behavior. The use of these preclinical models has also provided valuable opportunities for discovering new and novel therapeutic targets that may be useful in the treatment of alcohol use disorder (AUD). While no single model can fully capture the complexities of AUD, the goal is to develop animal models that closely approximate characteristics of heavy alcohol drinking in humans to enhance their translational value and utility. A variety of experimental approaches have been employed to produce the desired phenotype of interest—robust and reliable excessive levels of alcohol drinking. Here we provide an updated review of five animal models that are commonly used. The models entail procedural manipulations of scheduled access to alcohol (time of day, duration, frequency), periods of time when access to alcohol is withheld, and history of alcohol exposure. Specially, the models involve (a) scheduled access to alcohol, (b) scheduled periods of alcohol deprivation, (c) scheduled intermittent access to alcohol, (d) scheduled-induced polydipsia, and (e) chronic alcohol (dependence) and withdrawal experience. Each of the animal models possesses unique experimental features that engender excessive levels of alcohol consumption. Both advantages and disadvantages of each model are described along with discussion of future work to be considered in developing more optimal models. Ultimately, the validity and utility of these models will lie in their ability to aid in the discovery of new and novel potential therapeutic targets

Curr Topics Behav Neurosci (2025) 71: 79–110
https://doi.org/10.1007/7854_2024_461
Published Online: 11 February 2024

as well as serve as a platform to evaluate treatment strategies that effectively reduce excessive levels of alcohol consumption associated with AUD.

Keywords Alcohol use disorder · Animal models · Exececssive alcohol drinking

1 Introduction

Alcohol use disorder (AUD) is a chronic relapsing disease that constitutes a significant public health problem. Heavy (excessive) levels of alcohol consumption over a prolonged period of time along with increased vulnerability to relapse represent hallmark features of AUD. The development of animal models that incorporate these key behavioral characteristics has played an important role in advancing our knowledge about biological underpinnings and environmental circumstances that engender such maladaptive behavior. These preclinical models are also crucial for identifying new potential therapeutic targets as well as providing a platform for evaluating the efficacy and safety of various treatment strategies.

We previously outlined numerous experimental approaches employed in developing rodent models of excessive alcohol self-administration (Becker 2013). As noted in that review, a major effort devoted to this endeavor has entailed devising experimental strategies that overcome the fact that rodents typically do not self-administer alcohol in sufficient amounts to produce overt signs of intoxication. Further, when given the opportunity to voluntarily drink alcohol, even under circumstances when access is unlimited, rodents rarely will consume alcohol in a manner that results in significant elevation in blood alcohol levels (above legal limits). In the past 2–3 decades, the field has progressed with the development of several new models and the resurrection and refinement of some older ones that demonstrate excessive and physiologically relevant levels of alcohol consumption. As detailed in our last review (Becker 2013), these models have generally entailed incorporating genetic manipulations (e.g., selective breeding for high alcohol drinking and preference), environmental manipulations that involve modifying scheduled access to alcohol and scheduled periods of alcohol deprivation, and linking drinking procedures with dependence models. Here we provide an updated review of these various models, again outlining procedural and translational advantages and disadvantages as well as addressing more contemporary work highlighting potential sex-related differences.

2 Models of Continuous Free-Choice Access to Alcohol

A common approach for studying voluntary alcohol consumption in rodents involves providing continuous (24-h) access to alcohol in a 2-bottle choice situation. Alcohol solutions of varying concentrations are presented in the home cage along with an alternative fluid (typically water) over a number of days. The main advantages of this approach are that it is relatively simple to implement, it enables relatively quick assessment of general avidity for alcohol, and it is a useful model for screening genetic determinants of the behavior (Ciccocioppo 2012; Crabbe et al. 2010). Indeed, unrestricted (24-h) daily access to alcohol in this 2-bottle choice model has been used to selectively breed lines of rats (e.g., P/NP, HAD/LAP, AA/ANA, and UChA/UChB) (McBride et al. 2014; Quintanille et al. 2006; Sommer et al. 2006) and mice (e.g., HAP/LAP) (Grahame et al. 1999) that display high vs. low alcohol preference and intake. The model has also been extensively employed to characterize alcohol consumption in unique recombinant inbred models of mice (e.g., BXD lines) (Gill et al. 1996; Phillips et al. 1994; Rodriguez et al. 1994, 1995) and different outbred strains of rats (Azarov and Woodward 2014a, b; Khanna et al. 1990; Priddy et al. 2017). The major disadvantage of this unlimited free-access model, however, is that it is difficult to determine whether alcohol intake reaches levels that are physiologically relevant (achieving significant blood alcohol levels that accompany behavioral signs of intoxication). Except for studies using selectively bred mice (Matson and Grahame 2013; Matson et al. 2014), most studies have not shown relevant levels of intoxication using this standard continuous access 2-bottle choice procedure. Also, since the main dependent variable is the cumulative amount of alcohol consumed each day (24-h period), the model does not allow for more refined analyses of temporal patterns and structure of drinking (e.g., bout frequency and duration). Restricting access to alcohol for shorter periods of time is a convenient way to relate alcohol consumption more precisely to resultant blood alcohol levels. Further, since rodents are nocturnal, providing scheduled access to alcohol during the dark phase of their circadian cycle (when eating, drinking, and general activity are at the highest levels) facilitates greater alcohol consumption.

3 Models Involving Scheduled Access to Alcohol

A mouse model that involves limited access to alcohol restricted to the dark phase of the circadian cycle was developed to model binge-like drinking (Crabbe et al. 2011a). The model, commonly referred to as "drinking-in-the-dark" (DID), was designed to yield a high level of alcohol intake over a defined period of time so as to produce blood alcohol concentrations above the US legal limit of intoxication ($\geq$0.08 g/dL)—thereby satisfying the clinical criteria for binge-like drinking (NIAAA 2004). Typically, the procedure entails offering mice a single bottle of alcohol (20% v/v) for 2 h starting 3 h after the dark phase begins for 3 consecutive days. This is followed by a 4th day when access is extended to 4 h. This scheduled alcohol access produced significant consumption on the final day of this 4-day procedure in C57BL/6 mice, with resultant blood alcohol levels typically reaching ≥0.10 g/dL (Rhodes et al. 2005; Thiele et al. 2014; Thiele and Navarro 2014). Not surprisingly, alcohol intake in this model differed substantially across genotypes (inbred strains and recombinant mouse lines), and, importantly, drinking in C57BL/6 mice produced observable signs of intoxication as indexed by measures of motor incoordination (Rhodes et al. 2007). When the model incorporated a 2-bottle choice situation (water available as the alternative fluid), reduced alcohol intake and resultant blood alcohol levels have been reported (Phillips et al. 2010; Rhodes et al. 2007). However, another study with mice that had a history of drinking sucrose showed that alcohol intake in this 2-bottle choice situation (water as the alternative fluid) achieved significantly elevated blood alcohol levels (≥0.80 g/dL) (Giardino and Ryabinin 2013). Overall, the DID model has proven to reliably produce high levels of alcohol consumption in a short period of time, and this binge-like alcohol drinking does not appear related to motivation for obtaining calories contained in the alcohol (Lyons et al. 2008). In general, alcohol consumption in the DID model has been reported to be greater in female compared to male C57BL/6J mice (Levine et al. 2021; Younis et al. 2019).

A modified version of the 4-day DID model was employed by Crabbe and his colleagues to generate selectively bred lines of mice that drink substantial amounts of alcohol that produce behavioral signs of intoxication. In this work, mice from a genetically heterogeneous stock (HS/Npt) were tested in a 2-day single-bottle (20% alcohol) paradigm (2-h access the first day and then 4-h access the next day, both during the early part of the dark cycle). Over several generations, average blood alcohol levels registered immediately following the 4-h drinking session increased from an initial value of approximately 0.03 g/dL (prior to selective breeding) to 0.10 g/dL (Crabbe et al. 2009). This high drinking-in-the-dark (HDID-1) selected line also consumed significantly more alcohol than the control line from which they were selected, even though the selection was based on blood alcohol levels (not the amount of alcohol consumed). Interestingly, HDID-1 mice from the 13-17th selected generations consumed similar amounts of alcohol and other tastants (sucrose, saccharin, and quinine) as the control line when the solutions were presented under continuous (24-h) access conditions. However, greater intake was noted in the HDID-1 mice when preference testing was extended for several days under limited access conditions (Crabbe et al. 2011b). A second replicate line (HDID-2) displayed average blood alcohol levels that increased to 0.10 g/dL in 19 generations (Barkley-Levenson and Crabbe 2014). These HDID selected lines do not show differences in alcohol metabolism, but intake and blood alcohol levels were reduced when water was included as an alternative solution in the 4-day DID model (Barkley-Levenson and Crabbe 2014). A relatively modest preference for alcohol was displayed by male and female HDID lines when access was extended continuously (24-h/day) in the standard 2-bottle choice situation and intake remained relatively stable over several weeks (Crabbe et al. 2022). This suggests there are some distinct genetic influences that shape binge-like drinking when access is restricted versus those governing consumption when alcohol access is unlimited. Other studies have shown that HDID mice display increased sensitivity to some acute alcohol effects (locomotor stimulant effects) but reduced sensitivity to sedative/ataxia effects (Barkley-Levenson and Crabbe 2014). In studies using operant conditioning procedures, male and female HDID-2 mice (but not the HDID-1 line) showed increased oral self-administration (under both FR-1 and FR-3 schedules), but there was no apparent difference in progressive ratio responding (breakpoint) or cue-induced reinstatement of alcohol seeking behavior for either selected line compared to the progenitor line (Jensen et al. 2021; Savarese et al. 2021). These results align with findings indicating no differences in expression of conditioned place preference (Barkley-Levenson et al. 2015) and suggest some independence of genetic influences for motivational/rewarding effects of alcohol versus those underlying selection for binge-like drinking. On the other hand, the fact that both HDID lines show reduced alcohol-induced conditioned taste aversion suggests that reduced sensitivity to alcohol-related aversion may play a permissive role in elevated drinking in these mice (Barkley-Levenson et al. 2015).

Several studies have provided alcohol under limited access conditions (2-h/day) for an extended period of time. For example, presenting alcohol alone 2-h/day for 14 days produced faster rates of consumption (more drinking during the early portion of the drinking sessions) and tolerance to the ataxic effects of alcohol (Linsenbardt et al. 2011). This model has also been effectively used to study consequences of alcohol binge-like exposure in utero (Boehm et al. 2008) and during adolescence (Metten et al. 2011). Other studies have provided scheduled daily access to alcohol (2-h/day) over several weeks, but alcohol was presented along with water. Using this model, selectively bred high alcohol-preferring (HAP) lines of mice were shown to "front-load" (accelerate drinking during the early portion of the limited access sessions), and the emergence of this binge-like drinking pattern was associated with rewarding effects of alcohol (Ardinger et al. 2020). Using male and female C57BL/6 mice, studies have shown that 2–3 weeks of drinking alcohol under these conditions also lead to apparent habitual drinking, as defined by resistance to alcohol reward devaluation produced by adulteration of the alcohol solution with quinine (i.e., persistence of alcohol drinking in the face of aversion related to the bitter taste of quinine) (Bauer et al. 2021; Schuh et al. 2022; Sneddon et al. 2019, 2021). Another model of limited access drinking that entails simultaneously offering several concentrations of alcohol (5–40%) for 2 weeks was shown to demonstrate measures of negative affect related to alcohol withdrawal that varied depending on sex, age, and procedural conditions (Lee et al. 2017; Szumlinski et al. 2019).

Models involving scheduled access to alcohol have also been used in rats. Male rats selectively bred for high alcohol preference (P rats) consumed more alcohol in a 2-bottle free-choice situation (10% alcohol vs. water) when access was scheduled over four 1-h periods (each separated by 2 h) during the dark cycle compared to when the alcohol was available continuously for the equivalent 4-h period (Murphy et al. 1986). Building on these results, more recent studies have examined the effect of offering P rats concurrent access to three fluids (water vs. 15% alcohol vs. 30% alcohol) over three 1-h access periods during the dark phase of the circadian cycle.

Over several weeks alcohol consumption in this model was shown to register significant blood alcohol levels ($\geq$ 0.08 g/dL) as well as behavioral signs of intoxication (motor impairment), with effects more robust in females compared to males (Bell et al. 2011; McBride et al. 2010). Similar results were observed with selectively bred lines of high alcohol preference/drinking (P, HAD) rats when access to alcohol was limited to 2–3 h (Bell et al. 2014). Using a protocol that involved limited access (1-h/day) to multiple alcohol solutions in male Sardinian preferring (sP) rats, it was observed that randomizing the time of alcohol access each day produced higher levels of alcohol intake (and resultant blood alcohol levels) compared to consistently scheduled alcohol access (Colombo et al. 2014, 2015, 2017).

4 Models Involving Scheduled Alcohol Deprivation

Animals with a long history of daily access to alcohol display a transient yet robust increase in voluntary alcohol consumption and preference when alcohol is reintroduced after a period of deprivation. This alcohol deprivation effect was first formally described in rats (Sinclair and Senter 1968) but has also been demonstrated in mice (Salimov and Salimova 1993; Salimov et al. 2000; Tambour et al. 2008). Most studies have examined the phenomenon in rats using 2-bottle choice continuous access models. Increased alcohol drinking has been noted after relatively brief periods of deprivation (~24 h) as well as following longer (several weeks) deprivation intervals (Sinclair and Li 1989). The alcohol deprivation effect has also been demonstrated using limited access operant self-administration procedures in rats (Heyser et al. 1997; Holter et al. 1997) and mice (Sparta et al. 2009). However, there were no effects of deprivation on alcohol intake reported in a study using a modified (sipper-tube) self-administration procedure (Samson and Chappell 2001).

The alcohol deprivation effect has been demonstrated in outbred rat strains such as Wistar (Vengeliene et al. 2003) and Long-Evans (Sinclair and Tiihonen 1988). Similarly, an alcohol deprivation effect has been reported in rats selectively bred for high alcohol preference (P rats) under free-choice continuous access and limited access operant paradigms (McKinzie et al. 1998; Sinclair and Li 1989; Vengeliene et al. 2003). However, a robust increase in alcohol consumption following a period of deprivation has not been reliably observed in other rat lines selectively bred for high alcohol preference, including the Alko alcohol-accepting (AA) rats (Sinclair and Li 1989; Sinclair and Tiihonen 1988) and the Indiana high alcohol drinking (HAD) rats (Rodd-Henricks et al. 2000a). The Sardinian P (sP) rats, which were generated using the same selection criteria as the Indiana P rats, showed a fairly modest increase in alcohol intake that was very brief in duration (Agabio et al. 2000) or no increase after deprivation (Serra et al. 2003). Collectively, these data do not support a consistent relationship between selection for high alcohol preference/intake and expression of a robust alcohol deprivation effect.

Although the alcohol deprivation effect has been viewed as a model for alcohol relapse and craving, there are some drawbacks related to the model. One concern relates to the specificity of the phenomenon, since exaggerated intake of other rewarding tastants (e.g., sucrose and saccharin) can be demonstrated in rats following a period of deprivation (Avena et al. 2005; Wayner et al. 1972). As noted above, the increase in alcohol intake after short or long periods of deprivation is typically short-lived, with intake returning to baseline (pre-deprivation) levels in a few days. However, when P rats are given concurrent access to several alcohol concentrations (10, 20, and 30%) along with water, the alcohol deprivation effect was shown to be more robust and more durable (Rodd-Henricks et al. 2001). Further, this same experimental paradigm was reported to produce an alcohol deprivation effect in HAD rats even though these animals do not show such an effect when a single alcohol concentration is offered in a free-choice situation (Rodd et al. 2004).

While enhanced alcohol intake following a single deprivation period has been shown to be a transient effect, repeated deprivation experience produces longer lasting increases in alcohol consumption. For example, after long-term free access to several alcohol solutions, repeated "forced" abstinence periods resulted in progressively greater increases in alcohol intake, a shift in preference for higher alcohol concentrations, and longer lasting deprivation effects in Wistar rats (Spanagel and Holter 1999, 2000) and P rats (Rodd-Henricks et al. 2000b, 2001). Additionally, concurrent access to multiple concentrations of alcohol along with exposure to repeated cycles of deprivation produced significant increases in alcohol consumption in HAD rats, a genotype that does not readily exhibit an alcohol deprivation effect following a single period of deprivation (Rodd et al. 2009; Rodd-Henricks et al. 2000a). Using a similar experimental strategy involving multiple alcohol concentrations (0, 5, 10, and 15%) and several cycles of deprivation, increased alcohol consumption was demonstrated over repeated episodes of re-exposure to alcohol in rats selectively bred for low alcohol preference and drinking (NP and LAD rats) (Bell et al. 2004). This suggests that genetic selection for low alcohol preference/consumption can be overcome by experimental parameters that ordinarily engender expression of a more robust alcohol deprivation effect. Interestingly, offering several alcohol concentrations and repeated cycles of deprivation did not alter the magnitude or duration of a relatively brief and modest alcohol deprivation effect in sP rats (Serra et al. 2003).

In addition to enhancing the alcohol deprivation effect under 24-h free-choice conditions, repeated episodes of deprivation augmented and prolonged oral alcohol self-administration using operant conditioning procedures in Wistar, P, and HAD rats (Oster et al. 2006; Rodd et al. 2003; Spanagel and Holter 2000). Further, this effect was shown to be accompanied by an apparent enhancement of the reinforcing efficacy of alcohol, as indexed by higher breakpoint values under progressive ratio testing procedures (Oster et al. 2006; Rodd et al. 2003; Spanagel and Holter 2000). In a long-term drinking model involving several months of free-choice alcohol access and multiple episodes of deprivation, Wistar rats not only increased alcohol intake and demonstrated a progressive shift in preference for higher previously less preferred alcohol concentrations, but these rats also exhibited less sensitivity to the otherwise unfavorable adulteration of alcohol with quinine (Spanagel et al. 1996). This latter effect has been suggested to reflect more compulsive aspects of drinking that develops as a function of long-term access to alcohol with repeated intervening periods of abstinence (deprivation) (Spanagel 2009).

Fewer studies have systematically studied the alcohol deprivation model in mice (Vengeliene et al. 2014). In one study, the effect of repeated deprivation cycles on alcohol intake in a 2-bottle choice (10% alcohol vs. water) continuous access situation differed in substrains of C57BL/6 mice (Khisti et al. 2006). Repeated 4-day deprivation periods initially produced a robust alcohol deprivation effect in C57BL/6NCrl mice, but the transient increase in intake diminished in magnitude over successive deprivation cycles. In contrast, alcohol consumption did not significantly change following single or multiple cycles of deprivation in C57BL/6J mice. In a modified version of the alcohol deprivation effect, C57BL/6J mice showed increased alcohol intake following repeated weekly deprivation periods of 6 days (alcohol was reinstated 1 day each week). However, this effect was abolished with a longer (2-week) deprivation period (Melendez et al. 2006). Another study using male C57BL/6N mice also failed to show an effect of repeated alcohol deprivation periods (Vengeliene et al. 2014). In a recent study, male and female mice of the selectively bred HDID-1 and HDID-2 lines were given unlimited access to alcohol in their home cage continuously for 46 weeks. This was followed by five 2-week deprivation periods, with 2 weeks of resumed drinking after each deprivation period. Mice showed a stable pattern of intake over the initial 46 weeks that was not affected by repeated deprivation periods compared to mice that continued having uninterrupted access to alcohol (Crabbe et al. 2022).

Although relatively few studies have examined the alcohol deprivation effect in mice, single or multiple deprivation periods have not reliably produced enhanced alcohol drinking when alcohol is offered in the home cage under limited access conditions. A study in male C57BL/6J mice using the DID model examined alcohol intake over 6 weeks, with each of the 4-day drinking opportunities separated by 3 days off. A lickometer system was used in the study to show that mice gradually develop higher levels of intake during the first 15-min of access to the alcohol bottle (front-loading). In addition, increased alcohol intake was observed with repeated DID cycles (Wilcox et al. 2014). Another study using this procedure showed that repeated experience with weekly DID cycles (up to 10) did not result in significant signs of anxiety-like behavior but favored a subsequent increase in alcohol intake using a 24-h access protocol. However, mice did not show an increase in alcohol intake during the repeated cycles of DID (Cox et al. 2013).

5 Models Involving Scheduled Intermittent Alcohol Access

Another model that engenders a high level of alcohol consumption involves chronic intermittent access to alcohol. In this model, inherent in the scheduled intermittency of free access to alcohol are repeated periods of abstinence. Although the model is similar to the paradigm described above involving repeated periods of deprivation, in this case, the periods of alcohol access and deprivation are relatively short (days rather than weeks), thereby accelerating the pace at which excessive levels of alcohol intake can be established. This chronic intermittent access procedure was first described to produce increased drinking in rats when alcohol was provided on a continuous basis for 2 days with intervening 2-day abstinence periods (Wayner et al. 1972) or for 24-h every other day (Wise 1973). More recently, free access to 20% alcohol was offered in a 2-bottle choice situation (with water) for 24 h, 3 days a week (with no more than 2 days of abstinence between access days). Within 5–6 drinking sessions, alcohol consumption increased from baseline levels of about 2 g/kg/24-h to approximately 5–6 g/kg/24-h in Long-Evans rats (Simms et al. 2008). A similar outcome was reported in another study where Long-Evans rats exposed to the same procedure displayed progressively increased consumption and preference for 20% alcohol over 20 drinking sessions (Carnicella et al. 2009). This escalation of drinking along with an increased preference for alcohol was also demonstrated in Wistar rats (Simms et al. 2008), although another study using a 3-bottle choice situation (water vs. 5% vs. 20%) reported a two- to threefold difference in the change in alcohol intake and preference depending on the supplier of Wistar rats (Palm et al. 2011). The escalation of intake in Long-Evans and Wistar rats produced significantly elevated blood alcohol levels in samples taken after the first 30 min of the drinking sessions, with several subjects attaining levels above 0.08 g/dL (Carnicella et al. 2009; Simms et al. 2008). In another study, male Wistar rats that had 24-h access to alcohol in the home cage every other day not only displayed increased alcohol intake but also showed impaired working memory during acute (but not protracted) periods of abstinence (George et al. 2012). Increased alcohol consumption has also been noted in Sprague-Dawley rats following the 2-bottle (water vs. 20% alcohol) intermittent access paradigm (Bito-Onon et al. 2011), but the effect may only be observed in a portion of the animals (Moorman and Aston-Jones 2009).

In addition to home-cage drinking, this intermittent alcohol access model has also been extended to oral alcohol self-administration behavior using operant conditioning procedures. For example, Long-Evans rats were shown to vigorously respond to self-administered 20% alcohol when operant sessions scheduled every other day were gradually reduced from overnight to 30 min in duration (Simms et al. 2010). The increased amount of alcohol self-administered resulted in significant elevation of blood alcohol levels following the 30-min session, with average values of ~0.06 g/dL and several rats registering blood alcohol levels above 0.10 g/dL (Simms et al. 2010). In another study, prolonging the intermittent access schedule for several months not only increased home-cage alcohol drinking but also transferred to increased operant oral alcohol self-administration in Wistar rats (Hopf et al. 2010). Further, rats maintained on the intermittent access schedule to 20% alcohol for 3–4 months demonstrated resistance to quinine adulteration of alcohol in home-cage drinking and operant responding, but this effect was not observed in rats with a history of intermittent alcohol access for only 1.5 months (Hopf et al. 2010).

A few studies have examined drinking in this intermittent access model in rats selectively bred for high alcohol preference. For instance, P rats were shown to exhibit increased alcohol intake under conditions in which 24-h free-choice (20% alcohol vs. water) access was given every other day. However, this increase in alcohol consumption from an average baseline level of 4–5 g/kg/24-h) to 6–7 g/kg/24-h over 20 drinking sessions was relatively modest compared to the escalation of intake exhibited in Long-Evans and Wistar rats reported in the same study (Simms et al. 2008). In contrast, using a similar 2-bottle choice (20% alcohol vs. water), every other day scheduled access paradigm, the Sardinian P (sP) rats showed robust escalation of drinking (nearly a twofold increase in alcohol intake over 20 drinking sessions (Loi et al. 2010). This increase was also noted during the first hour of access during the dark phase, with intake rising from baseline levels of ~0.5 to 1.5–2.0 g/kg. Alcohol consumption in sP rats given intermittent access significantly exceeded intake registered in sP rats that were given the same alcohol solution (20% vs. water), but in a continuous access pattern. After 10 drinking sessions, consumption in the intermittent access group produced behavioral signs of intoxication, as measured by motor impairment in a rotarod task. Additionally, these rats exhibited resistance to the effects of quinine adulteration of alcohol as well as competing effects of concurrent access to saccharin (Loi et al. 2010). It is interesting that sP rats are very responsive to this chronic intermittent access procedure in which relatively short periods of access and abstinence (deprivation) are repeatedly alternated, while the Indiana P rats (but not sP rats) display robust escalation of drinking in a model of repeated deprivations where access and deprivation periods are longer in duration (Rodd-Henricks et al. 2001; Serra et al. 2003). An explanation for this discrepancy is not readily apparent at present (Loi et al. 2010).

Recent studies have suggested that sex and environmental factors (e.g., housing conditions) may modulate escalation of drinking in this model. Female Wistar rats that were pair housed (separated by a divider) did not show an increase in drinking compared to single-housed females or males independent of housing conditions (Moench and Logrip 2021). In another study, female Long-Evans rats showed significantly higher levels of intake compared to males and a gradual increase in alcohol preference, but neither males nor females showed an increase in alcohol intake over 7 weeks of intermittent access to 20% alcohol (vs. water) (Pirino et al. 2022). Intermittent access to alcohol in the home cage has also been used to “prime” rats before training them to respond to alcohol in an operant self-administration protocol. In these studies, it was observed that male P, Lister Hooded, and Long-Evans rats showed the expected gradual increase in alcohol intake over days of intermittent access while Wistar rats did not (Hernandez and Moorman 2020; McCane et al. 2021; Smeets et al. 2022). Overall, the emergence of escalated alcohol intake in the rats (and mice; see below) using this model depends on various factors including sex, genotype, and housing conditions (Carnicella et al. 2014; Spear 2020).

Similar studies have been conducted in mice. For example, Melendez (2011) reported that adult male C57BL/6J mice provided 24-h access to alcohol in a 2-bottle choice situation (15% alcohol vs. water) consumed significantly more alcohol when it was presented every other day in comparison to mice that received continuous access to alcohol every day. Specifically, initial alcohol intake (6–7 g/kg/24-h) increased to 14–15 g/kg/24-h over 7 drinking sessions in the intermittent access group while intake increased to 8–9 g/kg/24-h in the continuous access group. A large portion of the alcohol consumed occurred within the first 6 h when it was presented during the dark phase of the circadian cycle, and the increased level of drinking in the intermittent group reverted to lower (baseline) levels of intake when a continuous access schedule was implemented (Melendez 2011). In another study, C57BL/6J mice were first acclimated to increasing concentrations of alcohol and then maintained on a 24-h 2-bottle choice (20% alcohol vs. water) regimen, with access scheduled either every other day or continuously every day. Over the course of 4 weeks, alcohol consumption increased to ~20 g/kg/24-h in the intermittent access group compared to ~16 g/kg/24-h for the continuous access group (Hwa et al. 2011). This increase in alcohol consumption as a function of intermittent access was more robust in female (30 g/kg/day) compared to male (20 g/kg/day) C57BL/6J mice, and extending intermittent access for 16 weeks in the male subjects resulted in mild expression of withdrawal-related hyperexcitability. Also, intake over the first 2-h in a single-bottle test with 20% alcohol was greater in mice with intermittent compared to continuous access, and this greater intake resulted in higher blood alcohol levels (Hwa et al. 2011). However, using similar procedures, others have not observed this large a difference in intake between mice offered alcohol in an intermittent versus continuous fashion (Crabbe et al. 2012). The intermittent access model has been used with a variety of mouse genotypes including HDID, C3H/Hej, and C57BL/6J, among others). Although increased alcohol intake was observed in all the genotypes, it was not related to previous baseline levels of intake under continuous access conditions or subsequent withdrawal symptoms (Rosenwasser et al. 2013). Other recent studies have shown that female but not male C57BL/6J mice show escalation of alcohol intake in this model (Bloch et al. 2020; Cannady et al. 2020). Thus, in both rats and mice, females appear more likely to demonstrate elevated alcohol consumption when it is presented in an intermittent fashion. An explanation for this possible sex-related difference awaits further investigation.

6 Models Involving Schedule-Induced Polydipsia

Animals have been shown to engage in excessive drinking behavior when delivery of food reinforcement is scheduled in an intermittent fashion (typically a fixed time interval) that is not under the animal's control (Falk 1961). This adjunctive behavior (excessive drinking) is displayed as a consequence of and in relation to another behavior that is evoked by environmental change (eating small amounts of food delivered in a scheduled manner that is not determined by the animal). The term schedule-induced polydipsia refers to the excessive nature of adjunctive drinking under these conditions, which greatly exceeds the fluid intake that would occur if the same total amount of food was presented all at once.

When an alcohol solution is the available fluid, this schedule-induced polydipsia results in excessive levels of alcohol consumption (10–14 g/kg/24-h) in rats that leads to dependence, as evidenced by overt signs of withdrawal when the alcohol is removed (Falk and Samson 1975; Falk et al. 1972). Alcohol consumption during daily 3-h sessions over several months was reported to be sufficient to produce dependence (Tang and Falk 1983). In a more recent study, a schedule-induced polydipsia procedure was used to assess alcohol consumption in rats selectively bred for high and low alcohol preference (Gilpin et al. 2008a). Across a number of alcohol concentrations, P rats and one of the replicate lines of HAD rats showed greater water and alcohol intake compared to their non-preferring counterparts (NP and LAD-2 rats). In all cases, blood alcohol levels were positively correlated with alcohol intake after the 1-h sessions, with many rats registering levels >0.08 g/dL (Gilpin et al. 2008a). This procedure has also been used to induce high levels of alcohol intake in adolescent male and female Sprague-Dawley rats, with intake (10% alcohol mixed with chocolate Boost®) yielding blood alcohol levels exceeding 0.08 g/dL in 30 min in males (0.086 ± 0.013 g/dL) and females (0.075 ± 0.010 g/dL) (Hosova and Spear 2017).

Schedule-induced polydipsia procedures have also been used to examine alcohol consumption in mice. In an early study involving outbred (ICR-DUB) female mice, four daily 1-h sessions (each separated by 6-h) produced high levels of drinking in mice given access to 6% alcohol (14–20 g/kg/day) or 10% alcohol (17–25 g/kg/day). In both cases, this level of intake over 7 days was not sufficient to produce significant signs of withdrawal following the scheduled access phase of the study (Ogata et al. 1972). Over 20 daily 1-h sessions, the alcohol-preferring C57BL/6J inbred strain consumed a substantial amount of 5% alcohol (~5 g/kg) relative to their initial intake (~1 g/kg). In contrast, the non-preferring DBA/2J inbred strain showed only a very modest increase in alcohol consumption under the same schedule conditions (Mittleman et al. 2003). However, in another study using a fixed-time schedule of food delivery (as opposed to a variable schedule used in the Mittelman et al. (2003) study, both C57BL/6J and DBA/2J mouse strains were shown to exhibit high levels of alcohol intake and signs of intoxication (Ford et al. 2013). These results highlight the importance of environmental factors (e.g., schedule of food presentation) interacting with genotype in governing alcohol consumption in this model.

Advantages of this model are that animals consume large quantities of alcohol orally and on a voluntary basis (Falk and Tang 1988). Disadvantages of this approach include lack of specificity of the effect since polydipsia can be seen when other fluids are made available (including water) and the fact that animals are typically maintained on a food-restricted diet. This latter issue raises concern about whether motivation to drink alcohol is related to its pharmacological effects or its caloric content. Of note, challenging the non-specific nature of this polydipsia model, a recent study showed that a subset of male Sprague-Dawley rats exhibited heightened alcohol consumption but no change in water intake under the same schedule-induced polydipsia experimental parameters (Fouyssac et al. 2021).

Another shortcoming of this model to consider is that when the schedule of intermittent reinforcement is relaxed, alcohol consumption reverts to control levels in rats (Tang et al. 1982). That is, elevated alcohol drinking does not endure under free-choice conditions even though the animals consumed large amounts of alcohol when it was available under intermittent schedules of food reinforcement (Ford 2014). However, inasmuch as such schedules that induce adjunctive behaviors are stressful (Falk 1971; Lopez-Grancha et al. 2006), it may be that studies in rodents have not utilized experimental parameters that are optimal for establishing the negative reinforcing effects of alcohol. That is, while schedule-induced polydipsia procedures are effective in establishing the positive reinforcing effects of alcohol (Meisch 1975), experimental conditions that facilitate association of alcohol consumption with stress relief (escape from the onerous nature of the intermittent, response non-contingent schedule of food delivery) may be required for producing long-lasting elevated drinking. A study conducted with high- and low-drinking mouse strains (C57BL/6J and DBA2/J, respectively) found that schedule-induced polydipsia results in high levels of alcohol intake and intoxication in both strains. Increases in blood alcohol levels were also associated with elevations in circulating levels of corticosterone due to the schedule restrictions (Ford et al. 2013; see also studies with monkeys in Jimenez et al. 2017). A more recent study conducted with male Sprague-Dawley identified a subpopulation of subjects that failed to drink a high level of water but consumed a high level of alcohol under the same schedule that induced polydipsia. This subpopulation was identified as "alcohol copers" for their avidity to drink alcohol to cope with the stress of this schedule. This same group of "alcohol coper" rats showed higher resistance to reduce drinking when the alcohol solution contained quinine (Fouyssac et al. 2021).

While this chapter mainly focuses on rodent models of excessive alcohol drinking, it is noteworthy that the schedule-induced polydipsia paradigm has been effectively used in nonhuman primates to demonstrate sustained high levels of alcohol intake (Grant et al. 2008). Further, the pattern of drinking during the induction phase of this model was shown to predict the degree of heavy drinking once the schedule-induced polydipsia regimen was relaxed. That is, cynomolgus monkeys that reached levels of intake that produced blood alcohol levels above 0.08 g/dL during the induction phase were classified as "gulpers" (as opposed to "sippers") and showed higher levels of alcohol consumption during a subsequent 12-month continuous free-choice access period. Excessive alcohol consumption during this free-access period produced behavioral signs of intoxication in many of the subjects. Additionally, extending the open-access period to more than 2 years along with intervening periods of abstinence not only produced sustained excessive levels of alcohol consumption but also resulted in functional (synaptic) and morphological adaptations in the brain (putamen) (Cuzon Carlson et al. 2011). In separate studies, the level of aggressive temperament displayed by male and female rhesus macaques during late adolescence was shown to predict the level of alcohol intake in the schedule-induced polydipsia model; i.e., subjects displaying higher aggressive behavior also showed greater levels of alcohol intake and resultant blood alcohol levels (McClintick and Grant 2016). In another study, dominance hierarchy (dominant or subordinate status) did not relate to the levels of intake under schedule-induced polydipsia in adult male cynomolgus monkeys. However, after the induction phase of the study, when the schedule was relaxed, subordinate monkeys showed higher levels of alcohol intake than dominants (Galbo et al. 2022). Recently, Grant and her colleagues have shown that alcohol intake following induction in the schedule-induced polydipsia paradigm can be modulated by chemogenetic inhibition of the putamen in male and female rhesus monkeys. Specifically, inhibiting this area associated with habitual responding produced higher water and alcohol drinking (no sex-related differences were observed). However, during subsequent sessions alcohol intake reverted to baseline levels and water drinking was reduced to below baseline levels (Grant et al. 2022). Thus, the schedule-induced polydipsia procedure has proven to be effective and integral to this monkey model of heavy drinking that captures many of the features of alcohol use disorder.

7 Models Involving Alcohol Dependence and Withdrawal

Over the past two decades, rodent models involving chronic alcohol exposure producing dependence have been successfully linked with self-administration procedures to demonstrate excessive levels of alcohol intake (Becker 2008, 2014; Becker and Lopez 2016; Becker and Ron 2014). Indeed, numerous studies involving mice and rats have demonstrated escalated alcohol consumption using home-cage free-choice models and operant conditioning procedures (Griffin 2014; Lopez and Becker 2014; Vendruscolo and Roberts 2014). In most cases, dependence has been induced by administering alcohol vapor via inhalation chambers. For example, rats exposed to chronic alcohol vapor treatment consumed significantly more alcohol than non-dependent controls under free-choice unlimited (24 h/day) access conditions (Rimondini et al. 2002, 2003; Sommer et al. 2008). Similar results have been reported in mice following chronic alcohol vapor exposure, with voluntary alcohol consumption assessed using a limited access (2 h/day) schedule (Becker and Lopez 2004; Dhaher et al. 2008; Finn et al. 2007; Huitron-Resendiz et al. 2018; Lopez and Becker 2005; Lopez et al. 2017). Additionally, studies using operant conditioning procedures have demonstrated increased alcohol self-administration in mice (Chu et al. 2007; Lopez et al. 2014) and rats (C. K. Funk and Koob 2007; Gilpin et al. 2008b, c, 2009; Meinhardt and Sommer 2015; O'Dell et al. 2004; Richardson et al. 2008; Roberts et al. 1996, 2000) with a history of chronic alcohol vapor experience.

A key feature of this model that yields robust and reliable escalated alcohol responding/intake is the delivery of chronic alcohol exposure in an intermittent pattern such that multiple withdrawal episodes are experienced (Lopez and Becker 2005; O'Dell et al. 2004). This point highlights the importance of establishing the negative reinforcing effects of alcohol in driving enhanced motivation to imbibe (Becker 2014; Koob 2021, 2022). Additionally, the intensity of repeated chronic intermittent ethanol (CIE) exposure cycles (producing high and sustained blood alcohol levels) was shown to be critical in favoring escalation of alcohol consumption in the model (Griffin et al. 2009a). Further, the effect appears specific to alcohol because repeated cycles of CIE exposure did not produce alterations in water intake or consumption of highly palatable fluids such as sucrose and saccharin (Becker and Lopez 2004; Lopez et al. 2012). This suggests that the increase in alcohol consumption is not a non-specific effect related to a general need to hydrate with fluids or increase caloric intake.

Using this approach, enhanced alcohol responding/intake has been shown to be durable, evident in dependent animals well beyond acute withdrawal. Indeed, with an increased number of CIE exposure cycles, upregulated alcohol intake was shown to be not only further augmented but also sustained for a longer period of time (several weeks) following final withdrawal compared to intake in a separate group of non-dependent mice (Lopez and Becker 2005). Further, analysis of the temporal pattern of alcohol consumption revealed that dependent mice not only consumed more alcohol than non-dependent animals over the entire 2-h access period, but the rate of consumption was faster and progressively increased over successive withdrawal test periods (Griffin et al. 2009b; see also Robinson and McCool 2015 with rats).

In both mice and rats, escalation of alcohol self-administration following repeated cycles of CIE exposure was reported to be associated with significantly higher resultant blood alcohol levels compared to that achieved by more modest and stable levels of intake in non-dependent animals (Becker and Lopez 2004; Roberts et al. 2000). Additionally, the faster rate of alcohol intake and greater overall amount consumed exhibited by dependent mice have been shown to result in significantly higher peak and more sustained alcohol concentrations measured in the brain compared to levels achieved from consumption of alcohol in non-dependent animals (Griffin et al. 2009b). Moreover, greater voluntary alcohol consumption in dependent mice produced brain alcohol concentrations that approximated those levels experienced during chronic intermittent alcohol exposure that rendered the subjects dependent in the first place. While it is tempting to speculate that CIE-exposed animals display increased voluntary alcohol drinking behavior to attain blood and brain alcohol levels in a range consistent with sustaining dependence, the extent to which resultant brain alcohol concentrations play a role in driving as well as perpetuating enhanced alcohol drinking in dependent animals remains to be determined.

Despite the growing and convergent body of evidence indicating that rodent models of dependence involving CIE exposure produce robust escalation of voluntary alcohol consumption, the mechanisms underlying enhanced motivation to imbibe in the context of dependence are not fully understood. As noted above, mechanisms that govern the regulation of drinking behavior involve complex and dynamic processes (Koob and Le Moal 2008; Koob and Volkow 2016). An interplay among numerous biological and environmental factors influence the motivational effects of alcohol, and these may change as the subject gains more experience with the drug (Cunningham et al. 2000). Alcohol dependence may be characterized as an allostatic state fueled by progressive dysregulation of motivational processes and neural circuitry controlling intake (Becker 2008; Heilig et al. 2010; Koob 2003; Koob and Le Moal 2008; Koob and Schulkin 2019). Such neuroadaptations may play a role in enhancing the rewarding effects of alcohol, thereby fostering the transition from regulated alcohol use to uncontrolled, excessive levels of drinking. Additionally, the potential for alcohol to alleviate negative affect and other symptoms of withdrawal serves as a powerful motivational force that likely promotes and sustains high levels of drinking (Becker 2008; Heilig et al. 2010; Koob 2021, 2022).

Studies involving CIE exposure have provided evidence for enhanced rewarding effects of alcohol. For example, studies employing operant self-administration procedures have demonstrated augmented motivation to self-administer alcohol (increased responding and consumption) in alcohol-dependent mice (Chu et al. 2007; Lopez et al. 2014) and rats (Gilpin et al. 2008c, 2009; O'Dell et al. 2004; Roberts et al. 1996, 2000). Further, employing progressive ratio schedules, it was demonstrated that the amount of work rats were willing to expend in order to receive alcohol reinforcement was significantly increased following repeated cycles of CIE exposure (Brown et al. 1998). Another study reported that CIE-exposed rats displayed greater resistance to extinction of responding to alcohol reward, perhaps reflecting greater persistence in alcohol seeking behavior despite the fact that alcohol was no longer available (Gass et al. 2017). Also, animals with a history of CIE exposure were shown to exhibit exaggerated sensitivity to events that trigger alcohol relapse, i.e., presentation of alcohol-related cues and stress exposure (Funk et al. 2019; Gehlert et al. 2007; Liu and Weiss 2002; Sommer et al. 2008). These findings suggest that the reinforcing value of alcohol may be enhanced, and subjects may be rendered more vulnerable to relapse as a consequence of experiencing repeated opportunities to self-administer alcohol in the context of chronic intermittent exposure to the drug.

At the same time, another factor that could contribute to excessive drinking is the development of tolerance to the aversive effects of alcohol. Tolerance has long been viewed as playing an important role in the regulation of alcohol self-administration behavior (Deitrich et al. 1996; Elvig et al. 2021; Kalant 1996, 1998; Rigter and Crabbe 1980; Suwaki et al. 2001). In this vein, the development of tolerance to the aversive effects of alcohol (which ordinarily temper the amount consumed) may serve as a permissive factor, enabling higher levels of drinking. Recent evidence indicates that repeated cycles of CIE exposure in mice not only produces escalation of voluntary drinking but also reduced sensitivity (tolerance) to the aversive effects of alcohol in the same subjects, as determined by a conditioned taste aversion procedure (Lopez et al. 2012). This reduced sensitivity to alcohol-induced conditioned taste aversion could not be attributed to pharmacokinetic factors, and it could not simply be explained by a general learning deficit since both dependent and non-dependent mice exhibited a similar learned aversion to a non-alcohol noxious stimulus (lithium chloride). In another study, rats with a history of repeated cycles of CIE exposure were reported to exhibit long-lasting tolerance to the sedative/hypnotic effects of alcohol (Rimondini et al. 2008). Additionally, using operant discrimination procedures, it was found that the ability to detect (perceive) the subjective cues associated with alcohol intoxication was diminished during withdrawal from chronic alcohol exposure, and this tolerance effect was greater in mice that experienced multiple withdrawals during the course of the chronic alcohol treatment (Becker and Baros 2006). Thus, reduced sensitivity to feedback about the intoxicating effects of alcohol along with reduced sensitivity to the aversive effects of the drug may serve a permissive role in enabling greater alcohol consumption associated with dependence.

Another important mechanism to consider is the shift from goal-directed to habitual responding/drinking that could underlie higher levels of consumption in alcohol-dependent subjects (Vandaele and Janak 2018). Several studies have evaluated this possibility using different procedures. In some studies, alcohol reward was devalued using contingency degradation (Barker et al. 2020), satiation (Renteria et al. 2020), or associating alcohol reward with an aversive unconditioned stimulus (Lopez et al. 2014). Using these diverse strategies, it has been shown that alcohol-dependent mice are more resistant to the devaluation of alcohol reward, as indicated by persistence in working (responding) to obtain the drug. Other studies have evaluated the habitual or "compulsive" nature of alcohol drinking by adding quinine to the alcohol solution (den Hartog et al. 2016; Gioia and Woodward 2021; Russo et al. 2018) or by contingently delivering foot shock along with alcohol as a reinforcer (Radke et al. 2017). In these studies, mice that experienced repeated CIE exposure were more resistant to these manipulations demonstrating more habitual alcohol seeking and drinking. Collectively, these data support the notion that with prolonged alcohol exposure, the relative balance between rewarding/reinforcing and aversive properties of alcohol is shifted away from aversion in favor of reward/reinforcement. Thus, the combination of enhanced rewarding effects (through both positive and negative reinforcement) along with reduced sensitivity (tolerance) to the aversive qualities of alcohol intoxication may, in large part, drive excessive drinking associated with dependence. Elucidating neurobiological mechanisms underlying changes in sensitivity to both the rewarding and the aversive effects of alcohol is key to understanding motivational processes that are critical for regulating and controlling alcohol consumption, as well as adaptations in such processes that mediate transition to uncontrolled, harmful levels of drinking characteristic of dependence.

Finally, compromised cognitive function may be an important contributing factor that promotes increased vulnerability to relapse and impaired ability to exert control over drinking (Le Berre et al. 2017). Indeed, repeated cycles of chronic alcohol exposure and withdrawal experience have been shown to produce significant cognitive deficits. For example, CIE-exposed mice that displayed elevated alcohol consumption also exhibited deficits in performance in attention set-shifting and novel object recognition tasks (Hu et al. 2015; Pradhan et al. 2018). Similarly, studies conducted in mice (Badanich et al. 2011) and rats (Meinhardt et al. 2021) indicate that CIE exposure leads to deficits in behaviors mediated by the prefrontal cortex such as reversal learning and delay discounting. These studies open future avenues of investigation that probe mechanisms and circuits that link alcohol-induced alterations in cognition and motivation, which ultimately govern decision-making and behavioral control regarding alcohol consumption.

8 Summary and Future Challenges

This chapter reviews a number of animal models that have been established to study excessive alcohol consumption in rodents. In all cases, the common experimental strategy has been focused on utilizing procedures that effectively overcome the natural tendency of rodents to either avoid alcohol or consume it in limited amounts that typically do not produce overt signs of intoxication. A corollary to this is the increased recognition that recording blood alcohol levels achieved following alcohol consumption in each of the models is critical for validation of their physiological (and clinical) relevance. This point cannot be overstated. In many instances, experimental manipulations have been shown to produce statistically significant changes in drinking behavior, but physiological and behavioral relevance can best be realized when consumption yields significant changes in circulating alcohol concentrations.

The six models described in this chapter incorporate several procedural variables that engender excessive levels of alcohol intake along with resultant blood alcohol concentrations that exceed the legal limit of intoxication (0.08 g/dL) (Table 1). This includes manipulating scheduled access to alcohol (time of day, duration, frequency), periods of time when access to alcohol is withheld, and history of alcohol exposure. As detailed above, each model possesses unique experimental characteristics that confer both advantages and disadvantages. Of course, no single approach can claim to capture all the complexities that define problem drinking in humans. Nevertheless, development of these rodent models of excessive alcohol drinking has proven to be extremely valuable in advancing our knowledge about the biological and environmental contingencies that bear on this complex behavior.

At the same time, there is the opportunity and a need to further optimize the translational impact of these animal models. In updating this body of work since our last review (Becker 2013), we have not only emphasized findings from more recent studies but also highlight aspects of the studies that bear on relevance to drinking in humans. For example, most studies described in this chapter focused on experimental manipulations that minimize variance in study outcomes, with reliance on group averages for alcohol intake. This is understandable from the standpoint of wanting to utilize a model that produces an overall robust and reliable phenotype to probe mechanisms with sophisticated neurobiological tools and approaches. However, this is done at the expense of highlighting factors that contribute to individual differences in drinking (amount and pattern). Given the known heterogeneity of AUD and the burgeoning area of personalized medicine in relation to treatment strategies, this point has been more greatly appreciated in recent years. This is a subject that is deserving of more attention in studies on animal models of excessive alcohol drinking.

A related issue regards the influence of genetic factors. It is well known that genetic background has a significant effect on alcohol consumption in various rodent models. As described above, genotype (species, strain) and genetic factors related to selective breeding procedures have been shown to influence alcohol intake under some conditions. Additional investigations into the modulating effects of genetic

Table 1 Advantages and disadvantages of models that engender excessive levels of alcohol consumption

Model	Advantages	Disadvantages
Continuous free choice	• Easy to implement • Simple method for assessing propensity to voluntarily drink • Provides measures of general preference for and consumption of alcohol • Extensively used for investigating genetic determinants of alcohol preference and consumption	• Difficult to determine whether alcohol consumption results in intoxication (significant elevation in blood alcohol levels) • Temporal pattern of alcohol intake is typically not ascertained • Limited utility for studies probing mechanisms underlying motivational aspects of drinking
Scheduled limited access	• Mimics "binge-like" pattern of intake when access is appropriately scheduled • Engenders high levels of alcohol consumption • Affords opportunity to more accurately correlate intake with resultant blood alcohol levels • Enables examination of relationship between alcohol intake and measures of intoxication	• High levels of drinking during limited access conditions may not predict high drinking when access is not restricted • Limited utility for studies probing mechanisms underlying motivational aspects of drinking
Scheduled deprivation	• Models alcohol relapse and craving • Can be studied using home-cage drinking and operant self-administration procedures • Translational value as repeated cycles of deprivation lead to more robust and durable effect	• Lack of specificity: Exaggerated intake following deprivation observed for other rewards (e.g., sucrose) • High propensity for alcohol preference/intake may not predict robust expression of alcohol deprivation effect • Increased intake following deprivation periods may be transient
Intermittent access	• Home-cage drinking procedure easy to implement • Escalation of alcohol intake observed in home-cage drinking and operant self-administration procedures	• Genetic models of high alcohol preference/intake may limit escalation effect • Observation of escalated alcohol intake highly dependent on various biological and experimental factors (e.g., age, sex, housing conditions)
Schedule-induced polydipsia	• Leads to high levels of intake that can result in alcohol dependence (evidenced by withdrawal signs) • Effectively used in nonhuman primates to produce long-lasting elevated alcohol intake	• Lack of specificity: same procedure leads to excessive intake of any fluid (including water) • Requires food deprivation, clouding interpretation of motivation • Once schedule is relaxed, alcohol intake reverts to control levels (in rodent models)
Alcohol dependence	• Leads to escalation of voluntary intake that results in significant elevated	• Requires unique setup for delivery of alcohol vapor

(continued)

Table 1 (continued)

Model	Advantages	Disadvantages
and withdrawal	blood and brain alcohol levels • Can be employed to study both physical signs of withdrawal and motivational factors that influence alcohol consumption • Escalation of alcohol intake observed in home-cage drinking and operant self-administration procedures • Increased intake is specific to alcohol (consumption of other palatable substances not altered)	• Induction of alcohol dependence involves experimenter-administered alcohol • Some unique methodological issues need to be considered in studies involving rats vs. mice

factors on regulation of alcohol drinking in these models are certainly warranted. Extending this work to focus on genetically based risk factors that are predictive of excessive drinking phenotypes in these models has important clinical relevance. The emerging interest and evidence of epigenetic factors that influence alcohol consumption may be of importance to this area of research.

Most studies described in this chapter have been predominantly conducted using male subjects. In recent years, a growing body of evidence indicates sex-related differences in alcohol consumption and the mechanisms that regulate this behavior. Thus, there is a critical need for these models of excessive drinking to incorporate females in the study designs. Filling this relative void in information will enhance our understanding of potential sex-related differences in mechanisms that govern the amount and pattern of drinking. This, in turn, may have important implications for tailored treatment strategies for tempering excessive levels of alcohol intake in males and females.

In the quest to further improve the relevance of these preclinical models in relation to the human condition, other contributing factors that deserve more consideration in these models include the role of initial sensitivity to alcohol as well as acute and chronic tolerance (i.e., changes in response to alcohol as the subject gains more experience and exposure to the drug). This includes procedures that enable assessment of changes in the positive reinforcing (rewarding) effects of alcohol as well as the emergence of motivation to drink to alleviate a negative emotional state associated with chronic alcohol exposure (negative reinforcing effects of alcohol). Further, additional work is needed to probe cognitive (learning/memory) factors that guide decisions about initiating and terminating drinking behavior, as well as studies focused on distinguishing circumstances in which environmental factors such as cues, stress, and timing and predictability of access exert different effects on propensity to drink. Through increased refinement and more detailed characterization of procedures and factors that engender excessive alcohol drinking, the overall goal in developing these animal models is to advance our understanding of biological underpinnings and environmental influences that drive increased motivation for alcohol seeking and consumption. This enhances the translational value of this preclinical work and facilitates the ability of the models to better inform the clinical condition (AUD). Ultimately, the validity and utility of these models will lie in their ability to aid in the discovery of new and novel potential therapeutic targets as well as serve as a platform to evaluate treatment strategies that effectively reduce excessive levels of alcohol consumption.

Acknowledgments Supported by NIH grants P50 AA010761 (HCB), R01 AA026536 (HCB), U01 AA014095 (HCB), and U24 AA020929 (MFL) and a grant from the Department of Veterans Affairs (BLRD BX000813) (HCB).

References

Agabio R, Carai MA, Lobina C, Pani M, Reali R, Vacca G et al (2000) Development of short-lasting alcohol deprivation effect in Sardinian alcohol-preferring rats. Alcohol 21(1):59–62. https://doi.org/10.1016/s0741-8329(00)00072-0

Ardinger CE, Grahame NJ, Lapish CC, Linsenbardt DN (2020) High alcohol-preferring mice show reaction to loss of ethanol reward following repeated binge drinking. Alcohol Clin Exp Res 44(9):1717–1727. https://doi.org/10.1111/acer.14419

Avena NM, Long KA, Hoebel BG (2005) Sugar-dependent rats show enhanced responding for sugar after abstinence: evidence of a sugar deprivation effect. Physiol Behav 84(3):359–362. https://doi.org/10.1016/j.physbeh.2004.12.016

Azarov AV, Woodward DJ (2014a) Early ethanol and water consumption: accumulating experience differentially regulates drinking pattern and bout parameters in male alcohol preferring (P) vs. Wistar and Sprague Dawley rats. Physiol Behav 123:20–32. https://doi.org/10.1016/j.physbeh.2013.09.015

Azarov AV, Woodward DJ (2014b) Early ethanol and water intake: choice mechanism and total fluid regulation operate in parallel in male alcohol preferring (P) and both Wistar and Sprague Dawley rats. Physiol Behav 123:11–19. https://doi.org/10.1016/j.physbeh.2013.09.011

Badanich KA, Becker HC, Woodward JJ (2011) Effects of chronic intermittent ethanol expsoure on orbitofrontal and medial prefrontal cortex-dependent behaviors in mice. Behav Neurosci 125(6): 879–891. https://doi.org/10.1037/a0025922

Barker JM, Bryant KG, Montiel-Ramos A, Goldwasser B, Chandler LJ (2020) Selective deficits in contingency-driven ethanol seeking following chronic ethanol exposure in male mice. Alcohol Clin Exp Res 44(9):1896–1905. https://doi.org/10.1111/acer.14418

Barkley-Levenson AM, Crabbe JC (2014) High drinking in the dark mice: a genetic model of drinking to intoxication. Alcohol 48(3):217–223. https://doi.org/10.1016/j.alcohol.2013.10.007

Barkley-Levenson AM, Cunningham CL, Smitasin PJ, Crabbe JC (2015) Rewarding and aversive effects of ethanol in high drinking in the dark selectively bred mice. Addict Biol 20(1):80–90. https://doi.org/10.1111/adb.12079

Bauer MR, McVey MM, Boehm SL 2nd (2021) Three weeks of binge alcohol drinking generates increased alcohol front-loading and robust compulsive-like alcohol drinking in male and female C57BL/6J mice. Alcohol Clin Exp Res 45(3):650–660. https://doi.org/10.1111/acer.14563

Becker HC (2008) Alcohol dependence, withdrawal, and relapse. Alcohol Res Health 31(4): 348–361. Retrieved from https://www.ncbi.nlm.nih.gov/pubmed/23584009

Becker HC (2013) Animal models of excessive alcohol consumption in rodents. Curr Top Behav Neurosci 13:355–377. https://doi.org/10.1007/7854_2012_203

Becker HC (2014) Alcohol dependence, withdrawal, and relapse. In: Noronha A, Cui C, Harris RA, Crabbe JC (eds) Neurobiology of alcohol dependence. Academic Press (Elsevier), Waltham, MA, pp 377–410

Becker HC, Baros AM (2006) Effect of duration and pattern of chronic ethanol exposure on tolerance to the discriminative stimulus effects of ethanol in C57BL/6J mice. J Pharmacol Exp Ther 319(2):871–878. https://doi.org/10.1124/jpet.106.108795

Becker HC, Lopez MF (2004) Increased ethanol drinking after repeated chronic ethanol exposure and withdrawal experience in C57BL/6 mice. Alcohol Clin Exp Res 28(12):1829–1838. https://doi.org/10.1097/01.alc.0000149977.95306.3a

Becker HC, Lopez MF (2016) An animal model of alcohol dependence to screen medications for treating alcoholism. Int Rev Neurobiol 126:157–177. https://doi.org/10.1016/bs.irn.2016.02.006

Becker HC, Ron D (2014) Animal models of excessive alcohol consumption: recent advances and future challenges. Alcohol 48(3):205–208. https://doi.org/10.1016/j.alcohol.2014.04.001

Bell RL, Rodd ZA, Boutwell CL, Hsu CC, Lumeng L, Murphy JM et al (2004) Effects of long-term episodic access to ethanol on the expression of an alcohol deprivation effect in low alcohol-consuming rats. Alcohol Clin Exp Res 28(12):1867–1874. https://doi.org/10.1097/01.alc.0000148101.20547.0a

Bell RL, Rodd ZA, Smith RJ, Toalston JE, Franklin KM, McBride WJ (2011) Modeling binge-like ethanol drinking by peri-adolescent and adult P rats. Pharmacol Biochem Behav 100(1):90–97. https://doi.org/10.1016/j.pbb.2011.07.017

Bell RL, Rodd ZA, Engleman EA, Toalston JE, McBride WJ (2014) Scheduled access alcohol drinking by alcohol-preferring (P) and high-alcohol-drinking (HAD) rats: modeling adolescent and adult binge-like drinking. Alcohol 48(3):225–234. https://doi.org/10.1016/j.alcohol.2013.10.004

Bito-Onon JJ, Simms JA, Chatterjee S, Holgate J, Bartlett SE (2011) Varenicline, a partial agonist at neuronal nicotinic acetylcholine receptors, reduces nicotine-induced increases in 20% ethanol operant self-administration in Sprague-Dawley rats. Addict Biol 16(3):440–449. https://doi.org/10.1111/j.1369-1600.2010.00309.x

Bloch S, Rinker JA, Marcus MM, Mulholland PJ (2020) Absence of effects of intermittent access to alcohol on negative affective and anxiety-like behaviors in male and female C57BL/6J mice. Alcohol 88:91–99. https://doi.org/10.1016/j.alcohol.2020.07.011

Boehm SL 2nd, Moore EM, Walsh CD, Gross CD, Cavelli AM, Gigante E, Linsenbardt DN (2008) Using drinking in the dark to model prenatal binge-like exposure to ethanol in C57BL/6J mice. Dev Psychobiol 50(6):566–578. https://doi.org/10.1002/dev.20320

Brown G, Jackson A, Stephens DN (1998) Effects of repeated withdrawal from chronic ethanol on oral self-administration of ethanol on a progressive ratio schedule. Behav Pharmacol 9(2):149–161. Retrieved from https://www.ncbi.nlm.nih.gov/pubmed/10065934

Cannady R, Nimitvilai-Roberts S, Jennings SD, Woodward JJ, Mulholland PJ (2020) Distinct region- and time-dependent functional cortical adaptations in C57BL/6J mice after short and prolonged alcohol drinking. eNeuro 7(3). https://doi.org/10.1523/ENEURO.0077-20.2020

Carnicella S, Amamoto R, Ron D (2009) Excessive alcohol consumption is blocked by glial cell line-derived neurotrophic factor. Alcohol 43(1):35–43. https://doi.org/10.1016/j.alcohol.2008.12.001

Carnicella S, Ron D, Barak S (2014) Intermittent ethanol access schedule in rats as a preclinical model of alcohol abuse. Alcohol 48(3):243–252. https://doi.org/10.1016/j.alcohol.2014.01.006

Chu K, Koob GF, Cole M, Zorrilla EP, Roberts AJ (2007) Dependence-induced increases in ethanol self-administration in mice are blocked by the CRF1 receptor antagonist antalarmin and by CRF1 receptor knockout. Pharmacol Biochem Behav 86(4):813–821. https://doi.org/10.1016/j.pbb.2007.03.009

Ciccocioppo R (2012) Genetically selected alcohol preferring rats to model human alcoholism. In: Sommer W, Spanagel R (eds) Behavioral neurobiology of alcohol addiction. Current topics in behavioral neurosciences, vol 13. Springer, Berlin, pp 251–269

Colombo G, Maccioni P, Acciaro C, Lobina C, Loi B, Zaru A et al (2014) Binge drinking in alcohol-preferring sP rats at the end of the nocturnal period. Alcohol 48(3):301–311. https://doi.org/10.1016/j.alcohol.2014.02.001

Colombo G, Lobina C, Maccioni P, Carai MA, Lorrai I, Zaru A et al (2015) Anxiety-like behaviors at the end of the nocturnal period in sP rats with a "history" of unpredictable, limited access to alcohol. Alcohol 49(7):707–712. https://doi.org/10.1016/j.alcohol.2015.04.010

Colombo G, Lobina C, Lorrai I, Acciaro C, Maccioni P, Gessa GL (2017) Binge drinking and anxiety at the end of the nocturnal period in alcohol-preferring sP rats. Alcohol 63:27–32. https://doi.org/10.1016/j.alcohol.2017.04.002

Cox BR, Olney JJ, Lowery-Gionta EG, Sprow GM, Rinker JA, Navarro M et al (2013) Repeated cycles of binge-like ethanol (EtOH)-drinking in male C57BL/6J mice augments subsequent voluntary EtOH intake but not other dependence-like phenotypes. Alcohol Clin Exp Res 37(10): 1688–1695. https://doi.org/10.1111/acer.12145

Crabbe JC, Metten P, Rhodes JS, Yu CH, Brown LL, Phillips TJ, Finn DA (2009) A line of mice selected for high blood ethanol concentrations shows drinking in the dark to intoxication. Biol Psychiatry 65(8):662–670. https://doi.org/10.1016/j.biopsych.2008.11.002

Crabbe JC, Phillips TJ, Belknap JK (2010) The complexity of alcohol drinking: studies in rodent genetic models. Behav Genet 40(6):737–750. https://doi.org/10.1007/s10519-010-9371-z

Crabbe JC, Harris RA, Koob GF (2011a) Preclinical studies of alcohol binge drinking. Ann N Y Acad Sci 1216:24–40. https://doi.org/10.1111/j.1749-6632.2010.05895.x

Crabbe JC, Spence SE, Brown LL, Metten P (2011b) Alcohol preference drinking in a mouse line selectively bred for high drinking in the dark. Alcohol 45(5):427–440. https://doi.org/10.1016/j.alcohol.2010.12.001

Crabbe JC, Harkness JH, Spence SE, Huang LC, Metten P (2012) Intermittent availability of ethanol does not always lead to elevated drinking in mice. Alcohol Alcohol 47(5):509–517. https://doi.org/10.1093/alcalc/ags067

Crabbe JC, Hack WR, Ozburn AR, Savarese AM, Metten P (2022) Long-term alcohol drinking in high drinking in the dark mice is stable for many months and does not show alcohol deprivation effects. Addict Biol 27(1):e13074. https://doi.org/10.1111/adb.13074

Cunningham CL, Fidler TL, Hill KG (2000) Animal models of alcohol's motivational effects. Alcohol Res Health 24(2):85–92. Retrieved from https://www.ncbi.nlm.nih.gov/pubmed/11199282

Cuzon Carlson VC, Seabold GK, Helms CM, Garg N, Odagiri M, Rau AR et al (2011) Synaptic and morphological neuroadaptations in the putamen associated with long-term, relapsing alcohol drinking in primates. Neuropsychopharmacology 36(12):2513–2528. https://doi.org/10.1038/npp.2011.140

Deitrich RA, Radcliffe R, Erwin VG (1996) Pharmacological effects in the development of physiological tolerance and physical dependence. In: Begleiter H, Kissin B (eds) The pharmacology of alcohol and alcohol dependence. Oxford University Press, New York, pp 431–476

den Hartog C, Zamudio-Bulcock P, Nimitvilai S, Gilstrap M, Eaton B, Fedarovich H et al (2016) Inactivation of the lateral orbitofrontal cortex increases drinking in ethanol-dependent but not non-dependent mice. Neuropharmacology 107:451–459. https://doi.org/10.1016/j.neuropharm.2016.03.031

Dhaher R, Finn D, Snelling C, Hitzemann R (2008) Lesions of the extended amygdala in C57BL/6J mice do not block the intermittent ethanol vapor-induced increase in ethanol consumption. Alcohol Clin Exp Res 32(2):197–208. https://doi.org/10.1111/j.1530-0277.2007.00566.x

Elvig SK, McGinn MA, Smith C, Arends MA, Koob GF, Vendruscolo LF (2021) Tolerance to alcohol: a critical yet understudied factor in alcohol addiction. Pharmacol Biochem Behav 204: 173155. https://doi.org/10.1016/j.pbb.2021.173155

Falk JL (1961) Production of polydipsia in normal rats by an intermittent food schedule. Science 133(3447):195–196. https://doi.org/10.1126/science.133.3447.195

Falk JL (1971) The nature and determinants of adjunctive behavior. Physiol Behav 6(5):577–588. https://doi.org/10.1016/0031-9384(71)90209-5

Falk JL, Samson HH (1975) Schedule-induced physical dependence on ethanol. Pharmacol Rev 27(4):449–464. Retrieved from https://www.ncbi.nlm.nih.gov/pubmed/772699

Falk JL, Tang M (1988) What schedule-induced polydipsia can tell us about alcoholism. Alcohol Clin Exp Res 12(5):577–585. https://doi.org/10.1111/j.1530-0277.1988.tb00246.x

Falk JL, Samson HH, Winger G (1972) Behavioral maintenance of high concentrations of blood ethanol and physical dependence in the rat. Science 177(4051):811–813. https://doi.org/10.1126/science.177.4051.811

Finn DA, Snelling C, Fretwell AM, Tanchuck MA, Underwood L, Cole M et al (2007) Increased drinking during withdrawal from intermittent ethanol exposure is blocked by the CRF receptor antagonist D-Phe-CRF(12-41). Alcohol Clin Exp Res 31(6):939–949. https://doi.org/10.1111/j.1530-0277.2007.00379.x

Ford MM (2014) Applications of schedule-induced polydipsia in rodents for the study of an excessive ethanol intake phenotype. Alcohol 48(3):265–276. https://doi.org/10.1016/j.alcohol.2014.01.005

Ford MM, Steele AM, McCracken AD, Finn DA, Grant KA (2013) The relationship between adjunctive drinking, blood ethanol concentration and plasma corticosterone across fixed-time intervals of food delivery in two inbred mouse strains. Psychoneuroendocrinology 38(11):2598–2610. https://doi.org/10.1016/j.psyneuen.2013.06.011

Fouyssac M, Puaud M, Ducret E, Marti-Prats L, Vanhille N, Ansquer S et al (2021) Environment-dependent behavioral traits and experiential factors shape addiction vulnerability. Eur J Neurosci 53(6):1794–1808. https://doi.org/10.1111/ejn.15087

Funk CK, Koob GF (2007) A CRF(2) agonist administered into the central nucleus of the amygdala decreases ethanol self-administration in ethanol-dependent rats. Brain Res 1155:172–178. https://doi.org/10.1016/j.brainres.2007.04.009

Funk D, Coen K, Tamadon S, Le AD (2019) Effect of chronic alcohol vapor exposure on reinstatement of alcohol seeking induced by U50,488. Neuropharmacology 148:210–219. https://doi.org/10.1016/j.neuropharm.2019.01.017

Galbo LK, Davenport AT, Epperly PM, Daunais JB, Stinson BT, Czoty PW (2022) Social dominance in monkeys: lack of effect on ethanol self-administration during schedule induction. Alcohol 98:1–7. https://doi.org/10.1016/j.alcohol.2021.10.001

Gass JT, McGonigal JT, Chandler LJ (2017) Deficits in the extinction of ethanol-seeking behavior following chronic intermittent ethanol exposure are attenuated with positive allosteric modulation of mGlu5. Neuropharmacology 113(Pt A):198–205. https://doi.org/10.1016/j.neuropharm.2016.10.005

Gehlert DR, Cippitelli A, Thorsell A, Le AD, Hipskind PA, Hamdouchi C et al (2007) 3-(4-Chloro-2-morpholin-4-yl-thiazol-5-yl)-8-(1-ethylpropyl)-2,6-dimethyl-imidazo[1,2-b]pyridazine: a novel brain-penetrant, orally available corticotropin-releasing factor receptor 1 antagonist with efficacy in animal models of alcoholism. J Neurosci 27(10):2718–2726. https://doi.org/10.1523/JNEUROSCI.4985-06.2007

George O, Sanders C, Freiling J, Grigoryan E, Vu S, Allen CD, Crawford E, Mandyam CD, Koob GF (2012) Recruitment of medial prefrontal cortex neurons during alcohol withdrawal predicts cognitive impairment and excessive alcohol drinking. Proc Natl Acad Sci U S A 109(44):18156–18161. https://doi.org/10.1073/pnas.1116523109

Giardino WJ, Ryabinin AE (2013) CRF1 receptor signaling regulates food and fluid intake in the drinking-in-the-dark model of binge alcohol consumption. Alcohol Clin Exp Res 37(7):1161–1170. https://doi.org/10.1111/acer.12076

Gill K, Liu Y, Deitrich RA (1996) Voluntary alcohol consumption in BXD recombinant inbred mice: relationship to alcohol metabolism. Alcohol Clin Exp Res 20(1):185–190. https://doi.org/10.1111/j.1530-0277.1996.tb01063.x

Gilpin NW, Badia-Elder NE, Elder RL, Stewart RB (2008a) Schedule-induced polydipsia in lines of rats selectively bred for high and low ethanol preference. Behav Genet 38(5):515–524. https://doi.org/10.1007/s10519-008-9224-1

Gilpin NW, Richardson HN, Koob GF (2008b) Effects of CRF1-receptor and opioid-receptor antagonists on dependence-induced increases in alcohol drinking by alcohol-preferring

(P) rats. Alcohol Clin Exp Res 32(9):1535–1542. https://doi.org/10.1111/j.1530-0277.2008.00745.x

Gilpin NW, Richardson HN, Lumeng L, Koob GF (2008c) Dependence-induced alcohol drinking by alcohol-preferring (P) rats and outbred Wistar rats. Alcohol Clin Exp Res 32(9):1688–1696. https://doi.org/10.1111/j.1530-0277.2008.00678.x

Gilpin NW, Smith AD, Cole M, Weiss F, Koob GF, Richardson HN (2009) Operant behavior and alcohol levels in blood and brain of alcohol-dependent rats. Alcohol Clin Exp Res 33(12): 2113–2123. https://doi.org/10.1111/j.1530-0277.2009.01051.x

Gioia DA, Woodward JJ (2021) Altered activity of lateral orbitofrontal cortex neurons in mice following chronic intermittent ethanol exposure. eNeuro 8(2). https://doi.org/10.1523/ENEURO.0503-20.2021

Grahame NJ, Li TK, Lumeng L (1999) Selective breeding for high and low alcohol preference in mice. Behav Genet 29(1):47–57. https://doi.org/10.1023/a:1021489922751

Grant KA, Leng X, Green HL, Szeliga KT, Rogers LS, Gonzales SW (2008) Drinking typography established by scheduled induction predicts chronic heavy drinking in a monkey model of ethanol self-administration. Alcohol Clin Exp Res 32(10):1824–1838. https://doi.org/10.1111/j.1530-0277.2008.00765.x

Grant KA, Newman NN, Gonzales SW, Cuzon Carlson VC (2022) Impact of putamen inhibition by DREADDs on schedule-induced drinking in rhesus monkeys. J Exp Anal Behav 117(3): 493–504. https://doi.org/10.1002/jeab.761

Griffin WC 3rd, Lopez MF, Becker HC (2009a) Intensity and duration of chronic ethanol exposure is critical for subsequent escalation of voluntary ethanol drinking in mice. Alcohol Clin Exp Res 33(11):1893–1900. https://doi.org/10.1111/j.1530-0277.2009.01027.x

Griffin WC 3rd, Lopez MF, Yanke AB, Middaugh LD, Becker HC (2009b) Repeated cycles of chronic intermittent ethanol exposure in mice increases voluntary ethanol drinking and ethanol concentrations in the nucleus accumbens. Psychopharmacology (Berl) 201(4):569–580. https://doi.org/10.1007/s00213-008-1324-3

Griffin WC 3rd. (2014) Alcohol dependence and free-choice drinking in mice. Alcohol 48(3): 287–293. https://doi.org/10.1016/j.alcohol.2013.11.006

Heilig M, Egli M, Crabbe JC, Becker HC (2010) Acute withdrawal, protracted abstinence and negative affect in alcoholism: are they linked? Addict Biol 15(2):169–184. https://doi.org/10.1111/j.1369-1600.2009.00194.x

Hernandez JS, Moorman DE (2020) Orbitofrontal cortex encodes preference for alcohol. eNeuro 7(4). https://doi.org/10.1523/ENEURO.0402-19.2020. ENEURO.0402

Heyser CJ, Schulteis G, Koob GF (1997) Increased ethanol self-administration after a period of imposed ethanol deprivation in rats trained in a limited access paradigm. Alcohol Clin Exp Res 21(5):784–791. Retrieved from https://www.ncbi.nlm.nih.gov/pubmed/9267526

Holter SM, Landgraf R, Zieglgansberger W, Spanagel R (1997) Time course of acamprosate action on operant ethanol self-administration after ethanol deprivation. Alcohol Clin Exp Res 21(5): 862–868. Retrieved from https://www.ncbi.nlm.nih.gov/pubmed/9267536

Hopf FW, Chang SJ, Sparta DR, Bowers MS, Bonci A (2010) Motivation for alcohol becomes resistant to quinine adulteration after 3 to 4 months of intermittent alcohol self-administration. Alcohol Clin Exp Res 34(9):1565–1573. https://doi.org/10.1111/j.1530-0277.2010.01241.x

Hosova D, Spear LP (2017) Voluntary binge consumption of ethanol in a sweetened, chocolate-flavored solution by male and female adolescent Sprague Dawley rats. Alcohol Clin Exp Res 41(3):541–550. https://doi.org/10.1111/acer.13315

Hu W, Morris B, Carrasco A, Kroener S (2015) Effects of acamprosate on attentional set-shifting and cellular function in the prefrontal cortex of chronic alcohol-exposed mice. Alcohol Clin Exp Res 39(6):953–961. https://doi.org/10.1111/acer.12722

Huitron-Resendiz S, Nadav T, Krause S, Cates-Gatto C, Polis I, Roberts AJ (2018) Effects of withdrawal from chronic intermittent ethanol exposure on sleep characteristics of female and male mice. Alcohol Clin Exp Res 42(3):540–550. https://doi.org/10.1111/acer.13584

Hwa LS, Chu A, Levinson SA, Kayyali TM, DeBold JF, Miczek KA (2011) Persistent escalation of alcohol drinking in C57BL/6J mice with intermittent access to 20% ethanol. Alcohol Clin Exp Res 35(11):1938–1947. https://doi.org/10.1111/j.1530-0277.2011.01545.x

Jensen BE, Townsley KG, Grigsby KB, Metten P, Chand M, Uzoekwe M et al (2021) Ethanol-related behaviors in mouse lines selectively bred for drinking to intoxication. Brain Sci 11(2). https://doi.org/10.3390/brainsci11020189

Jimenez VA, Porcu P, Morrow AL, Grant KA (2017) Adaptations in basal and hypothalamic-pituitary-adrenal-activated deoxycorticosterone responses following ethanol self-administration in Cynomolgus monkeys. Front Endocrinol (Lausanne) 8:19. https://doi.org/10.3389/fendo.2017.00019

Kalant H (1996) Current state of knowledge about the mechanisms of alcohol tolerance. Addict Biol 1(2):133–141. https://doi.org/10.1080/1355621961000124756

Kalant H (1998) Research on tolerance: what can we learn from history? Alcohol Clin Exp Res 22(1):67–76. https://doi.org/10.1111/j.1530-0277.1998.tb03618.x

Khanna JM, Kalant H, Shah G, Sharma H (1990) Comparison of sensitivity and alcohol consumption in four outbred strains of rats. Alcohol 7(5):429–434. https://doi.org/10.1016/0741-8329(90)90027-a

Khisti RT, Wolstenholme J, Shelton KL, Miles MF (2006) Characterization of the ethanol-deprivation effect in substrains of C57BL/6 mice. Alcohol 40(2):119–126. https://doi.org/10.1016/j.alcohol.2006.12.003

Koob GF (2003) Alcoholism: allostasis and beyond. Alcohol Clin Exp Res 27(2):232–243. https://doi.org/10.1097/01.ALC.0000057122.36127.C2

Koob GF (2021) Drug addiction: Hyperkatifeia/negative reinforcement as a framework for medications development. Pharmacol Rev 73(1):163–201. https://doi.org/10.1124/pharmrev.120.000083

Koob GF (2022) Anhedonia, Hyperkatifeia, and negative reinforcement in substance use disorders. Curr Top Behav Neurosci 58:147–165. https://doi.org/10.1007/7854_2021_288

Koob GF, Le Moal M (2008) Review. Neurobiological mechanisms for opponent motivational processes in addiction. Philos Trans R Soc Lond B Biol Sci 363(1507):3113–3123. https://doi.org/10.1098/rstb.2008.0094

Koob GF, Schulkin J (2019) Addiction and stress: an allostatic view. Neurosci Biobehav Rev 106:245–262. https://doi.org/10.1016/j.neubiorev.2018.09.008

Koob GF, Volkow ND (2016) Neurobiology of addition: a neurocircuitry analysis. Lancet Psychiatry 3(8):760–773. https://doi.org/10.1016/S2215-0366(16)00104-8

Le Berre A-P, Farna R, Sullivan EV (2017) Executive functions, memory, and social cognitive deficits and recovery in chronic alcoholism: a critical review to inform future research. Alcohol Clin Exp Res 41(8):1432–1443. https://doi.org/10.1111/acer.13431

Lee KM, Coehlo MA, Solton NR, Szumlinski KK (2017) Negative affect and excessive alcohol intake incubate during protracted withdrawal from binge-drinking in adolescent, but not adult, mice. Front Psychol 8:1128. https://doi.org/10.3389/fpsyg.2017.01128

Levine OB, Skelly MJ, Miller JD, Rivera-Irizarry JK, Rowson SA, DiBerto JF et al (2021) The paraventricular thalamus provides a polysynaptic brake on limbic CRF neurons to sex-dependently blunt binge alcohol drinking and avoidance behavior in mice. Nat Commun 12(1):5080. https://doi.org/10.1038/s41467-021-25368-y

Linsenbardt DN, Moore EM, Griffin KD, Gigante ED, Boehm SL 2nd (2011) Tolerance to ethanol's ataxic effects and alterations in ethanol-induced locomotion following repeated binge-like ethanol intake using the DID model. Alcohol Clin Exp Res 35(7):1246–1255. https://doi.org/10.1111/j.1530-0277.2011.01459.x

Liu X, Weiss F (2002) Additive effect of stress and drug cues on reinstatement of ethanol seeking: exacerbation by history of dependence and role of concurrent activation of corticotropin-releasing factor and opioid mechanisms. J Neurosci 22(18):7856–7861. https://doi.org/10.1523/JNEUROSCI.22-18-07856.2002

Loi B, Lobina C, Maccioni P, Fantini N, Carai MA, Gessa GL, Colombo G (2010) Increase in alcohol intake, reduced flexibility of alcohol drinking, and evidence of signs of alcohol intoxication in Sardinian alcohol-preferring rats exposed to intermittent access to 20% alcohol. Alcohol Clin Exp Res 34(12):2147–2154. https://doi.org/10.1111/j.1530-0277.2010.01311.x

Lopez MF, Becker HC (2005) Effect of pattern and number of chronic ethanol exposures on subsequent voluntary ethanol intake in C57BL/6J mice. Psychopharmacology (Berl) 181(4): 688–696. https://doi.org/10.1007/s00213-005-0026-3

Lopez MF, Becker HC (2014) Operant ethanol self-administration in ethanol dependent mice. Alcohol 48(3):295–299. https://doi.org/10.1016/j.alcohol.2014.02.002

Lopez MF, Griffin WC 3rd, Melendez RI, Becker HC (2012) Repeated cycles of chronic intermittent ethanol exposure leads to the development of tolerance to aversive effects of ethanol in C57BL/6J mice. Alcohol Clin Exp Res 36(7):1180–1187. https://doi.org/10.1111/j.1530-0277.2011.01717.x

Lopez MF, Becker HC, Chandler LJ (2014) Repeated episodes of chronic intermittent ethanol promote insensitivity to devaluation of the reinforcing effect of ethanol. Alcohol 48(7):639–645. https://doi.org/10.1016/j.alcohol.2014.09.002

Lopez MF, Miles MF, Williams RW, Becker HC (2017) Variable effects of chronic intermittent ethanol exposure on ethanol drinking in a genetically diverse mouse cohort. Alcohol 58:73–82. https://doi.org/10.1016/j.alcohol.2016.09.003

Lopez-Grancha M, Lopez-Crespo G, Venero C, Canadas F, Sanchez-Santed F, Sandi C, Flores P (2006) Differences in corticosterone level due to inter-food interval length: implications for schedule-induced polydipsia. Horm Behav 49(2):166–172. https://doi.org/10.1016/j.yhbeh.2005.05.019

Lyons AM, Lowery EG, Sparta DR, Thiele TE (2008) Effects of food availability and administration of orexigenic and anorectic agents on elevated ethanol drinking associated with drinking in the dark procedures. Alcohol Clin Exp Res 32(11):1962–1968. https://doi.org/10.1111/j.1530-0277.2008.00784.x

Matson LM, Grahame NJ (2013) Pharmacologically relevant intake during chronic, free-choice drinking rhythms in selectively bred high alcohol-preferring mice. Addict Biol 18(6):921–929. https://doi.org/10.1111/j.1369-1600.2011.00412.x

Matson LM, Kasten CR, Boehm SL 2nd, Grahame NJ (2014) Selectively bred crossed high-alcohol-preferring mice drink to intoxication and develop functional tolerance, but not locomotor sensitization during free-choice ethanol access. Alcohol Clin Exp Res 38(1):267–274. https://doi.org/10.1111/acer.12216

McBride WJ, Kimpel MW, Schultz JA, McClintick JN, Edenberg HJ, Bell RL (2010) Changes in gene expression in regions of the extended amygdala of alcohol-preferring rats after binge-like alcohol drinking. Alcohol 44(2):171–183. https://doi.org/10.1016/j.alcohol.2009.12.001

McBride WJ, Rodd ZA, Bell RL, Lumeng L, Li TK (2014) The alcohol-preferring (P) and high-alcohol-drinking (HAD) rats--animal models of alcoholism. Alcohol 48(3):209–215. https://doi.org/10.1016/j.alcohol.2013.09.044

McCane AM, Auterson CD, DeLory MJ, Lapish CC, Czachowski CL (2021) Differential effects of quinine adulteration of alcohol on seeking and drinking. Alcohol 92:73–80. https://doi.org/10.1016/j.alcohol.2021.01.003

McClintick MN, Grant KA (2016) Aggressive temperament predicts ethanol self-administration in late adolescent male and female rhesus macaques. Psychopharmacology (Berl) 233(23–24): 3965–3976. https://doi.org/10.1007/s00213-016-4427-2

McKinzie DL, Nowak KL, Yorger L, McBride WJ, Murphy JM, Lumeng L, Li TK (1998) The alcohol deprivation effect in the alcohol-preferring P rat under free-drinking and operant access conditions. Alcohol Clin Exp Res 22(5):1170–1176. Retrieved from https://www.ncbi.nlm.nih.gov/pubmed/9726292

Meinhardt M, Sommer WH (2015) Postdependent state in rats as a model for medication development in alcoholism. Addict Biol 20(1):1–21. https://doi.org/10.1111/adb.12187

Meinhardt MW, Pfarr S, Fouquet G, Rohleder C, Meinhardt ML, Barroso-Flores J, Hoffman R, Jeanblanc J, Paul E, Wagner K, Hansson AC, Kohr G, Meier N, Halbach OB, Bell RL, Endepols H, Neumaier B, Schlonig K, Bartsch D, Naassila M, Spanagel R, Sommer WH (2021) Sci Adv 7(47):eabh2399. https://doi.org/10.1126/sciadv.abh2399

Meisch RA (1975) The function of schedule-induced polydipsia in establishing ethanol as a positive reinforcer. Pharmacol Rev 27(4):465–473. Retrieved from https://www.ncbi.nlm.nih.gov/pubmed/1223913

Melendez RI (2011) Intermittent (every-other-day) drinking induces rapid escalation of ethanol intake and preference in adolescent and adult C57BL/6J mice. Alcohol Clin Exp Res 35(4): 652–658. https://doi.org/10.1111/j.1530-0277.2010.01383.x

Melendez RI, Middaugh LD, Kalivas PW (2006) Development of an alcohol deprivation and escalation effect in C57BL/6J mice. Alcohol Clin Exp Res 30(12):2017–2025. https://doi.org/10.1111/j.1530-0277.2006.00248.x

Metten P, Brown LL, Crabbe JC (2011) Limited access ethanol drinking in the dark in adolescent and adult mice. Pharmacol Biochem Behav 98(2):279–285. https://doi.org/10.1016/j.pbb.2011.01.003

Mittleman G, Van Brunt CL, Matthews DB (2003) Schedule-induced ethanol self-administration in DBA/2J and C57BL/6J mice. Alcohol Clin Exp Res 27(6):918–925. https://doi.org/10.1097/01.ALC.0000071930.48632.AE

Moench KM, Logrip ML (2021) Housing condition differentially impacts escalation of alcohol intake, relapse-like drinking, anxiety-like behavior, and stress history effects by sex. Alcohol Clin Exp Res 45(2):480–489. https://doi.org/10.1111/acer.14540

Moorman DE, Aston-Jones G (2009) Orexin-1 receptor antagonism decreases ethanol consumption and preference selectively in high-ethanol--preferring Sprague--Dawley rats. Alcohol 43(5): 379–386. https://doi.org/10.1016/j.alcohol.2009.07.002

Murphy JM, Gatto GJ, Waller MB, McBride WJ, Lumeng L, Li TK (1986) Effects of scheduled access on ethanol intake by the alcohol-preferring (P) line of rats. Alcohol 3(5):331–336. https://doi.org/10.1016/0741-8329(86)90010-8

NIAAA (2004) National Institute on Alcohol Abuse and Alcoholism Council approves definition of binge drinking. Winter 2004. NIAAA Newslett:3

O'Dell LE, Roberts AJ, Smith RT, Koob GF (2004) Enhanced alcohol self-administration after intermittent versus continuous alcohol vapor exposure. Alcohol Clin Exp Res 28(11): 1676–1682. https://doi.org/10.1097/01.alc.0000145781.11923.4e

Ogata H, Ogato F, Mendelson JH, Mello NK (1972) A comparison of techniques to induce alcohol dependence and tolerance in the mouse. J Pharmacol Exp Ther 180(2):216–230. Retrieved from https://www.ncbi.nlm.nih.gov/pubmed/5062298

Oster SM, Toalston JE, Kuc KA, Pommer TJ, Murphy JM, Lumeng L et al (2006) Effects of multiple alcohol deprivations on operant ethanol self-administration by high-alcohol-drinking replicate rat lines. Alcohol 38(3):155–164. https://doi.org/10.1016/j.alcohol.2006.06.001

Palm S, Roman E, Nylander I (2011) Differences in voluntary ethanol consumption in Wistar rats from five different suppliers. Alcohol 45(6):607–614. https://doi.org/10.1016/j.alcohol.2010.11.005

Phillips TJ, Crabbe JC, Metten P, Belknap JK (1994) Localization of genes affecting alcohol drinking in mice. Alcohol Clin Exp Res 18(4):931–941. https://doi.org/10.1111/j.1530-0277.1994.tb00062.x

Phillips TJ, Reed C, Burkhart-Kasch S, Li N, Hitzemann R, Yu CH et al (2010) A method for mapping intralocus interactions influencing excessive alcohol drinking. Mamm Genome 21(1–2):39–51. https://doi.org/10.1007/s00335-009-9239-9

Pirino BE, Martin CR, Carpenter BA, Curtis GR, Curran-Alfaro CM, Samels SB et al (2022) Sex-related differences in pattern of ethanol drinking under the intermittent-access model and its impact on exploratory and anxiety-like behavior in Long-Evans rats. Alcohol Clin Exp Res 46(7):1282–1293. https://doi.org/10.1111/acer.14853

Pradhan G, Melugin PR, Wu F, Fang HM, Weber R, Kroener S (2018) Calcium chloride mimics the effects of acamprosate on cognitive deficits in chronic alcohol-exposed mice. Psychopharmacology (Berl) 235(7):2027–2040. https://doi.org/10.1007/s00213-018-4900-1

Priddy BM, Carmack SA, Thomas LC, Vendruscolo JC, Koob GF, Vendruscolo LF (2017) Sex, strain, and estrous cycle influences on alcohol drinking in rats. Pharmacol Biochem Behav 152: 61–67. https://doi.org/10.1016/j.pbb.2016.08.001

Quintanille ME, Israel Y, Sapag A, Tampier L (2006) The UChA and UChB rat lines: metabolic and genetic differences influencing ethanol intake. Addict Biol 11(3–4):310–323. https://doi.org/10.1111/j.1369-1600.2006.00030.x

Radke AK, Jury NJ, Kocharian A, Marcinkiewcz CA, Lowery-Gionta EG, Pleil KE et al (2017) Chronic EtOH effects on putative measures of compulsive behavior in mice. Addict Biol 22(2): 423–434. https://doi.org/10.1111/adb.12342

Renteria R, Cazares C, Gremel CM (2020) Habitual ethanol seeking and licking microstructure of enhanced ethanol self-administration in ethanol-dependent mice. Alcohol Clin Exp Res 44(4): 880–891. https://doi.org/10.1111/acer.14302

Rhodes JS, Best K, Belknap JK, Finn DA, Crabbe JC (2005) Evaluation of a simple model of ethanol drinking to intoxication in C57BL/6J mice. Physiol Behav 84(1):53–63. https://doi.org/10.1016/j.physbeh.2004.10.007

Rhodes JS, Ford MM, Yu CH, Brown LL, Finn DA, Garland T Jr, Crabbe JC (2007) Mouse inbred strain differences in ethanol drinking to intoxication. Genes Brain Behav 6(1):1–18. https://doi.org/10.1111/j.1601-183X.2006.00210.x

Richardson HN, Lee SY, O'Dell LE, Koob GF, Rivier CL (2008) Alcohol self-administration acutely stimulates the hypothalamic-pituitary-adrenal axis, but alcohol dependence leads to a dampened neuroendocrine state. Eur J Neurosci 28(8):1641–1653. https://doi.org/10.1111/j.1460-9568.2008.06455.x

Rigter H, Crabbe JC (1980) Alcohol tolerance and dependence. Elsevier/North-Holland, New York, NY

Rimondini R, Arlinde C, Sommer W, Heilig M (2002) Long-lasting increase in voluntary ethanol consumption and transcriptional regulation in the rat brain after intermittent exposure to alcohol. FASEB J 16(1):27–35. https://doi.org/10.1096/fj.01-0593com

Rimondini R, Sommer W, Heilig M (2003) A temporal threshold for induction of persistent alcohol preference: behavioral evidence in a rat model of intermittent intoxication. J Stud Alcohol 64(4): 445–449. https://doi.org/10.15288/jsa.2003.64.445

Rimondini R, Sommer WH, Dall'Olio R, Heilig M (2008) Long-lasting tolerance to alcohol following a history of dependence. Addict Biol 13(1):26–30. https://doi.org/10.1111/j.1369-1600.2007.00079.x

Roberts AJ, Cole M, Koob GF (1996) Intra-amygdala muscimol decreases operant ethanol self-administration in dependent rats. Alcohol Clin Exp Res 20(7):1289–1298. https://doi.org/10.1111/j.1530-0277.1996.tb01125.x

Roberts AJ, Heyser CJ, Cole M, Griffin P, Koob GF (2000) Excessive ethanol drinking following a history of dependence: animal model of allostasis. Neuropsychopharmacology 22(6):581–594. https://doi.org/10.1016/S0893-133X(99)00167-0

Robinson SL, McCool BA (2015) Microstructural analysis of rat ethanol and water drinking patterns using a modified operant self-administration model. Physiol Behav 149:119–130. https://doi.org/10.1016/j.physbeh.2015.05.034

Rodd ZA, Bell RL, Kuc KA, Murphy JM, Lumeng L, Li TK, McBride WJ (2003) Effects of repeated alcohol deprivations on operant ethanol self-administration by alcohol-preferring (P) rats. Neuropsychopharmacology 28(9):1614–1621. https://doi.org/10.1038/sj.npp.1300214

Rodd ZA, Bell RL, Sable HJ, Murphy JM, McBride WJ (2004) Recent advances in animal models of alcohol craving and relapse. Pharmacol Biochem Behav 79(3):439–450. https://doi.org/10.1016/j.pbb.2004.08.018

Rodd ZA, Bell RL, Kuc KA, Murphy JM, Lumeng L, McBride WJ (2009) Effects of concurrent access to multiple ethanol concentrations and repeated deprivations on alcohol intake of high-alcohol-drinking (HAD) rats. Addict Biol 14(2):152–164. https://doi.org/10.1111/j.1369-1600.2008.00140.x

Rodd-Henricks ZA, McKinzie DL, Murphy JM, McBride WJ, Lumeng L, Li TK (2000a) The expression of an alcohol deprivation effect in the high-alcohol-drinking replicate rat lines is dependent on repeated deprivations. Alcohol Clin Exp Res 24(6):747–753. Retrieved from https://www.ncbi.nlm.nih.gov/pubmed/10888060

Rodd-Henricks ZA, McKinzie DL, Shaikh SR, Murphy JM, McBride WJ, Lumeng L, Li TK (2000b) Alcohol deprivation effect is prolonged in the alcohol preferring (P) rat after repeated deprivations. Alcohol Clin Exp Res 24(1):8–16. Retrieved from https://www.ncbi.nlm.nih.gov/pubmed/10656186

Rodd-Henricks ZA, Bell RL, Kuc KA, Murphy JM, McBride WJ, Lumeng L, Li TK (2001) Effects of concurrent access to multiple ethanol concentrations and repeated deprivations on alcohol intake of alcohol-preferring rats. Alcohol Clin Exp Res 25(8):1140–1150. Retrieved from https://www.ncbi.nlm.nih.gov/pubmed/11505045

Rodriguez LA, Plomin R, Blizard DA, Jones BC, McClearn GE (1994) Alcohol acceptance, preference, and sensitivity in mice. I. Quantitative genetic analysis using BXD recombinant inbred strains. Alcohol Clin Exp Res 18(6):1416–1422. https://doi.org/10.1111/j.1530-0277.1994.tb01444.x

Rodriguez LA, Plomin R, Blizard DA, Jones BC, McClearn GE (1995) Alcohol acceptance, preference, and sensitivity in mice. II. Quantitative trait loci mapping analysis using BXD recombinant inbred strains. Alcohol Clin Exp Res 19(2):367–373. https://doi.org/10.1111/j.1530-0277.1995.tb01517.x

Rosenwasser AM, Fixaris MC, Crabbe JC, Brooks PC, Ascheid S (2013) Escalation of intake under intermittent ethanol access in diverse mouse genotypes. Addict Biol 18(3):496–507. https://doi.org/10.1111/j.1369-1600.2012.00481.x

Russo M, Funk D, Loughlin A, Coen K, Le AD (2018) Effects of alcohol dependence on discrete choice between alcohol and saccharin. Neuropsychopharmacology 43(9):1859–1866. https://doi.org/10.1038/s41386-018-0101-1

Salimov RM, Salimova NB (1993) The alcohol-deprivation effect in hybrid mice. Drug Alcohol Depend 32(2):187–191. https://doi.org/10.1016/0376-8716(93)80012-4

Salimov RM, Salimova NB, Shvets LN, Maisky AI (2000) Haloperidol administered subchronically reduces the alcohol-deprivation effect in mice. Alcohol 20(1):61–68. https://doi.org/10.1016/s0741-8329(99)00057-9

Samson HH, Chappell A (2001) Effects of alcohol deprivation on alcohol consumption using a sipper-tube procedure. Alcohol Clin Exp Res 25(5):680–686. Retrieved from https://www.ncbi.nlm.nih.gov/pubmed/11371717

Savarese AM, Ozburn AR, Barkley-Levenson AM, Metten P, Crabbe JC (2021) The impact of drinking in the dark (DID) procedural manipulations on ethanol intake in high drinking in the dark (HDID) mice. Alcohol 93:45–56. https://doi.org/10.1016/j.alcohol.2021.02.001

Schuh KM, Sneddon EA, Nader AM, Muench MA, Radke AK (2022) Orbitofrontal cortex subregion inhibition during binge-like and aversion-resistant alcohol drinking. Alcohol 99:1–8. https://doi.org/10.1016/j.alcohol.2021.11.004

Serra S, Brunetti G, Vacca G, Lobina C, Carai MA, Gessa GL, Colombo G (2003) Stable preference for high ethanol concentrations after ethanol deprivation in Sardinian alcohol-preferring (sP) rats. Alcohol 29(2):101–108. https://doi.org/10.1016/s0741-8329(03)00003-x

Simms JA, Steensland P, Medina B, Abernathy KE, Chandler LJ, Wise R, Bartlett SE (2008) Intermittent access to 20% ethanol induces high ethanol consumption in Long-Evans and Wistar rats. Alcohol Clin Exp Res 32(10):1816–1823. https://doi.org/10.1111/j.1530-0277.2008.00753.x

Simms JA, Bito-Onon JJ, Chatterjee S, Bartlett SE (2010) Long-Evans rats acquire operant self-administration of 20% ethanol without sucrose fading. Neuropsychopharmacology 35(7):1453–1463. https://doi.org/10.1038/npp.2010.15

Sinclair JD, Li TK (1989) Long and short alcohol deprivation: effects on AA and P alcohol-preferring rats. Alcohol 6(6):505–509. https://doi.org/10.1016/0741-8329(89)90059-1

Sinclair JD, Senter RJ (1968) Development of an alcohol-deprivation effect in rats. Q J Stud Alcohol 29(4):863–867. Retrieved from https://www.ncbi.nlm.nih.gov/pubmed/5705408

Sinclair JD, Tiihonen K (1988) Lack of alcohol-deprivation effect in AA rats. Alcohol 5(1):85–87. https://doi.org/10.1016/0741-8329(88)90048-1

Smeets JAS, Minnaard AM, Ramakers GMJ, Adan RAH, Vanderschuren L, Lesscher HMB (2022) On the interrelation between alcohol addiction-like behaviors in rats. Psychopharmacology (Berl) 239(4):1115–1128. https://doi.org/10.1007/s00213-021-06059-4

Sneddon EA, White RD, Radke AK (2019) Sex differences in binge-like and aversion-resistant alcohol drinking in C57BL/6J mice. Alcohol Clin Exp Res 43(2):243–249. https://doi.org/10.1111/acer.13923

Sneddon EA, Schuh KM, Frankel JW, Radke AK (2021) The contribution of medium spiny neuron subtypes in the nucleus accumbens core to compulsive-like ethanol drinking. Neuropharmacology 187:108497. https://doi.org/10.1016/j.neuropharm.2021.108497

Sommer W, Hyytia P, Kiianmaa K (2006) The alcohol-preferring AA and alcohol avoiding ANA rats: neurobiology of the regulation of alcohol drinking. Addict Biol 11(3–4):289–309. https://doi.org/10.1111/j.1369-1600.2006.00037.x

Sommer WH, Rimondini R, Hansson AC, Hipskind PA, Gehlert DR, Barr CS, Heilig MA (2008) Upregulation of voluntary alcohol intake, behavioral sensitivity to stress, and amygdala crhr1 expression following a history of dependence. Biol Psychiatry 63(2):139–145. https://doi.org/10.1016/j.biopsych.2007.01.010

Spanagel R (2009) Alcoholism: a systems approach from molecular physiology to addictive behavior. Physiol Rev 89(2):649–705. https://doi.org/10.1152/physrev.00013.2008

Spanagel R, Holter SM (1999) Long-term alcohol self-administration with repeated alcohol deprivation phases: an animal model of alcoholism? Alcohol Alcohol 34(2):231–243. https://doi.org/10.1093/alcalc/34.2.231

Spanagel R, Holter SM (2000) Pharmacological validation of a new animal model of alcoholism. J Neural Transm (Vienna) 107(6):669–680. https://doi.org/10.1007/s007020070068

Spanagel R, Holter SM, Allingham K, Landgraf R, Zieglgansberger W (1996) Acamprosate and alcohol: I. Effects on alcohol intake following alcohol deprivation in the rat. Eur J Pharmacol 305(1–3):39–44. https://doi.org/10.1016/0014-2999(96)00174-4

Sparta DR, Ferraro FM 3rd, Fee JR, Knapp DJ, Breese GR, Thiele TE (2009) The alcohol deprivation effect in C57BL/6J mice is observed using operant self-administration procedures and is modulated by CRF-1 receptor signaling. Alcohol Clin Exp Res 33(1):31–42. https://doi.org/10.1111/j.1530-0277.2008.00808.x

Spear LP (2020) Timing eclipses amount: the critical importance of intermittency in alcohol exposure effects. Alcohol Clin Exp Res 44(4):806–813. https://doi.org/10.1111/acer.14307

Suwaki H, Kalant H, Higuchi S, Crabbe JC, Ohkuma S, Katsura M et al (2001) Recent research on alcohol tolerance and dependence. Alcohol Clin Exp Res 25(5 Suppl ISBRA):189S–196S. https://doi.org/10.1097/00000374-200105051-00031

Szumlinski KK, Coelho MA, Lee KM, Tran T, Sern KR, Bernal A, Kippin TE (2019) DID it or DIDn't it? Exploration of a failure to replicate binge-like alcohol-drinking in C57BL/6J mice. Pharmacol Biochem Behav 178:3–18. https://doi.org/10.1016/j.pbb.2018.12.002

Tambour S, Brown LL, Crabbe JC (2008) Gender and age at drinking onset affect voluntary alcohol consumption but neither the alcohol deprivation effect nor the response to stress in mice. Alcohol Clin Exp Res 32(12):2100–2106. https://doi.org/10.1111/j.1530-0277.2008.00798.x

Tang M, Falk JL (1983) Production of physical dependence on ethanol by a short drinking episode each day. Pharmacol Biochem Behav 19(1):53–55. https://doi.org/10.1016/0091-3057(83)90311-8

Tang M, Brown C, Falk JL (1982) Complete reversal of chronic ethanol polydipsia by schedule withdrawal. Pharmacol Biochem Behav 16(1):155–158. https://doi.org/10.1016/0091-3057(82)90028-4

Thiele TE, Navarro M (2014) "Drinking in the dark" (DID) procedures: a model of binge-like ethanol drinking in non-dependent mice. Alcohol 48(3):235–241. https://doi.org/10.1016/j.alcohol.2013.08.005

Thiele TE, Crabbe JC, Boehm SL 2nd (2014) "Drinking in the Dark" (DID): a simple mouse model of binge-like alcohol intake. Curr Protoc Neurosci 68., 9 49 41-49 49 12. https://doi.org/10.1002/0471142301.ns0949s68

Vandaele Y, Janak PH (2018) Defining the place of habit in substance use disorders. Prog Neuropsychopharmacol Biol Psychiatry 87(Pt A):22–32. https://doi.org/10.1016/j.pnpbp.2017.06.029

Vendruscolo LF, Roberts AJ (2014) Operant alcohol self-administration in dependent rats: focus on the vapor model. Alcohol 48(3):277–286. https://doi.org/10.1016/j.alcohol.2013.08.006

Vengeliene V, Siegmund S, Singer MV, Sinclair JD, Li TK, Spanagel R (2003) A comparative study on alcohol-preferring rat lines: effects of deprivation and stress phases on voluntary alcohol intake. Alcohol Clin Exp Res 27(7):1048–1054. https://doi.org/10.1097/01.ALC.0000075829.81211.0C

Vengeliene V, Bilbao A, Spanagel R (2014) The alcohol deprivation effect model for studying relapse behavior: a comparison between rats and mice. Alcohol 48(3):313–320. https://doi.org/10.1016/j.alcohol.2014.03.002

Wayner MJ, Greenberg I, Tartaglione R, Nolley D, Fraley S, Cott A (1972) A new factor affecting the consumption of ethyl alcohol and other sapid fluids. Physiol Behav 8(2):345–362. https://doi.org/10.1016/0031-9384(72)90383-6

Wilcox MV, Cuzon Carlson VC, Sherazee N, Sprow GM, Bock R, Thiele TE et al (2014) Repeated binge-like ethanol drinking alters ethanol drinking patterns and depresses striatal GABAergic transmission. Neuropsychopharmacology 39(3):579–594. https://doi.org/10.1038/npp.2013.230

Wise RA (1973) Voluntary ethanol intake in rats following exposure to ethanol on various schedules. Psychopharmacologia 29(3):203–210. https://doi.org/10.1007/BF00414034

Younis RM, Wolstenholme JT, Bagdas D, Bettinger JC, Miles MF, Damaj MI (2019) Adolescent but not adult ethanol binge drinking modulates ethanol behavioral effects in mice later in life. Pharmacol Biochem Behav 184:172740. https://doi.org/10.1016/j.pbb.2019.172740

Modeling Brain Gene Expression in Alcohol Use Disorder with Genetic Animal Models

Robert Hitzemann, Angela R. Ozburn, Denesa Lockwood, and Tamara J. Phillips

Contents

Abstract Animal genetic models have and will continue to provide important new information about the behavioral and physiological adaptations associated with alcohol use disorder (AUD). This chapter focuses on two models, ethanol preference and drinking in the dark (DID), their usefulness in interrogating brain gene expression data and the relevance of the data obtained to interpret AUD-related GWAS and TWAS studies. Both the animal and human data point to the importance for AUD of changes in synaptic transmission (particularly glutamate and GABA transmission), of changes in the extracellular matrix (specifically including collagens, cadherins and protocadherins) and of changes in neuroimmune processes. The implementation of new technologies (e.g., cell type-specific gene expression) is expected to further enhance the value of genetic animal models in understanding AUD.

R. Hitzemann (✉), A. R. Ozburn, and D. Lockwood
Department of Behavioral Neuroscience, Portland Alcohol Research Center, Oregon Health and Science University, Portland, OR, USA
e-mail: hitzeman@ohsu.edu

T. J. Phillips
Department of Behavioral Neuroscience, Portland Alcohol Research Center, Oregon Health and Science University, Portland, OR, USA

Veterans Affairs Portland Health Care System, Portland, OR, USA

Curr Topics Behav Neurosci (2025) 71: 111–144
https://doi.org/10.1007/7854_2023_455
Published Online: 21 November 2023

Keywords Alcohol use disorder · Animal genetic models · Brain gene expression · Drinking in the dark · Ethanol preference · GWAS · RNA sequencing · TWAS

1 Introduction

Crabbe et al. (2013) in a chapter entitled "Modeling the Diagnostic Criteria for Alcohol Dependence with Genetic Animal Models" focused on the following question: Do the animal genetic models best support the idea of single or multiple genetic risk factors for alcohol dependence? It was concluded, "that there is modest support in the animal literature that alcohol tolerance and withdrawal reflect distinct genetic risk factors, in agreement with our human data."

A decade has passed since the chapter appeared. Alcohol abuse and alcohol dependence have been combined in DSM-V for the diagnosis of alcohol use disorder (AUD) (*Diagnostic and Statistical Manual of Mental Disorders (5th Ed., Text Revision)*, 2022). Legal problems were dropped as a diagnostic criterion and alcohol craving was added. Mild, moderate, and severe AUD requires, respectively, 2–3, 4–5, and 6 or more of the 11 DSM-V diagnostic criteria. Alcohol tolerance and alcohol dependence/withdrawal, key elements of the Crabbe et al. (2013) analyses, were retained in DSM-V. Thus, from a certain perspective substituting AUD for alcohol dependence has not been a great change. We are now (2023) in a better position to ask if the data generated from the animal genetic models, even those models that appear somewhat distant from AUD, are yielding relevant results. For example, are the model data similar to those extracted from genome-wide association studies (GWAS) and transcriptome-wide association studies (TWAS)? Have the animal genetic models pointed to the development of new therapies? In addition, how have the animal genetic models, informed us about the somewhat perplexing but repeated observation that many individuals at high genetic risk for AUD, abstain from excessive alcohol consumption? Finally, how have the animal genetic models informed us about the important sex × risk interaction? This chapter will focus on these and related issues with an emphasis on new gene expression data and the animal models used to collect these data.

In what is something of a chapter update, it is useful to review the previous recommendations for future directions. We quote here again from Crabbe et al. (2013):

> "The existing data on gene expression differences, either predisposing to alcohol responses or consequent to exposure to alcohol, have unfortunately rarely characterized either tolerance or withdrawal phenotypes. A systematic characterization of the gene expression networks predisposing to and invoked by ethanol tolerance could be compared with the networks seen after initiating dependence, and during withdrawal. A maximally informative line of mice for such a genetic experiment would be the heterogeneous stock – collaborative cross (HS-CC) outbred stock, which was developed by members of the Collaborative Cross consortium specifically to display maximal allelic diversity (Churchill et al. 2004)." (*Note:*

The HS-CC population is now extinct; the Diversity Outbred (DO) population is a similar alternative.)

Are there new data (over the past 10 years) that align with these recommendations? To our knowledge, there has been no substantial effort to develop new genetic models of functional ethanol tolerance. It could be argued that some aspects of functional tolerance are captured in the animal models that are routinely used, i.e., preference consumption, chronic intermittent exposure (CIE), and drinking in the dark (DID). However, parsing out the tolerance component would be difficult if not impossible. There are some new genetic/gene expression data. For example, Radcliffe et al. (2020) collected brain gene expression data in 40 Long Sleep × Short Sleep, or LXS, recombinant inbred (RI) strains under identical conditions to those previously used to detect and map acute functional tolerance (see Bennett et al. 2015). The DE genes associated with acute functional tolerance were enriched in ontologies associated with neuron-to-neuron synaptic transmission and potassium transport.

The situation for withdrawal is similar in that no new animal genetic models have been developed. The withdrawal sensitive prone and withdrawal sensitive resistant (WSP/WSR) mouse lines have been retired to cryopreservation. However, a new selection of lines similar to the WSP/WSR seems relevant for several reasons. First, the withdrawal phenotype remains the animal model most closely related to a characteristic of AUD. Secondly, from the gene expression perspective, the past 10 years have witnessed some remarkable developments in terms of collecting gene expression data on the scale necessary to accurately form gene expression networks. Third, any new selection could use highly diverse rodent populations such as the diversity outbred (DO) mouse (Churchill et al. 2012) or the N/NIH rat (Hansen and Spuhler 1984; Solberg Woods et al. 2010). The rationale for using these populations, all of which are more genetically diverse than humans, has been discussed elsewhere (see, e.g., Hitzemann et al. 2014; Solberg Woods et al. 2010). From the perspective of translation, these diverse populations offer the prospect of detecting new genetic features of withdrawal and related phenotypes that in turn could be used to direct clinical studies.

2 Animal Models: General

For animal models of neuropsychiatric disorders such as AUD, there will probably always be a conversation on whether the models are relevant to understanding the human condition, i.e., are the models translationally relevant? For alcohol research, this is hardly a new concern. McClearn (1979) commented on this issue >40 years ago and the comments remain topical. As the article is not freely available, we directly quote here two passages. The first focuses more on general issues.

"I would like to make a few comments about the issue of animal models in the study of alcoholism. It has been contended by some that animal models in biomedical research should be isomorphic in all respects with the disease being modeled. By this standard, an animal model of hypertension, for example, should have the same causal etiology, the same symptomatology, the same prognosis, and the same response to medication as human hypertension. I have been confused by this requirement of perfection, because it implies that adequacy of a model cannot be established unless one is in possession of so much information that the model is no longer needed. At least, it would not be needed for heuristic or hypothesis-testing purposes, even though it might be useful in routine clinical laboratory applications. It would seem to me that the most useful models are those in which some known features of the target phenomenon are incorporated, but some are not yet known, so that future exploration of the model provides surprises. That is to say, in a useful model, one gets back more than is put into it. If not, one is in possession of a summary statement, not a model. As soon as we back off from the counsel of perfection, the question arises as to which aspects of the target phenomenon are to be included in the model."

The second passage focuses on issues more relevant to alcohol models.

"....it has been suggested that criteria for a proper animal model of alcoholism include oral ingestion of alcohol without food deprivation and with competing fluids available, ingestion to the point of intoxication sustained over a long period, performance of work to obtain alcohol, occurrence of a withdrawal syndrome, and reacquisition of drinking to intoxication after abstinence. Many critics believe that no animal model has yet met these criteria. I freely confess that my own work has not. There is no question that a model meeting these or other comprehensive criteria would be extremely valuable. However, while awaiting the development of such a model, I think our time can be well spent in research that utilizes partial models. It is extremely unlikely that all of the information we accumulate on a model that is limited to voluntary ingestion of ethanol is automatically and entirely irrelevant to the problem of human alcohol consumption and its sequelae. Thus, while we aspire to comprehensive models and struggle to achieve them, use of those that are less adequate and available can contribute to the pool of knowledge from which better understanding of the human condition will be derived."

We agree that animal models can contribute to a better understanding of the human condition. A recent example of convergent results across rodent and human studies is described by Farokhnia et al. (2022) for the effect of the mineralocorticoid receptor antagonist, spironolactone, on alcohol drinking. Similar arguments can be made for the development of mifepristone, a glucocorticoid receptor antagonist (Vendruscolo et al. 2015), and for the phosphodiesterase type 4 inhibitor, apremilast (see below) as potential AUD pharmacotherapies.

3 Animal Models: Preference

Although not mentioned, the model at the heart of the above argument is ethanol preference. This model is important since the majority of the animal model genomic data collected over the past decade was obtained using the preference model or some model derivative. The usual format for the preference model is 24/7, two-bottle choice (water vs ethanol, usually 10%, although sometimes higher concentrations are used) in which both the preference for consuming fluid from the ethanol bottle is measured, as well as the amount of ethanol consumed on a g ethanol/kg body weight basis. The model has a long history in animal alcohol genetics research. Borruto et al. (2021) reviewed the use of rat preference models and selective breeding (which is often associated with the large number of animals required for genetic/genomic studies). These authors noted that "the first selective breeding program was initiated at Universidad de Chile (UCh), where high alcohol-drinking (UChB) and low alcohol-drinking (UChA) lines were developed (Mardones et al. 1953; Mardones and Segovia-Riquelme 1984; Quintanilla et al. 2006). Eriksson and colleagues later initiated another selective breeding program at Alko Research Laboratories in Finland, where the Alko Alcohol high alcohol-preferring (AA) and Alko Non-Alcohol low alcohol-preferring (ANA) rat lines were generated (Eriksson 1968; Sommer et al. 2006). After the success of these two initiatives, additional selective breeding programs were initiated in several laboratories. These programs resulted in the development of alcohol preferring (P) and non-preferring (NP) rat lines at Indiana University/Purdue University Indianapolis (Indianapolis, IN, USA; Li et al. 1979) and the Sardinian alcohol-preferring (sP) and non-preferring (sNP) rats in Cagliari, Italy (Ciccocioppo et al. 2006; Colombo et al. 2006; Li et al. 1979). Years later, starting from selectively bred rats from the N/NIH founder stock, a new bidirectional breeding program was launched that resulted in the generation of High Alcohol Drinking (HAD) and Low Alcohol Drinking (LAD) rat lines (Hansen and Spuhler 1984; Murphy et al. 2002)." The concern is frequently raised that with the preference model, the animals either do not or only briefly, exceed intoxicating blood ethanol concentrations (BECs) of 80 mg/dl. However, this is not always the case. For example, McBride et al. (2014) asked how well the P and HAD replicate 1 and 2 rats matched the criteria for an animal model of AUD suggested by Lester and Freed (1973) and Cicero (1979) (described above by McClearn 1979). For six key criteria, including obtaining relevant BECs, the rat selected lines matched reasonably well (see Table 1 in McBride et al. 2014).

Although there may be earlier reports, the use of mice in the context of a genetic model is usually thought to have begun with the observations by McClearn and Rodgers (1959) that among a small panel of inbred mouse strains, C57BL/6J (B6) mice consistently prefer ethanol over water. This experiment, with numerous variations, has certainly been repeated hundreds of times, often with very large panels of inbred strains (e.g., Yoneyama et al. 2008). Until recently, the B6 strain was always found to have the highest ethanol preference and consumption which in turn has led to something of a B6 monoculture in ethanol research. However, Bagley et al. (2021) examined ethanol preference in the eight founder strains of the Collaborative Cross (CC) (Churchill et al. 2004) and found that the PWK/PhJ (PWK) strain had an ethanol preference and total consumption that exceeded that of the B6 strain. The question of whether the B6 and PWK strains prefer ethanol for similar or different reasons remains to be determined although Anderson et al. (2022) have suggested that the B6 and PWK strains are most similar when compared to the other six CC founder strains across dimensions related to neuroimmune function. The incorporation of the PWK strain into ethanol research will be somewhat problematic. First, the PWK strain is wild derived and this offers challenges for handling the animals, as they are truly "wild." Secondly, the current cost and availability of the animals prohibit their routine use.

Although the focus was on cocaine addiction, Saul et al. (2020) made several salient points about the use of genetic animal models that are relevant to the current discussion. These authors noted that . . ."Concerns about external validity of rodent models and translation of findings across species are often based on narrow investigations of populations with limited diversity. Sources of individual variation – including genetics and sex – are only infrequently encompassed in model organism studies. . . Explicit inclusion of individual differences in rodent research may reveal conserved phenotypes and molecular systems relevant to human addiction." Primarily focusing on the 8 CC founder strains these authors observed that individual differences explained a substantial proportion of variance in cocaine responsive or cocaine response-predictive behavioral and physiological phenotypes. Further, they observed that striatum transcriptional responses to cocaine were also highly dependent upon strain and sex differences. When they compared the mouse transcriptional data to human data the similarity between species was highly dependent upon mouse genetic background and sex. Specifically, they observed that male WSB/EiJ mice and female NOD/ShiLtJ mice exhibited the greatest degree of neural transcriptional consilience with humans with cocaine use disorder. Thus, the consilience between the mouse and human data was detected with mouse strains not normally used in drug and alcohol research.

A key derivative of the inbred strain panel approach has been the use of recombinant inbred (RI) strains (both mouse and rat) to extract genetic information; e.g., quantitative trait loci (QTL) and genetically correlated traits related to excessive ethanol consumption (e.g., Gora-Maslak et al. 1991; Lusk et al. 2018; Metten et al. 1998; Philip et al. 2010, p. 2010; Phillips et al. 1994; Saba et al. 2015; Vanderlinden et al. 2015). The rapid incorporation of the RI panels into alcohol research follows from the long history of genetic research, some of which are noted above. These reference panels were and remain an ideal mechanism for determining the genetic overlap of the animal models. For example, is there a genetic overlap between the withdrawal and tolerance models? To date most of the RI data relevant to alcohol research has been extracted from the B6 × DBA/2J (D2) or BXD RI panel; these data have been collected in an easily usable form at the Gene Network web site. One caveat to the use of these data is that the B6 and D2 strains capture only a small slice of the genetic diversity available in *Mus musculus* (Roberts et al. 2007) which may limit the general relevance of the results. Nonetheless, the BXD RI panel continues to be used largely because of the unique features of the progenitor strains across several alcohol phenotypes including preference and withdrawal; the B6 strain is characterized by high preference and low withdrawal whereas the D2 strain has the reciprocal phenotypes. Thus, the preference distribution among the RI strains ranges more or less normally from the B6 to D2 phenotype (Phillips et al. 1994). The original RI panel is now more than 50 years old (Berek et al. 1976) and many of the original 40 strains have become extinct. However, Williams and colleagues have developed a newer and larger set of RI strains that greatly increases the available statistical power (Williams and Williams 2017). Finally, it should be noted that the preference and withdrawal QTLs first detected in the RI panel have been repeatedly confirmed (see, e.g., Belknap and Atkins 2001; Kozell et al. 2020). Two quantitative trait genes for withdrawal have emerged: *Mpdz* and *Kcnj* (Kozell et al. 2018; Kruse et al. 2014; Milner et al. 2015; Shirley et al. 2004).

There have been several ethanol consumption or preference selective breeding studies in mice over the past decade. These studies have a somewhat outsized influence on the integration of preference with gene expression, given that mouse populations can be easily scaled to the numbers required for genome-wide analyses. One problem with such studies, including studies using the rat selected lines (see above), is that the number of families is usually limited to no more than 20 which in turn means that there are significant kinship issues that need to be considered in the data analysis (Parker et al. 2014). Grahame and colleagues selectively bred replicate lines of High Alcohol Preference (HAP) and Low Alcohol Preference (LAP) mice from HS/Ibg founders and then crossed the replicate HAP1 and HAP2 lines to form the cHAP mouse (Matson et al. 2013; Matson and Grahame 2013). These authors found that cHAP mice routinely escalate drinking to an excess of 25 g/kg/day, achieve mean BECs in excess of 250 mg/dl, and sustain pharmacologically relevant BECs throughout the active (dark) portion of their light-dark cycle following chronic (3 weeks) ethanol access. In addition, with chronic consumption the mice demonstrate tolerance to the effects of alcohol on motor coordination; there was no evidence of pharmacokinetic tolerance or sensitization to alcohol-induced locomotion (reviewed in Kippin 2014). There are many possibilities as to the mechanisms that combined to produce the cHAP mouse. We favor the view that in genetically diverse populations such as the HS/Ibg, there are multiple selection funnels that can be engaged to produce a high-preference animal. Hopefully, the funnels all lead to a small number of final common pathways that can be detected and thus provide targets for pharmacological manipulation. There is a long-standing concern that preference selection in mice only works because selection is simply a process leading to an enrichment in B6 alleles and thus, selection line testing is redundant to testing in B6 mice. However, the ethanol intake in the cHAP mice considerably exceeds that of the B6, indicating effects of non-B6-derived genes on intake.

The B6 bias (or from some perspectives, the B6 menace to alcohol and drug research) was frequently discussed at the planning meetings (2001–2003) for the development of the Collaborative Cross (CC). The CC was designed to be a very large panel (>1,000) of RI strains derived from eight founder strains that captured ~90% of *Mus musculus* genetic diversity. It was assumed that the increase in genetic diversity over existing mouse populations (see Roberts et al. 2007) would mitigate the B6 bias. For a complex set of reasons, the very large panel of CC RI strains was not developed, although there are ~100 CC strains that have proven useful in a variety of contexts (Bagley et al. 2021; Hackett et al. 2022; Schoenrock et al. 2020; Scoggin et al. 2022; Tryndyak et al. 2022). That said, the existing CC RI strains have not been widely used in alcohol research. Parallel to the development of the RI strains, the founder strains were interbred to establish an outbred population: the HS-CC, mentioned above (Iancu et al. 2010). Subsequently, and using early generation CC RI strains, the above-mentioned DO population was created (Churchill et al. 2012). The breeding of the HS-CC began in 2005, using a pseudo-random breeding design. The offspring from the first full crossing of all eight strains were tested for ethanol preference, with 20% showing a preference >0.5. This was at least double the preference we and others have observed in another 8-strain heterogeneous stock, the HS/NPT mice (see, e.g., Crabbe et al. 2010). We have also observed high preference in the offspring from 48 random breeding pairs of DO mice (unpublished observation). In retrospect, we should have strongly considered and investigated that in addition to the B6, one or more of the other 7 CC founder strains had a high ethanol preference. However, as noted above, it was 15 years before this discovery was made.

Colville et al. (2017, 2018) used HS-CC founders for the selective breeding of high and low-preference lines; the goal was to detect new transcriptional features associated with preference. The gene expression data are discussed below. Here we note some of the caveats associated with selective breeding. Selective breeding is best accomplished when the phenotypes of interest (here high and low preference) diverge quickly so as to minimize inbreeding and genetic drift. To some degree, this can be accomplished by maintaining a large number of breeding families and not breeding related individuals (usually those with common grandparents). For Colville et al. (2017), there were 24 families for the high and low lines; to our knowledge, this is about the maximum that has ever been used. Colville et al. (2017) also used a short-term selective breeding design (see Belknap et al. 1997; Metten et al. 1998). In this design, one sets a limit on the number of generations of selective breeding or for the difference between the divergent lines, e.g., 3 or 4 standard deviations. Once this limit is reached, the experiment ends even though if the experiment had continued, it is likely that further divergence would have occurred and for preference, the lines would have differed more along the features noted above. It was assumed that the genes associated with high and low preference would be the first to segregate and would not be confounded by the effects of genetic drift. Thus, one is looking for the early risk signals.

A desired feature of any animal genetic model would be the escalation of ethanol consumption with chronic use (see, e.g., Matson et al. 2014). Hitzemann et al. (2020) examined ethanol preference in 56 female and 58 male HS-CC mice over a 3-month period. Overall, preference increased (>50%) from week 1 to week 13 in female but not male mice. Figure 1 taken from Hitzemann et al. (2020) shows that the patterns of escalation appeared to be different among individuals but the numbers are too small to draw significant conclusions. However, these data suggest that the preference model can detect significant sex × individual variation × escalation interactions (Fig. 1).

Preference remains a widely used animal genetic model largely (a) because selection generally proceeds rapidly and is easily scalable to the numbers needed for genetic studies and (b) because preference and consumption are highly correlated ($r > 0.9$); i.e., preference is a surrogate for consumption. Thus, it seems likely that genetic studies of preference will necessarily also detect genes and gene networks associated with excessive alcohol intake. However, intake over time (days/weeks/months/years) is not a criterion for AUD given that the amount consumed will be

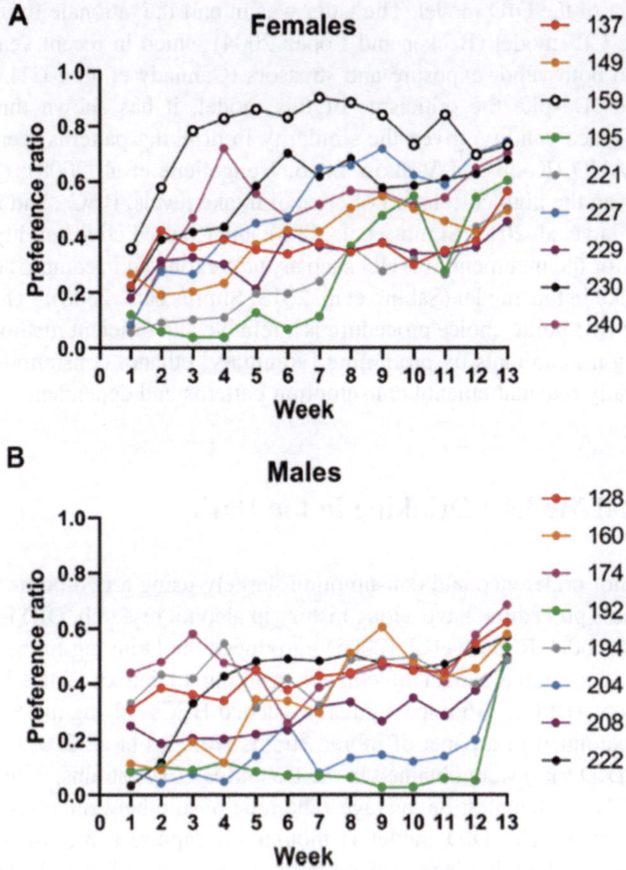

Fig. 1 Escalation of ethanol consumption in HS-CC mice (from Hitzemann et al. 2020). Shown are individual preference ratios for (**a**) females and (**b**) males that had initial preference ratios <0.5 during the first week of drinking that increased to ratios ≥0.5 in the final week. Subject numbers are included for those interested in reviewing the original data

biased by a number of factors including sex, body type, rate of alcohol elimination, age, cultural factors, and so on. That said, there is some agreement as to what can be considered heavy alcohol use that is reliably seen in AUD. NIAAA has defined heavy alcohol use as follows: for men, consuming more than 4 drinks on any day or more than 14 drinks per week; for women, consuming more than 3 drinks on any day or more than 7 drinks per week. SAMHSA defines heavy alcohol use as binge drinking on 5 or more days in the past month.

Nieto et al. (2021) in an article entitled "Translational opportunities in animal and human models to study alcohol use disorder" have commented on the disadvantages and advantages of the preference model. Several criticisms regarding this model have been raised, particularly with regard to its ability to capture binge drinking or dependence-driven ethanol intake. The former will be discussed below in the development of the DID model. The latter was in part the rationale for the development of the CIE model (Becker and Lopez 2004) which in recent years has been expanded to both vapor exposure and stressors (Cannady et al. 2021). Nieto et al. (2021) note "Despite the criticisms of this model, it has shown three types of validity: (1) face validity: given the similarity in drinking patterns seen in humans with mild AUD (Koob and Volkow 2016; Vengeliene et al. 2008); (2) construct validity: given the high correlation of ethanol intake levels, BECs, and neuroplastic effects (Sabino et al. 2013; Simms et al. 2008); and (3) predictive validity: given that drugs used for the treatment of AUD such as, naltrexone and acamprosate, suppress alcohol intake in this model (Sabino et al. 2013; Simms et al. 2008)." These findings indicate the two-bottle choice procedure is a reliable and efficient method of alcohol administration in animals by promoting [voluntary] ethanol consumption that may yield clinically relevant ethanol consumption patterns and dependence.

4 Animal Models: Drinking in the Dark

While ethanol preference and consumption, largely using a two-bottle choice continuous access procedure, have a long history in alcohol research, DID is a relatively new animal model. Rhodes et al. (2005) introduced the Drinking In the Dark (DID) procedure as a simple model of ethanol drinking to intoxicating blood ethanol concentrations (BECs). B6 mice regularly reached BECs >1 mg/ml. Subsequently, DID was examined in a panel of inbred strains (Rhodes et al. 2007). The highest BECs (4 h DID trial) were obtained in the B6 and BALB/c strains, with a <0.2-mg/ml higher BEC in females, on average. The relationship between consumption and BEC was strong. The DID model is thought to capture some aspects of binge consumption. Although binge consumption is not one of the AUD criteria, it frequently aligns with AUD (Ehlers et al. 2022; Gowin et al. 2021; Patrick et al. 2023).

Crabbe et al. (2009) reported on the selection of the High DID-1 (HDID-1) line; HS/NPT mice were the founders; six of the eight strains combined to form the HS/NPT (Hitzemann et al. 1994) were the same as those used for form the HS/Ibg used by Grahame and colleagues (Matson et al. 2013; Matson and Grahame 2013) for selective breeding of the HAP and LAP mice (see above). The selection phenotype for the HDID mice was BEC at the end of a 4-h DID trial from ingestion of a 20% ethanol solution. After 11 generations of selection, the BEC increased from 0.30 to 1.07 mg/ml. A replicate HDID selection (HDID-2) followed the HDID-1 selection. The selection response was largely identical (see Fig. 3 in Crabbe et al. 2010). However, the microstructure of drinking in the HDID-1 and -2 lines is different. HDID-1 mice drink in larger ethanol bouts than the HS founders, whereas HDID-2 mice drink in more frequent bouts (Barkley-Levenson and Crabbe 2012). These data compliment the arguments noted above for the cHAP mouse. Seemingly identical selections can yield similar outcomes, due to a different underlying cause. Further evidence of this point came from the genotype data. Iancu et al. (2013) used

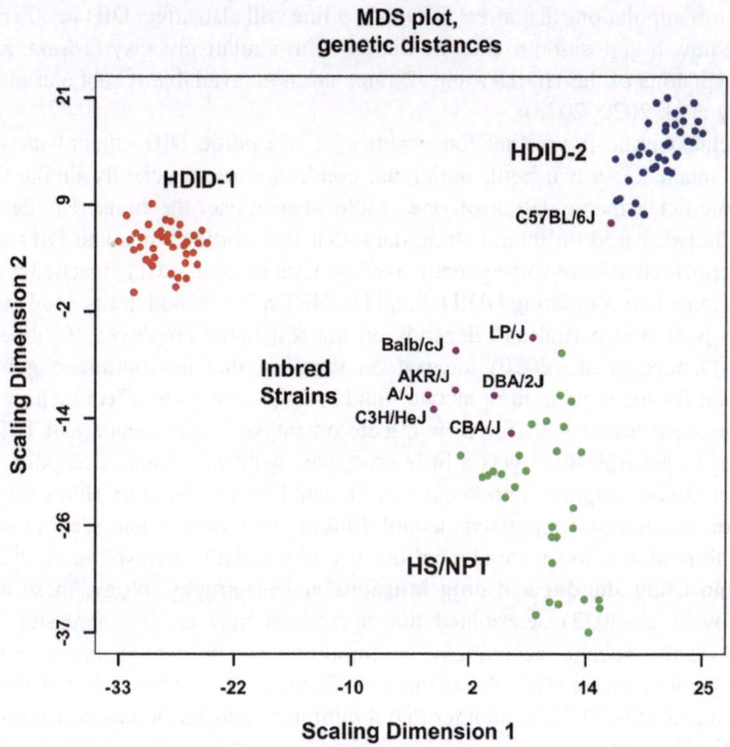

Fig. 2 Genome-wide genetic distances between the HDID-1 and -2 selected lines, HS/NPT animals and the inbred strains used to form the HS/NPT (from Iancu et al. 2013). Data are presented as a multidimensional scaling (MDS) plot. Note the difference between the HDID-1 and HDID-2 animals across scaling dimension 1

the Mouse Universal Genotyping Array (MUGA) to genotype HDID-1, HDID-2, and HS/NPT mice ($N = 48$/group balanced for sex). The MUGA contained 7,851 SNP markers, with an average spacing of 325 ± 191 kb. After the elimination of non-polymorphic or low frequency (below 2.5%) SNPs, the data contained 3,683 markers that were further analyzed using a marker-by-marker approach. The genotype data extracted (Fig. 2) illustrate two important points. One, compared to the HS/NPT founders, the genetic variance was strikingly reduced in both of the selected lines, presumably the result of the inbreeding that occurs when using a relatively small number of families for selection. Two, the genotype data illustrated that the selected lines were genetically distinct. The subsequent QTL analysis confirmed this point. The adjusted LOD threshold was set high (10.6) to correct for genetic drift. Five unique QTLs exceeding the threshold of 10.6 were found in the HDID-1 line and three unique QTLs were found in the HDID-2 line. In addition, there were three common QTLs on chromosomes 4, 14, and 16, each of which were mapped to relatively good (<5 Mbp) resolution. These and related data beg the question as to how many unique selections are possible? However, the more relevant question is whether manipulations that affect DID in one line will also affect DID in other lines? We are now in a position to address this issue in a relatively easy fashion, as fully inbred versions of the HDID-1 and -2 mice are now available (Crabbe et al. 2019; Grigsby et al. 2022, 2023).

A related issue is whether the genetics of one-bottle DID ethanol intake and ethanol intake under two-bottle preference conditions are sufficiently similar that we could predict a manipulation of one would also impact the other. Rhodes et al. (2007) noted, based on inbred strain data, that two-bottle choice and DID ethanol intake appeared to have some genetic overlap. Crabbe et al. (2011) reached a similar conclusion when comparing HDID-1 and HS/NPT mice for both traits. However, the overlap is at best partial and depends on the test being employed (Crabbe et al. 2011). Ozburn et al. (2020) focused on the idea that neuroimmune genes are important for the regulation of alcohol intake; drugs known to affect neuroimmune function were tested both for their effects on intake in preference and DID procedures. In general, there was a little crossover with the notable exception of the PDE4 inhibitor, rolipram. Importantly, in a recent human laboratory study, the PDE4 inhibitor, apremilast, suppressed alcohol drinking in non-treatment-seeking individuals with moderate to severe alcohol use disorder (AUD) (Grigsby et al. 2023). In mice and using site-directed drug infusions and electrophysiology, these authors (Blednov et al. 2022) determined that apremilast may act by increasing neural activity in the nucleus accumbens, an important alcohol-related brain region, to reduce alcohol intake. Of related interest, Franklin et al. (2015) found that both rolipram and RO--20-1724, another PDE4 inhibitor, reduced alcohol consumption in P and HAD-1 rats.

5 Animal Genetic Models: Transcriptional Data

Since Crabbe et al. (2013) the transcriptional tool kit that can be used to assess gene expression in animal genetic models has significantly changed. RNA-Seq, with its numerous advantages (see, e.g., Hitzemann et al. 2013, 2014) has largely replaced microarrays for the assessment of gene expression. This shift has been particularly important for gene co-expression analysis, as RNA-Seq data produce higher-quality network modules (Iancu et al. 2012). Single cell, single nuclei, and assay for transposase-accessible chromatin (ATAC) sequencing are now routinely used to detect the neuronal populations affected by the manipulation of interest. For example, Dilly et al. (2022) found that alcohol withdrawal effects occur in a specific population of central nucleus of the amygdala (CeA) GABAergic neurons, namely those enriched in protein kinase C delta, in Wistar rats. To our knowledge, the single cell approach has not been used in the analysis of alcohol genetic models, although we are aware that such studies are in progress.

Iancu et al. (2013) examined striatal gene expression in HDID-1. -2 and HS/NPT animals. There were marked differences between the HDID-1 vs. HS/NPT and HDID-2 vs. HS/NPT in terms of the number of differentially expressed (DE;

FDR < 0.1) transcripts (1,430 vs. 301). One hundred and four transcripts were DE in both comparisons; 94 of these had the same directionality. A majority of the DE transcripts (85 out of 94) were found among the gray-network module, which is reserved for poorly connected transcripts. Gene Ontology (GO) annotation of the DE genes revealed significant enrichments in extracellular region part and the extracellular matrix (ECM). The weighted gene co-expression network analysis (WGCNA; Langfelder and Horvath 2008) was implemented to examine the effects of selection on co-expression network structure. The nuances of this analysis are discussed in Iancu et al. (2013) and Hitzemann et al. (2021). Despite the genetic differences noted above, two of the co-expression modules (black and magenta; color has no meaning) were significantly disrupted (see Fig. 3). Module disruption may be either a significant increase or decrease in module connectivity. Both modules were highly enriched in neuronal genes (black module – $p < 3 \times 10^{-27}$; magenta module – $p < 3 \times 10^{-5}$). GO annotation of the black module revealed significant enrichments in the neurological system process, glutamate secretion, and neurotransmitter transport. GO annotation of the magenta module revealed significant enrichments in neuropeptide hormone activity, peptide receptor activity, and postsynaptic membrane. The progressive effects of selection on *Dgkz*, a gene found in the black module and known to be associated with glutamate neurotoxicity and brain trauma, are illustrated in Fig. 4. Gene module connectivity was increased in the HDID-2 animals and further increased in the HDID-1.

We bring two points to the readers' attention. The first is that both selections affected a subgroup of GABA- and glutamate-related genes; this is a familiar observation. The second point is the observation that selection affected the neuropeptide Y system. Manipulation of the neuropeptide Y system affects both DID and ethanol preference consumption (see Barkley-Levenson et al. 2016) and references therein). There is some evidence, at least for ethanol preference, that these effects may be genotype dependent (Thiele et al. 2000).

Ferguson et al. (2019) compared gene expression in the HDID-1 and HS/NPT founders across 7 brain regions. The data obtained from the extended amygdala were integrated with similar data obtained from human post-mortem analyses. The data overlap is illustrated in Fig. 5. At the GO level, key areas of overlap included glutamatergic signaling, dopaminergic signaling, CREB signaling, NFkB signaling, and axonal guidance signaling. Ferguson et al. (2018) used these gene expression data to query the LINCS-L1000 (Library of Integrated Network-Based Cellular Signatures) database to predict candidate drugs with the greatest potential to decrease alcohol consumption. The top 2 candidates were validated in vivo as a proof-of-concept. Terreic acid (a Bruton's tyrosine kinase inhibitor) and pergolide (a dopamine and serotonin receptor agonist) robustly reduced alcohol intake and BECs in HDID-1 mice.

Iancu et al. (2018) asked whether the HDID-2 selection affected males and females differently. Transcriptional data were collected from the ventral striatum. In addition to DE, the data analysis focused on the differential variance (DV) and differential wiring (DW) metrics introduced by Colville et al. (2017). The DV statistic has proven to be a computationally simple way to identify those genes

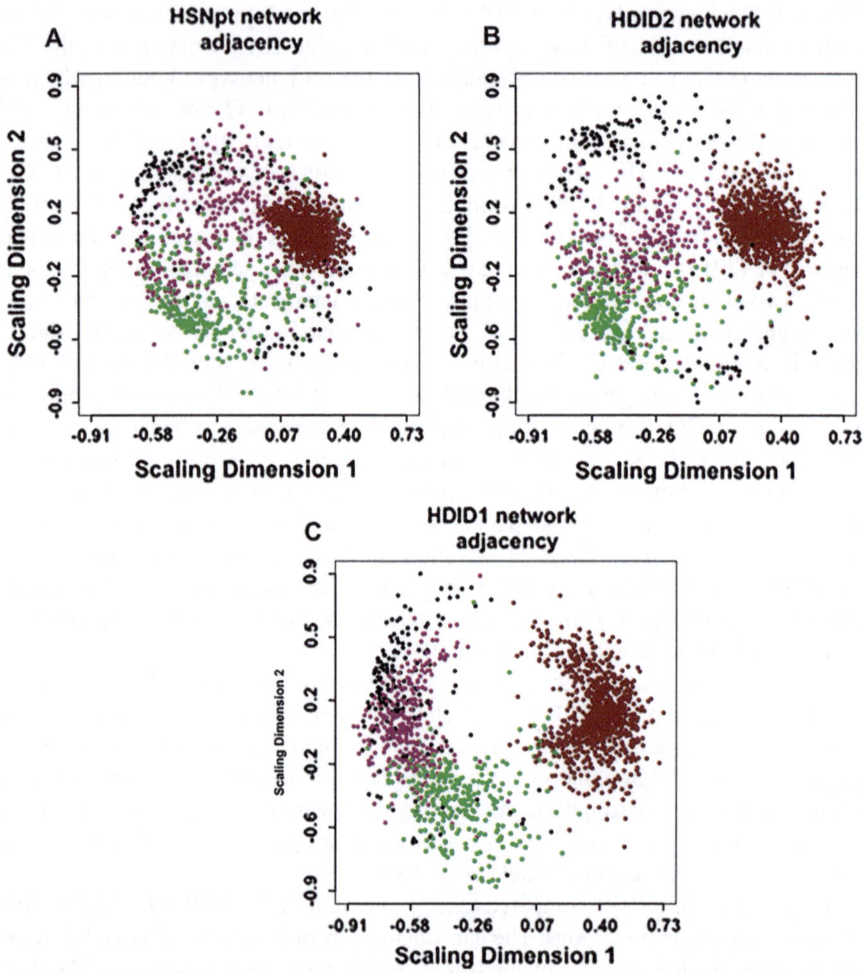

Fig. 3 Multidimensional scaling plots of the co-expression networks in HS/NPT (**a**), HDID-2 (**b**) and HDID-1 (**c**) subjects (from Iancu et al. 2013). For visual clarity, only the 4 modules most consistently affected by DID selection ("black," "magenta," "dark-red," and "green") are depicted. Each dot represents a transcript, with colors corresponding to module assignments. In particular, note how selection increased the dispersion of the dark-red module and increased compaction of the magenta module

that will show a difference in network co-expression analyses, e.g., the WGCNA. The overlap of the male vs female DE, DV, and DW genes was negligible (see Fig. 6c in Iancu et al. 2018). However, at the level of GO, the overlap was significant ($p < 10^{-3}$). The overlap is illustrated in Fig. 6 in the form of a Word cloud representation of the common GO categories. Note again that some of the most prominent GO categories are those associated with the ECM. Alcohol and other addictive drugs can have marked effects on ECM constituents (reviewed in

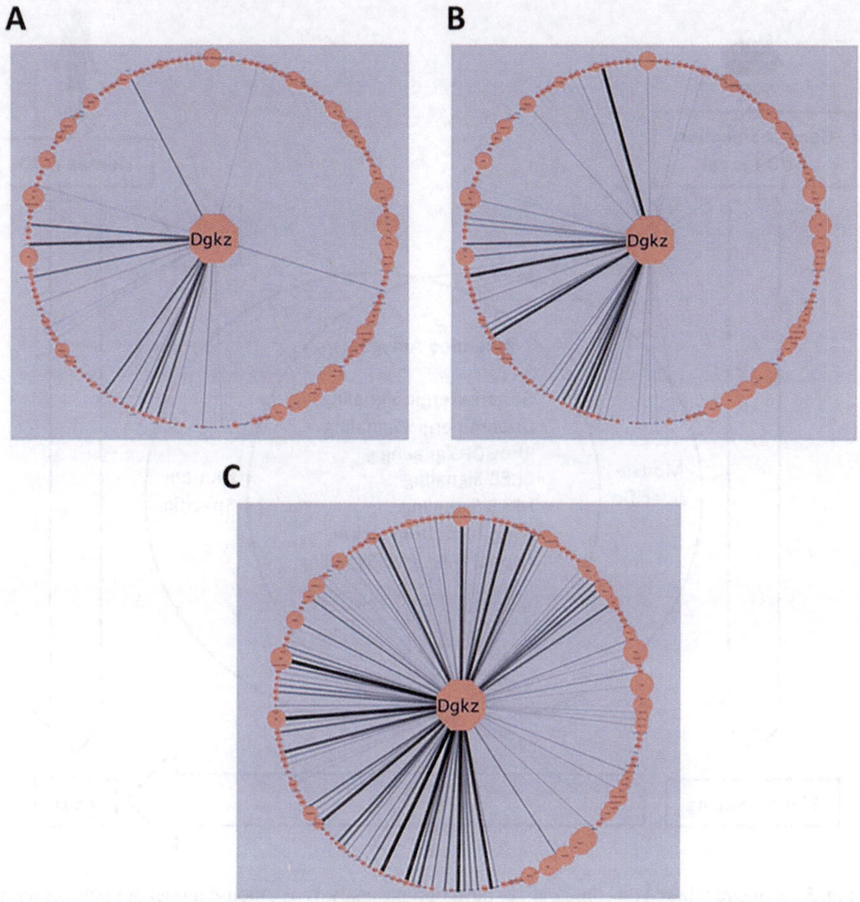

Fig. 4 The effects of selection on intra-modular connectivity for *Dgkz* which is found is found in the "black" module (see Fig. 3) (from Iancu et al. 2013). Edge thickness and opacity are proportional to network adjacency between *Dgkz* and other module transcripts. The intra-modular connectivity of the other module genes is reflected in the node size. A: HS/NPT network connectivity. B: HDID-2 network connectivity. C: HDID-1 network connectivity. Note the more pronounced increase in connectivity in the HDID-1 as compared to the HDID-2 animals

Hitzemann et al. 2020; Lasek 2016; Lubbers et al. 2014). Ethanol has been found to affect the brain expression of tPA (Bahi and Dreyer 2012; Pawlak et al. 2005), of Mmp-9 (Wright et al. 2003), Bcan & Ncan (Coleman et al. 2014), and Tsp2 & Tsp4 (Risher et al. 2015). Some data indicate that all elements of the brain ECM – the basement membrane, the interstitial ECM, and the perineuronal nets – are affected by acute and/or chronic ethanol treatment (Lasek 2016). The evidence that changes in the brain ECM are associated with risk of developing an AUD is less compelling. However, polymorphisms have been detected in MMP-9M, TNC, and TNR in individuals with AUD (Samochowiec et al. 2010; Zuo et al. 2012). GWAS studies

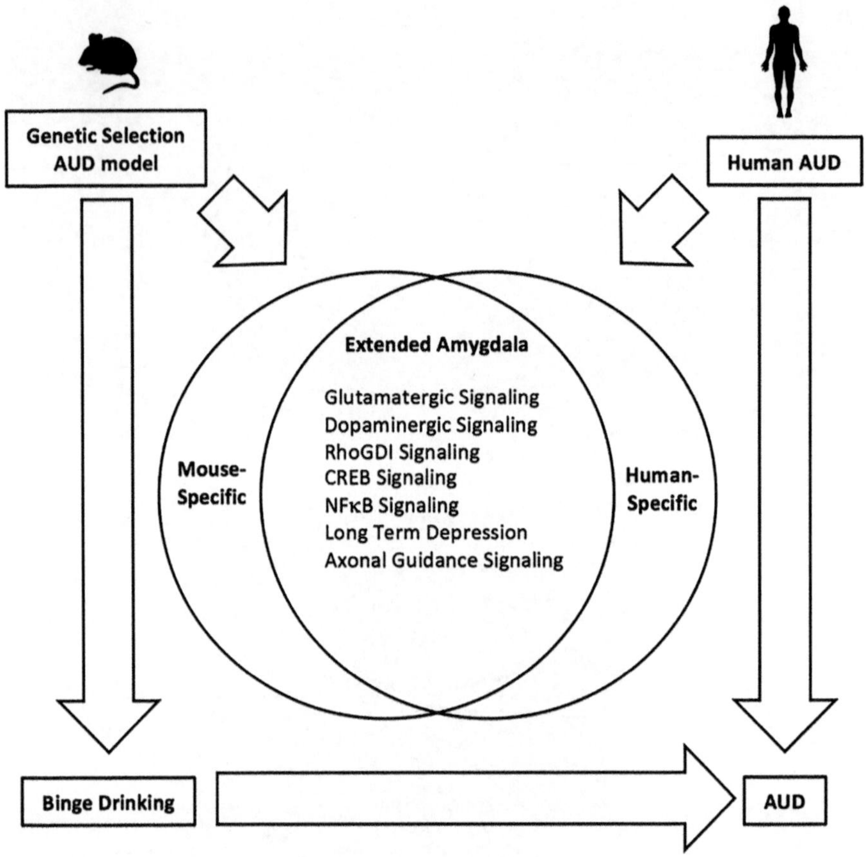

Fig. 5 A diagram based on comparative transcriptome analysis of a mouse model of binge drinking and people diagnosed with AUD (from Ferguson et al. 2019). The analyses revealed gene expression patterns that are conserved^ between alcoholics and HDID-1 mice, suggesting molecular functions driving AUD risk

have revealed a polymorphism in COL6A3 associated with AUD (Adkins et al. 2017); a gene that encodes a collagen that is an essential part of the ECM (Lamandé and Bateman 2018). Our data illustrate that HDID risk is associated with ECM-associated genes in both males and females. We found that the FDA-approved drug doxycycline (a broad spectrum MMPI with anti-inflammatory effects), as well as other tetracycline derivatives (minocycline and tigecycline) reduce binge-like drinking and BECs in HDID mice (Crabbe et al. 2020). Prior experiments from other groups have shown that tetracycline derivatives (doxycycline, minocycline, and tigecycline), reduce binge drinking in inbred C57BL/6J (B6J) mice (see references in Crabbe et al. 2020).

A common observation in both basic science and clinical populations is that substantial individual variation is retained even in groups at high risk for excessive

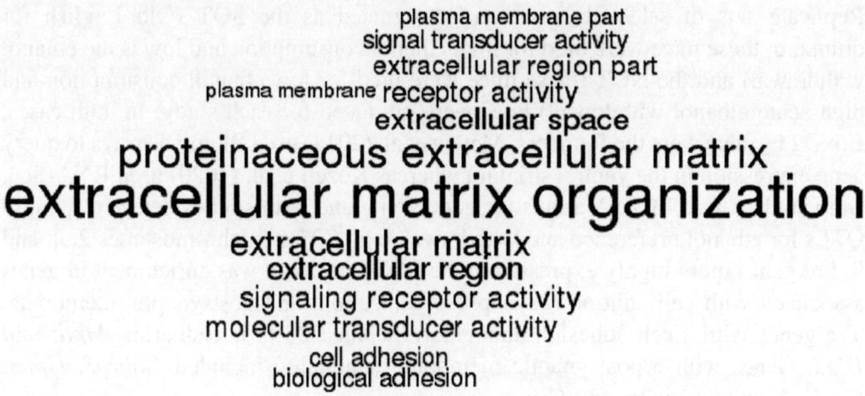

Fig. 6 Word cloud representation of the common GO categories between HDID-2 males and females (from Iancu et al. 2018). Especially note the strong overlap of ontologies associated with the ECM

ethanol consumption. Interestingly, this individual variation is seen even within inbred mouse strains such as the B6 (see Mulligan et al. 2011). We asked whether the genes associated with individual variation in HDID-1 mice are different from those associated with selection (risk) (Hitzemann et al. 2017). Thirty-five HDID-1 mice (18 males and 17 females) phenotyped for their BECs at the end of a standard 4-day DID trial, were euthanized 3 weeks later. RNA-Seq was used to analyze the striatal transcriptome. Pearson correlations were used to assess the relationships between gene expression and BEC. Five hundred and fifty-seven genes (375 positive vs. 182 negative) met the criteria for inclusion in the gene set enrichment analysis. The most significant (FDR < 0.01) annotation enrichments were for the positively correlated genes (see Table 2 in Hitzemann et al. 2017). Broadly, the enriched gene categories were associated with the regulation of synaptic function. Genes associated with the category included *Grik5, Syn1, Stxbp1, Stx1a, Rims4, Rims1, and Stx1b Camk2g, Chrm3, Crhbp, Gria3, Grin1, Strn4, Syngap1 and Syt2*. Notably, individual variation was not associated with ECM-related genes.

From 2013 forward and paralleling the HDID studies noted above, there have been several new studies examining the transcriptional features associated with genetic preference in animal models. Hoffman et al. (2014) examined gene expression in HAP3 and LAP3 animals derived from HS/Ibg mice (Oberlin et al. 2011). These authors focused on the DE genes between the two lines (see their Table 2) and in particular on *Gnb1* expression; the authors provided confirming evidence that *Gnb1* has a key role in the regulation of ethanol preference.

Metten et al. (2014) and Kozell et al. (2020) are two studies that from the animal genetic model perspective are essentially identical. Both studies used a dual selection protocol for high preference/low withdrawal and vice versa. These two traits have been found to be negatively genetically related; thus, high preference risk is associated with low withdrawal susceptibility (Hitzemann et al. 2009; Metten et al. 1998).

Replicate sets of selected lines were designated as the SOT (Old English for drunkard; these mice were bred for high ethanol consumption and low acute ethanol withdrawal) and the NOT (these mice were bred for low ethanol consumption and high acute ethanol withdrawal) lines were produced for each study. In both cases, B6xD2 F_2 mice were the founders. Metten et al. (2014) used Illumina arrays to query gene expression in the ventral striatum whereas Kozell et al. (2020) used RNA-Seq. Both studies used MUGA arrays for genotyping and both studies detected familiar QTLs for ethanol preference and withdrawal, e.g., QTLs on chromosomes 2, 4, and 9. For genes more highly expressed in the SOT line, there was enrichment in genes associated with cell adhesion, synapse organization, and postsynaptic membrane. The genes with a cell adhesion annotation included 23 protocadherins, *Mpdz and Dlg2*. Genes with a postsynaptic membrane annotation included *Gabrb3, Gphn, Grid1, Grin2b, Grin2c, and Grm3*.

Colville et al. (2018) used RNA-Seq in high and low-ethanol preference selected lines (24 h/7 d, 10% ethanol vs. water) bred from HS-CC founders to examine the transcriptional changes across three brain regions. The three brain regions were the nucleus accumbens shell, the prelimbic cortex, and the CeA. Sample sizes were moderate ($N = \sim 30$/region/line). The selection protocol was short-term, terminated after four generations of selection, to curtail inbreeding. In the "High" line, ethanol preference more than doubled to ~0.5 whereas in the "Low" line preference was <0.1. As expected (see Contet 2012), there were a large number of transcriptional changes, unique to each brain region. Here, we focus on the changes that were common to all three regions (see Fig. 3 in Colville et al. 2018). *5730455P16Rik, Gdi2, Skiv2, Tsr1, and Glod4* were the only common DE genes. There were 30 common DV genes and this grouping was significantly enriched in genes associated with cell-to-cell signaling. Genes with this GO annotation included *Dlg2, Egr3, Gabbr2, Lnpep, Pcdhgb2, Pcdhac2, Sstr4, and Syt10*. The common DV genes were enriched in a common network module that differed in size across the three regions but shared common annotations. The three modules also shared 183 common genes. These common genes included several receptors, *Adra1a, Chrna7, Grin2b, Htr2a, Oprd1, and Sstr4* and17 protocadherins including 14 of the 22 known clustered protocadherins. Common hub nodes across regions included *Dlg2, Gatad2b, Pcdhac2, Tnks, Usp29, and Usp9x*. Figure 7 illustrates the co-expression and physical interaction partners for *Dlg2*. Key partners include a number of glutamate-related genes: e.g., *Grin2b, Grid1 Dlg1, Dlg4, and Dlgap1*. These data extend the observations of Bell et al. (2016) who noted when comparing ethanol naïve P and NP rats, there were a number of differences in glutamate signaling genes. Further, clinical studies have found that in family history-positive individuals there is an altered response to the NMDA antagonist ketamine (Joslyn et al. 2010; Petrakis et al. 2004). A statistic added by Colville et al. (2018) was differential wiring (DW). DW was restricted to searching for correlations between individual genes that differed by >0.5. This general procedure has been used to quantify network rewiring in both genomic (Wright et al. 2003) and neural imaging studies (Coleman et al. 2014). Colville et al. (2018) identified for each gene, the number of changed edges (each changed edge is a statistically significant change in

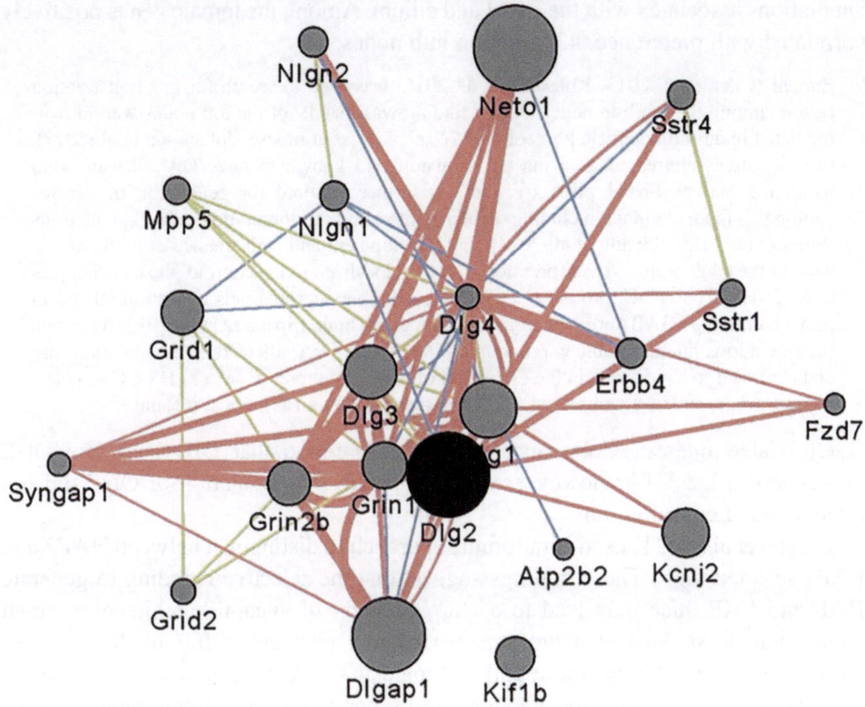

Fig. 7 Interaction partners for *Dlg2* extracted using Gene Mania (Warde-Farley et al. 2010) which was accessed as a Cytoscape plugin with default settings. Depicted are top 20 genes related to *Dlg2* through physical interactions, colocalizations, or sharing protein domains. *Dlg2* which encodes for PSD93, interacts with a number of genes and gene products associated with glutamate receptor activity including *Dlg4, Syngap1, Neto, Grin1, Grin2b, Dlgap1,* and *Dlg3*

the Pearson correlation) and then inquired as to whether some genes had a disproportionately high number of changed edges. For the latter, a binomial test was used to test for significance. There were 72 significant DW genes common to all three brain regions and this grouping included *Chrna7, Als2, Pppir9a, Strn, Kcna4, Kif1a, and Slc1a2. Slc1a2,* which encodes for the excitatory amino acid transporter 2 (EAAT2); the inhibition of EAAT2 has been reported to reduce ethanol consumption (Sari et al. 2016).

Hitzemann et al. (2020) examined in HS-CC mice the effects of chronic (13 weeks) ethanol consumption (24 h/7 d 2-bottle choice) on CeA gene expression. Here we focus on the correlation between individual gene expression and week 13 ethanol preference. For females, the enriched annotations were associated with cilium organization, extracellular region, and collagen-containing ECM. For males, there were no significant annotation enrichments. The majority (70%) of female genes correlated with preference were found in a single WGCNA network module. This module was enriched ($p < 0.0001$) in genes with an astrocyte annotation and in annotations associated with the ECM and cilium. Among the female genes positively correlated with preference, 43 were top hub nodes:

> Enrichr (Chen et al. 2013; Kuleshov et al. 2016) was used to search for key transcription factors among the top hub nodes. A key finding was that 19 of the top nodes were down-regulated in an orthodenticle homeobox 2 (*Otx2*) knockout mouse (Johansson et al. 2013). *Otx2* is often referred to as a master regulator, and known to have key roles in brain patterning and post-natal plasticity. *Otx2* is further required for generation of various neuronal subpopulations, including ocular motor and midbrain dopaminergic neurons (Sherf et al. 2015; Tripathi et al. 2014), and development and maintenance of perineuronal nets. In the adult brain, *Otx2* expression is largely localized to the choroid plexus (Planques et al. 2019). The OTX2 protein is captured by the perineuronal nets and accumulated in parvalbumin type GABAergic neurons throughout the brain (Spatazza et al. 2013). Our data indicate a low, but detectable expression of Otx2 in the CeA, affected by ethanol exposure and predicted to have a role in the escalation of ethanol preference seen in HS-CC females, but not males, and in the observed sex differences in the transcriptional response.

Of related interest, Coles and Lasek (2021) found that DID increased Otx2 expression in the VTA; however, viral-mediated downregulation of Otx2 did not affect ethanol consumption.

Grecco et al. (2021) used a multiomic approach to distinguish between HAP2 and LAP2 selected lines. These findings suggest that the selective breeding to generate HAP and LAP mice may lead to a rearrangement of synaptic architecture which could alter dorsal striatal neurotransmission and plasticity differentially between mouse lines. Importantly, these authors noted that selection affected a number of phosphorylated proteins associated with post synaptic specializations. Rich data sets, like those described here, will serve as excellent resources to inform future studies on how inherited differences in gene, protein, and phosphorylated protein expression contribute to AUD-related phenotypes.

6 Human Genome-Wide Transcriptomics and GWAS

This chapter focuses, in part, on the question of whether the animal genetic model data described above have enhanced our understanding of human data obtained from post-mortem studies and/or GWAS/TWAS investigations. Crabbe et al. (2013) described and commented on the early attempts at data integration through the microarray era. For those interested in the more recent post-mortem studies, Warden and Mayfield (2017) would be a useful starting place. These authors reviewed nine different post-mortem studies. Broad transcriptional features that differentiated control brains from brains obtained from individuals with AUD included epigenetic regulation, miRNA regulation, long non-coding RNAs, ion channels and signal transduction, immune and stress responses, and metabolism. Dissecting such results is difficult. Some of these changes will be associated with the risk of developing AUD, others will reflect the consequences of chronic consumption (including the amount consumed) and others will strongly reflect other gene-by-environment interactions (where environment can be quite variable). The overlap of ion channels and synaptic transmission between the animal genetic models and the post-mortem studies cannot be unexpected and probably would be most scientists' first good guess. Again not unexpectedly, the emphasis here is on GABA and glutamate neurotransmission.

Reports subsequent to Warden and Mayfield (2017) have broadly confirmed their observations, with many emphasizing the integration of animal models, post-mortem, and GWAS studies. Farris et al. (2018) reviewed the results of a symposium titled "Cross-species molecular dissection across alcohol behavioral domains." Included in the symposium were non-human primate (NHP) data (see Iancu et al. 2018; Walter et al. 2020, 2021). These data are important for several reasons including the fact the animals in these studies were chosen for being genetically unrelated. The data focused on the effects of chronic ethanol consumption within the context of a preference/choice design. Further, these data emphasized a biopsy/necropsy design in which small samples of tissue were removed from Area 46 (dorsolateral prefrontal cortex), prior to ethanol exposure; contralateral samples of a similar size were taken at necropsy (pre-perfusion). Walter et al. (2020) stated, "A total of 675 genes were significantly down-regulated following [ethanol] consumption [12 months]; these were functionally enriched for immune response, cell adhesion, plasma membrane, and [ECM]. A total of 567 genes that were up-regulated following [ethanol] consumption were enriched in microRNA target sites and included target sites associated with Toll-like receptor pathways." Importantly, these data align with both the post-mortem data and aspects of the animal model data. We focus on two aspects of the data. The animal model, human, and NHP data suggest that changes in ECM-related genes are associated with both the risk for and the consequences of excessive ethanol consumption. The precise pathways involved are not clear but some data suggest the involvement of the perineuronal nets (Lasek et al. 2018). Riley (reviewed in Adkins et al. 2017) detected a signal for *ColA63* associated with DSM-IV alcohol dependence; although the signal was not replicated, animal studies have confirmed that *Col6A3*, an ECM collagen, (Lamandé and Bateman 2018, p. 201) has a role in ethanol consumption.

We also emphasize the effects on neuroimmune genes. There is now substantial evidence that neuroimmune mechanisms have a role in AUD and not only in the neurotoxicity associated with AUD (Erickson et al. 2019). However immune mechanisms are notably missing or underemphasized in the animal genetic model data in the absence of ethanol exposure. This is not to imply that immune mechanisms are not associated with the risk of developing excessive ethanol consumption. Rather the data suggest that the bulk RNA analysis across the usual regions of interest, e.g., the CeA, has not detected a signal. In sub-populations of cells, perhaps within the domain of microglial cells, there may well be a neuroimmune signal. In dependency-driven animal models such as CIE, the neuroimmune features are clear (see, e.g., Roberto and colleagues). In an NHP study of repeated abstinence (Hitzemann, Grant et al. personal communication) the abstinence-associated increase in consumption, especially among heavy drinkers, has a clear neuroimmune signal. Importantly, both inflammatory and anti-inflammatory genes were affected suggesting that the balance between these genes is critical.

- Neuroimmune mechanisms associated with AUD may be sex-dependent. Ferrer et al. (2020) conducted a systematic review and meta-analysis of human alcohol dependence transcriptomics studies in which both males and females were represented. To summarize, for female AUD patients, there were associations of decreased tissue regeneration, embryo malformations, altered intracellular transport, and increased rate of RNA and protein replacement. For male AUD patients, increased inflammatory response and blood pressure, and a reduction in DNA repair capabilities were found. Overall, we believe it could be argued that one of the most important advances over the past decade has been the repeated observations for both animal genetic models and human studies that sex is an important determinant of the transcriptional and GWAS signatures associated with excessive ethanol consumption and AUD. For readers, especially interested in the sex by transcriptome interaction, we direct them to a recent special issue of *Biological Psychiatry* (volume 91, 2022).

It was noted above that ethanol preference is highly correlated with consumption, but the amount consumed per se is not a criterion for AUD. Sanchez-Roige et al. (2020) have begun to address this issue. These authors noted that AUDs are defined by multiple symptom criteria, each of which can be dissected at the genetic level. We now have a more complete understanding of the genetic factors influencing alcohol use and AUD; but there are hundreds of loci implicated in different features of use. There has been a definitive replication of associations with alcohol metabolizing enzymes (ADH1B, ALDH2), with some aspects of alcohol consumption appearing to be more strongly linked to differences in metabolism. A number of novel associations have also been found. Overall, these authors indicated that the genetic architecture of alcohol consumption and AUD are only partially overlapping.

The hundredth anniversary of RA Fisher's classic publication entitled, "The correlation between relatives on the supposition of Mendelian inheritance" (Fisher 1918) was in 2018. This important work was followed in 1919 by the publication, "The causes of human variability" (Fisher 1919). The ideas put forward in these publications are foundational to our understanding of complex trait genetics. The basic premise is that when heritable phenotypes are normally distributed, the effects of individual genes on the phenotype of interest must be small. However, it is unlikely that Fisher appreciated the number of genes involved and how small their effect size(s) even for one of his favorite phenotypes' height. For those interested, Visscher and Goddard (2019) provide an excellent overview of Fisher's work. In the last paragraph of this chapter, the authors note that the development of new tools, which we assume includes the need for very large sample sizes, has greatly facilitated our understanding of human complex trait genetics.

For those interested in contemporary GWAS/TWAS of AUD and related disorders, Zhou et al. (2020) would be a useful starting place. These authors conducted a proxy-phenotype meta-analysis of problematic alcohol usage that combined AUD and problematic drinking in over 400,000 individuals of European ancestry. Twenty-nine independent risk variants were identified, of which 19 were novel. Eleven of these 19 novel variants were located in or near the following genes: *PDE4B,*

THSD7B, CADM2, ADH1B, DPP6, SLC39A13, TMX2, ARID4A, C14orf2, TNRC6A, and *FUT2*. The remaining ten previously detected variants were located in or near the following genes: *GCKR, SIX3, KLB, ADH1B, ADH1C, SLC39A8, DRD2,* and *FTO*. The detection of *PDE4B* (phosphodiesterase 4B) and *CADM2* (cell adhesion molecule M2) are particularly relevant to the animal genetic model data discussed above. The TWAS detected 34 genes whose expression was likely to be affected by problematic alcohol usage; these genes included *ADH1B, ADH4, ADH5, C1QTNF4, GCKR,* and *DRD2*. As many if not most readers are aware, *DRD2*, the dopamine receptor, is a familiar target for association with alcohol and drug addiction.

Rao et al. (2021) noted that it is common for GWAS of complex traits, of which AUD is one, to identify non-coding region variants, but such analyses, in and of themselves, are not able to distinguish whether the associated variants have a functional role in the trait or are in linkage disequilibrium with one or more functional variants. To address this, allele-specific expression data from RNA-seq and GWAS were integrated. Data were from post-mortem sample for four brain regions from 30 control individuals and 30 individuals with AUD. Eighty-eight genes were identified with differential allele-specific expression. In vitro confirmation assays were conducted and the effects of ethanol were examined; i.e., the effects of ethanol on cis-acting elements. One of the genes identified in these analyses is *PCDHB16* (protocadherin HB16). The PCDHB16 protein is a potential calcium-dependent cell adhesion element that may be involved in the establishment and maintenance of specific neuronal connections in the brain. As noted above, the expression of clustered protocadherins has been found to be strongly linked with selection for ethanol preference in mice (Colville et al. 2017, 2018).

Kapoor et al. (2019) examined the prefrontal cortex transcriptome for 65 individuals with AUD and 73 controls, all of European descent. This study is important for several reasons. (1) The sample sizes are large for this type of study, which means the co-expression analysis (here the WGCNA) will produce highly reliable gene modules; (2) Twenty percent of the subjects were female; although not sufficient for a male/female comparison, it was an important step toward building a female post-mortem AUD database; and (3) The authors attempted to integrate the gene expression and GWAS data. The authors emphasized the significant correlations of alcohol-related traits to two of the identified gene modules (thistle 2 and brown4; the names of the modules have no meaning). Differentially expressed genes in the thistle2 module mapped to networks involved in G-protein coupled receptor signaling, calcium signaling, and opioid signaling; genes in this module were generally downregulated. Interestingly, the thistle2 module did not show an enrichment of GWAS association. Pathway analysis for differentially expressed genes in the brown4 module showed significant enrichment for growth arrest and DNA damage (GADD45) signaling and for biological processes related to the inflammatory response. In contrast to the thistle2 module, the brown4 module was enriched for GWAS association.

Dai et al. (2020) characterized multidimensional evidence of genetic, epigenetic, and transcriptomic data in AUD. Their workflow is a good example of how to link together seemingly diverse datasets. In the end, they identified 206 AUD-related genes. Figure 4 summarizes the functional enrichment of the AUD gene set. Of relevance to the current discussion, the AUD gene set was enriched in the overlapping ontologies of ECM/Collagen-related ECM and in inhibitory synaptic transmission (GABAergic).

7 Conclusions

The past decade has seen remarkable advances in our ability to align brain gene function with behavior. These advances have included the rapidly increasing efficiency and rapidly decreasing cost of RNA-Seq, the emergence of CRISPR-*Cas9*-based technologies to modify gene function and expression, the emergence of single cell, single nuclei, and ATAC sequencing to more fully understand gene expression within discrete cell populations, the recognition that very large samples are needed to fully implement GWAS/TWAS experimental designs, the necessity to investigate the causes of individual variation, the recognition that sex is an important factor in the outcomes, and significant improvements in the algorithms used to analyze genetic and genomic data. Animal genetic models, including those described in this chapter, have contributed to and benefitted from these advances. However, skepticism remains as to whether the animal genetic models will significantly contribute to our understanding of psychiatric disorders. This chapter is unlikely to change the skeptics to believers. Nonetheless, we argue that the alcohol animal genetic models have and will continue to contribute to our understanding of AUD in ways that models of other psychiatric disorders such as schizophrenia and depression will be less predictive. We conclude with the following observations:

- The genetics of preference/consumption and binge consumption (DID) appear to be substantially different. Further, the data indicate show that there are likely to be multiple genetic pathways that increase the risk for binge consumption. Human data (see, e.g., Kapoor et al. 2019) confirm that the alignment of problem alcohol use with gene expression data depends on the details of the alcohol phenotype.
- The ECM has emerged as a risk factor for excessive ethanol consumption in both animal genetic models and human GWAS. How this will inform the development of new therapies remains unclear?
- The animal model data has confirmed the clinical observation that the escalation of ethanol consumption is more rapid in females, compared to males. This may suggest that different approaches are needed for early intervention to prevent AUD in females and males.
- The integration of the animal model and human (and NHP) data suggest that neuroimmune mechanisms emerge in AUD as a consequence of excessive consumption. This of course does not mean that neuroimmune factors are not latent risk mechanisms.

- The animal genetic model data has suggested new targets for therapeutic intervention, e.g., the primary cilium.

We conclude with a quotation from Gerry McClearn (1979):

> Thus, while we aspire to comprehensive models and struggle to achieve them, use of those that are less adequate and available can contribute to the pool of knowledge from which better understanding of the human condition will be derived.

Acknowledgments National Institutes of Health National Institute on Alcohol Abuse and Alcoholism grants R01AA13484 (R.H.) and U01AA013519 (A.O.) and P60AA010760 (R.H. and A. O.), National Institutes of Health National Institute on Drug Abuse grants R01DA046081 (T.P.), the Department of Veterans Affairs grants I01BX002106 (T.P.) and IO1BX004699 (A.O.), Department of Veterans Affairs Research Career Scientist grant 15F-RCS-009 (T.P.) and a gift from the John R. Andrews family (A.O.). Note that Tamara Phillips (T.P.) is funded under the name Tamara Richards.

References

Adkins AE, Hack LM, Bigdeli TB, Williamson VS, McMichael GO, Mamdani M, Edwards AC, Aliev F, Chan RF, Bhandari P, Raabe RC, Alaimo JT, Blackwell GG, Moscati A, Poland RS, Rood B, Patterson DG, Walsh D, Collaborative Study of the Genetics of Alcoholism Consortium et al (2017) Genome wide association study of alcohol dependence identifies risk loci altering ethanol-response behaviors in model organisms. Alcohol Clin Exp Res 41(5): 911–928. https://doi.org/10.1111/acer.13362

Anderson JQ, Darakjian P, Hitzemann R, Lockwood DR, Phillips TJ, Ozburn AR (2022) Brain gene expression differences related to ethanol preference in the collaborative cross founder strains. Front Behav Neurosci 16:992727. https://doi.org/10.3389/fnbeh.2022.992727

Bagley JR, Chesler EJ, Philip VM, Center for the Systems Genetics of Addiction, Jentsch JD (2021) Heritability of ethanol consumption and pharmacokinetics in a genetically diverse panel of collaborative cross mouse strains and their inbred founders. Alcohol Clin Exp Res. https://doi.org/10.1111/acer.14582

Bahi A, Dreyer J-L (2012) Involvement of tissue plasminogen activator "tPA" in ethanol-induced locomotor sensitization and conditioned-place preference. Behav Brain Res 226(1):250–258. https://doi.org/10.1016/j.bbr.2011.09.024

Barkley-Levenson AM, Crabbe JC (2012) Ethanol drinking microstructure of a high drinking in the dark selected mouse line. Alcohol Clin Exp Res 36(8):1330–1339. https://doi.org/10.1111/j.1530-0277.2012.01749.x

Barkley-Levenson AM, Ryabinin AE, Crabbe JC (2016) Neuropeptide Y response to alcohol is altered in nucleus accumbens of mice selectively bred for drinking to intoxication. Behav Brain Res 302:160–170. https://doi.org/10.1016/j.bbr.2016.01.015

Becker HC, Lopez MF (2004) Increased ethanol drinking after repeated chronic ethanol exposure and withdrawal experience in C57BL/6 mice. Alcohol Clin Exp Res 28(12):1829–1838. 00000374-200412000-00007 [pii]

Belknap JK, Atkins AL (2001) The replicability of QTLs for murine alcohol preference drinking behavior across eight independent studies. Mamm Genome 12(12):893–899

Belknap JK, Richards SP, O'Toole LA, Helms ML, Phillips TJ (1997) Short-term selective breeding as a tool for QTL mapping: ethanol preference drinking in mice. Behav Genet 27(1): 55–66

Bell RL, Hauser SR, McClintick J, Rahman S, Edenberg HJ, Szumlinski KK, McBride WJ (2016) Ethanol-associated changes in glutamate reward neurocircuitry: a minireview of clinical and preclinical genetic findings. Prog Mol Biol Transl Sci 137:41–85. https://doi.org/10.1016/bs.pmbts.2015.10.018

Bennett B, Larson C, Richmond PA, Odell AT, Saba LM, Tabakoff B, Dowell R, Radcliffe RA (2015) Quantitative trait locus mapping of acute functional tolerance in the LXS recombinant inbred strains. Alcohol Clin Exp Res 39(4):611–620. https://doi.org/10.1111/acer.12678

Berek C, Taylor BA, Eichmann K (1976) Genetics of the idotype of BALB/c myeloma S117: multiple chromosomal loci for Vh genes encoding specificity for group A streptococcal carbohydrate. J Exp Med 144(5):1164–1174. https://doi.org/10.1084/jem.144.5.1164

Blednov YA, Da Costa A, Mason S, Mayfield J, Moss SJ, Messing RO (2022) Apremilast-induced increases in acute ethanol intoxication and decreases in ethanol drinking in mice involve PKA phosphorylation of GABAA β3 subunits. Neuropharmacology 220:109255. https://doi.org/10.1016/j.neuropharm.2022.109255

Borruto AM, Stopponi S, Li H, Weiss F, Roberto M, Ciccocioppo R (2021) Genetically selected alcohol-preferring msP rats to study alcohol use disorder: anything lost in translation? Neuropharmacology 186:108446. https://doi.org/10.1016/j.neuropharm.2020.108446

Cannady R, Nguyen T, Padula AE, Rinker JA, Lopez MF, Becker HC, Woodward JJ, Mulholland PJ (2021) Interaction of chronic intermittent ethanol and repeated stress on structural and functional plasticity in the mouse medial prefrontal cortex. Neuropharmacology 182:108396. https://doi.org/10.1016/j.neuropharm.2020.108396

Chen EY, Tan CM, Kou Y, Duan Q, Wang Z, Meirelles GV, Clark NR, Ma'ayan, A. (2013) Enrichr: interactive and collaborative HTML5 gene list enrichment analysis tool. BMC Bioinformatics 14:128. https://doi.org/10.1186/1471-2105-14-128

Churchill GA, Airey DC, Allayee H, Angel JM, Attie AD, Beatty J, Beavis WD, Belknap JK, Bennett B, Berrettini W, Bleich A, Bogue M, Broman KW, Buck KJ, Buckler E, Burmeister M, Chesler EJ, Cheverud JM, Clapcote S et al (2004) The collaborative cross, a community resource for the genetic analysis of complex traits. Nat Genet 36(11):1133–1137. https://doi.org/10.1038/ng1104-1133

Churchill GA, Gatti DM, Munger SC, Svenson KL (2012) The diversity outbred mouse population. Mamm Genome 23(9–10):713–718. https://doi.org/10.1007/s00335-012-9414-2

Ciccocioppo R, Economidou D, Cippitelli A, Cucculelli M, Ubaldi M, Soverchia L, Lourdusamy A, Massi M (2006) Genetically selected Marchigian Sardinian alcohol-preferring (msP) rats: an animal model to study the neurobiology of alcoholism. Addict Biol 11(3–4):339–355. https://doi.org/10.1111/j.1369-1600.2006.00032.x

Cicero TJ, Meyer ER, Bell RD (1979) Effects of ethanol on the hypothalamic-pituitary-luteinizing hormone axis and testicular steroidogenesis. J Pharmacol Exp Ther 208(2):210–215

Coleman LG, Liu W, Oguz I, Styner M, Crews FT (2014) Adolescent binge ethanol treatment alters adult brain regional volumes, cortical extracellular matrix protein and behavioral flexibility. Pharmacol Biochem Behav 116:142–151. https://doi.org/10.1016/j.pbb.2013.11.021

Coles C, Lasek AW (2021) Binge-like ethanol drinking increases Otx2, Wnt1, and Mdk gene expression in the ventral tegmental area of adult mice. Neurosci Insights 16: 26331055211009850. https://doi.org/10.1177/26331055211009850

Colombo G, Lobina C, Carai MAM, Gessa GL (2006) Phenotypic characterization of genetically selected Sardinian alcohol-preferring (sP) and -non-preferring (sNP) rats. Addict Biol 11(3–4): 324–338. https://doi.org/10.1111/j.1369-1600.2006.00031.x

Colville AM, Iancu OD, Oberbeck DL, Darakjian P, Zheng CL, Walter NA, Harrington CA, Searles RP, McWeeney S, Hitzemann RJ (2017) Effects of selection for ethanol preference on gene expression in the nucleus accumbens of HS-CC mice. Genes Brain Behav 16(4):462–471. https://doi.org/10.1111/gbb.12367

Colville AM, Iancu OD, Lockwood DR, Darakjian P, McWeeney SK, Searles R, Zheng C, Hitzemann R (2018) Regional differences and similarities in the brain transcriptome for mice selected for ethanol preference from HS-CC founders. Front Genet 9:300. https://doi.org/10.3389/fgene.2018.00300

Contet C (2012) Gene expression under the influence: transcriptional profiling of ethanol in the brain. Curr Psychopharmacol 1(4):301–314

Crabbe JC, Metten P, Rhodes JS, Yu C-H, Brown LL, Phillips TJ, Finn DA (2009) A line of mice selected for high blood ethanol concentrations shows drinking in the dark to intoxication. Biol Psychiatry 65(8):662–670. https://doi.org/10.1016/j.biopsych.2008.11.002

Crabbe JC, Phillips TJ, Belknap JK (2010) The complexity of alcohol drinking: studies in rodent genetic models. Behav Genet 40(6):737–750. https://doi.org/10.1007/s10519-010-9371-z

Crabbe JC, Spence SE, Brown LL, Metten P (2011) Alcohol preference drinking in a mouse line selectively bred for high drinking in the dark. Alcohol 45(5):427–440. https://doi.org/10.1016/j.alcohol.2010.12.001

Crabbe JC, Kendler KS, Hitzemann RJ (2013) Modeling the diagnostic criteria for alcohol dependence with genetic animal models. Curr Top Behav Neurosci 13:187–221. https://doi.org/10.1007/7854_2011_162

Crabbe JC, Metten P, Savarese AM, Ozburn AR, Schlumbohm JP, Spence SE, Hack WR (2019) Ethanol conditioned taste aversion in high drinking in the dark mice. Brain Sci 9(1):2. https://doi.org/10.3390/brainsci9010002

Crabbe JC, Ozburn AR, Hitzemann RJ, Spence SE, Hack WR, Schlumbohm JP, Metten P (2020) Tetracycline derivatives reduce binge alcohol consumption in high drinking in the dark mice. Brain Behav Immun Health 4:100061. https://doi.org/10.1016/j.bbih.2020.100061

Dai Y, Hu R, Pei G, Zhang H, Zhao Z, Jia P (2020) Diverse types of genomic evidence converge on alcohol use disorder risk genes. J Med Genet 57(11):733–743. https://doi.org/10.1136/jmedgenet-2019-106490

Dilly GA, Kittleman CW, Kerr TM, Messing RO, Mayfield RD (2022) Cell-type specific changes in PKC-delta neurons of the central amygdala during alcohol withdrawal. Transl Psychiatry 12(1):289. https://doi.org/10.1038/s41398-022-02063-0

Ehlers CL, Wills D, Karriker-Jaffe KJ, Gilder DA (2022) Extreme binge drinking during adolescence: associations with subsequent substance use disorders in American Indian and Mexican American young adults. J Addict Med 16(1):33–40. https://doi.org/10.1097/ADM.0000000000000815

Erickson EK, Grantham EK, Warden AS, Harris RA (2019) Neuroimmune signaling in alcohol use disorder. Pharmacol Biochem Behav 177:34–60. https://doi.org/10.1016/j.pbb.2018.12.007

Eriksson K (1968) Genetic selection for voluntary alcohol consumption in the albino rat. Science 159(3816):739–741. https://doi.org/10.1126/science.159.3816.739

Farokhnia M, Rentsch CT, Chuong V, McGinn MA, Elvig SK, Douglass EA, Gonzalez LA, Sanfilippo JE, Marchette RCN, Tunstall BJ, Fiellin DA, Koob GF, Justice AC, Leggio L, Vendruscolo LF (2022) Spironolactone as a potential new pharmacotherapy for alcohol use disorder: convergent evidence from rodent and human studies. Mol Psychiatry 27(11):4642–4652. https://doi.org/10.1038/s41380-022-01736-y

Farris SP, Riley BP, Williams RW, Mulligan MK, Miles MF, Lopez MF, Hitzemann R, Iancu OD, Colville A, Walter NAR, Darakjian P, Oberbeck DL, Daunais JB, Zheng CL, Searles RP, McWeeney SK, Grant KA, Mayfield RD (2018) Cross-species molecular dissection across alcohol behavioral domains. Alcohol 72:19–31. https://doi.org/10.1016/j.alcohol.2017.11.036

Ferrer FC, Pascual M, Hidalgo MR, Malmierca-Merlo P, Guerri C, Garcia-Garcia F (2020) Unveiling sex-based differences in the effects of alcohol abuse: a comprehensive functional meta-analysis of transcriptomic studies. Genes 11(9). https://doi.org/10.3390/genes11091106

Ferguson LB, Ozburn AR, Ponomarev I, Metten P, Reilly M, Crabbe JC, Harris RA, Mayfield RD (2018) Genome-wide expression profiles drive discovery of novel compounds that reduce binge drinking in mice. Neuropsychopharmacology 43(6):1257–1266. https://doi.org/10.1038/npp.2017.301

Ferguson LB, Zhang L, Kircher D, Wang S, Mayfield RD, Crabbe JC, Morrisett RA, Harris RA, Ponomarev I (2019) Dissecting brain networks underlying alcohol binge drinking using a systems genomics approach. Mol Neurobiol 56(4):2791–2810. https://doi.org/10.1007/s12035-018-1252-0

Fisher RA (1918) The correlation between relatives on the supposition of Mendelian inheritance. https://digital.library.adelaide.edu.au/dspace/handle/2440/15097

Fisher RA (1919) The causes of human variability. Eugen Rev 10(4):213–220

Franklin KM, Hauser SR, Lasek AW, McClintick J, Ding Z-M, McBride WJ, Bell RL (2015) Reduction of alcohol drinking of alcohol-preferring (P) and high-alcohol drinking (HAD1) rats by targeting phosphodiesterase-4 (PDE4). Psychopharmacology (Berl) 232(13):2251–2262. https://doi.org/10.1007/s00213-014-3852-3

Gora-Maslak G, McClearn GE, Crabbe JC, Phillips TJ, Belknap JK, Plomin R (1991) Use of recombinant inbred strains to identify quantitative trait loci in psychopharmacology. Psychopharmacology (Berl) 104(4):413–424

Gowin JL, Sloan ME, Morris JK, Schwandt ML, Diazgranados N, Ramchandani VA (2021) Characteristics associated with high-intensity binge drinking in alcohol use disorder. Front Psychol 12:750395. https://doi.org/10.3389/fpsyg.2021.750395

Grecco GG, Haggerty DL, Doud EH, Fritz BM, Yin F, Hoffman H, Mosley AL, Simpson E, Liu Y, Baucum AJ, Atwood BK (2021) A multi-Omic analysis of the dorsal striatum in an animal model of divergent genetic risk for alcohol use disorder. J Neurochem 157(4):1013–1031. https://doi.org/10.1111/jnc.15226

Grigsby K, Ledford C, Batish T, Kanadibhotla S, Smith D, Firsick E, Tran A, Townsley K, Reyes K-AV, LeBlanc K, Ozburn A (2022) Targeting the maladaptive effects of binge drinking on circadian gene expression. Int J Mol Sci 23(19):11084. https://doi.org/10.3390/ijms231911084

Grigsby KB, Mangieri RA, Roberts AJ, Lopez MF, Firsick EJ, Townsley KG, Beneze A, Bess J, Eisenstein TK, Meissler JJ, Light JM, Miller J, Quello S, Shadan F, Skinner MH, Aziz HC, Metten P, Morissett RA, Crabbe JC et al (2023) Pre-clinical and clinical evidence for suppression of alcohol intake by apremilast. J Clin Invest:e159103. https://doi.org/10.1172/JCI159103

Hackett JB, Glassbrook JE, Muñiz MC, Bross M, Fielder A, Dyson G, Movahhedin N, McCasland J, McCarthy-Leo C, Gibson HM (2022) A diversity outbred F1 mouse model identifies host-intrinsic genetic regulators of response to immune checkpoint inhibitors. Onco Targets Ther 11(1):2064958. https://doi.org/10.1080/2162402X.2022.2064958

Hansen C, Spuhler K (1984) Development of the National Institutes of Health genetically heterogeneous rat stock. Alcohol Clin Exp Res 8(5):477–479

Hitzemann B, Dains K, Kanes S, Hitzemann R (1994) Further studies on the relationship between dopamine cell density and haloperidol-induced catalepsy. J Pharmacol Exp Ther 271(2): 969–976

Hitzemann R, Edmunds S, Wu W, Malmanger B, Walter N, Belknap J, Darakjian P, McWeeney S (2009) Detection of reciprocal quantitative trait loci for acute ethanol withdrawal and ethanol consumption in heterogeneous stock mice. Psychopharmacology (Berl) 203(4):713–722. https://doi.org/10.1007/s00213-008-1418-y

Hitzemann R, Bottomly D, Darakjian P, Walter N, Iancu O, Searles R, Wilmot B, McWeeney S (2013) Genes, behavior and next-generation RNA sequencing. Genes Brain Behav 12(1):1–12. https://doi.org/10.1111/gbb.12007

Hitzemann R, Bottomly D, Iancu O, Buck K, Wilmot B, Mooney M, Searles R, Zheng C, Belknap J, Crabbe J, McWeeney S (2014) The genetics of gene expression in complex mouse crosses as a tool to study the molecular underpinnings of behavior traits. Mamm Genome 25(1–2):12–22. https://doi.org/10.1007/s00335-013-9495-6

Hitzemann R, Oberbeck D, Iancu O, Darakjian P, McWeeney S, Spence S, Schlumbohm J, Metten P, Crabbe J (2017) Alignment of the transcriptome with individual variation in animals selectively bred for high drinking-in-the-dark (HDID). Alcohol 60:115–120. https://doi.org/10.1016/j.alcohol.2017.02.176

Hitzemann R, Phillips TJ, Lockwood DR, Darakjian P, Searles RP (2020) Phenotypic and gene expression features associated with variation in chronic ethanol consumption in heterogeneous stock collaborative cross mice. Genomics. https://doi.org/10.1016/j.ygeno.2020.08.004

Hitzemann R, Lockwood DR, Ozburn AR, Phillips TJ (2021) On the use of heterogeneous stock mice to map transcriptomes associated with excessive ethanol consumption. Front Psych 12: 725819. https://doi.org/10.3389/fpsyt.2021.725819

Hoffman PL, Saba LM, Flink S, Grahame NJ, Kechris K, Tabakoff B (2014) Genetics of gene expression characterizes response to selective breeding for alcohol preference. Genes Brain Behav 13(8):743–757. https://doi.org/10.1111/gbb.12175

Iancu OD, Darakjian P, Walter NA, Malmanger B, Oberbeck D, Belknap J, McWeeney S, Hitzemann R (2010) Genetic diversity and striatal gene networks: focus on the heterogeneous stock-collaborative cross (HS-CC) mouse. BMC Genomics 11:585. https://doi.org/10.1186/1471-2164-11-585

Iancu OD, Kawane S, Bottomly D, Searles R, Hitzemann R, McWeeney S (2012) Utilizing RNA-Seq data for de novo coexpression network inference. Bioinformatics 28(12): 1592–1597. https://doi.org/10.1093/bioinformatics/bts245

Iancu OD, Oberbeck D, Darakjian P, Metten P, McWeeney S, Crabbe JC, Hitzemann R (2013) Selection for drinking in the dark alters brain gene coexpression networks. Alcohol Clin Exp Res 37(8):1295–1303. https://doi.org/10.1111/acer.12100

Iancu OD, Colville AM, Wilmot B, Searles R, Darakjian P, Zheng C, McWeeney S, Kawane S, Crabbe JC, Metten P, Oberbeck D, Hitzemann R (2018) Gender-specific effects of selection for drinking in the dark on the network roles of coding and noncoding RNAs. Alcohol Clin Exp Res. https://doi.org/10.1111/acer.13777

Johansson PA, Irmler M, Acampora D, Beckers J, Simeone A, Götz M (2013) The transcription factor Otx2 regulates choroid plexus development and function. Development 140(5): 1055–1066. https://doi.org/10.1242/dev.090860

Joslyn G, Ravindranathan A, Brush G, Schuckit M, White RL (2010) Human variation in alcohol response is influenced by variation in neuronal signaling genes. Alcohol Clin Exp Res 34(5): 800–812. ACER1152 [pii]. https://doi.org/10.1111/j.1530-0277.2010.01152.x

Kapoor M, Wang J-C, Farris SP, Liu Y, McClintick J, Gupta I, Meyers JL, Bertelsen S, Chao M, Nurnberger J, Tischfield J, Harari O, Zeran L, Hesselbrock V, Bauer L, Raj T, Porjesz B, Agrawal A, Foroud T et al (2019) Analysis of whole genome-transcriptomic organization in brain to identify genes associated with alcoholism. Transl Psychiatry 9(1):89. https://doi.org/10.1038/s41398-019-0384-y

Kippin TE (2014) Adaptations underlying the development of excessive alcohol intake in selectively bred mice. Alcohol Clin Exp Res 38(1):36–39. https://doi.org/10.1111/acer.12327

Koob GF, Volkow ND (2016) Neurobiology of addiction: a neurocircuitry analysis. Lancet Psychiatry 3(8):760–773. https://doi.org/10.1016/S2215-0366(16)00104-8

Kozell LB, Denmark DL, Walter NAR, Buck KJ (2018) Distinct roles for two chromosome 1 loci in ethanol withdrawal, consumption, and conditioned place preference. Front Genet 9:323. https://doi.org/10.3389/fgene.2018.00323

Kozell LB, Lockwood D, Darakjian P, Edmunds S, Shepherdson K, Buck KJ, Hitzemann R (2020) RNA-Seq analysis of genetic and transcriptome network effects of dual-trait selection for ethanol preference and withdrawal using SOT and NOT genetic models. Alcohol Clin Exp Res 44(4):820–830. https://doi.org/10.1111/acer.14312

Kruse LC, Walter NA, Buck KJ (2014) Mpdz expression in the caudolateral substantia nigra pars reticulata is crucially involved in alcohol withdrawal. Genes Brain Behav. https://doi.org/10.1111/gbb.12171

Kuleshov MV, Jones MR, Rouillard AD, Fernandez NF, Duan Q, Wang Z, Koplev S, Jenkins SL, Jagodnik KM, Lachmann A, McDermott MG, Monteiro CD, Gundersen GW, Ma'ayan, A. (2016) Enrichr: a comprehensive gene set enrichment analysis web server 2016 update. Nucleic Acids Res 44(W1):W90–W97. https://doi.org/10.1093/nar/gkw377

Lamandé SR, Bateman JF (2018) Collagen VI disorders: insights on form and function in the extracellular matrix and beyond. Matrix Biol 71–72:348–367. https://doi.org/10.1016/j.matbio.2017.12.008

Langfelder P, Horvath S (2008) WGCNA: an R package for weighted correlation network analysis. BMC Bioinformatics 9:559. https://doi.org/10.1186/1471-2105-9-559

Lasek AW (2016) Effects of ethanol on brain extracellular matrix: implications for alcohol use disorder. Alcohol Clin Exp Res 40(10):2030–2042. https://doi.org/10.1111/acer.13200

Lasek AW, Chen H, Chen W-Y (2018) Releasing addiction memories trapped in Perineuronal nets. Trends Genet 34(3):197–208. https://doi.org/10.1016/j.tig.2017.12.004

Lester D, Freed EX (1973) Criteria for an animal model of alcoholism. Pharmacol Biochem Behav 1(1):103–107. https://doi.org/10.1016/0091-3057(73)90062-2

Li TK, Lumeng L, McBride WJ, Waller MB (1979) Progress toward a voluntary oral consumption model of alcoholism. Drug Alcohol Depend 4(1–2):45–60. https://doi.org/10.1016/0376-8716(79)90040-1

Lubbers BR, Smit AB, Spijker S, van den Oever MC (2014) Neural ECM in addiction, schizophrenia, and mood disorder. Prog Brain Res 214:263–284. https://doi.org/10.1016/B978-0-444-63486-3.00012-8

Lusk R, Saba LM, Vanderlinden LA, Zidek V, Silhavy J, Pravenec M, Hoffman PL, Tabakoff B (2018) Unsupervised, statistically based systems biology approach for unraveling the genetics of complex traits: a demonstration with ethanol metabolism. Alcohol Clin Exp Res 42(7): 1177–1191. https://doi.org/10.1111/acer.13763

Mardones J, Segovia-Riquelme N (1984) Effects of different diets on voluntary consumption of ethanol in UChA and UChB rats. Acta Physiol Pharmacol Latinoam 34(1):25–30

Mardones J, Segovia N, Hederra A (1953) Heredity of experimental alcohol preference in rats. II. Coefficient of heredity. Q J Stud Alcohol 14(1):1–2

Matson LM, Grahame NJ (2013) Pharmacologically relevant intake during chronic, free-choice drinking rhythms in selectively bred high alcohol-preferring mice. Addict Biol 18(6):921–929. https://doi.org/10.1111/j.1369-1600.2011.00412.x

Matson L, Liangpunsakul S, Crabb D, Buckingham A, Ross RA, Halcomb M, Grahame N (2013) Chronic free-choice drinking in crossed high alcohol preferring mice leads to sustained blood ethanol levels and metabolic tolerance without evidence of liver damage. Alcohol Clin Exp Res 37(2):194–201. https://doi.org/10.1111/j.1530-0277.2012.01873.x

Matson LM, Kasten CR, Boehm SL, Grahame NJ (2014) Selectively bred crossed high-alcohol-preferring mice drink to intoxication and develop functional tolerance, but not locomotor sensitization during free-choice ethanol access. Alcohol Clin Exp Res 38(1):267–274. https://doi.org/10.1111/acer.12216

McBride WJ, Rodd ZA, Bell RL, Lumeng L, Li T-K (2014) The alcohol-preferring (P) and high-alcohol-drinking (HAD) rats – animal models of alcoholism. Alcohol 48(3):209–215. https://doi.org/10.1016/j.alcohol.2013.09.044

McClearn GE (1979) Genetics and alcoholism simulacra. Alcohol Clin Exp Res 3(3):255–258. https://doi.org/10.1111/j.1530-0277.1979.tb05310.x

McClearn GE, Rodgers DA (1959) Differences in alcohol preference among inbred strains of mice. Q J Stud Alcohol 20(4):691–695. https://doi.org/10.15288/qjsa.1959.20.691

Metten P, Phillips TJ, Crabbe JC, Tarantino LM, McClearn GE, Plomin R, Erwin VG, Belknap JK (1998) High genetic susceptibility to ethanol withdrawal predicts low ethanol consumption. Mamm Genome 9(12):983–990. https://doi.org/10.1007/s003359900911

Metten P, Iancu OD, Spence SE, Walter NAR, Oberbeck D, Harrington CA, Colville A, McWeeney S, Phillips TJ, Buck KJ, Crabbe JC, Belknap JK, Hitzemann RJ (2014) Dual-trait selection for ethanol consumption and withdrawal: genetic and transcriptional network effects. Alcohol Clin Exp Res 38(12):2915–2924. https://doi.org/10.1111/acer.12574

Milner LC, Shirley RL, Kozell LB, Walter NA, Kruse LC, Komiyama NH, Grant SGN, Buck KJ (2015) Novel MPDZ/MUPP1 transgenic and knockdown models confirm Mpdz's role in ethanol withdrawal and support its role in voluntary ethanol consumption. Addict Biol 20(1): 143–147. https://doi.org/10.1111/adb.12087

Mulligan MK, Rhodes JS, Crabbe JC, Mayfield RD, Harris RA, Ponomarev I (2011) Molecular profiles of drinking alcohol to intoxication in C57BL/6J mice. Alcohol Clin Exp Res 35(4): 659–670. https://doi.org/10.1111/j.1530-0277.2010.01384.x

Murphy JM, Stewart RB, Bell RL, Badia-Elder NE, Carr LG, McBride WJ, Lumeng L, Li T-K (2002) Phenotypic and genotypic characterization of the Indiana University rat lines selectively bred for high and low alcohol preference. Behav Genet 32(5):363–388. https://doi.org/10.1023/a:1020266306135

Nieto SJ, Grodin EN, Aguirre CG, Izquierdo A, Ray LA (2021) Translational opportunities in animal and human models to study alcohol use disorder. Transl Psychiatry 11(1):496. https://doi.org/10.1038/s41398-021-01615-0

Oberlin B, Best C, Matson L, Henderson A, Grahame N (2011) Derivation and characterization of replicate high- and low-alcohol preferring lines of mice and a high-drinking crossed HAP line. Behav Genet 41(2):288–302. https://doi.org/10.1007/s10519-010-9394-5

Ozburn AR, Metten P, Potretzke S, Townsley KG, Blednov YA, Crabbe JC (2020) Effects of pharmacologically targeting neuroimmune pathways on alcohol drinking in mice selectively bred to drink to intoxication. Alcohol Clin Exp Res 44(2):553–566. https://doi.org/10.1111/acer.14269

Parker CC, Carbonetto P, Sokoloff G, Park YJ, Abney M, Palmer AA (2014) High-resolution genetic mapping of complex traits from a combined analysis of F2 and advanced intercross mice. Genetics 198(1):103–116. https://doi.org/10.1534/genetics.114.167056

Patrick ME, Pang YC, Jang BJ, Arterberry BJ, Terry-McElrath YM (2023) Alcohol use disorder symptoms reported during midlife: results from the monitoring the future study among US adults at modal ages 50, 55, and 60. Subst Use Misuse:1–9. https://doi.org/10.1080/10826084.2022.2161826

Pawlak R, Melchor JP, Matys T, Skrzypiec AE, Strickland S (2005) Ethanol-withdrawal seizures are controlled by tissue plasminogen activator via modulation of NR2B-containing NMDA receptors. Proc Natl Acad Sci U S A 102(2):443–448. https://doi.org/10.1073/pnas.0406454102

Petrakis IL, Limoncelli D, Gueorguieva R, Jatlow P, Boutros NN, Trevisan L, Gelernter J, Krystal JH (2004) Altered NMDA glutamate receptor antagonist response in individuals with a family vulnerability to alcoholism. Am J Psychiatry 161(10):1776–1782. https://doi.org/10.1176/ajp.161.10.1776

Philip VM, Duvvuru S, Gomero B, Ansah TA, Blaha CD, Cook MN, Hamre KM, Lariviere WR, Matthews DB, Mittleman G, Goldowitz D, Chesler EJ (2010) High-throughput behavioral phenotyping in the expanded panel of BXD recombinant inbred strains. Genes Brain Behav 9(2):129–159. https://doi.org/10.1111/j.1601-183X.2009.00540.x

Phillips TJ, Crabbe JC, Metten P, Belknap JK (1994) Localization of genes affecting alcohol drinking in mice. Alcohol Clin Exp Res 18(4):931–941. https://doi.org/10.1111/j.1530-0277.1994.tb00062.x

Planques A, Oliveira Moreira V, Dubreuil C, Prochiantz A, Di Nardo AA (2019) OTX2 signals from the choroid plexus to regulate adult neurogenesis. ENeuro 6(2). https://doi.org/10.1523/ENEURO.0262-18.2019

Quintanilla ME, Israel Y, Sapag A, Tampier L (2006) The UChA and UChB rat lines: metabolic and genetic differences influencing ethanol intake. Addict Biol 11(3–4):310–323. https://doi.org/10.1111/j.1369-1600.2006.00030.x

Radcliffe RA, Dowell R, Odell AT, Richmond PA, Bennett B, Larson C, Kechris K, Saba LM, Rudra P, Wen S (2020) Systems genetics analysis of the LXS recombinant inbred mouse strains: Genetic and molecular insights into acute ethanol tolerance. PloS One 15(10):e0240253. https://doi.org/10.1371/journal.pone.0240253

Rao X, Thapa KS, Chen AB, Lin H, Gao H, Reiter JL, Hargreaves KA, Ipe J, Lai D, Xuei X, Wang Y, Gu H, Kapoor M, Farris SP, Tischfield J, Foroud T, Goate AM, Skaar TC, Mayfield RD et al (2021) Allele-specific expression and high-throughput reporter assay reveal functional genetic variants associated with alcohol use disorders. Mol Psychiatry 26(4):1142–1151. https://doi.org/10.1038/s41380-019-0508-z

Rhodes JS, Best K, Belknap JK, Finn DA, Crabbe JC (2005) Evaluation of a simple model of ethanol drinking to intoxication in C57BL/6J mice. Physiol Behav 84(1):53–63. https://doi.org/10.1016/j.physbeh.2004.10.007

Rhodes JS, Ford MM, Yu C-H, Brown LL, Finn DA, Garland T, Crabbe JC (2007) Mouse inbred strain differences in ethanol drinking to intoxication. Genes Brain Behav 6(1):1–18. https://doi.org/10.1111/j.1601-183X.2006.00210.x

Risher M-L, Sexton HG, Risher WC, Wilson WA, Fleming RL, Madison RD, Moore SD, Eroglu C, Swartzwelder HS (2015) Adolescent intermittent alcohol exposure: dysregulation of thrombospondins and synapse formation are associated with decreased neuronal density in the adult hippocampus. Alcohol Clin Exp Res 39(12):2403–2413. https://doi.org/10.1111/acer.12913

Roberts A, Pardo-Manuel de Villena F, Wang W, McMillan L, Threadgill DW (2007) The polymorphism architecture of mouse genetic resources elucidated using genome-wide resequencing data: implications for QTL discovery and systems genetics. Mamm Genome 18(6–7):473–481. https://doi.org/10.1007/s00335-007-9045-1

Saba LM, Flink SC, Vanderlinden LA, Israel Y, Tampier L, Colombo G, Kiianmaa K, Bell RL, Printz MP, Flodman P, Koob G, Richardson HN, Lombardo J, Hoffman PL, Tabakoff B (2015) The sequenced rat brain transcriptome – its use in identifying networks predisposing alcohol consumption. FEBS J 282(18):3556–3578. https://doi.org/10.1111/febs.13358

Sabino V, Kwak J, Rice KC, Cottone P (2013) Pharmacological characterization of the 20% alcohol intermittent access model in Sardinian alcohol-preferring rats: A model of binge-like drinking. Alcohol Clin Exp Res 37(4):635–643. https://doi.org/10.1111/acer.12008

Samochowiec A, Grzywacz A, Kaczmarek L, Bienkowski P, Samochowiec J, Mierzejewski P, Preuss UW, Grochans E, Ciechanowicz A (2010) Functional polymorphism of matrix metalloproteinase-9 (MMP-9) gene in alcohol dependence: family and case control study. Brain Res 1327:103–106. https://doi.org/10.1016/j.brainres.2010.02.072

Sanchez-Roige S, Palmer AA, Clarke T-K (2020) Recent efforts to dissect the genetic basis of alcohol use and abuse. Biol Psychiatry 87(7):609–618. https://doi.org/10.1016/j.biopsych.2019.09.011

Sari Y, Toalston JE, Rao PSS, Bell RL (2016) Effects of ceftriaxone on ethanol, nicotine or sucrose intake by alcohol-preferring (P) rats and its association with GLT-1 expression. Neuroscience 326:117–125. https://doi.org/10.1016/j.neuroscience.2016.04.004

Saul MC, Bagley JR, Bailey LS, Datta U, Dickson PE, Dodd R, Gagnon LH, Hugett SB, Kimble VM, Leonardo M, Kim S-M, Olson A, Roy T, Schoenrock SA, Wilcox T, Jentsch JD, Logan RW, McClung CA, Palmer RHC et al (2020) Consideration of genetic and sex effects in mice enhances consilience with human addiction studies. bioRxiv:2020.02.14.949784. https://doi.org/10.1101/2020.02.14.949784

Schoenrock SA, Kumar P, Gómez-A A, Dickson PE, Kim S-M, Bailey L, Neira S, Riker KD, Farrington J, Gaines CH, Khan S, Wilcox TD, Roy TA, Leonardo MR, Olson AA, Gagnon LH, Philip VM, Valdar W, de Villena FP-M et al (2020) Characterization of genetically complex collaborative cross mouse strains that model divergent locomotor activating and reinforcing properties of cocaine. Psychopharmacology (Berl) 237(4):979–996. https://doi.org/10.1007/s00213-019-05429-3

Scoggin K, Lynch R, Gupta J, Nagarajan A, Sheffield M, Elsaadi A, Bowden C, Aminian M, Peterson A, Adams LG, Kirby M, Threadgill DW, Andrews-Polymenis HL (2022) Genetic background influences survival of infections with Salmonella enterica serovar typhimurium in the collaborative cross. PLoS Genet 18(4):e1010075. https://doi.org/10.1371/journal.pgen.1010075

Sherf O, Nashelsky Zolotov L, Liser K, Tilleman H, Jovanovic VM, Zega K, Jukic MM, Brodski C (2015) Otx2 requires Lmx1b to control the development of mesodiencephalic dopaminergic neurons. PloS One 10(10):e0139697. https://doi.org/10.1371/journal.pone.0139697

Shirley RL, Walter NAR, Reilly MT, Fehr C, Buck KJ (2004) Mpdz is a quantitative trait gene for drug withdrawal seizures. Nat Neurosci 7(7):699–700. https://doi.org/10.1038/nn1271

Simms JA, Steensland P, Medina B, Abernathy KE, Chandler LJ, Wise R, Bartlett SE (2008) Intermittent access to 20% ethanol induces high ethanol consumption in long-Evans and Wistar rats. Alcohol Clin Exp Res 32(10):1816–1823. https://doi.org/10.1111/j.1530-0277.2008.00753.x

Solberg Woods LC, Stelloh C, Regner KR, Schwabe T, Eisenhauer J, Garrett MR (2010) Heterogeneous stock rats: a new model to study the genetics of renal phenotypes. Am J Physiol Renal Physiol 298(6):F1484–F1491. https://doi.org/10.1152/ajprenal.00002.2010

Sommer W, Hyytiä P, Kiianmaa K (2006) The alcohol-preferring AA and alcohol-avoiding ANA rats: neurobiology of the regulation of alcohol drinking. Addict Biol 11(3–4):289–309. https://doi.org/10.1111/j.1369-1600.2006.00037.x

Spatazza J, Lee HHC, Di Nardo AA, Tibaldi L, Joliot A, Hensch TK, Prochiantz A (2013) Choroid-plexus-derived Otx2 homeoprotein constrains adult cortical plasticity. Cell Rep 3(6):1815–1823. https://doi.org/10.1016/j.celrep.2013.05.014

Thiele TE, Miura GI, Marsh DJ, Bernstein IL, Palmiter RD (2000) Neurobiological responses to ethanol in mutant mice lacking neuropeptide Y or the Y5 receptor. Pharmacol Biochem Behav 67(4):683–691. https://doi.org/10.1016/s0091-3057(00)00413-5

Tripathi PP, Di Giovannantonio LG, Sanguinetti E, Acampora D, Allegra M, Caleo M, Wurst W, Simeone A, Bozzi Y (2014) Increased dopaminergic innervation in the brain of conditional mutant mice overexpressing Otx2: effects on locomotor behavior and seizure susceptibility. Neuroscience 261:173–183. https://doi.org/10.1016/j.neuroscience.2013.12.045

Tryndyak VP, Willett RA, Avigan MI, Sanyal AJ, Beland FA, Rusyn I, Pogribny IP (2022) Non-alcoholic fatty liver disease-associated DNA methylation and gene expression alterations in the livers of collaborative cross mice fed an obesogenic high-fat and high-sucrose diet. Epigenetics:1–15. https://doi.org/10.1080/15592294.2022.2043590

Vanderlinden LA, Saba LM, Bennett B, Hoffman PL, Tabakoff B (2015) Influence of sex on genetic regulation of "drinking in the dark" alcohol consumption. Mammalian Genome 26(1–2):43–56. https://doi.org/10.1007/s00335-014-9553-8

Vendruscolo LF, Estey D, Goodell V, Macshane LG, Logrip ML, Schlosburg JE, McGinn MA, Zamora-Martinez ER, Belanoff JK, Hunt HJ, Sanna PP, George O, Koob GF, Edwards S, Mason BJ (2015) Glucocorticoid receptor antagonism decreases alcohol seeking in alcohol-dependent individuals. J Clin Invest 125(8):3193–3197. https://doi.org/10.1172/JCI79828

Vengeliene V, Bilbao A, Molander A, Spanagel R (2008) Neuropharmacology of alcohol addiction. Br J Pharmacol 154(2):299–315. https://doi.org/10.1038/bjp.2008.30

Visscher PM, Goddard ME (2019) From R.A. Fisher's 1918 paper to GWAS a century later. Genetics 211(4):1125–1130. https://doi.org/10.1534/genetics.118.301594

Walter NAR, Zheng CL, Searles RP, McWeeney SK, Grant KA, Hitzemann R (2020) Chronic voluntary ethanol drinking in Cynomolgus macaques elicits gene expression changes in prefrontal cortical area 46. Alcohol Clin Exp Res 44(2):470–478. https://doi.org/10.1111/acer.14259

Walter N, Cervera-Juanes R, Zheng C, Darakjian P, Lockwood D, Cuzon-Carlson V, Ray K, Fei S, Conrad D, Searles R, Grant K, Hitzemann R (2021) Effect of chronic ethanol consumption in rhesus macaques on the nucleus accumbens core transcriptome. Addict Biol 26(5):e13021. https://doi.org/10.1111/adb.13021

Warde-Farley D, Donaldson SL, Comes O, Zuberi K, Barrawi R, Chao P, Franz M, Grouios C, Kazi F, Lopes CT, Maitland A, Mostafavi S, Montojo J, Shao Q, Wright G, Bader GD, Morris Q (2010) The GeneMANIA prediction server: biological network integration for gene prioritization and predicting gene function. Nucleic Acids Res (38). https://doi.org/10.1093/nar/gkq537

Warden AS, Mayfield RD (2017) Gene expression profiling in the human alcoholic brain. Neuropharmacology 122:161–174. https://doi.org/10.1016/j.neuropharm.2017.02.017

Williams RW, Williams EG (2017) Resources for systems genetics. Methods Mol Biol 1488:3–29. https://doi.org/10.1007/978-1-4939-6427-7_1

Wright JW, Masino AJ, Reichert JR, Turner GD, Meighan SE, Meighan PC, Harding JW (2003) Ethanol-induced impairment of spatial memory and brain matrix metalloproteinases. Brain Res 963(1–2):252–261

Yoneyama N, Crabbe JC, Ford MM, Murillo A, Finn DA (2008) Voluntary ethanol consumption in 22 inbred mouse strains. Alcohol 42(3):149–160. https://doi.org/10.1016/j.alcohol.2007.12.006

Zhou H, Sealock JM, Sanchez-Roige S, Clarke T-K, Levey DF, Cheng Z, Li B, Polimanti R, Kember RL, Smith RV, Thygesen JH, Morgan MY, Atkinson SR, Thursz MR, Nyegaard M, Mattheisen M, Børglum AD, Johnson EC, Justice AC et al (2020) Genome-wide meta-analysis of problematic alcohol use in 435,563 individuals yields insights into biology and relationships with other traits. Nat Neurosci 23(7):809–818. https://doi.org/10.1038/s41593-020-0643-5

Zuo L, Zhang F, Zhang H, Zhang X-Y, Wang F, Li C-SR, Lu L, Hong J, Lu L, Krystal J, Deng H-W, Luo X (2012) Genome-wide search for replicable risk gene regions in alcohol and nicotine co-dependence. Am J Med Genet B Neuropsychiatr Genet 159B(4):437–444. https://doi.org/10.1002/ajmg.b.32047

From Natural Behavior to Drug Screening: Invertebrates as Models to Study Mechanisms Associated with Alcohol Use Disorders

Henrike Scholz

Contents

Abstract Humans consume ethanol-containing beverages, which may cause an uncontrollable or difficult-to-control intake of ethanol-containing liquids and may result in alcohol use disorders. How the transition at the molecular level from "normal" ethanol-associated behaviors to addictive behaviors occurs is still unknown. One problem is that the components contributing to normal ethanol intake and their underlying molecular adaptations, especially in neurons that regulate behavior, are not clear. The fruit fly *Drosophila melanogaster* and the earthworm *Caenorhabditis elegans* show behavioral similarities to humans such as signs of intoxication, tolerance, and withdrawal. Underlying the phenotypic similarities, invertebrates and vertebrates share mechanistic similarities. For example in *Drosophila melanogaster*, the dopaminergic neurotransmitter system regulates the positive reinforcing properties of ethanol and in *Caenorhabditis elegans,* serotonergic neurons regulate feeding behavior. Since these mechanisms are fundamental molecular mechanisms and are highly conserved, invertebrates are good models for uncovering the basic principles of neuronal adaptation underlying the behavioral response to ethanol. This review will focus on the following aspects that might shed light on the mechanisms underlying normal ethanol-associated behaviors. First, the

---

H. Scholz (✉)
Department of Biology, Institute for Zoology, University of Köln, Köln, Germany
e-mail: henrike.scholz@uni-koeln.de

Curr Topics Behav Neurosci (2025) 71: 145–167
https://doi.org/10.1007/7854_2022_413
Published Online: 05 January 2023

current status of what is required at the behavioral and cellular level to respond to naturally occurring levels of ethanol is summarized. Low levels of ethanol delay the development and activate compensatory mechanisms that in turn might be beneficial for some aspects of the animal's physiology. Repeated exposure to ethanol however might change brain structures involved in mediating learning and memory processes. The smell of ethanol is already a key component in the environment that is able to elicit behavioral changes and molecular programs. Minimal networks have been identified that regulate normal ethanol consumption. Other environmental factors that influence ethanol-induced behaviors include the diet, dietary supplements, and the microbiome. Second, the molecular mechanisms underlying neuronal adaptation to the cellular stressor ethanol are discussed. Components of the heat shock and oxidative stress pathways regulate adaptive responses to low levels of ethanol and in turn change behavior. The adaptive potential of the brain cells is challenged when the organism encounters additional cellular stressors caused by aging, endosymbionts or environmental toxins or excessive ethanol intake. Finally, to underline the conserved nature of these mechanisms between invertebrates and higher organisms, recent approaches to identify drug targets for ethanol-induced behaviors are provided. Already approved drugs regulate ethanol-induced behaviors and they do so in part by interfering with cellular stress pathways. In addition, invertebrates have been used to identify new compounds targeting molecules involved in the regulation in ethanol withdrawal-like symptoms. This review primarily highlights the advances of the last 5 years concerning *Drosophila melanogaster*, but also provides intriguing examples of *Caenorhabditis elegans* and *Apis mellifera* in support.

Keywords Acamprosate · *Apis mellifera* · Approach · Benzodiazepine · *Caenorhabditis elegans* · Cellular stress response · Decision · *Drosophila melanogaster* · Ethanol

Abbreviations

ADH	Alcohol dehydrogenase
ALDH	Aldehyde dehydrogenase
AUD	Alcohol use disorder
EPA	Eicosapentaenoic acid
ESR1	Estrogen-related receptor 1
GABA	Gamma-aminobutyric acid
Gdp1L	Glycerol-3-phosphate dehydrogenase 1 like
GstO1	Glutathione transferase omega 1
hGlyRA	Human glycine receptor
hSLO	Human BK channel
MC4R	Melanocortin receptor 4 receptor
MeHg	Metal methylmercury
NPY	Neuropeptide Y

NPF	Neuropeptide F
Nrf2	Nuclear factor erythroid-derived 2-like
PECR	Peroxisomal trans-2-enoyl-CoA reductase
PDE4d	Phosphodiesterase 4D
PPP2R2B	Protein phosphatase 2 regulatory subunit B beta
ROS	Reactive oxygen species
rSK	Rat calcium-sensitive potassium channel
Rsu1	Ras suppressor 1
SLO1	Slowpoke 1
TSPO	Translocator protein

1 Invertebrate Models for Studying Alcohol Use Disorders

Caenorhabditis elegans (C. elegans) and *Drosophila melanogaster (Drosophila)* are two well-established genetic model systems for studying the mechanisms underlying behaviors related to alcohol use disorders (AUD) and mechanisms underlying the action of ethanol (Scholz 2019). The behaviors of worms and flies under the influence of ethanol resemble those of intoxicated humans such as ethanol increased locomotor activity, ethanol-induced sedation, the development of ethanol tolerance, and withdrawal-like symptoms (Scholz and Mustard 2013). Some of the advantages of using *C. elegans* and *Drosophila* as model systems include the control of the genetic background, the high number of individuals grown under very similar, well-controlled environmental conditions, and the genetic toolboxes available to change gene function and monitor cellular changes in response to ethanol. It is possible to raise worms and flies at relatively low cost and without regulatory restrictions. This offers the opportunity to answer basic research questions and act in accordance with the animal protection act of test animals that aims to reduce, replace, and refined the use of animals for scientific purpose (European Directive 2010/63/EU). Research over the last 20 years has shown that *Drosophila* and *C. elegans* can be used to uncover genes underlying the action of ethanol at the cellular level or underlying the behaviors associated with AUD. Various methods, ranging from classical genetic screens, candidate gene analysis, genome-wide association studies, transcriptome, micro-RNA analysis, and genome-wide histone acetylation surveys have been employed to identify genes (for review, see Lathen et al. 2020; Engel et al. 2019; Park et al. 2017; Grotewiel and Bettinger 2015). This review focuses on the following three aspects. First, mechanisms underlying the natural and ecologically relevant behaviors related to ethanol consumption will be discussed. To understand the dysregulation of behavior, it is important to understand how normal behavior is regulated, and what kind of mechanisms underlying behaviors are related or induced by ethanol that are still considered well-adapted. In an organism, how is the balance maintained between the benefits and the deteriorating damage from ethanol? What can the system or organism do before it becomes dysregulated? Ethanol is a challenging molecule for the organism. Ethanol is naturally occurring and low levels of ethanol are produced as by-products of metabolism in every cell – for example, during fatty-acid synthesis (Holmes 1994). In humans, low levels of ethanol are also found in the gut. Here, fungi or bacteria in the microbiome produce ethanol in the presence of high-carbohydrate diets (Bayoumy et al. 2021). Animals might take up higher concentrations of naturally occurring ethanol by feeding on fermenting fruits when foraging for food. Yeast is the organism responsible for ethanol production. In the presence of carbohydrates, yeast is converted into ethanol under anaerobic conditions. In natural environments such as orchards, fermentation can generate ethanol concentrations of up to 2% in plums and up to 7% in grapes (McKenzie and McKechnie 1979; Gibson and Wilks 1988). Within the organism, low levels of ethanol are normally efficiently degraded. A class of enzymes required for the catabolism of ethanol includes the alcohol dehydrogenase (ADH). ADH is found in all living animals and is highly conserved between different clades, including vertebrates such as humans and invertebrates, such as *Drosophila melanogaster*, *C. elegans*, and *Apis mellifera* (Holmes 1994; Ashburner 1998; Martins et al. 1977). Due to the presence of ADH, the ingestion of low levels of ethanol is not cytotoxic to the organism. In addition to the cellular level, invertebrates but also other species have adapted to low ethanol concentrations at the behavioral level. What are the benefits of these adaptations for animals? Possibilities include that if ethanol's toxicity is taken out of action, ethanol might serve as energy source and as important environmental cue, and ethanol may protect from predators.

The second aspect will focus on the cellular stress response to ethanol in neurons. After ethanol intake, all cells, including neurons and glia, are exposed to ethanol. In these cells, ethanol interferes with cellular mechanisms that have evolved to protect cells from toxic effects. In neurons, the mechanisms might be neuroprotective and ensure the maintenance of neuronal function. Interestingly, mechanisms identified in the cellular stress response to ethanol, such as the oxidative stress response, are also involved in other neurodegenerative diseases such as Alzheimer's disease, Parkinson's disease, and amyotrophic lateral sclerosis (Peng et al. 2020).

The third aspect will shed light on the identification of drug targets and understanding the action of drugs used for the treatment of AUD. Some of the drug targets impinge on the cellular stress system, and others might do so indirectly and therefore are included. This review will use examples that are mainly derived from studies that were conducted in the last 5 years in the invertebrates in *Drosophila*, *C. elegans*, and *Apis mellifera*.

2 Natural Behaviors Associated with Ethanol

2.1 Natural Levels of Ethanol Influence Behaviors

Animals exposed to increasing ethanol concentrations throughout development or as adults react to these external stimuli in a manner that ensures their internal energy demands and/or survival. However, in early embryogenesis, even a short exposure to ethanol can cause long-term changes in neuronal function and behavior. During embryogenesis, a brief pulse of 2% ethanol results in axonal pathfinding and neuron morphology defects in flies; in larvae, it results in defects in sensory perception (Scepanovic and Stewart 2019). Rearing embryos in the presence of 7% ethanol reduced food intake during the larval stage and adulthood. In the brain, the distribution of neuropeptide F (NPF) is changed. NPF is a peptide that normally regulates food intake and is considered to be the functional orthologue of neuropeptide Y in higher organisms (Guevara et al. 2018). Developmental exposure to ethanol also impairs locomotor activity in adult flies. Normally, ethanol-naïve embryos show an age-dependent decline in locomotor activity. In ethanol-raised embryos, this decline was reduced. Genes that encode antioxidant enzymes are upregulated in response to ethanol exposure. Despite the neurological defects caused by ethanol, the compensatory mechanism against the damage is neuroprotective and might delay aging (Belhorma et al. 2021). Protective mechanisms are also upregulated in response to ethanol in vertebrate neurons. In hippocampal neuronal cultures of rats, exposure to ethanol in a binge-like fashion stimulates the melanocortin receptor 4 receptor (MC4R) and increases the expression of the nuclear factor (erythroid-derived 2)-like 2 (Nrf-2) transcription factor, the "master regulator of the antioxidant defense and gatekeeper of species longevity" (Lewis et al. 2010; Quintanilla et al. 2020).

In developing *C. elegans* larvae, exposure to ethanol triggers mechanisms that positively influence the survival of animal. *C. elegans* larvae can survive long periods of food deprivation and extreme environmental conditions by entering a state called "dauer." In this state, the larvae are restricted to their internal energy supply. Internal energy supply influences larval survival. Exposure to ethanol in this dauer state increases enzymes involved in ethanol metabolism, such as ADH, and results in an extended lifespan (Kaptan et al. 2020). In addition to changes in metabolism, ethanol increases stress resistance and lipid storage in animals, and the mitochondria deteriorate less. Thus, in the dauer state ethanol induced a switch of the metabolism and serves as energy source for the larvae (Kaptan et al. 2020). In *Drosophila* larvae, ethanol also switches the metabolism. Ethanol-reared larvae have more lipid droplets and accumulate more lipids. Ethanol changes the transcription of genes involved in fatty-acid biosynthesis and reduces the expression of insulin-like peptides (DILPs) and their receptors. The transcriptional changes in *Drosophila* larvae might account for developmental delay and increased lethality (Logan-Garbisch et al. 2014).

If the organism has a choice and can move toward or away from an ethanol-containing source, the first step for the organism is to recognize that ethanol is present – it must be able to distinguish between an ethanol-containing food source and a non-ethanol-containing food source. The developing *Drosophila* larvae recognize the odorant ethanol and are attracted to its smell. Attractiveness does not require previous experience with ethanol and varies slightly throughout different developmental stages. Moderate, naturally occurring ethanol concentrations of up to 4% increase the fitness of the larvae, as measured by a higher likelihood of survival.

In addition, ethanol increases the attractiveness of associated odorants (Schumann et al. 2021).

Adult flies are attracted to ethanol-containing food odors (Ogueta et al. 2010). Both flying and walking flies respond to ethanol as a single odorant, moving toward it (van Breugel et al. 2018). In nature, ethanol is rarely present as a single odorant, and in artificial environments, in ethanol-containing food such as a Pilsner-type beer, 76 odor-active compounds can be identified (Kishimoto et al. 2018). If ethanol is crucial for the fly to elicit a specific behavior, the fly needs to recognize ethanol within a complex odor bouquet. Ethanol is recognized as a key odorant in complex odor mixtures (Giang et al. 2017). At the molecular level, increased serotonin signaling in the neurons of the olfactory pathway enhances the detection of ethanol. A second set of serotonergic neurons can suppress the attractiveness of ethanol by integrating the internal states. Olfactory attraction can override inhibitory mechanisms (Xu et al. 2016). The experiments indicate that information on external ethanol-related cues can override more internal antagonizing mechanisms and highlight the importance of external cues in eliciting behavior. Another example of the power of smell of ethanol is the observation that the smell of yeast increases the locomotor activity of flies. The smell extended the survival rate during starvation and reduced the decline in triacylglyceride. The component in the odor of yeast that elicits these behavioral changes is ethanol (Luo et al. 2021).

The experience of ethanol-associated odorants might result in positive or negative memories depending on the ethanol dose, the length of ethanol experience, and the spacing between the events when ethanol is presented (Nunez et al. 2018). Other insect model systems for AUD, such as honeybees, also learn to associate odorants with different concentrations of ethanol (Varnon et al. 2018). In adult flies, multiple experiences of ethanol-related situations change the pattern of activated neurons in the brain. Ethanol concentrations that normally elicit an approach to an odorant paired with ethanol activate different sets of dopaminergic neurons when they are experienced more often (Scaplen et al. 2020).

The presentation of the odorant ethanol may also influence other behaviors. Male flies became more aggressive in the presence of food containing 5% ethanol. Here, ethanol increases sensitivity to the male aggression-promoting pheromone 11-cis-Vaccenyl acetate (cVa) (Park et al. 2020). In female flies, food containing 2–6% ethanol is an important cue for egg-laying (Zhu and Fry 2015). Females were selective for the right ethanol concentration. They lay eggs in 10% ethanol-containing food when an alternative nearby does not contain ethanol. When the alternative is further away, they prefer non-ethanol-containing foods. The larvae that hatched on the 10% ethanol-containing food left the substrate. The idea here is that the eggs in the presence of 10% ethanol are protected from predators such as parasitoid wasps, but for the larvae, exposure to 10% ethanol might be damaging. A food source without ethanol might be a good alternative. Thus, ethanol has different advantages for the organism during development, and the influence of ethanol on behavior can be dimorphic (Sumethasorn and Turner 2016).

Ethanol in the environment elicits some of the behavioral changes described above. Before the animal can react to ethanol, a decision must be made to do so. The octopaminergic neurotransmitter system is required for olfactory ethanol attraction. By changing the activation of different sets of octopaminergic neurons, the decision to approach or avert the ethanol-containing food source can be switched. These findings uncover a decision-making system involved in the selection of behavioral outcomes in response to ethanol (Claßen and Scholz 2018). Dysregulation of the decision-making process might increase the likelihood of approaching ethanol-containing food. Finally, the decision to favor ethanol over non-ethanol increases the likelihood of ingesting ethanol-containing food. This raises the question of whether there is a connection between odorant attraction and ethanol consumption. At the molecular level, there seems to be a partial overlap in genes associated with the regulation of olfactory ethanol attraction and ethanol consumption preference, supporting the idea that some mechanisms might regulate both (Sekhon et al. 2016). Normally, the preference for consuming ethanol develops (Gonzalez et al. 2018). In contrast to this, the loss of the small GTPase Arf6, and the Arf6 guanine exchange factor Efa6, results in an innate preference for consuming ethanol-enriched food, suggesting that both molecules act as negative regulators of ethanol consumption preference. Polymorphisms associated with the human orthologues of Arf6 and Efa6 (PSD1–4) are associated with an increased frequency of drinking and binge-drinking episodes in adolescents. In a European sample, polymorphisms correlated with increased alcohol dependence. Thus, in flies and humans, similar genes regulate ethanol consumption (Gonzalez et al. 2018). Genes that might share a function in ethanol consumption in flies and humans also belong to the Ras suppressor 1 (Rsu1) (Ojelade et al. 2015). Rac1 together with cofilin regulates the experience-dependent preference to consume alcohol in the mushroom bodies of flies, a center involved in the regulation of learning and memory. The mushroom bodies are functionally similar to the mammalian hippocampus. Repetitive ethanol consumption results in structural changes in the neurons involved in learning and memory (Butts et al. 2019).

Do other insects prefer to consume ethanol? Flower nectar contains ethanol. Some flowers even smell fermented and contain ethanol-enriched floral nectar generated from yeast (Goodrich et al. 2006). Honeybees normally choose the closest food source and do not distinguish between yeast and non-yeast containing nectar (Kevan et al. 1988). However, they can be trained to forage on ethanol-containing foods (Sokolowski et al. 2012). The taste of ethanol is aversive to honeybees, and adding sucrose to the solution overcomes the aversion. Honeybees can develop a mild preference to consume 1–5% ethanol-containing sucrose solution over a regular sucrose solution (Mustard et al. 2019). When worker bees were fed for a prolonged time with 1% ethanol-containing sucrose, they even increased their consumption after a period of abstinence, suggesting that bees show withdrawal-like symptoms (Ostap-Chec et al. 2021).

In *C. elegans*, ethanol promotes feeding behavior, and a part of a neural circuit that mediates this behavior has been identified. Specific sensory neurons – the BAG neurons – respond to ethanol stimulation with changes in intracellular calcium levels. The calcium response required guanylate cyclases GCY-31 and GCY-33. Ethanol-stimulated feeding behavior also requires the release of serotonin from the downstream NSM motor neurons, showing a function for serotonin in regulating ethanol-induced feeding behavior (Wang et al. 2021). It will be interesting to see how a minimal network appears to regulate ethanol-induced feeding behavior in mammals.

2.2 Other Environmental Factors that Influence Ethanol-Induced Behaviors

Environmental factors influence the behavior elicited by ethanol. One factor is the diet. In male and female flies, yeast as a protein source delays ethanol-induced sedation. If the yeast diet is changed to a non-yeast containing diet, flies sedate normally again. Inactivation of serotonergic neurons in yeast-fed flies also results in normal sedation, suggesting that serotonergic neurons regulate the yeast-induced delay of ethanol-induced sedation (Schmitt et al. 2020). Dietary medium also influences the preference for consuming ethanol-enriched food in flies. Diets with high nutrient content suppress the preference for consuming ethanol. The ethanol intake of the fly is further limited by the total amount of food consumed. The total amount of alcohol consumed and the total volume of food consumed are better indicators of ethanol consumption preference than calories in the food (Park and Ja 2020).

A translation approach of the results obtained in *C. elegans* and mice showed that nutrients provided in the diet might influence ethanol-induced locomotion in both worms and rodents. Eicosapentaenoic acid (EPA), a component of the mammalian diet, belongs to the class of omega-3 fatty acids. Worms lacking EPA fail to develop acute tolerance, and in controls, EPA supplementation enhanced acute functional tolerance (Raabe et al. 2014). Depending on the genetic background, mice fed with omega-3 fatty acids show differences in their ethanol-induced locomotor activity, supporting the finding that diet influences the effect of ethanol in rodents (Wolstenholme et al. 2018).

In addition to diet, the microbiome influences ethanol-induced behaviors. The fungus *Candida albicans* has a special ADH that catalyzes the conversion of acetaldehyde to ethanol. In the fungus, the loss of ADH results in decreased fungal levels in the kidneys and livers of infected mice and worms. The presence of mutated *Candida albicans* also extends the lifespan of mice and worms. Normally, the ADH of the fungus increases oxidative phosphorylation in the host by increasing reactive oxygen species, thereby increasing the pathogenicity of *Candida albicans* (Song et al. 2019).

2.3 Adaptations to Ethanol-Enriched Environments

One specific characteristic of *Drosophila melanogaster* in comparison with other *Drosophilidae* is that they are well-adapted to ethanol-enriched environments. The adaption is still beneficial for the species and allows conquering ecological niches in which other less well-adapted *Drosophilidae* cannot live. The neuronal bases of this adaptation might provide insight into the range of ethanol that a normal system might tolerate before it becomes dysregulated. One idea of how an organism might cope better with ethanol-rich environments is that they metabolize ethanol more efficiently, for example, due to a better functioning ADH. To address whether *Drosophila melanogaster* has a better functioning ADH in comparison with less well-adapted *Drosophilidae,* the sequence of the ADH gene of *Drosophila melanogaster* and the closely related species *Drosophila simulans* were compared, and a progenitor of ADH was derived. The activity and turnover of ethanol in the ancestral ADH and the ADH of *Drosophila melanogaster* were indistinguishable in vitro and in vivo. Therefore, a better functioning ADH does not necessarily account for better adaptation to ethanol-enriched environments (Siddiq et al. 2017).

Besides the functional level of ADH, other enzymes important for ethanol metabolism might influence how an organism copes with environmental ethanol. For example, polymorphisms in aldehyde dehydrogenase (ALDH) can increase the ability of animals to adapt to ethanol. The ALDH in *Drosophila melanogaster* has more than one substrate, and polymorphisms of ALDH allow for increased activity for one substrate in favor of another, when the environmental conditions change. One substrate of ALDH is acetaldehyde, which is a product of ethanol metabolism, whereas the second class of substrates includes larger aldehydes, which are by-products of oxidative phosphorylation. Natural occurring ALDH variants are differentially selective for the substrates. One variant of ALDH shows a higher turnover rate of acetaldehydes than for larger aldehydes and is found in populations that encounter higher levels of ethanol. The heterogeneity of different ALDH variants in a population allows for better adaptation to changing ethanol concentration in the environment (Chakraborty and Fry 2016).

What other mechanisms exist that allow an organism to adapt better to a changing environment? Mutations in genes that normally have little or no effect on the phenotype might be advantageous for survival in changing environments that contain ethanol. When new species conquer a habitat, the polymorphisms in the genome might increase survival chances, and the genetic trait that is best adapted to the new environment might be dominantly inherited. Flies with this variant were dominant in such a habitat. In *Drosophila simulans*, a non-ethanol-adapted *Drosophila species,* a broader set of splice variants was found in comparison with the ethanol-adapted species *Drosophila melanogaster.* The cost of adaptation to an ethanol-enriched environment is the loss of cryptic genetic variation (Signor 2020). Consistent with the idea that there is a correlation between splicing and ethanol, ethanol exposure changes transcript levels and alternative splicing in a pattern that changes over time (Signor and Nuzhdin 2018). It would be interesting to test whether the amount of cryptic genetic variation differs between individuals that tolerate more (or less) ethanol, and whether these differences might predict how good an organism is adapted to ethanol.

Other factors that influence the adaptation to ethanol-enriched environments and promote the survival of a population is a higher mating success in the presence of ethanol (Zhu and Fry 2018). Experiences with environmental ethanol exposure not only influence the behavior of animals that directly encounter ethanol, but also influence the behavior of the next generation. The offspring of adult flies exposed repetitively to ethanol showed increased resistance to the sedating effect of ethanol and developed increased tolerance (Bonilla et al. 2021). The offspring of the surviving *Drosophila melanogaster* larvae after a predatory wasp attack showed a preference to lay their eggs in ethanol-enriched food for five generations after the attack. This preference was correlated with low maternally derived NPF levels in a subset of neurons (Bozler et al. 2019). Thus, the beneficial effects of ethanol exposure in a very specific situation might be strong enough to be carried over to the next generation, even when the parental generation is no longer present.

In summary, the exposure to ethanol results in developmental changes that might result in behavioral changes of the adult organism. The effects of ethanol are – depending on the ethanol dose – beneficial or detrimental. In addition to ethanol, other environmental factors potentiate or reduce the effects of ethanol. In invertebrates, adaptive mechanisms have evolved that protect the organism from increasing ethanol concentrations. Some of the adaptation might be beneficial under the same environmental conditions. However, whether the adaptations are sufficient to cope with increasing ethanol exposures or whether the adaptations reinforce mechanisms resulting in addictive behaviors needs to be further analyzed.

3 Cellular Stress Responses to Ethanol

The next section focuses on the cellular stress response to ethanol to better understand the basic adaptive potential of every cell in an organism. Exposure to ethanol affects all cells in the brain equally. Ethanol is lipid soluble, migrates through membranes, and affects structures in the membranes, cytoplasm, and nucleus. In addition to migrating through the membrane, there is also evidence that ethanol directly influences membrane integrity. In *C. elegans* exposed to 7% ethanol, there was increased membrane permeability and increased ethanol-induced immobility. The increase depends on the patched-related family member-6 protein, a protein that contains 11 transmembrane domains and a sterol-sensing domain. The requirement of a patched-related protein in regulating membrane integrity indicates that the passage of ethanol through the membrane depends on the composition of the membrane (Choi et al. 2016). Within the cell, the effects of ethanol are highly selective and are mediated by specific receptor systems. Cells perceive ethanol as a cellular stressor. Different protective mechanisms against this stressor have evolved. As mentioned above, even single cellular organisms already contain ADH that mediates ethanol catabolism (Holmes 1994; Ashburner 1998). Metabolites that improve ADH activity also improve the defects caused by ethanol in developing *Drosophila* larvae. The metabolite alpha-ketoglutarate is an intermediate of the Krebs cycle, increases ADH activity, and improves the negative effect of ethanol on development. In addition, alpha-ketoglutarate functions as an antioxidant (Bayliak et al. 2016).

Oxidative stress is a form of cellular stress induced by ethanol. For example, in *C. elegans*, ethanol exposure results in fragmentation of mitochondria – a process that depends on dynamin-related protein 1 and its receptor at the outer mitochondrial membrane – and ethanol-damaged mitochondria increase the oxidative stress in the cell (Oh et al. 2020). In *Drosophila*, mitochondrial function influences the degree of stress resistance on the toxic effects of ethanol during development. *Drosophila* larvae with a greater antioxidant capacity in response to elevated mitochondrial reactive oxygen radicals (ROS) pupate sooner when grown in the presence of ethanol (Towarnicki et al. 2020). Given that ethanol influences mitochondrial function, it is not surprising that ethanol influences other neurological diseases that are associated with dysfunctional mitochondria, such as Alzheimer's disease (Morton et al. 2021). In a model of Alzheimer's disease, in which flies express pan-neuronal constructs of $A\beta_{42}$ or tau^{R406W}, flies increased their sensitivity to ethanol and showed a rise in their ROS levels (Nikookar et al. 2021).

Molecules that regulate oxidative stress include glutathione transferase omega 1 (GstO1) (Board and Menon 2016). In flies, ethanol alters the expression of GstO1, and loss of GstO1 in dopaminergic neurons results in earlier ethanol-induced sedation (Choi et al. 2019). This links the regulation of oxidative stress directly to the dopaminergic neurotransmitter system. The dopaminergic neurotransmitter system regulates a broad range of ethanol-induced behaviors. For example, the increase in the voluntary ethanol intake of flies is regulated by dopamine receptor 1 (Kanno et al. 2021). Sensitization to the ethanol-induced courtship of flies is mediated by the G-protein-coupled dopamine ecdysone receptor (Aranda et al. 2017); the same receptor also regulates ethanol-induced sedation (Petruccelli et al. 2016). In *C. elegans*, ethanol enhances dopamine release in dopamine receptor 2 like autoreceptor mutants, suggesting that the D2-like receptor normally represses dopamine release upon ethanol exposure. In turn, the increased release of dopamine activates the further downstream NLP12 releasing interneuron that expresses the excitatory dopamine D1-like receptor (Pandey et al. 2021). In *Drosophila* and mice, the D1-like dopamine receptor has been shown to promote ethanol-induced hyperactivity (Kong et al. 2010; Abrahao et al. 2011).

The accumulation of damage caused by oxidative stress is one mechanism underlying the aging of the organism. Ethanol reduces lifespan. Flies inbred for a longer lifespan are more resistant to the lifespan-reducing effect of ethanol. They showed a higher oxidative defense, lower reactive oxygen species levels, enhanced alcohol dehydrogenase activity, and better motor skills (Towarnicki et al. 2020). There are sex-specific differences in lifespan and oxidative stress resistance. Female flies live longer than males. Females, compared to age-matched males, showed reduced levels of ROS and increased activity of antioxidant enzymes such as superoxide dismutase and catalase. Ethanol-exposed females are more resistant to ethanol-induced mortality and ethanol-induced loss of climbing activity compared to males. Thus, females with a higher antioxidant defense have a higher resistance to ethanol (Niveditha et al. 2017). Resistance to oxidative stress caused by ethanol may serve as a predictor of longevity.

Protection against oxidative stress caused by ethanol might also arise from the interaction of bacteria in the microbiome of the animal. In mice and flies, the human gut bacteria *Lactobacillus rhamnosus GG* activates the expression of the nuclear factor erythroid-derived 2-like (Nrf2) in the liver and the respective liver-like structure – the fat bodies – in flies. Induction of Nrf2 protects against ethanol-induced liver and fat body damage. The presence of specific gut bacteria modulates hepatic susceptibility to damage caused by ethanol (Saeedi et al. 2020). Whether bacteria or the microbiome in flies can also influence neuronal function underlying ethanol-induced behaviors remains to be determined.

Environmental toxins that cause cellular stress may further enhance the effects of ethanol. Heavy metal methylmercury (MeHg) is a toxic environmental pollutant. Exposure to methylmercury causes an increase in ROS production and lipid peroxidation. Ethanol increased ROS production even further. MeHG has a direct impact on ethanol catabolism by inhibiting the activity of ADH, thereby increasing the concentration of ethanol in the fly. Therefore, the influence of MeHG on ethanol-induced behavior is not surprising. Flies exposed to ethanol recover later from sedation in the presence of MeHG, probably due to the inhibition of ADH (Chauhan and Chauhan 2016).

In addition to oxidative stress mechanisms, neurons contain cellular stress responses that are mechanistically similar to heat shock stress responses that can result in changes in neuronal plasticity and behavior (Scholz et al. 2005: Fig. 2). In *Drosophila*, the nuclear RNA-binding protein Hangover might achieve these changes in neuronal plasticity by changing the transcript composition of the neurons. One prominent example that is changed includes transcripts of the phosphodiesterase 4d orthologue *dunce*. The *dunce* gene encodes a phosphodiesterase that specifically hydrolyzes cAMP and terminates the action of the cAMP secondary messenger system. These results link the regulation of cellular stress responses to cellular cAMP levels (Ruppert et al. 2017). In *C. elegans*, the small heat shock protein HSF-1 is required in IL2 neurons to regulate ethanol-induced increased locomotor activity. The observation that ethanol activates a $G_{\alpha s}$-cAMP-protein kinase A signaling pathway in these cells supports a connection between the regulation of the cellular heat shock/ethanol response to the regulation of cellular cAMP levels (Johnson et al. 2016, 2017).

Taken together, ethanol causes cellular stress responses involving components of the cellular heat shock response and the oxidative stress response. Neurons are protected against low levels of cellular stress. The mechanisms of activation change neuronal plasticity and behavior. These adaptive changes are challenged when organisms encounter additional forms of oxidative stress caused by aging, endosymbionts, or environmental toxins. These neuroprotective mechanisms might protect organisms from damage caused by oxidative stress exposure. This might be due to natural processes such as aging, endosymbionts, or environmental toxins. However, excessive cellular stress is detrimental.

4 Model to Test Pharmacological Agents

The third aspect illuminates how invertebrate model systems are used to identify new pharmacological tools and identify the molecular mechanisms of approved drugs. The studies performed in invertebrates also suggest that some drugs already used for the treatment of AUD might convey their action by interfering with the cellular stress response. In the USA, only three drugs have been approved for the treatment of AUD, including the oral alcohol-aversive drug disulfiram, the opioid antagonist naltrexone, and the homotaurine acamprosate. In addition, the opioid antagonist nalmefene has been approved in Europe (Mason and Heyser 2021). The relapse rate of patients with AUD is still relatively high. For example, after 6 months of withdrawal and treatment with acamprosate 63.9% of patients relapse instead of 76.6% when treated with placebos (Mann et al. 2004). Invertebrate test systems might help improve the effectiveness of treatment by uncovering the mechanism underlying the effects of approved drugs.

The first example focuses on an elegant study conducted on *C. elegans*. In this study, worms were used to identify a new pharmacological ligand for a known molecular target of ethanol, the well-studied large-conductance calcium- and voltage-activated potassium channels (BK or SLO-1 channel). The function of the BK/SLO1 channel is important for the regulation of ethanol-induced behaviors in rodents, *C. elegans*, and flies (for review, see Dopico et al. 2016). For example, the SLO-1 channel is required for ethanol-induced withdrawal-like symptoms in *C. elegans* (Scott et al. 2017), and the clustering of SLO-1 channels in the active zone is required for the regulation of resistance to the locomotor pressing effect of ethanol (Oh and Kim 2019). At first glance, the BK channel does not seem to be involved in cellular stress pathways (such as oxidative stress pathways). There is emerging evidence that oxidative stress modulates the function of BK channels (Hermann et al. 2015), further increasing the effect of ethanol directly in the BK channel. To identify molecules that selectively alter BK channel function in response to ethanol, a high-throughput phagemid-display screen expression of 30 million peptide sequences was combined with behavioral experiments using *C. elegans*. The screen was based on three HEK293 lines that expressed different channels. The human ethanol-sensitive BK channel (hSLO) was used for positive selection, and the human glycine receptor (hGlyRa1) – another ethanol-targeted receptor – and the calcium-sensitive potassium channel (rSK2) were used for negative selection. Phagemids containing 30 million unique peptide sequences were sequentially exposed to HEK293 cells containing the rSK2 and hGLYRA1 channels. Afterwards, the supernatant was cleared of the molecular targets that bind to rSK2, and hGLYRA1 was transferred to HEK293 cells stably transfected with the human SLO channel. After pre-incubation with other channels, the peptides bound to the BK channel were isolated and sequenced. A set of 20 candidate peptides were isolated. To validate the effect of these peptides on the BK channel, the peptides were applied to worms, and their ethanol-induced behavior was analyzed. The crawling speeds of controls and SLO-1 mutant worms were measured before and after ethanol exposure. Of the tested peptides that all affected normal crawling speed, only five did not alter the resistance of SLO-1 mutants to the locomotor repressing effect of ethanol, suggesting that they normally interact with the BK channel. Next, the remaining five peptides were tested in a humanized worm model expressing the human SLO1-channel in SLO1 mutants. The peptide LS10 selectively altered the physiological effects of ethanol without affecting the basal function of BK in the regulation of locomotion (Scott et al. 2018a). The peptide can now be further validated in mouse models of AUD.

Another promising identification of a drug in *C. elegans* that targets a molecule involved in the regulation of ethanol withdrawal is the ligand JVW-1034 of the transmembrane receptor sigma-2 receptor. The sigma-2 receptor regulates the homeostasis of cholesterol and lipids. In *C. elegans*, the sigma-2 receptor is required to regulate the symptoms of ethanol withdrawal. JVW-1034 and three additional ligands improved the withdrawal-like symptoms in *C. elegans*. The function of JVW-1034 depends on the sigma-2 receptor, since worms lacking the receptor do not improve their withdrawal-like symptoms (Scott et al. 2018b). Based on these results, the function of the ligand was analyzed in mice. Administration of the selective ligand JVW-1034 specifically reduced ethanol intake and preference in mice (Quadir et al. 2021).

The identification of another putative drug for the treatment of AUD in *C. elegans* is based on the observation that traditional herbal medicine extracts or the plant *Schisandra chinensis* is used to treat various neurological diseases. It is thought that the plant is useful for treatment because of its antioxidant capacity (Zhou et al. 2021). In *C. elegans*, the water extract of *Schisandra chinensis* fruit delays the reduction of the lifespan and increases the mobility of worms exposed to ethanol. In addition, the extract reduces internal ethanol concentrations and the free radical content and blocks the continuous dopamine release caused by ethanol. The plant contains molecules that reduce the neurotoxic effects of ethanol and may have a neuroprotective function in dopaminergic neurons (Wu et al. 2020).

The effects and functions of pharmacological agents on ethanol-induced behavior have also been investigated in *Drosophila*. The agonist SKF 97541 of the GABA-B receptor increased sensitivity to the sedating effect of ethanol, whereas the antagonist CGP54626 GABA-B receptor decreased sensitivity. The agonist resulted in a higher degree of tolerance to the sedative effect of ethanol, whereas administration of the antagonist did not change the degree of tolerance (Ranson et al. 2020). A previous study showed that the reduction of GABA-B receptor 1 function and administration of the same antagonist reduced the locomotor impairing effect of ethanol (Dzitoyeva et al. 2003). Another class of drugs that interacts with the GABA receptor function includes benzodiazepines, which bind to $GABA_A$ and thereby exert inhibitory effects on the central nervous system, similar to ethanol (Gravielle 2018). Benzodiazepines interfere with the GABA receptor and alter the function of the mitochondrial outer membrane translocator protein 18 kDa (TSPO). Reduction of dTSPO in adult male flies increases sensitivity to the sedating effect of ethanol. The reduction in dTSPO also increased ROS and blocked caspase activity. In addition to the effect of sedation, the reduction of dTSPO results in a loss of tolerance; however, neurons may not be responsible for this effect. Female flies are more sensitive than males to the sedating effect of ethanol. dTSPO not only regulates the effect of benzodiazepines, but also ethanol. These animals could be used as models to screen for the effectiveness of benzodiazepines (Lin et al. 2015).

To uncover the mechanisms underlying the action of approved drugs, the drugs should induce similar behavioral effects in invertebrates. This is the case with naltrexone and acamprosate. In flies, naltrexone reduced the ethanol consumption preference. The effect of naltrexone is reversible, and naltrexone can suppress the ethanol-induced activity of protein kinase C (Koyyada et al. 2018). Feeding flies with acamprosate reduced ethanol consumption preference, suggesting that acamprosate has a neuronal target in flies, similar to what causes a reduction in ethanol consumption in humans. In flies, the preference for ethanol consumption is also reduced in estrogen-related receptor (ERR/ESR1) mutants and peroxisomal trans-2-enoyl-CoA reductase (dPECR/PECR) mutants (Fig. 1). In addition to their function in the regulation of ethanol consumption, ERR/ESR1 and dPECR/PECR1 are required to regulate the response to oxidative stress in neurons. The observation that dPECR/PECR and ERR/ESR1 regulate the oxidative stress response and

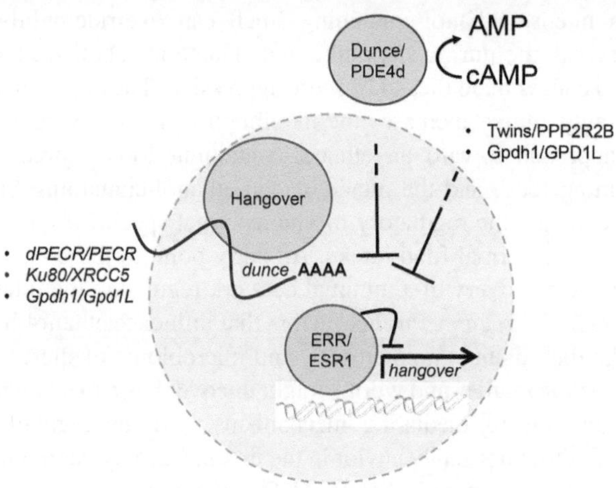

Fig. 1 Cellular stress pathway regulating ethanol-associated behaviors. The phenotypic analysis of the *hangover* mutants uncovered a cellular stress response required for the development of ethanol tolerance (Scholz et al. 2005). The Hangover protein is a neuronal nuclear RNA-binding protein (Scholz et al. 2005; Ruppert et al. 2017). One target transcript is a specific splice variant of the *dunce* gene. Other transcripts that are regulated negatively by the Hangover protein include *dPECR/PECR, Ku80/XRCC5,* and *Gpdh1/Gdp1L* (Velo Escarcena et al. 2021). The *dunce* gene encodes the cAMP-specific phosphodiesterase PDE4d homolog of vertebrates. Dunce is required for the development of ethanol tolerance and for the regulation of a cellular stress response. Dunce or elevated cAMP levels in addition to Twins/PPP2R2B, Gpdh1/GPD1L, and the ERR/ESR1 negatively regulate *hangover* transcripts (Ruppert et al. 2017; Velo Escarcena et al. 2021). Hangover, ERR and PECR are required for the reinstatement of ethanol consumption preference after periods of abstinence and the cellular stress response (Velo Escarcena et al. 2021). Great round circles indicate proteins, italic letters transcripts

ethanol consumption suggests that cellular stress mechanisms might also modulate the preference for ethanol (Velo Escarcena et al. 2021). Evidence that this mechanism might also regulate ethanol intake in humans comes from the observation that polymorphism in the human orthologue of estrogen receptor 1 and peroxisomal trans-2-enoyl-CoA reductase has been implicated in a GWAS for AUD (Treutlein et al. 2009, Frank et al. 2012; Fig. 1).

5 Conclusions

Under natural condition organisms such as humans, flies, and worms are exposed to external and internal levels of ethanol. During evolution mechanisms at the behavioral and cellular level have evolved that allow the organisms to adapt to changes of ethanol concentration and protect the organism from the toxicity of ethanol. These mechanisms appear to be dysregulated in AUDs resulting in uncontrollable intake of ethanol and cellular damage.

At the behavioral level, ethanol functions as a key odorant. Ethanol triggers an attraction to the source of ethanol. Mechanisms that integrate internal states suppress attraction. An intense ethanol-containing smell can override inhibition and can extend the survival rate during starvation and change metabolism. Exposure to the odorant ethanol causes male flies to be more aggressive. The decision to approach an ethanol-containing source increases the likelihood of intake. Similar sets of genes regulate the approach toward an ethanol-containing food source, the intake of ethanol-containing food, and the intake of non-ethanol-containing food. The challenge is to disentangle the regulatory mechanism that specifically regulates ethanol intake by regulating normal food intake. An entry point to these analyses might be provided from the discovery of a minimal network regulating the intake of ethanol-containing food in *C. elegans*. Further factors that influence ethanol-induced behaviors include the diet, dietary supplements, and microbiome of flies. Comparison of knowledge gained from flies and worms might uncover basic mechanisms regulating decision-making, reward mediating mechanisms, and the regulation of ethanol consumption. Without normal behavior in the first place, a dysregulation of behavior resulting in addiction, more specifically AUD, cannot evolve.

At the cellular level, already short developmental exposure to ethanol in flies and worms increases the transcription of genes involved in the regulation of the oxidative stress response and induces metabolic switches that prevent the organism from damage. In contrast to this, increased ethanol exposure damages brain structures involved in learning and memory or shorten lifespan. Increased environmental ethanol also induces changes that alter the efficiency of ethanol metabolism and regulate oxidative stress responses. Factors that increase oxidative damage, such as the process of aging or environmental pollutants, increase the effects of ethanol on the organism. Factors that decrease oxidative stress, such as specific microorganisms in the microbiome of animals, reduce the effect of ethanol. The discovery of molecules in *C. elegans* that interfere with cellular stress responses and are neuroprotective might reduce ethanol-induced damages in the brain. Already approved drugs used to treat AUDs such as acamprosate and benzodiazepines might assert their action through genes involved in the regulation of oxidative stress responses.

Taken together increasing evidences from research over the last 5 years in flies and worms suggest that some aspects of AUDs might be caused by dysregulation of the oxidative balance at the cellular level and the discovery of new compounds interfering with cellular stress response pathways might protect against neuronal damage or excessive ethanol intake.

Acknowledgments The present work was supported by DFG 656-10-1 and a grant from the Hetzler Foundation to HS.

References

Abrahao KP, Quadros IM, Souza-Formigoni ML (2011) Nucleus accumbens dopamine D_1 receptors regulate the expression of ethanol-induced behavioural sensitization. Int J Neuropsychopharmacol 14(2):175–185. https://doi.org/10.1017/S1461145710000441

Aranda GP, Hinojos SJ, Sabandal PR, Evans PD, Han KA (2017) Behavioral sensitization to the disinhibition effect of ethanol requires the dopamine/ecdysone receptor in Drosophila. Front Syst Neurosci 11:56. https://doi.org/10.3389/fnsys.2017.00056

Ashburner M (1998) Speculations on the subject of alcohol dehydrogenase and its properties in Drosophila and other flies. BioEssays 20(11):949–954. https://doi.org/10.1002/(SICI)1521-1878(199811)20:11<949::AID-BIES10>3.0.CO;2-0

Bayliak MM, Shmihel HV, Lylyk MP, Storey KB, Lushchak VI (2016) Alpha-ketoglutarate reduces ethanol toxicity in Drosophila melanogaster by enhancing alcohol dehydrogenase activity and antioxidant capacity. Alcohol 55:23–33. https://doi.org/10.1016/j.alcohol.2016.07.009

Bayoumy AB, Mulder CJJ, Mol JJ, Tushuizen ME (2021) Gut fermentation syndrome: a systematic review of case reports. United European Gastroenterol J 9(3):332–342. https://doi.org/10.1002/ueg2.12062

Belhorma K, Darwish N, Benn-Hirsch E, Duenas A, Gates H, Sanghera N, Wu J, French RL (2021) Developmental ethanol exposure causes central nervous system dysfunction and may slow the aging process in a Drosophila model of fetal alcohol spectrum disorder. Alcohol 94:65–73. https://doi.org/10.1016/j.alcohol.2021.03.006

Board PG, Menon D (2016) Structure, function and disease relevance of omega-class glutathione transferases. Arch Toxicol 90(5):1049–1067. https://doi.org/10.1007/s00204-016-1691-1

Bonilla M, McPherson M, Coreas J, Boulos M, Chavol P, Alrabadi RI, Loza-Coll M (2021) Repeated ethanol intoxications of Drosophila melanogaster adults increases the resistance to ethanol of their progeny. Alcohol Clin Exp Res 45(7):1370–1382. https://doi.org/10.1111/acer.14647

Bozler J, Kacsoh BZ, Bosco G (2019) Transgenerational inheritance of ethanol preference is caused by maternal NPF repression. eLife 8:e45391. https://doi.org/10.7554/eLife.45391

Butts AR, Ojelade SA, Pronovost ED, Seguin A, Merrill CB, Rodan AR, Rothenfluh A (2019) Altered actin filament dynamics in the Drosophila mushroom bodies lead to fast acquisition of alcohol consumption preference. J Neurosci 39(45):8877–8884. https://doi.org/10.1523/JNEUROSCI.0973-19.2019

Chakraborty M, Fry JD (2016) Evidence that environmental heterogeneity maintains a detoxifying enzyme polymorphism in Drosophila melanogaster. Curr Biol 26(2):219–223. https://doi.org/10.1016/j.cub.2015.11.049

Chauhan V, Chauhan A (2016) Effects of methylmercury and alcohol exposure in Drosophila melanogaster: potential risks in neurodevelopmental disorders. Int J Dev Neurosci 51:36–41. https://doi.org/10.1016/j.ijdevneu.2016.04.010

Choi HJ, Cha SJ, Kim K (2019) Glutathione transferase modulates acute ethanol-induced sedation in Drosophila neurones. Insect Mol Biol 28(2):246–252. https://doi.org/10.1111/imb.12547

Choi MK, Son S, Hong M, Choi MS, Kwon JY, Lee J (2016) Maintenance of membrane integrity and permeability depends on a patched-related protein in Caenorhabditis elegans. Genetics 202(4):1411–1420. https://doi.org/10.1534/genetics.115.179705

Claßen G, Scholz H (2018) Octopamine shifts the behavioral response from indecision to approach or aversion in Drosophila melanogaster. Front Behav Neurosci 12:131. https://doi.org/10.3389/fnbeh.2018.00131

Dopico AM, Bukiya AN, Kuntamallappanavar G, Liu J (2016) Modulation of BK channels by ethanol. Int Rev Neurobiol 128:239–279. https://doi.org/10.1016/bs.irn.2016.03.019

Dzitoyeva S, Dimitrijevic N, Manev H (2003) Gamma-aminobutyric acid B receptor 1 mediates behavior-impairing actions of alcohol in Drosophila: adult RNA interference and pharmacological evidence. Proc Natl Acad Sci U S A 100(9):5485–5490. https://doi.org/10.1073/pnas.0830111100

Engel GL, Taber K, Vinton E, Crocker AJ (2019) Studying alcohol use disorder using Drosophila melanogaster in the era of 'Big Data'. Behav Brain Funct 15(1):7. https://doi.org/10.1186/s12993-019-0159-x

Frank J, Cichon S, Treutlein J, Ridinger M, Mattheisen M, Hoffmann P, Herms S, Wodarz N, Soyka M, Zill P, Maier W, Mössner R, Gaebel W, Dahmen N, Scherbaum N, Schmäl C, Steffens M, Lucae S, Ising M, Müller-Myhsok B, Nöthen MM, Mann K, Kiefer F, Rietschel M (2012) Genome-wide significant association between alcohol dependence and a variant in the ADH gene cluster. Addict Biol 17(1):171–180. https://doi.org/10.1111/j.1369-1600.2011.00395.x

Giang T, He J, Belaidi S, Scholz H (2017) Key odorants regulate food attraction in Drosophila melanogaster. Front Behav Neurosci 11:160. https://doi.org/10.3389/fnbeh.2017.00160

Gibson JB, Wilks AV (1988) The alcohol dehydrogenase polymorphism of Drosophila melanogaster in relation to environmental ethanol, ethanol tolerance and alcohol dehydrogenase activity. Heredity 60(3):403–414. https://doi.org/10.1038/hdy.1988.58

Gonzalez DA, Jia T, Pinzón JH, Acevedo SF, Ojelade SA, Xu B, Tay N, Desrivières S, Hernandez JL, Banaschewski T, Büchel C, Bokde ALW, Conrod PJ, Flor H, Frouin V, Gallinat J, Garavan H, Gowland PA, Heinz A, Ittermann B, Lathrop M, Martinot JL, Paus T, Smolka MN, IMAGEN Consortium, Rodan AR, Schumann G, Rothenfluh A (2018) The Arf6 activator Efa6/PSD3 confers regional specificity and modulates ethanol consumption in Drosophila and humans. Mol Psychiatry 23(3):621–628. https://doi.org/10.1038/mp.2017.112

Goodrich KR, Zjhra ML, Ley CA, Raguso RA (2006) When flowers smell fermented: the chemistry and ontogeny of yeasty floral scent in pawpaw (Asimina triloba: Annonaceae). Int J Plant Sci 167(1):33–46. https://doi.org/10.1086/498351

Gravielle MC (2018) Regulation of GABAA receptors by prolonged exposure to endogenous and exogenous ligands. Neurochem Int 118:96–104. https://doi.org/10.1016/j.neuint.2018.05.015

Grotewiel M, Bettinger JC (2015) Drosophila and Caenorhabditis elegans as discovery platforms for genes involved in human alcohol use disorder. Alcohol Clin Exp Res 39(8):1292–1311. https://doi.org/10.1111/acer.12785

Guevara A, Gates H, Urbina B, French R (2018) Developmental ethanol exposure causes reduced feeding and reveals a critical role for neuropeptide F in survival. Front Physiol 9:237. https://doi.org/10.3389/fphys.2018.00237

Hermann A, Sitdikova GF, Weiger TM (2015) Oxidative stress and maxi calcium-activated potassium (BK) channels. Biomol Ther 5(3):1870–1911. https://doi.org/10.3390/biom5031870

Holmes RS (1994) Alcohol dehydrogenases: a family of isozymes with differential functions. Alcohol Alcohol Suppl 2:127–130

Johnson JR, Edwards MR, Davies H, Newman D, Holden W, Jenkins RE, Burgoyne RD, Lucas RJ, Barclay JW (2017) Ethanol stimulates locomotion via a $G_{\alpha s}$-signaling pathway in IL2 neurons in Caenorhabditis elegans. Genetics 207(3):1023–1039. https://doi.org/10.1534/genetics.117.300119

Johnson JR, Rajamanoharan D, McCue HV, Rankin K, Barclay JW (2016) Small heat shock proteins are novel common determinants of alcohol and nicotine sensitivity in Caenorhabditis elegans. Genetics 202(3):1013–1027. https://doi.org/10.1534/genetics.115.185025

Kanno M, Hiramatsu S, Kondo S, Tanimoto H, Ichinose T (2021) Voluntary intake of psychoactive substances is regulated by the dopamine receptor Dop1R1 in Drosophila. Sci Rep 11(1):3432. https://doi.org/10.1038/s41598-021-82813-0

Kaptan D, Penkov S, Zhang X, Gade VR, Raghuraman BK, Galli R, Sampaio JL, Haase R, Koch E, Shevchenko A, Zaburdaev V, Kurzchalia TV (2020) Exogenous ethanol induces a metabolic switch that prolongs the survival of Caenorhabditis elegans dauer larva and enhances its resistance to desiccation. Aging Cell 19(10):e13214. https://doi.org/10.1111/acel.13214

Kevan PG, Eisikowitch D, Fowle S, Thomas K (1988) Yeast-contaminated nectar and its effects on bee foraging. J Apic Res 27(1):26–29. https://doi.org/10.1080/00218839.1988.11100777

Kishimoto T, Noba S, Yako N, Kobayashi M, Watanabe T (2018) Simulation of pilsner-type beer aroma using 76 odor-active compounds. J Biosci Bioeng 126(3):330–338. https://doi.org/10.1016/j.jbiosc.2018.03.015

Kong EC, Woo K, Li H, Lebestky T, Mayer N, Sniffen MR, Heberlein U, Bainton RJ, Hirsh J, Wolf FW (2010) A pair of dopamine neurons target the D1-like dopamine receptor DopR in the central complex to promote ethanol-stimulated locomotion in Drosophila. PLoS One 5(4): e9954. https://doi.org/10.1371/journal.pone.0009954

Koyyada R, Latchooman N, Jonaitis J, Ayoub SS, Corcoran O, Casalotti SO (2018) Naltrexone reverses ethanol preference and protein kinase C activation in Drosophila melanogaster. Front Physiol 9:175. https://doi.org/10.3389/fphys.2018.00175

Lathen DR, Merrill CB, Rothenfluh A (2020) Flying together: Drosophila as a tool to understand the genetics of human alcoholism. Int J Mol Sci 21(18):6649. https://doi.org/10.3390/ijms21186649

Lewis KN, Mele J, Hayes JD, Buffenstein R (2010) Nrf2, a guardian of healthspan and gatekeeper of species longevity. Integr Comp Biol 50(5):829–843. https://doi.org/10.1093/icb/icq034

Lin R, Rittenhouse D, Sweeney K, Potluri P, Wallace DC (2015) TSPO, a mitochondrial outer membrane protein, controls ethanol-related behaviors in Drosophila. PLoS Genet 11(8): e1005366. https://doi.org/10.1371/journal.pgen.1005366

Logan-Garbisch T, Bortolazzo A, Luu P, Ford A, Do D, Khodabakhshi P, French RL (2014) Developmental ethanol exposure leads to dysregulation of lipid metabolism and oxidative stress in Drosophila. G3 (Bethesda) 5(1):49–59. https://doi.org/10.1534/g3.114.015040

Luo Y, Johnson JC, Chakraborty TS, Piontkowski A, Gendron CM, Pletcher SD (2021) Yeast volatiles double starvation survival in Drosophila. Sci Adv 7(20):eabf8896. https://doi.org/10.1126/sciadv.abf8896

Mann K, Lehert P, Morgan MY (2004) The efficacy of acamprosate in the maintenance of abstinence in alcohol-dependent individuals: results of a meta-analysis. Alcohol Clin Exp Res 28(1):51–63. https://doi.org/10.1097/01.ALC.0000108656.81563.05

Martins E, Mestriner MA, Contel EP (1977) Alcohol dehydrogenase polymorphism in Apis mellifera. Biochem Genet 15(3–4):357–366. https://doi.org/10.1007/BF00484466

Mason BJ, Heyser CJ (2021) Alcohol use disorder: the role of medication in recovery. Alcohol Res 41(1):07. https://doi.org/10.35946/arcr.v41.1.07

McKenzie JA, McKechnie SW (1979) A comparative study of resource utilization in natural populations of Drosophila melanogaster and D. simulans. Oecologia 40(3):299–309. https://doi.org/10.1007/BF00345326

Morton H, Kshirsagar S, Orlov E, Bunquin LE, Sawant N, Boleng L, George M, Basu T, Ramasubramanian B, Pradeepkiran JA, Kumar S, Vijayan M, Reddy AP, Reddy PH (2021) Defective mitophagy and synaptic degeneration in Alzheimer's disease: focus on aging, mitochondria and synapse. Free Radic Biol Med 172:652–667. https://doi.org/10.1016/j.freeradbiomed.2021.07.013

Mustard JA, Oquita R, Garza P, Stoker A (2019) Honey bees (Apis mellifera) show a preference for the consumption of ethanol. Alcohol Clin Exp Res 43(1):26–35. https://doi.org/10.1111/acer.13908

Nikookar H, Haddadi M, Haghi M, Masoudi R (2021) DNT1 downregulation and increased ethanol sensitivity in transgenic Drosophila Models of Alzheimer's disease. Arch Gerontol Geriatr 94:104355. https://doi.org/10.1016/j.archger.2021.104355

Niveditha S, Deepashree S, Ramesh SR, Shivanandappa T (2017) Sex differences in oxidative stress resistance in relation to longevity in Drosophila melanogaster. J Comp Physiol B 187(7):899–909. https://doi.org/10.1007/s00360-017-1061-1

Nunez KM, Azanchi R, Kaun KR (2018) Cue-induced ethanol seeking in Drosophila melanogaster is dose-dependent. Front Physiol 9:438. https://doi.org/10.3389/fphys.2018.00438

Ogueta M, Cibik O, Eltrop R, Schneider A, Scholz H (2010) The influence of Adh function on ethanol preference and tolerance in adult Drosophila melanogaster. Chem Senses 35(9):813–822. https://doi.org/10.1093/chemse/bjq084

Oh KH, Kim H (2019) BK channel clustering is required for normal behavioral alcohol sensitivity in C. elegans. Sci Rep 9(1):10224. https://doi.org/10.1038/s41598-019-46615-9

Oh KH, Sheoran S, Richmond JE, Kim H (2020) Alcohol induces mitochondrial fragmentation and stress responses to maintain normal muscle function in Caenorhabditis elegans. FASEB J 34(6):8204–8216. https://doi.org/10.1096/fj.201903166R

Ojelade SA, Jia T, Rodan AR, Chenyang T, Kadrmas JL, Cattrell A, Ruggeri B, Charoen P, Lemaitre H, Banaschewski T, Büchel C, Bokde AL, Carvalho F, Conrod PJ, Flor H, Frouin V, Gallinat J, Garavan H, Gowland PA, Heinz A, Ittermann B, Lathrop M, Lubbe S, Martinot JL, Paus T, Smolka MN, Spanagel R, O'Reilly PF, Laitinen J, Veijola JM, Feng J, Desrivières S, Jarvelin MR, IMAGEN Consortium, Schumann G, Rothenfluh A (2015) Rsu1 regulates ethanol consumption in Drosophila and humans. Proc Natl Acad Sci U S A 112(30):E4085–E4093. https://doi.org/10.1073/pnas.1417222112

Ostap-Chec M, Opalek M, Stec D, Miler K (2021) Discontinued alcohol consumption elicits withdrawal symptoms in honeybees. Biol Lett 17(6):20210182. https://doi.org/10.1098/rsbl.2021.0182

Pandey P, Singh A, Kaur H, Ghosh-Roy A, Babu K (2021) Increased dopaminergic neurotransmission results in ethanol dependent sedative behaviors in Caenorhabditis elegans. PLoS Genet 17(2):e1009346. https://doi.org/10.1371/journal.pgen.1009346

Park A, Ghezzi A, Wijesekera TP, Atkinson NS (2017) Genetics and genomics of alcohol responses in Drosophila. Neuropharmacology 122:22–35. https://doi.org/10.1016/j.neuropharm.2017.01.032

Park A, Tran T, Scheuermann EA, Smith DP, Atkinson NS (2020) Alcohol potentiates a pheromone signal in flies. eLife 9:e59853. https://doi.org/10.7554/eLife.59853

Park SJ, Ja WW (2020) Absolute ethanol intake predicts ethanol preference in Drosophila melanogaster. J Exp Biol 223(11):jeb224121. https://doi.org/10.1242/jeb.224121

Peng B, Yang Q, Joshi RB, Liu Y, Akbar M, Song BJ, Zhou S, Wang X (2020) Role of alcohol drinking in Alzheimer's disease, Parkinson's disease, and amyotrophic lateral sclerosis. Int J Mol Sci 21(7):2316. https://doi.org/10.3390/ijms21072316

Petruccelli E, Li Q, Rao Y, Kitamoto T (2016) The unique dopamine/ecdysteroid receptor modulates ethanol-induced sedation in Drosophila. J Neurosci 36(16):4647–4657. https://doi.org/10.1523/JNEUROSCI.3774-15.2016

Quadir SG, Tanino SM, Rohl CD, Sahn JJ, Yao EJ, Cruz LDR, Cottone P, Martin SF, Sabino V (2021) The Sigma-2 receptor / transmembrane protein 97 (σ2R/TMEM97) modulator

JVW-1034 reduces heavy alcohol drinking and associated pain states in male mice. Neuropharmacology 184:108409. https://doi.org/10.1016/j.neuropharm.2020.108409

Quintanilla RA, Pérez MJ, Aranguiz A, Tapia-Monsalves C, Mendez G (2020) Activation of the melanocortin-4 receptor prevents oxidative damage and mitochondrial dysfunction in cultured hippocampal neurons exposed to ethanol. Neurotox Res 38(2):421–433. https://doi.org/10.1007/s12640-020-00204-1

Raabe RC, Mathies LD, Davies AG, Bettinger JC (2014) The omega-3 fatty acid eicosapentaenoic acid is required for normal alcohol response behaviors in C. elegans. PLoS One 9(8):e105999. https://doi.org/10.1371/journal.pone.0105999

Ranson DC, Ayoub SS, Corcoran O, Casalotti SO (2020) Pharmacological targeting of the GABAB receptor alters Drosophila's behavioural responses to alcohol. Addict Biol 25(2):e12725. https://doi.org/10.1111/adb.12725

Ruppert M, Franz M, Saratsis A, Velo Escarcena L, Hendrich O, Gooi LM, Schwenkert I, Klebes A, Scholz H (2017) Hangover links nuclear RNA Signaling to cAMP regulation via the phosphodiesterase 4d ortholog dunce. Cell Rep 18(2):533–544. https://doi.org/10.1016/j.celrep.2016.12.048

Saeedi BJ, Liu KH, Owens JA, Hunter-Chang S, Camacho MC, Eboka RU, Chandrasekharan B, Baker NF, Darby TM, Robinson BS, Jones RM, Jones DP, Neish AS (2020) Gut-resident lactobacilli activate hepatic Nrf2 and protect against oxidative liver injury. Cell Metab 31(5):956–968.e5. https://doi.org/10.1016/j.cmet.2020.03.006

Scaplen KM, Talay M, Nunez KM, Salamon S, Waterman AG, Gang S, Song SL, Barnea G, Kaun KR (2020) Circuits that encode and guide alcohol-associated preference. eLife 9:e48730. https://doi.org/10.7554/eLife.48730

Scepanovic G, Stewart BA (2019) Analysis of Drosophila nervous system development following an early, brief exposure to ethanol. Dev Neurobiol 79(8):780–793. https://doi.org/10.1002/dneu.22718

Schmitt RE, Messick MR, Shell BC, Dunbar EK, Fang HF, Shelton KL, Venton BJ, Pletcher SD, Grotewiel M (2020) Dietary yeast influences ethanol sedation in Drosophila via serotonergic neuron function. Addict Biol 25(4):e12779. https://doi.org/10.1111/adb.12779

Scholz H (2019) Unraveling the mechanisms of behaviors associated with AUD using flies and worms. Alcohol Clin Exp Res 43(11):2274–2284. https://doi.org/10.1111/acer.14199

Scholz H, Franz M, Heberlein U (2005) The hangover gene defines a stress pathway required for ethanol tolerance development. Nature 436(7052):845–847. https://doi.org/10.1038/nature03864

Scholz H, Mustard JA (2013) Invertebrate models of alcoholism. Curr Top Behav Neurosci 13:433–457. https://doi.org/10.1007/7854_2011_128

Schumann I, Berger M, Nowag N, Schäfer Y, Saumweber J, Scholz H, Thum AS (2021) Ethanol-guided behavior in Drosophila larvae. Sci Rep 11(1):12307. https://doi.org/10.1038/s41598-021-91677-3

Scott LL, Davis SJ, Yen RC, Ordemann GJ, Nordquist SK, Bannai D, Pierce JT (2017) Behavioral deficits following withdrawal from chronic ethanol are influenced by SLO channel function in Caenorhabditis elegans. Genetics 206(3):1445–1458. https://doi.org/10.1534/genetics.116.193102

Scott LL, Iyer S, Philpo AE, Avalos MN, Wu NS, Shi T, Prakash BA, Nguyen TT, Mihic SJ, Aldrich RW, Pierce JT (2018a) A novel peptide restricts ethanol modulation of the BK channel in vitro and in vivo. J Pharmacol Exp Ther 367(2):282–290. https://doi.org/10.1124/jpet.118.251918

Scott LL, Sahn JJ, Ferragud A, Yen RC, Satarasinghe PN, Wood MD, Hodges TR, Shi T, Prakash BA, Friese KM, Shen A, Sabino V, Pierce JT, Martin SF (2018b) Small molecule modulators of σ2R/Tmem97 reduce alcohol withdrawal-induced behaviors. Neuropsychopharmacology 43(9):1867–1875. https://doi.org/10.1038/s41386-018-0067-z

Sekhon ML, Lamina O, Hogan KE, Kliethermes CL (2016) Common genes regulate food and ethanol intake in Drosophila. Alcohol 53:27–34. https://doi.org/10.1016/j.alcohol.2016.04.001

Siddiq MA, Loehlin DW, Montooth KL, Thornton JW (2017) Experimental test and refutation of a classic case of molecular adaptation in Drosophila melanogaster. Nat Ecol Evol 1(2):25. https://doi.org/10.1038/s41559-016-0025

Signor SA (2020) Evolution of plasticity in response to ethanol between sister species with different ecological histories (Drosophila melanogaster and D. simulans). Am Nat 196(5):620–633. https://doi.org/10.1086/710763

Signor S, Nuzhdin S (2018) Dynamic changes in gene expression and alternative splicing mediate the response to acute alcohol exposure in Drosophila melanogaster. Heredity 121(4):342–360. https://doi.org/10.1038/s41437-018-0136-4

Sokolowski MB, Abramson CI, Craig DP (2012) Ethanol self-administration in free-flying honeybees (Apis mellifera L.) in an operant conditioning protocol. Alcohol Clin Exp Res 36(9): 1568–1577. https://doi.org/10.1111/j.1530-0277.2012.01770.x

Song Y, Li S, Zhao Y, Zhang Y, Lv Y, Jiang Y, Wang Y, Li D, Zhang H (2019) ADH1 promotes Candida albicans pathogenicity by stimulating oxidative phosphorylation. Int J Med Microbiol 309(6):151330. https://doi.org/10.1016/j.ijmm.2019.151330

Sumethasorn M, Turner TL (2016) Oviposition preferences for ethanol depend on spatial arrangement and differ dramatically among closely related Drosophila species. Biol Open 5(11): 1642–1647. https://doi.org/10.1242/bio.019380

Towarnicki SG, Kok LM, Ballard JWO (2020) Yin and Yang of mitochondrial ROS in Drosophila. J Insect Physiol 122:104022. https://doi.org/10.1016/j.jinsphys.2020.104022

Treutlein J, Cichon S, Ridinger M, Wodarz N, Soyka M, Zill P, Maier W, Moessner R, Gaebel W, Dahmen N, Fehr C, Scherbaum N, Steffens M, Ludwig KU, Frank J, Wichmann HE, Schreiber S, Dragano N, Sommer WH, Leonardi-Essmann F, Lourdusamy A, Gebicke-Haerter-P, Wienker TF, Sullivan PF, Nöthen MM, Kiefer F, Spanagel R, Mann K, Rietschel M (2009) Genome-wide association study of alcohol dependence. Arch Gen Psychiatry 66(7):773–784. https://doi.org/10.1001/archgenpsychiatry.2009.83

van Breugel F, Huda A, Dickinson MH (2018) Distinct activity-gated pathways mediate attraction and aversion to CO2 in Drosophila. Nature 564(7736):420–424. https://doi.org/10.1038/s41586-018-0732-8

Varnon CA, Dinges CW, Black TE, Wells H, Abramson CI (2018) Failure to find ethanol-induced conditioned taste aversion in honey bees (Apis mellifera L.). Alcohol Clin Exp Res 42(7): 1260–1270. https://doi.org/10.1111/acer.13761

Velo Escarcena L, Neufeld M, Rietschel M, Spanagel R, Scholz H (2021) ERR and dPECR suggest a link between neuroprotection and the regulation of ethanol consumption preference. Front Psych 12:655816. https://doi.org/10.3389/fpsyt.2021.655816

Wang X, Zhang C, Chen Q, Ma Z, Liu H, Huang J (2021) Guanylate cyclases link serotoninergic signaling to modulate ethanol-induced food intake in C. elegans. Biochem Biophys Res Commun 567:29–34. https://doi.org/10.1016/j.bbrc.2021.06.006

Wolstenholme JT, Bowers MS, Pais AB, Pais AC, Poland RS, Poklis JL, Davies AG, Bettinger JC (2018) Dietary omega-3 fatty acids differentially impact acute ethanol-responsive behaviors and ethanol consumption in DBA/2J versus C57BL/6J mice. Alcohol Clin Exp Res 42(8): 1476–1485. https://doi.org/10.1111/acer.13780

Wu ZQ, Li K, Tian X, Zhou MX, Li ZJ (2020) Schisandra chinensis water extract protects ethanol-induced neurotoxicity in Caenorhabditis elegans. J Food Biochem 44(8):e13249. https://doi.org/10.1111/jfbc.13249

Xu L, He J, Kaiser A, Gräber N, Schläger L, Ritze Y, Scholz H (2016) A single pair of serotonergic neurons counteracts serotonergic inhibition of ethanol attraction in Drosophila. PLoS One 11(12):e0167518. https://doi.org/10.1371/journal.pone.0167518

Zhou Y, Men L, Sun Y, Wei M, Fan X (2021) Pharmacodynamic effects and molecular mechanisms of lignans from Schisandra chinensis Turcz. (Baill.), a current review. Eur J Pharmacol 892:173796. https://doi.org/10.1016/j.ejphar.2020.173796

Zhu J, Fry JD (2015) Preference for ethanol in feeding and oviposition in temperate and tropical populations of Drosophila melanogaster. Entomol Exp Appl 155(1):64–70. https://doi.org/10.1111/eea.12285

Zhu J, Fry JD (2018) Effects of a low dose of ethanol on mating success of Drosophila melanogaster males: implications for the evolution of ethanol resistance? Entomol Exp Appl 166(10): 801–809. https://doi.org/10.1111/eea.12714

Good Practice Guideline for Preclinical Alcohol Research: The STRINGENCY Framework

Marcus W. Meinhardt, Björn Gerlach, and Rainer Spanagel

Contents

M. W. Meinhardt (✉)
Institute of Psychopharmacology, Central Institute of Mental Health, Medical Faculty Mannheim, University of Heidelberg, Heidelberg, Germany

Department of Molecular Neuroimaging, Medical Faculty Mannheim, Central Institute of Mental Health, University of Heidelberg, Heidelberg, Germany
e-mail: marcus.meinhardt@zi-mannheim.de

B. Gerlach
Institute of Psychopharmacology, Central Institute of Mental Health, Medical Faculty Mannheim, University of Heidelberg, Heidelberg, Germany

Guarantors of EQIPD e.V., Heidelberg, Germany

PAASP GmbH, Heidelberg, Germany

R. Spanagel (✉)
Institute of Psychopharmacology, Central Institute of Mental Health, Medical Faculty Mannheim, University of Heidelberg, Heidelberg, Germany

German Center for Mental Health (DZPG), Mannheim, Heidelberg, Ulm, Germany
e-mail: rainer.spanagel@zi-mannheim.de

Curr Topics Behav Neurosci (2025) 71: 169–189
https://doi.org/10.1007/7854_2024_484
Published Online: 09 August 2024

Abstract Research in the field of preclinical alcohol research, but also science in general, has a problem: Many published scientific results cannot be repeated. As a result, findings from preclinical research often do not translate well to humans, causing increasing disappointment and calls for restructuring of preclinical research, that is, better reproducibility of preclinical research. However, the replication crisis is an inherent problem in biomedical research. Replication failures are not only due to small experimental variations but are often the result of poor methodology. In response to the replication crisis, numerous guidelines and recommendations have been proposed to promote transparency, rigor, and reproducibility in scientific research. What is missing today is a framework that integrates all the confusing information that results from all these guidelines and recommendations. Here we present STRINGENCY, an integrative approach to good practice guidelines for preclinical alcohol research, which can also apply to behavioral research in general and which aims to improve preclinical research to better prepare it for translation and minimize the "valley of death" in translational research. STRINGENCY includes systematic review and, when possible, meta-analysis prior to study design, sample size calculation, preregistration, multisite experiments, scientific data management (FAIR), reporting of data using ARRIVE, generalization of research data, and transparent publications that allow reporting of null results. We invite the scientific community to adopt STRINGENCY to improve the reliability and impact of preclinical alcohol research.

Keywords Guideline · Methodology · Preclinical research · Replication crisis · Reproducibility

1 Introduction

In recent years, the scientific community has been confronted with observations of low reproducibility of experimental findings (Freedman et al. 2015). However, much earlier, Crabbe and colleagues published observations in Science, which disrupted the field of behavioral phenotyping, causing doubts on the replicability also on alcohol-related research, and thereby already challenging reproducibility (Crabbe et al. 1999). John Crabbe from Portland together with colleagues from laboratories in Edmonton and Albany studied eight inbred strains of mice using various behavioral tests. All laboratory conditions—from the light-dark cycle to the manufacturer of the mouse food, etc.—were harmonized between the three laboratories. However, the results were strikingly different across laboratories. The study by Crabbe et al. (1999) published in Science caused a shock wave in the global scientific community because it indicated that even simple parameters such as motor activity and weight gain are site dependent (with the exception of gender differences) and thus not replicable. In 2005, the seminal publication authored by John P. A. Ioannidis made predictions for low reproducibility and is one of the most cited articles (Ioannidis

2005). It describes several theoretical assumptions why most research findings could be false. Six years later, in 2011, two observations from research groups at Bayer (Prinz et al. 2011) and Amgen (Begley and Ellis 2012) looked retrospectively how many of published studies they could reproduce and found very low numbers in the range of 11–25%.

In the preclinical alcohol research field, animal models advanced our understanding of brain mechanisms of addictive-like behavior and relapse, but these mechanistic gains did in many cases fail to translate to the human situation. The most prominent example is the translation failure of corticotropin-releasing hormone (CRH) 1 receptor antagonists as treatment for relapse prevention. Over two decades, numerous animal studies provided convergent evidence that CRH1 receptor antagonists ought to be promising drugs for human AUD (Heilig and Koob 2007). Accordingly, numerous pharmaceutical companies generated excellent compounds but all clinical trials failed (e.g., Kwako et al. (2015) and Schwandt et al. (2016)). Clearly, this as well as many other translation failures is an increasing source of disappointment and calls to regroup.

Translation failures are not unique to preclinical alcohol research but to preclinical research in general, and this has led to the widespread calls for actions to enhance research integrity and address the underlying causes of nonreproducible research. Numerous guidelines and recommendations have been proposed to promote transparency, rigor, and reproducibility in scientific research. Several institutions and activities such as the Alliance of European Academies (ALLEA) and the Research Integrity Promotion Plan (RIPP) from the SOP4RI consortium have provided guidance to address these concerns (The European Code of Conduct for Research Integrity[1] and Horbach et al. 2022). The *Enhancing Quality in Preclinical Data* (EQIPD) project aimed to provide practical guidance by offering a systematic approach on implementation of best practices in preclinical research labs (Bespalov et al. 2021). Recently, we introduced ten points of recommendation, especially for preclinical confirmatory studies but also to some degree for exploratory studies, that will ultimately—if applied—improve the reproducibility and translation of animal research (Spanagel 2022). These initiatives aim to provide researchers with the tools and principles required to conduct robust and replicable studies. However, despite these efforts, there remains a need for more specific detailed recommendations tailored to the nature of individual research fields. This article introduces the STRINGENCY framework (see Fig. 1), which has a focus on preclinical alcohol research and may apply to behavioral research in general intending to improve preclinical research to better prepare for translation and to minimize the notorious "valley of death" in translational research (Seyhan 2019). The STRINGENCY framework aims to provide an unified framework, which is easy to follow and implement in daily research practice in order to enhance the quality and replicability of preclinical research.

[1] http://www.doi.org/10.26356/ECOC

Fig. 1 Ten recommended criteria that form the STRINGENCY framework

S ystematic review prior to study plan

T ailord scietific data management

R egistration of study plan

I nception of study by calculated sample size

N etworks of centres for multi-site experiments

G eneralization of research data

E nunciation of data using ARRIVE

N ull-results should be published

C apturing DSM5/ICD11 criteria in animal models

Y Methodolg<u>Y</u> should be validated and under control

FRAMEWORK

In the following sections, we will introduce the STRINGENCY framework, explore their components, and give practical examples on how to implement them. By addressing the challenges of reproducibility and translational research, the framework offers a holistic approach to research integrity, and we invite the scientific community to adopt this framework to drive meaningful improvements in the reliability and impact of preclinical alcohol research.

2 Systematic Review Prior to Study Plan and Meta-Analysis

A literature search lays the foundation for any research project. Typically, this task is undertaken by the researcher, and for experienced researchers, it is a routine matter. However, with the staggering number of over two million articles published annually (Ioannidis et al. 2018), staying current or delving into a new research field becomes a near-impossible challenge. To gain a comprehensive understanding, it is highly advisable to conduct a systematic review (SR) of the literature. An SR is an unbiased, systematic, and well-documented process that follows a predefined protocol to extract information from the available literature. To enhance transparency and the value to the research community, this protocol should be preregistered. While the process of systematic review is well established in clinical research, it has only recently been described for preclinical research as well. A prime example demonstrating the utility of a prior SR or meta-analysis, which also challenged a prevailing hypothesis in the alcohol field, was the observation of dynamic changes in accumbal dopamine levels during abstinence with a hypodopaminergic state during acute withdrawal followed by a hyperdopaminergic state in protracted abstinence (Hirth et al. 2016).

A standardized protocol for animal studies has been published (de Vries et al. 2015) detailing how SRs should be planned and executed, along with a template for performing a SR. In brief, a SR is structured into three sections: (A) General Information, (B) Objectives, and (C) Methods. These sections are further subdivided into 50 items. The General Information section covers administrative aspects, including questions about the researcher's team and reflecting about own practices. The Objectives section focuses on defining the review's purpose and should consider the elements of the PICO mnemonic: Population (including the disease model), Intervention, Comparisons (control groups), and Outcome measures. These elements form the basis for the research question.

Section (C) Methods contains the most items and provides a comprehensive description of the study. It encompasses the definition of search terms in as much detail as possible, the criteria for selecting studies (inclusion/exclusion), and measures to mitigate bias. It also addresses crucial study aspects, such as study design, disease model, and outcome measures. Furthermore, it includes an assessment of the risk of bias and study quality, evaluating the external validity or quality of the original studies. This is based on a "risk of bias" (RoB) tool which was originally developed for clinical research (Higgins et al. 2011) and subsequently adapted for animal research (Hooijmans et al. 2014). This tool developed by Hooijmans and colleagues named the SYRCLE RoB tool (Hooijmans et al. 2014) is particularly valuable for a comprehensive understanding of different biases which might occur during a research study. For example, it investigates selection bias (e.g., were the groups similar at baseline or were they adjusted for confounders in the analysis?), performance bias (e.g., were the animals randomly housed during the experiment?), detection bias (e.g., was the outcome assessor blinded?), attrition bias (e.g., were incomplete outcome data adequately addressed?), and reporting bias (e.g., are reports of the study free of selective outcome reporting?).

Conducting a systematic review is a comprehensive process that offers an in-depth view of the literature on a specific topic. Preregistering an SR is highly recommended to enhance its value, and various platforms are available for preregistration of systematic reviews. The "Systematic Review Facility" (SyRF) is one such platform designed for planning and facilitating SRs, allowing for collaboration, registration, and publication of the review. However, in many cases, a comprehensive SR may be impractical, and a more streamlined "SR light" versions might be suitable. In such instances, the SR protocol can serve as a guide for a general overview of the field, offering a systematic approach to exploring the literature, thereby laying the groundwork for the development of new pharmacological targets.

In cases when full SRs are performed, it could be considered to also make a meta-analysis of the collected data. This formal and quantitative study design may provide additional and valuable knowledge about a research field to derive conclusions beyond the SR. For this analysis, research data from different groups are analyzed together to provide a more precise estimate of the effect of treatment, or other outcomes, than any individual study contributing to the pooled analysis. Vesterinen and colleagues described very well the methodology for performing a meta-analysis (Vesterinen et al. 2014). Briefly, the protocol should be described as detailed as possible before starting the analysis, followed by the data extraction. The core of the meta-analysis is the calculation of the effect sizes from the different studies, weighting them to be able to subsequently combine from similar outcome measures and finally calculate the summary effect size. However, since the devil is in the detail, the "Preclinical Systematic Review Wiki"[2] provides very good guidance also on meta-analysis, and in some cases, the support from experienced colleagues might be required.

3 Data Management

Organizing research data can be a complex endeavor, as data are diverse and acquired in various contexts. Typically, individual researchers excel in structuring their own data, especially when they have developed their own systems that they understand well. Consequently, accessing data within a researcher's personal system or database usually works effectively. However, challenges arise when colleagues attempt to retrieve data from such individualized systems. The complexity further intensifies when large research consortia collaborate on a shared research goal, generating substantial volumes of data.

To address these issues, it is highly advisable to document all data-related information in a comprehensive Data Management Plan (DMP). This plan encompasses data organization and storage processes, a detailed data description, and metadata. Numerous guidelines and tools are available for crafting a DMP, such as the DataStewardWizard[3] or the DMPTool.[4] These tools are free and open-source to help researchers create DMPs. These plans are now required by many funding agencies as part of the grant proposal submission process, and these tools comply with funder requirements.

[2] https://www.camarades.de.

[3] https://ds-wizard.org.

[4] https://dmptool.org/plans.

In this context, one of the most pivotal guiding principles for data integrity is the FAIR framework. FAIR, formulated during a 2015 workshop, stands for "findable," "accessible," "interoperable," and "reusable" (Wilkinson et al. 2016). The FAIR principles aim to enhance data integrity and ensure the reusability of research data, not only by humans but also by automated systems. These guidelines primarily focus on enriching data with metadata to enhance its utility over an extended period. This is especially crucial when transferring data to a public repository, as data can become practically useless without such information, even if they are publicly available.

While understanding these guiding principles theoretically is relatively straightforward, putting them into practice and persuading researchers to adhere to them is often a more challenging task. Nevertheless, it is an essential requirement since data should remain discoverable, even after a researcher leaves a laboratory. Hence, every tool and approach that proves effective in a research setting is invaluable.

One systematic approach to bolster data integrity involves implementing a quality system within a research laboratory. For clinical and regulated research, adhering to Good Laboratory Practice, which adheres to rigorous procedures, is a common practice. In preclinical nonregulated research, systematic approaches are also available, such as the EQIPD Quality System (Bespalov et al. 2021). This system is less reliant on extensive documentation and instead emphasizes commonsense practices to achieve optimal results. They provide guidance and structure for implementing measures, such as the use of unique study identifiers across the research lab and designating a responsible individual (such as a data steward) to support these efforts and address data integrity concerns.

These approaches should be established in collaboration with the entire research team. Dedicating time during regular lab meetings to discuss data integrity can be highly productive and foster a culture of heightened awareness. Particularly in larger research laboratories, a systematic approach to data storage should incorporate electronic solutions, such as electronic lab notebooks (ELNs) or comprehensive electronic lab management systems. ELNs are helpful tools to organize research data and allow for easier searchability than paper lab notebooks. They can be purchased from vendors but also several very good open-source solutions exist. The functionality is also quite different, for example, some offer a generic user interface and work like a general documentation systems such as Microsoft OneNote whereas others are tailored for specific environments (Gerlach et al. 2020). It is important that each research unit defines their specific needs and chooses an ELN accordingly. There is no doubt that a well-implemented ELN ensures data management on the highest quality level for research groups.

In conclusion, reading about these guidelines and their theoretical understanding is easy; however, putting them into practice and asking researchers to follow them is usually much harder. But it is essential to do so as one has to fulfill the requirements of funders, and it is critically important for the future success of a lab to have an up-to-date data management structure.

4 Experimental Study Protocols Preregistered

Preregistration is a relatively new approach to combat publication bias and, in recent times, has gained recognition as a best practice in preclinical research. The primary rationale behind preregistering studies lies in mitigating reporting bias and preventing practices such as HARKing and p-hacking. Moreover, it contributes to reducing reporting bias because the study plan is disclosed to the public before results are known. Consequently, any researcher can scrutinize the results of a specific study plan, whether the outcome is positive or negative.

As of March 2023, more than 300 journals are affiliated with the Open Science Framework,[5] offering support for a review process through registered reports. If a study plan is accepted, the journal commits to publishing the data independently of the outcome. Multiple platforms are available for preregistration, including PeclinicalTrials.eu,[6] Animal Study Registry,[7] OSF with the Registered Reports,[8] and PROSPERO.[9] Notably, there are no significant differences in their functionality. It is worth highlighting that all of these platforms allow for an embargo period during which the study plan remains confidential. The long-term funding and sustainability of these platforms are assured, often with government funding playing a key role (Bert et al. 2019).

A prevalent misconception is that preregistration impedes the emergence of new scientific insights, which often arise from exploratory studies. However, preregistration can indeed be beneficial, particularly in the context of exploratory studies, and remains serving as a catalyst for uncovering new pathways and formulating innovative hypotheses in biomedical research. Platforms such as the "Animal Study Registry" enable researchers to specify whether their registered study is exploratory or confirmatory, while also facilitating the registration of "Minimal Content" for exploratory studies (Heinl et al. 2022). This practice of generating minimal content registrations significantly enhances the quality of study planning for exploratory experiments, prompting researchers to consider critical factors such as the species, strain, and sex of the animals, along with fundamental principles of good scientific practice such as blinding and randomization.

It is important to distinguish between various types of exploratory research. Our recently published workflow (Spanagel 2022) provides guidance on navigating exploratory research endeavors. In Type I exploratory studies, where novel methodological approaches are pioneered, registering the study design may be of limited utility, as experimental parameters are likely to undergo iterative modifications until the new method accurately measures the intended outcome. Conversely,

---

[5] https://www.cos.io/initiatives/registered-reports.

[6] https://preclinicaltrials.eu/.

[7] https://www.animalstudyregistry.org/asr_web/index.action.

[8] https://osf.io/.

[9] https://www.crd.york.ac.uk/PROSPERO/.

preregistration is strongly recommended for Type II exploratory studies, which use established methods in a laboratory setting to conduct new exploratory experiments.

5 Sample Size Calculation and Statistics

Statistical methods are a crucial cornerstone of biomedical research, essential for data planning and making informed assumptions. Consequently, having a fundamental understanding of these methods is vital. If needed, consulting with professional statisticians to identify the appropriate statistical approaches is advisable. However, especially when planning a study and performing the necessary power analysis to determine the required sample size, statisticians often request more information from the researcher, sending them back to the drawing board to pinpoint the parameters necessary for power calculations.

The power calculation is, in essence, straightforward as it requires only four factors: the sample size, the effect size, the alpha error, and the power. Typically, the sample size is the parameter that needs calculation, and if the other three factors are known, it can be computed using an R script.[10] So, how do you determine the other three parameters? Let us begin with the effect size. The effect size can usually be estimated from historical experiments or derived from existing literature data, especially as a key outcome of a meta-analysis. A recent study in the field of habit formation in AUD provides an excellent example for a meta-analysis-derived determination of effect size (Giannone et al. 2024). The second parameter is alpha, representing the chance of false positives. This error is conventionally set at 0.05 but often lacks a firm basis. It can be worthwhile to review historical experiments or estimate a likelihood for this error and consider whether there are reasons to be more rigorous. The same principle applies to statistical power, which is commonly defaulted to 0.8. This value represents the likelihood that an experimental result can be reliably replicated. However, in situations where greater stringency is required, increasing the power can yield more robust data. In essence, the most crucial aspect of sample size calculation is the ability to provide a solid rationale for the estimated sample size and the capacity to defend this decision.

Several online tools are available for sample size calculations. Websites like R bloggers[11] or the EGAP network[12] offer various R scripts for power calculations, although they may require a deeper understanding of statistical methods. Additionally, when planning a study with the sophisticated Experimental Design Assistant (EDA), sample size calculations can be conducted once all parameters are defined (du Sert et al. 2017).

---

[10] https://paasp.net/power-analysis-for-two-sided-t-test/.

[11] https://www.r-bloggers.com/2021/05/power-analysis-in-statistics-with-r/.

[12] https://egap.org/resource/script-power-analysis-simulations-in-r/.

Ideally, the statistical analysis, which is based on a primary hypotheses and the testing procedures, should be outlined during the study planning phase and preregistered to reduce bias. This is especially critical in the context of knowledge-claiming or confirmatory research (Bespalov et al. 2021; Vollert et al. 2022). Defining the statistical plan in advance is a vital safeguard against "p-hacking," a practice where various statistical approaches are tried until a significant p-value is achieved. Unfortunately, this malpractice is possible because different methods may yield different results, and it is often feasible to justify a statistical analysis, even if it was not the originally intended one. Always keep in mind that "P values, in the absence of indicators of effect sizes, should not be reported because even a very small P value in isolation does not tell us whether an observed effect was large enough to be deemed biologically relevant" (Michel et al. 2020).

This is different when exploratory research is performed. If research is entitled as such research, it is permissible to explore different meaningful statistical approaches to test the primary hypothesis as well as secondary hypothesis. However, it is always recommended to consult statisticians to provide reasoning for the chosen approach.

6 Networks of Centers for Multisite Experiments

While the aforementioned study conducted by Crabbe et al. (1999) serves as an outstanding example of a multisite animal study (Crabbe et al. 1999), this methodology is still relatively underutilized. The primary reason for its limited adoption is the lack of comprehensive guidance on how to execute confirmatory multisite preclinical trials. Addressing this gap, a multistakeholder workshop report was recently published with the aim of offering a blueprint for planning preclinical multicenter trials, thus enhancing the translation from fundamental research to clinical applications (Drude et al. 2021).

Just as clinical studies benefit from being conducted at multiple sites to enhance external validity, the same principle applies to confirmatory preclinical research. Replicating experiments across different sites introduces variability in study execution, leading to more robust outcomes. Moreover, it mitigates biases associated with a single site, including researcher bias and biases related to subject handling, equipment, and data management. Therefore, multisite experiments should become a standard practice in preclinical research, particularly when the data are intended to inform decisions, such as whether to proceed with human trials. In a notable recent development, the German Ministry for Education and Research initiated a funding call to reevaluate promising findings in multisite confirmatory trials. Our team is participating in this funding call with the AhEAD project,[13] which aims to confirm the previously observed positive effects of mesenchymal cell-derived exosomes in an addiction model for alcohol (Ezquer et al. 2019). In this project, we are

[13] www.ahead-study.de.

collaborating with two sites at the University of Erlangen and the Charité in Berlin. Additionally, we coordinated another multicenter trial as part of a European research program, investigating the effects of psychedelic drugs in an animal model of alcohol addiction.[14] Although research in both consortia is ongoing, the finalized data will offer significant contributions to the field and support go/no-go decisions in clinical development.

It is worth noting that even though multisite preclinical testing is well powered and will likely enhance translation of animal studies to humans, it does not eliminate the challenges of heterogeneity observed in human populations. Preclinical studies are generally conducted on specific outbred or inbred rodent strains, which, while exhibiting some behavioral variation even in inbred strains, lack the full genetic and behavioral diversity seen in human studies. To address this limitation, one approach is to use heterogeneous stock (HS) rats. These animals show increased genetic diversity and have been created by crossing eight genetically diverse founder strains (Hansen and Spuhler 1984; Solberg et al. 2019), resulting in a diversity that more closely mirrors the variation found in the human population. Combined with behavioral profiling, this approach allows for genetic analysis of any measurable behavioral or psychopathological quantitative trait through whole-genome sequencing of each individual. By aggregating genetic data from hundreds of animals, a subsequent genome-wide association study (GWAS) can be conducted to identify gene variants contributing to a specific phenotype of interest. The disparities in outcomes between the utilization of inbred or outbred rodent strains as opposed to HS rats may not necessarily exhibit significant differences. Nonetheless, owing to their genetic diversity, the magnitudes of effects tend to be reduced, thereby bearing a closer resemblance to clinical outcomes. This mitigates the risk of overestimating effect sizes. Therefore, it is recommended to incorporate HS rats into confirmatory preclinical multisite studies to mimic the human scenario and offer additional genetic insights within tightly controlled environmental conditions. Overall, such multicenter projects should be considered an essential practice in knowledge-claiming research.

7 Generalization of Research Data

Data generalization implies that a particular observation is not limited to a single, specific observation but holds broader applicability. In essence, research data should exhibit robustness and withstand variations in experimental conditions. It should remain valid regardless of the environment or researcher responsible for making the observation. Generalization also extends to the universality of an observation across both sexes, different species, and various age groups within the observational units. However, it is important to note that this may not apply when the research hypothesis specifically incorporates one of these factors as a parameter under investigation.

[14] www.psialc.org.

Though preclinical and clinical evidence underscores the impact of sex on disease progression and treatment outcomes in individuals with alcohol use disorder (AUD) (Sanchis-Segura and Becker 2016), the preclinical alcohol research landscape has historically been skewed toward investigating males. It is only recently that research encompassing both sexes in preclinical studies has gained momentum, largely spurred by policy revisions by major funding bodies such as NIH[15] or DFG.[16] A notable instance in this regard was the discovery of sex-specific alterations in neuroadaptations within the oxytocin system in alcohol-dependent rodents and humans. While a significant decline in oxytocin levels and an increase in oxytocin receptor binding sites were observed in male alcohol-dependent rats and patients, no such changes were noted in female rats and women, underscoring the importance of sex-specific considerations in psychopharmacological interventions for alcoholism (Hansson and Spanagel 2021).The significance of generalization lies in its capacity to prevent observations from being confined to a singular situation. Even if an effect is genuine, limiting it to a specific context may restrict its broader implications and hinder its translation to other scenarios, including potential applications to human research. Generalization can also be bolstered by intentionally introducing variations in certain parameters, such as environmental factors or genetic diversity. This deliberate heterogenization serves to enhance data robustness, as illustrated in recent studies (Arroyo-Araujo et al. 2022; Würbel 2017), demonstrating that systematic variation of factors such as age, sex, body weight, and others improves reproducibility of animal experiments.

8 Reporting of Data Using ARRIVE

The ARRIVE (Animal Research Reporting of In Vivo Experiments) guidelines represent a well-established tool for enhancing the transparency of research methodology reporting. Accurate reporting of elements such as study design, blinding, animal age and gender, and other experimental factors is crucial for facilitating the reproducibility of future studies. Many scientific journals have already mandated that authors adhere to the ARRIVE 2.0 guidelines, as outlined by Percie du Sert et al. in 2020 (Percie du Sert et al. 2020).

Authors are required to include these guidelines as a checklist alongside their submitted manuscripts. This practice aims to improve the quality of reporting in animal experiments, offering transparency to both scientific journal editors and reviewers. For intricate studies, particularly in the context of confirmatory preclinical animal research, seeking professional assistance is advisable. This can be essential for subsequent FDA or EMA approval processes. Therefore, queries related to study

[15] https://grants.nih.gov/grants/guide/notice-files/not-od-15-102.html.

[16] https://www.dfg.de/resource/blob/174154/abe30fcc0aaa3527e6930f9a64ab7375/stellungnahme-data.pdf.

design, statistical procedures, and patent law implications should be addressed by qualified experts. By following the ARRIVE guidelines and seeking professional guidance, researchers can establish a clear methodology that facilitates other laboratories' comprehension and replication of their data. ARRIVE guidelines basically represent a checklist in order to ensure that all important information required to understand and replicate a study are mentioned. To facilitate the process of proper reporting using ARRIVE, researchers should consider using the PREAPRE guidelines (Smith et al. 2018) during the preparation phase of their studies. Since both guidelines complement each other, it is recommended to plan a study properly from day one, starting with PREPARE and ending with ARRIVE. Furthermore, it is beneficial to preregister study designs (also see section "Experimental Study Protocols Preregistered."

9 Null Results Published

Null results or negative findings often carry a misleadingly negative connotation due to their terminology. However, they do not inherently signify something unfavorable but merely indicate that the research hypothesis has not been confirmed. While this can be a source of disappointment for the researcher whose hypothesis was not validated by the experiment, these results still hold substantial value, particularly if the experiments were conducted rigorously. One example concerns a mechanistic study of the effect of the approved antirelapse drug acamprosate (calcium N-acetylhomotaurinate). Although there is ample evidence that acamprosate interferes with several neurotransmitter systems and the glutamate system, it has been shown in a convincing series of animal and human experiments that N-acetylhomotaurinate itself is not an active psychotropic molecule, suggesting that calcium is the active moiety of acamprosate (Spanagel et al. 2014). It was a difficult undertaking to publish these "negative results," but in the end, the study was highly valued by the research community, received multiple awards, and was repeated several times (e.g., Ademar et al. (2022), Pradhan et al. (2018), Schuster et al. (2021)).

Null findings can be beneficial for other researchers working within the same field who may have formulated identical or similar hypotheses. By making this data available to the research community, we can prevent needless repetition of the same study. The CERTAIN guidelines, as introduced by Bespalov et al. (2019), offer valuable guidance on determining whether an experiment or study should indeed be published. The authors have devised a practical decision tree to assist researchers in making the decision on whether to publish this data.

Once the decision to publish has been reached, the FIDDLE[17] tool (which stands for "file drawer data liberation effort") comes into play, offering support in locating an appropriate platform for sharing and disseminating this data.

10 Capturing DSM 5 Criteria in Animal Models

The choice of animal testing methods significantly influences the translatability of results to the human context, especially when attempting to model complex pathological behaviors observed in various psychiatric conditions like alcohol use disorder (AUD). The current psychiatric diagnostic system, DSM-5, is built upon clinical observations and patient-reported symptoms, inherently relying on anthropomorphic terms. As diagnoses are made according to these classification systems worldwide and since these classification systems will be used for the next decades to come, logic dictates that face validity of animal models of psychiatric disorders should also be based on DSM-5/ICD-11 criteria. However, modeling the full spectrum of a mental disorder in humans is not possible in animals due to the high level of complexity. Nevertheless, we can transfer anthropomorphic terminology to animal models with empirical, translatable, and measurable parameters and thus reliably examine at least some key criteria of a disease of interest in animal models. Excellent examples of DSM-based animal models can be found in addressing concepts such as craving for alcohol and relapse (Spanagel 2017).

One such model is the alcohol deprivation model, which explores long-term alcohol consumption in Wistar rats. In this model, animals experience a significant increase in voluntary alcohol intake (the alcohol deprivation effect or ADE) when granted renewed access to alcohol solutions after a period of deprivation spanning several days. Following repeated deprivation phases, the ADE is characterized by an increased demand for the substance, resembling a relapse situation observed in alcoholic patients (Spanagel and Holter 1999; Vengeliene et al. 2009). Consequently, the ADE model exhibits strong face validity for several aspects of the relapse process.

Additionally, in the realm of studying drug craving and seeking behavior, the post-dependent reinstatement model is commonly used. This model is considered a valuable tool for medication development in alcoholism (Meinhardt et al. 2021; Meinhardt and Sommer 2015). It investigates the resumption of behavior in an operant task when dependent and nondependent rats are reexposed to alcohol-associated environmental stimuli (Sanchis-Segura and Spanagel 2006). In the context of alcohol cue exposure leading to craving and relapse, this model demonstrates high face and construct validity (Meinhardt and Sommer 2015).

Both the ADE and postdependent models have showcased the effectiveness of pharmacological agents in reducing alcohol consumption and susceptibility to

[17] https://s-quest.bihealth.org/fiddle/.

relapse in humans, underscoring the predictive validity of both paradigms. Hence, it is argued that aligning our animal models with the DSM-5/ICD11 classification is essential to ultimately yield meaningful data.

11 Methodology Should Be Validated and Under Control

A methodological concern that deserves more attention is what we refer to here as "method hopping" (Spanagel 2022). Method hopping represents the pursuit of the latest technology, and it raises significant concerns, particularly in the context of replication. The field of biomedical research, especially in neuroscience, experiences an exceptionally rapid development and turnover rate of technology. What is now considered state-of-the-art may become outdated within a year. This issue is exacerbated by the policies of top-tier journals and numerous grant institutions, which prioritize the use of cutting-edge technology for addressing research questions.

One indication that method hopping contributes to replication failures is the fact that studies published in high-profile journals, which subsequently could not be replicated, receive 153 times more citations than those that were successfully replicated (Serra-Garcia and Gneezy 2021). This suggests that highly cited papers tend to produce more "interesting" results driven by new technologies, which, in turn, can lead to a subjectively more lenient peer-review process and, consequently, a negative correlation between replicability and citation count (Serra-Garcia and Gneezy 2021). As a result, the race for the latest technology often comes at a steep cost, as researchers may not have full control over these cutting-edge tools, and published measurements may, as a result, hinge on simple technical mishaps. Researchers that apply a new cutting edge technology should at least validate their methods by checking its specific parameters for selectivity and specificity, precision, accuracy, linearity, range, stability, and limit of detection (LOD)/limit of quantitation (LOQ) where applicable.

In general, method hopping is a systemic challenge within biomedical research that can only be addressed by modifying the policies of high-impact journals and the evaluation procedures for high-profile grant applications. However, individual scientists also bear a responsibility in this systemic issue, as each scientist has the freedom to choose whether to engage in this technological race or to rely on well-established methodologies. A skeptical perspective on method hopping should not stifle technological advancements in the biomedical field, but rather, a critical evaluation of the utility of new technologies can strike a balance between using established methods and incorporating cutting-edge technology.

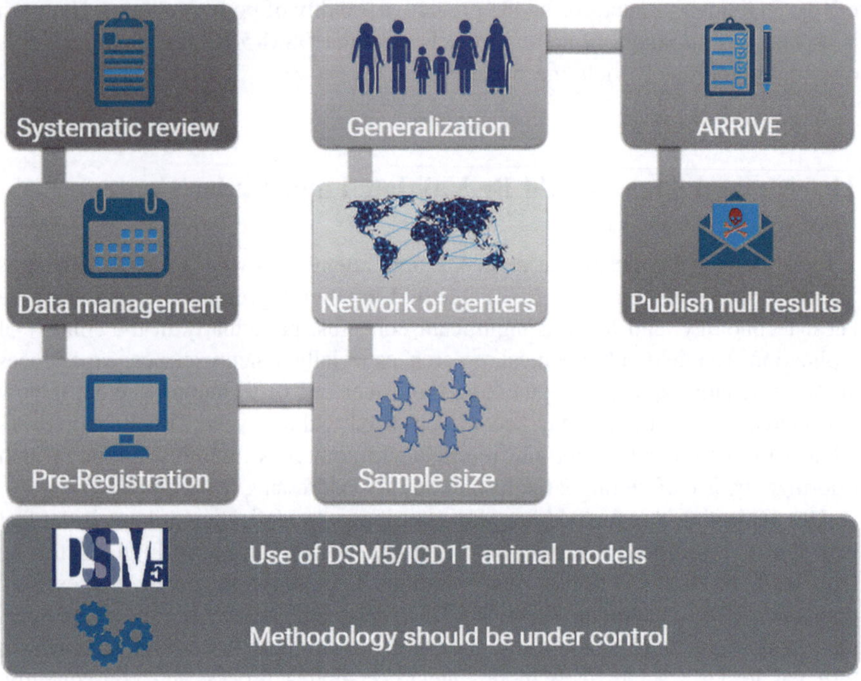

Fig. 2 Graphical representation of the ten recommendation points that are included in the STINGENCY framework

12 Conclusion

Science has long been considered "self-correcting" due to its foundation on replicating earlier research. However, in recent years, the safeguards that once ensured the integrity of scientific endeavors have faced challenges across the entire biomedical field. These challenges have compromised the ability of contemporary researchers to reproduce the findings of others (Collins et al. 2014) and have led to numerous failures in translation. This chapter presents the *Good Preclinical Practice Framework,* referred to as STRINGENCY, to enhance reproducibility and translation. Figures 1 and 2 summarize ten key recommendations aimed at improving the reproducibility and translation of animal experimentation. Many of these recommendations are also integral to the new 6R framework, which expands upon the widely accepted 3R framework for the ethical use of animals in research by adding three additional guiding principles, robustness, registration, and reporting, all aimed at safeguarding and enhancing the scientific value of animal research (Strech and Dirnagl 2019).

It is worth noting that even with the implementation of the STRINGENCY framework, translational failures can still occur. In such cases, researchers should inquire about the root causes of these failures. As already mentioned, a prominent example in the field of AUD is the development of CRH1 receptor antagonists for alcohol addiction and other stress-related psychiatric conditions. Despite compelling evidence from animal studies indicating that CRH activity plays a significant role in mediating stressor-induced relapse and the escalation of alcohol consumption in dependent rats (Heilig and Koob 2007), other preclinical data suggested that CRH1 receptor antagonists might not be effective in clinical trials (Molander et al. 2012; Refojo et al. 2011; Sillaber et al. 2002). Studies with negative or even conflicting results provided substantial evidence for the observed lack of effects of antagonists on stress-induced disorders in clinical trials (Kwako et al. 2015; Schwandt et al. 2016) and should have prompted a reconsideration of the initiation of costly clinical trials. This fact again emphasizes the need for in-depth and unbiased literature review prior study initiation using systematic reviews and meta-analysis approaches as stated above in the STRINGENY framework.

In conclusion, valuable lessons can be gleaned from translation errors. However, to do so, a critical dialogue among basic researchers, preclinicians, and clinicians is of utmost importance. A crisis only ensues if we fail to find an explanation for the lack of replication and/or a failure in translation. Furthermore, the integration of advanced statistical methods and possibly AI algorithms to analyze complex datasets could provide new insights into the factors influencing reproducibility. Encouraging collaborative networks and data sharing platforms will also facilitate the verification and validation of findings across different laboratories and research contexts. Ultimately, fostering a culture of transparency and rigorous methodology is essential for advancing the reliability and impact of preclinical research.

Funding and Support Financial support for this work was provided by the Ministry of Education and Research Baden-Württemberg for the 3R-Center Rhein-Neckar (www.3r-rn.de), the Bundesministerium für Bildung und Forschung (BMBF) funded SysMedSUDs consortium (01ZX1909A), AhEAD consortium (01KC2004A), Target-Oxy consortium (031L0190A), the ERA-NET consortium Psi-Alc (01EW1908; www.psialc.org), the Deutsche Forschungsgemeinschaft (DFG, German Research Foundation), and TRR 265 consortium (A05, B02; (Heinz et al. 2020)).

Declaration of Interest Statement

Marcus W. Meinhardt

Declaration of interest: None.

Björn Gerlach

Declaration of interest: Shareholder of PAASP GmbH and PAASP US, LLC; employee of Central Institute of Mental Health and PAASP GmbH and Guarantors of EQIPD e.V.; Member of the Executive Board at Guarantors of EQIPD e.V.

Rainer Spanagel

Declaration of interest: None.

References

Ademar K, Adermark L, Söderpalm B, Ericson M (2022) Sodium acamprosate and calcium exert additive effects on nucleus accumbens dopamine in the rat. Addict Biol 27(5):e13224. https://doi.org/10.1111/adb.13224

Arroyo-Araujo M, Voelkl B, Ment Laloux C, Novak J, Koopmans B, Waldron A-M, Seiffert I, Stirling H, Aulehner K, Janhunen SK, Ramboz S, Potschka H, Holappa J, Fine T, Loos M, Boulanger B, Würbel H, Kas MJ (2022) Systematic assessment of the replicability and generalizability of preclinical findings: impact of protocol harmonization across laboratory sites. PLoS Biol 20(11):e3001886. https://doi.org/10.1371/journal.pbio.3001886

Begley CG, Ellis LM (2012) Raise standards for preclinical cancer research. Nature 483(7391): 531–533. https://doi.org/10.1038/483531a

Bert B, Heinl C, Chmielewska J, Schwarz F, Grune B, Hensel A, Greiner M, Schönfelder G (2019) Refining animal research: the animal study registry. PLoS Biol 17(10):e3000463. https://doi.org/10.1371/journal.pbio.3000463

Bespalov A, Bernard R, Gilis A, Gerlach B, Guillén J, Castagné V, Lefevre IA, Ducrey F, Monk L, Bongiovanni S, Altevogt B, Arroyo-Araujo M, Bikovski L, de Bruin N, Castaños-Vélez E, Dityatev A, Emmerich CH, Fares R, Ferland-Beckham C et al (2021) Introduction to the EQIPD quality system. eLife 10. https://doi.org/10.7554/ELIFE.63294

Bespalov A, Steckler T, Skolnick P (2019) Be positive about negatives–recommendations for the publication of negative (or null) results. Eur Neuropsychopharmacol 29(12):1312–1320. https://doi.org/10.1016/j.euroneuro.2019.10.007

Collins FS, Tabak L, a. (2014) Policy: NIH plans to enhance reproducibility. Nature 505(7485): 612–613. https://doi.org/10.1038/505612a

Crabbe JC, Wahlsten D, Dudek BC (1999) Genetics of mouse behavior: interactions with laboratory environment. Science 284(5420):1670–1672. https://doi.org/10.1126/science.284.5420.1670

de Vries RBM, Hooijmans CR, Langendam MW, van Luijk J, Leenaars M, Ritskes-Hoitinga M, Wever KE (2015) A protocol format for the preparation, registration and publication of systematic reviews of animal intervention studies. Evidence-based Preclin Med 2(1):e00007. https://doi.org/10.1002/EBM2.7

Drude NI, Gamboa LM, Danziger M, Dirnagl U, Toelch U (2021) Science forum improving preclinical studies through replications. eLife 10:1–10. https://doi.org/10.7554/ELIFE.62101

du Sert NP, Bamsey I, Bate ST, Berdoy M, Clark RA, Cuthill IC, Fry D, Karp NA, Macleod M, Moon L, Stanford SC, Lings B (2017) The experimental design assistant. Nat Methods 14(11): 1024–1025. https://doi.org/10.1038/nmeth.4462

du Sert NP, Hurst V, Ahluwalia A, Alam S, Avey MT, Baker M, Browne WJ, Clark A, Cuthill IC, Dirnagl U, Emerson M, Garner P, Holgate ST, Howells DW, Karp NA, Lazic SE, Lidster K, MacCallum CJ, Macleod M et al (2020) The ARRIVE guidelines 2.0: updated guidelines for reporting animal research. PLoS Biol 18(7):e3000410. https://doi.org/10.1371/journal.pbio.3000410

Ezquer F, Quintanilla ME, Morales P, Santapau D, Ezquer M, Kogan MJ, Salas-Huenuleo E, Herrera-Marschitz M, Israel Y (2019) Intranasal delivery of mesenchymal stem cell-derived exosomes reduces oxidative stress and markedly inhibits ethanol consumption and post-deprivation relapse drinking. Addict Biol 24(5):994–1007. https://doi.org/10.1111/adb.12675

Freedman LP, Cockburn IM, Simcoe TS (2015) The economics of reproducibility in preclinical research. PLoS Biol 13(6):e1002165. https://doi.org/10.1371/journal.pbio.1002165

Gerlach B, Untucht C, Stefan A (2020) Electronic lab notebooks and experimental design assistants. Handb Exp Pharmacol 257:257–275. https://doi.org/10.1007/164_2019_287

Giannone F, Ebrahimi C, Endrass T, Hansson AC, Schlagenhauf F, Sommer WH (2024) Bad habits – good goals? Meta-analysis and translation of the habit construct to alcoholism. Transl Psychiatry. In Press

Hansen C, Spuhler K (1984) Development of the National Institutes of Health Genetically Heterogeneous Rat Stock. Alcohol Clin Exp Res 8(5):477–479. https://doi.org/10.1111/j.1530-0277.1984.tb05706.x

Hansson AC, Spanagel R (2021) No changes in the oxytocin system in alcohol-dependent female rodents and humans: towards a sex-specific psychopharmacology in alcoholism. Addict Biol 26(2):e12945. https://doi.org/10.1111/adb.12945

Heilig M, Koob GF (2007) A key role for corticotropin-releasing factor in alcohol dependence. Trends Neurosci 30(8):399–406. https://doi.org/10.1016/j.tins.2007.06.006

Heinl C, Scholman-Végh AMD, Mellor D, Schönfelder G, Strech D, Chamuleau S, Bert B (2022) Declaration of common standards for the preregistration of animal research-speeding up the scientific progress. PNAS Nexus 1(1):pgac016. https://doi.org/10.1093/pnasnexus/pgac016

Heinz A, Kiefer F, Smolka MN, Endrass T, Beste C, Beck A, Liu S, Genauck A, Romund L, Banaschewski T, Bermpohl F, Deserno L, Dolan RJ, Durstewitz D, Ebner-Priemer U, Flor H, Hansson AC, Heim C, Hermann D et al (2020) Addiction research consortium: losing and regaining control over drug intake (ReCoDe) – from trajectories to mechanisms and interventions. Addict Biol 25(2). https://doi.org/10.1111/adb.12866

Higgins JPT, Altman DG, Gotzsche PC, Juni P, Moher D, Oxman AD, Savovic J, Schulz KF, Weeks L, Sterne JAC (2011) The Cochrane Collaboration's tool for assessing risk of bias in randomised trials. BMJ 343(2):d5928–d5928. https://doi.org/10.1136/bmj.d5928

Hirth N, Meinhardt MW, Noori HR, Salgado H, Torres-Ramirez O, Uhrig S, Broccoli L, Vengeliene V, Roßmanith M, Perreau-Lenz S, Köhr G, Sommer WH, Spanagel R, Hansson AC (2016) Convergent evidence from alcohol-dependent humans and rats for a hyperdopaminergic state in protracted abstinence. Proc Natl Acad Sci U S A 113(11): 3024–3029. https://doi.org/10.1073/pnas.1506012113

Hooijmans CR, Rovers MM, de Vries RB, Leenaars M, Ritskes-Hoitinga M, Langendam MW (2014) SYRCLE's risk of bias tool for animal studies. BMC Med Res Methodol 14(1):43. https://doi.org/10.1186/1471-2288-14-43

Horbach SPJM, Bouter LM, Gaskell G, Hiney M, Kavouras P, Mejlgaard N, Allum N, Aubert Bonn N, Bendtsen A-K, Charitidis CA, Claesen N, Dierickx K, Domaradzka A, Elizondo AR, Föger N, Kaltenbrunner W, Konach T, Labib K, Marušić A et al (2022) Designing and implementing a research integrity promotion plan: recommendations for research funders. PLoS Biol 20(8):e3001773. https://doi.org/10.1371/journal.pbio.3001773

Ioannidis JPA (2005) Why most published research findings are false. PLoS Med 2:0696–0701. https://doi.org/10.1371/journal.pmed.0020124

Ioannidis JPA, Klavans R, Boyack KW (2018) Thousands of scientists publish a paper every five days. Nature 561(7722):167–169. https://doi.org/10.1038/d41586-018-06185-8

Kwako LE, Spagnolo PA, Schwandt ML, Thorsell A, George DT, Momenan R, Rio DE, Huestis M, Anizan S, Concheiro M, Sinha R, Heilig M (2015) The Corticotropin releasing hormone-1 (CRH1) receptor antagonist pexacerfont in alcohol dependence: a randomized controlled experimental medicine study. Neuropsychopharmacology 40(5):1053–1063. https://doi.org/10.1038/npp.2014.306

Meinhardt MW, Pfarr S, Fouquet G, Rohleder C, Meinhardt ML, Barroso-Flores J, Hoffmann R, Jeanblanc J, Paul E, Wagner K, Hansson AC, Köhr G, Meier N, von Bohlen und Halbach O, Bell RL, Endepols H, Neumaier B, Schönig K, Bartsch D et al (2021) Psilocybin targets a common molecular mechanism for cognitive impairment and increased craving in alcoholism. Sci Adv 7(47). https://doi.org/10.1126/sciadv.abh2399

Meinhardt MW, Sommer WH (2015) Postdependent state in rats as a model for medication development in alcoholism. Addict Biol 20(1):1–21. https://doi.org/10.1111/adb.12187

Michel MC, Murphy TJ, Motulsky HJ (2020) New author guidelines for displaying data and reporting data analysis and statistical methods in experimental biology. J Pharmacol Exp Ther 372(1):136–147. https://doi.org/10.1124/JPET.119.264143

Molander A, Vengeliene V, Heilig M, Wurst W, Deussing JM, Spanagel R (2012) Brain-specific inactivation of the Crhr1 gene inhibits post-dependent and stress-induced alcohol intake, but does not affect relapse-like drinking. Neuropsychopharmacology 37(4):1047–1056. https://doi.org/10.1038/npp.2011.297

Pradhan G, Melugin PR, Wu F, Fang HM, Weber R, Kroener S (2018) Calcium chloride mimics the effects of acamprosate on cognitive deficits in chronic alcohol-exposed mice. Psychopharmacology (Berl) 235(7):2027–2040. https://doi.org/10.1007/s00213-018-4900-1

Prinz F, Schlange T, Asadullah K (2011) Believe it or not: how much can we rely on published data on potential drug targets? Nat Rev Drug Discov 10(9):712. https://doi.org/10.1038/nrd3439-c1

Refojo D, Schweizer M, Kuehne C, Ehrenberg S, Thoeringer C, Vogl AM, Dedic N, Schumacher M, von Wolff G, Avrabos C, Touma C, Engblom D, Schütz G, Nave K-A, Eder M, Wotjak CT, Sillaber I, Holsboer F, Wurst W, Deussing JM (2011) Glutamatergic and dopaminergic neurons mediate anxiogenic and anxiolytic effects of CRHR1. Science 333(6051):1903–1907. https://doi.org/10.1126/science.1202107

Sanchis-Segura C, Becker JB (2016) Why we should consider sex (and study sex differences) in addiction research. Addict Biol 21:995–1006

Sanchis-Segura C, Spanagel R (2006) REVIEW: Behavioural assessment of drug reinforcement and addictive features in rodents: an overview. Addict Biol 11(1):2–38. https://doi.org/10.1111/j.1369-1600.2006.00012.x

Schuster R, Winkler M, Koopmann A, Bach P, Hoffmann S, Reinhard I, Spanagel R, Bumb JM, Sommer WH, Kiefer F (2021) Calcium carbonate attenuates withdrawal and reduces craving: a randomized controlled trial in alcohol-dependent patients. Eur Addict Res 27(5):332–340. https://doi.org/10.1159/000512763

Schwandt ML, Cortes CR, Kwako LE, George DT, Momenan R, Sinha R, Grigoriadis DE, Pich EM, Leggio L, Heilig M (2016) The CRF1 antagonist verucerfont in anxious alcohol-dependent women: translation of neuroendocrine, but not of anti-craving effects. Neuropsychopharmacology 41(12):2818–2829. https://doi.org/10.1038/npp.2016.61

Serra-Garcia M, Gneezy U (2021) Nonreplicable publications are cited more than replicable ones. Sci Adv 7(21). https://doi.org/10.1126/sciadv.abd1705

Seyhan AA (2019) Lost in translation: the valley of death across preclinical and clinical divide – identification of problems and overcoming obstacles. Transl Med Commun 4(1):18. https://doi.org/10.1186/s41231-019-0050-7

Sillaber I, Rammes G, Zimmermann S, Mahal B, Zieglgänsberger W, Wurst W, Holsboer F, Spanagel R (2002) Enhanced and delayed stress-induced alcohol drinking in mice lacking functional CRH1 receptors. Science 296(5569):931–933. https://doi.org/10.1126/science.1069836

Smith AJ, Clutton RE, Lilley E, Hansen KEA, Brattelid T (2018) PREPARE: guidelines for planning animal research and testing. Lab Anim 52(2):135–141. https://doi.org/10.1177/0023677217724823

Solberg BS, Zayats T, Posserud M-B, Halmøy A, Engeland A, Haavik J, Klungsøyr K (2019) Patterns of psychiatric comorbidity and genetic correlations provide new insights into differences between attention-deficit/hyperactivity disorder and autism Spectrum disorder. Biol Psychiatry 86(8):587–598. https://doi.org/10.1016/j.biopsych.2019.04.021

Spanagel R (2017) Animal models of addiction. Dialogues Clin Neurosci 19(3):247–258. 10.31887/DCNS.2017.19.3/rspanagel

Spanagel R (2022) Ten points to improve reproducibility and translation of animal research. Front Behav Neurosci 16. https://doi.org/10.3389/FNBEH.2022.869511

Spanagel R, Holter SM (1999) Long-term alcohol self-administration with repeated alcohol deprivation phases: an animal model of alcoholism? Alcohol 34(2):231–243. https://doi.org/10.1093/alcalc/34.2.231

Spanagel R, Vengeliene V, Jandeleit B, Fischer WN, Grindstaff K, Zhang X, Gallop MA, Krstew EV, Lawrence AJ, Kiefer F (2014) Acamprosate produces its anti-relapse effects via calcium. Neuropsychopharmacology 39(4):783–791. https://doi.org/10.1038/npp.2013.264

Strech D, Dirnagl U (2019) 3Rs missing: animal research without scientific value is unethical. BMJ Open Sci. https://doi.org/10.1136/bmjos-2018-000048

Vengeliene V, Celerier E, Chaskiel L, Penzo F, Spanagel R (2009) Research focus on compulsive behaviour in animals: Compulsive alcohol drinking in rodents. Addict Biol 14(4):384–396. https://doi.org/10.1111/j.1369-1600.2009.00177.x

Vesterinen HM, Sena ES, Egan KJ, Hirst TC, Churolov L, Currie GL, Antonic A, Howells DW, Macleod MR (2014) Meta-analysis of data from animal studies: a practical guide. J Neurosci Methods 221:92–102. https://doi.org/10.1016/j.jneumeth.2013.09.010

Vollert J, Macleod M, Dirnagl U, Kas MJ, Michel MC, Potschka H, Riedel G, Wever KE, Würbel H, Steckler T, Rice ASC (2022) The EQIPD framework for rigor in the design, conduct, analysis and documentation of animal experiments. Nat Methods 19(11):1334–1337. https://doi.org/10.1038/s41592-022-01615-y

Wilkinson MD, Dumontier M, Aalbersberg IJ, Appleton G, Axton M, Baak A, Blomberg N, Boiten JW, da Silva Santos LB, Bourne PE, Bouwman J, Brookes AJ, Clark T, Crosas M, Dillo I, Dumon O, Edmunds S, Evelo CT, Finkers R et al (2016) The FAIR guiding principles for scientific data management and stewardship. Sci Data 3(1):1–9. https://doi.org/10.1038/sdata.2016.18

Würbel H (2017) More than 3Rs: the importance of scientific validity for harm-benefit analysis of animal research. Lab Anim 46(4):164–166. https://doi.org/10.1038/laban.1220

Part II
Specific Mechanisms of Alcohol Use and Addiction

Role of Metabolism on Alcohol Preference, Addiction, and Treatment

María Elena Quintanilla and Yedy Israel

Contents

Abstract Studies presented in this chapter show that: (1) in the brain, ethanol is metabolized by catalase to acetaldehyde, which condenses with dopamine forming salsolinol; (2) acetaldehyde-derived salsolinol increases the release of dopamine

M. E. Quintanilla (✉)
Molecular and Clinical Pharmacology Program, Institute of Biomedical Sciences, Faculty of Medicine, University of Chile, Santiago, Chile
e-mail: equintanilla@uchile.cl

Y. Israel
Molecular and Clinical Pharmacology Program, Institute of Biomedical Sciences, Faculty of Medicine, University of Chile, Santiago, Chile

Centro de Medicina Regenerativa, ICM Clinica Alemana-Universidad de Desarrollo, Santiago, Chile

Curr Topics Behav Neurosci (2025) 71: 193–216
https://doi.org/10.1007/7854_2023_422
Published Online: 24 May 2023

mediating, via opioid receptors, the reinforcing effects of ethanol during the acquisition of ethanol consumption, while (3) brain acetaldehyde does not influence the maintenance of chronic ethanol intake, it is suggested that a learned cue-induced hyperglutamatergic system takes precedence over the dopaminergic system. However, (4) following a prolonged ethanol deprivation, the generation of acetaldehyde in the brain again plays a role, contributing to the increase in ethanol intake observed during ethanol re-access, called the alcohol deprivation effect (ADE), a model of relapse behavior; (5) naltrexone inhibits the high ethanol intake seen in the ADE condition, suggesting that acetaldehyde-derived salsolinol via opioid receptors also contributes to the relapse-like drinking behavior. The reader is referred to glutamate-mediated mechanisms that trigger the cue-associated alcohol-seeking and that also contribute to triggering relapse.

Keywords Acquisition of ethanol intake · Alcohol deprivation effect · Aldehyde dehydrogenase · Aminotriazole · Binge-like drinking · Brain acetaldehyde generation · Brain salsolinol · Catalase · Drug-seeking · Ethanol aversion · Gene therapy · Maintenance of ethanol intake · Naltrexone · Opioid receptor · Penicillamine · Reinforcing effect

1 Introduction

The conditions that lead to an excessive ethanol consumption by some individuals are complex as these involve interactions among genetic, psychosocial, environmental, and neurobiological factors (Weiss and Porrino 2002). To unravel the genetic factors involved in ethanol intake, rat lines/strains bred by their low or high voluntary ethanol consumption were developed in: Chile (Mardones and Segovia-Riquelme 1983; Quintanilla et al. 2006), the Unites States (Bell et al. 2006) Finland (Sommer et al. 2006), and Italy (Colombo et al. 2006). The two rat lines developed at the University of Chile consume ethanol at low (0.1–2 g ethanol/kg/day, UChA "alcohol abstainer rats") or high levels (consume over 5 g ethanol/kg/day, UChB "alcohol bibulous rats"), when these are offered both a 10% ethanol solution and water (Quintanilla et al. 2006).

The *Alcohol* Use Disorder (AUD) is a chronic and relapsing condition characterized by craving, compulsive ethanol use, and loss of control over limiting its intake; representing the final stage in the transition from the beginning of ethanol intake and the progression toward uncontrolled consumption (Volkow and Li 2005; Heilig and Egli 2006; Spanagel 2009; Koob and Volkow 2010). This review addresses the biological factors that influence: (1) the acquisition of ethanol intake; (2) the maintenance of chronic ethanol intake; and (3) ethanol relapse-like drinking behavior, and is based both on studies conducted in both heterogeneous (outbred) rats (Wistar, Sprague-Dawley) and rats selectively bred for their high or low voluntary ethanol consumption.

Noteworthy, this review does not address the biological factors involved in the development of negative reinforcement since rats do not develop overt dependence when 10% ethanol and water are offered on a 24-h/day basis. Indeed, the chronic ethanol consuming UChB rats fully cease their ethanol intake if bitter quinine (0.01%) is added to the ethanol solution (Quintanilla et al. 2012). Models of negative reinforcement require overt intoxication; for example, paradigms in which rats are placed in an ethanol vapor chamber to achieve blood ethanol levels approaching 200 mg/dl levels, which are often, repeated along interspaced ethanol withdrawal episodes (Rimondini et al. 2002; Gilpin et al. 2008; Meinhardt and Sommer 2015). In severe cases of human AUD, the withdrawal reaction is successfully treated with long-acting benzodiazepines (Long et al. 2017). On the other hand, ethanol relapse is seen months after ethanol abstinence, a condition that continues to constitute a pharmacological problem. We note that other authors suggest that negative reinforcement is the main condition that needs to be addressed in the field of alcoholism (Koob 2021). Such a view acquires special importance in jurisdictions where drinking to overt intoxication is the norm.

2 Hepatic Metabolism of Ethanol

Ethanol is absorbed from the stomach and intestine and is distributed throughout the body, reaching identical concentrations in the water of all tissues, including the brain. Ethanol is metabolized mainly in the liver by alcohol dehydrogenase (ADH), which has a Km for ethanol of about 1 mM (see Cederbaum 2012) and minimally in the brain mainly through the peroxidative activity of catalase, and to a minor extent by cytochrome P-4502E1 (Aragon and Amit 1992; Tampier et al. 1995; Zimatkin et al. 2006). *Acetaldehyde* is subsequently metabolized into acetate mainly by a mitochondrial aldehyde dehydrogenase (ALDH2), an enzyme with a Km for acetaldehyde in the micromolar range, known as the low-Km ALDH2 (Cederbaum and Rubin 1977), which is present in most organs. In individuals with an active ALDH2, very low ethanol-derived liver-generated acetaldehyde levels are found in the bloodstream, given that the levels range from undetectable to about 3 μM (Mizoi et al. 1994). The reaction catalyzed by ALDH2 involves the transfer of hydrogen atoms from acetaldehyde to nicotinamide adenine dinucleotide (NAD^+), resulting in the formation of reduced NAD (NADH). During this reaction, ALDH interacts with both acetaldehyde and NAD^+. In both humans and laboratory animals, researchers have identified several types of ALDHs (e.g., ALDH1, ALDH2, etc.), vide infra.

3 Aversive Effect of Liver-Generated Acetaldehyde

Some individuals of East Asian origin carry a point mutation in ALDH2 (gene *ALDH2*2*, E487K), which virtually abolishes the activity of this enzyme. Upon ethanol consumption, these individuals display marked elevations of blood acetaldehyde, which generates a dysphoric reaction (e.g., facial flushing, hypotension, headaches, and nausea) that deters individuals from drinking (Mizoi et al. 1994; Peng et al. 2007). Population studies have demonstrated that 20–40% of individuals of East Asian origin (e.g., Japan, China, Korea) carry the *ALDH2*2* allele and are protected between 66% (heterozygous *ALDH2*1/ALDH2*2)* and 99% (homozygous *ALDH2*2/ALDH2*2*) against heavy ethanol drinking and AUD (see Edenberg and McClintick 2018). Upon ethanol intake, individuals carrying only one *ALDH2*2* allele can display blood acetaldehyde over 10-times higher than the normal levels (Adachi et al. 1989; Peng et al. 2014). These findings are also observed in animal studies. Indeed, rats of the UChA line (virtually Abstainer) display a mutation in the *Aldh*2* gene (Gln67Arg) that codes for an enzyme with a higher Km for NAD^+ and a lower Vmax compared with the ALDH2 of the UChB drinker animals (Sapag et al. 2003; see Quintanilla et al. 2006), which results in transient elevations of blood acetaldehyde levels (40–50 uM) that deters their ethanol intake (Quintanilla et al. 2005a). The mutation in the *Aldh2* gene accounts for 50–60% of the low ethanol intake of UChA rats (Quintanilla et al. 2005b). Taken together, these findings support the idea that increased acetaldehyde in the periphery leads to ethanol aversion, which has been the hallmark of aversion therapy for AUD with disulfiram (Antabuse), an inhibitor of acetaldehyde *metabolism*.

4 Treatments to Generate an Aversion to Ethanol

Disulfiram (Antabuse®), a drug that non-specifically inhibits all the ALDH isozymes, is an effective medication in the treatment of alcoholism *when its daily intake is secured by another person* (Chick et al. 1992; see Jørgensen et al. 2011). However, its therapeutic efficacy varies widely, in part due to the fact that disulfiram is a prodrug that requires its transformation into an active form (Madan and Faiman 1995; Mays et al. 1995) and because it shows a wide range of secondary effects which often prevent the use of doses that ensure full therapeutic effectiveness (Brewer 1984; Christensen et al. 1991). Another source of variation may stem from the fact that there is tolerance to disulfiram induced by chronic ethanol intake. Tampier et al. (2008) showed that, while a single administration of a dose of disulfiram (12.5 or 25 mg/kg, i.p.) inhibited ethanol consumption (up to 60 to 70%) in rats that had been drinking ethanol for only 3 days, this same dose of disulfiram was completely inactive to inhibit ethanol intake in animals that had been drinking ethanol continuously for 30 days. Such a tolerance was seen despite that disulfiram induced virtually identical elevations of *acetaldehyde* in the blood of rats that had been drinking ethanol for only 3 days and chronic ethanol treated groups (Tampier et al. 2008). Tolerance to disulfiram could be due to the fact that acetaldehyde in the brain, unlike its effect in the periphery, is a reinforcing metabolite, since rats have been shown to self-administer acetaldehyde intracranially (Amit et al. 1977; Brown et al. 1979; Rodd-Henriks et al. 2002) and there is evidence that disulfiram inhibits not only ALDH2 in the liver, but also in the brain (Hellstrom and Tottmar 1982 Tampier and Quintanilla 2003). Thus, an elevation of brain acetaldehyde following disulfiram administration may, in some individuals, counteract the aversive effects of acetaldehyde in the periphery. These studies might also explain why some individuals who carry the *ALDH2*2* allele and continue to drink chronically become alcoholics.

The extent to which the placebo effect plays a role in the therapeutic effect of disulfiram was investigated (Skinner et al. 2014; Yoshimura et al. 2014). These studies showed that in double blind studies disulfiram was not superior to placebo, while reported or supervised use of disulfiram was indicated to be necessary for effectiveness.

4.1 Gene Therapy Approaches to Generate an Aversion to Ethanol: Preclinical Studies

4.1.1 Liver Aldehyde Dehydrogenase-2 (ALDH2)

Individuals carrying the virtually inactive *ALDH2*2* allele are protected by 66% against excessive ethanol consumption. Such a protection against *alcoholism* suggested that inhibiting hepatic ALDH2 synthesis by *treatment* with an adenoviral vector (AdV) coding for an anti-ALDH2 antisense oligonucleotide would generate an *aversion* to ethanol in animals after these have become chronic ethanol drinkers. Indeed, animals that intravenously received an adenoviral vector coding for an antisense mRNA against the *Aldh2* transcript, which lowered ALDH2 activity by 85% ($p < 0.02$), consumed 50% less ethanol (10%v/v) ($p < 0.001$) than that of controls animals, an effect that was maintained over 1 month (Ocaranza et al. 2008).

It should be noted that the AdV-vector (70 nm) administered intravenously is considered to be liver-directed due to its primary access to hepatocytes through the large liver fenestra (>200 nm), whereas most body capillaries only allow the passage of smaller particles (<20 nm) (vide infra). Guillot et al. (2009) reported similar results in mice after intravenously administering an adenoviral vector encoding an anti *Aldh2 shRNA*. These authors observed a significant inhibition of ethanol intake in shRNA-AdV treated mice, primarily when the animals were offered ethanol solutions in concentrations above 10%. Potential interaction problems have been reported (Kwon et al. 2014) as ALDH2-/-mice fed ethanol and/or carbon tetrachloride (CCl4) were more susceptible to liver damage than wild-type ones. Further, in recent *ALDH2* knock-in studies in 3T3 cells, cell injury was potentiated by the combination of ethanol and (novel human) genes coding for a low ALDH2 activity

(Chen et al. 2020). In such studies, injury was show following incubation with a very high concentration of ethanol (50 mM) for a prolonged period (48 h), which markedly increased the generation of reactive oxygen species. However, the epidemiological literature is at odds on the prevalence of liver cancer in subjects who carry the ALDH2*2 allele (reviewed by Chang et al. 2017). Indeed, Chang et al. (2017) conclude that studies to date have consistently shown that ALDH2*2 does not independently contribute to liver cancer risk. Thus, to answer the question of whether ALDH2*2 may contribute to the association between ethanol consumption and risk of liver cancer, further studies are required.

4.1.2 Liver Alcohol Dehydrogenase 1 (ADH1B*2)

Some people of East Asian and Caucasian descent carry a variant of alcohol dehydrogenase, (ADH1B*2; Arg47His), which is 100 times more active in to generate acetaldehyde than the more common enzyme ADH1B*1 (Hurley et al. 1990, 1991). Several studies have reported that carriers of ADH1B*2 show marked protection (~50%) against the development of alcoholism (Chambers et al. 2002; Kim et al. 2008; Chen et al. 2009; Li et al. 2011). Thus, a gene therapy study mimicking the protective hepatic ADH1B*2 phenotype was conducted. Toward such end, the wild-type rat cDNA encoding rADH-47Arg was mutated to encode rADH-47His, thus mimicking the more active human mutation. The rADH-47His mutated DNA was subsequently incorporated into an adenoviral vector and administered intravenously to alcohol-preferring rats. Importantly, the endogenous rADH-47Arg gene was not deleted. Animals transduced with rAdh-47His showed a 90% ($p < 0.01$) increase in liver ADH activity and a 50% reduction ($p < 0.001$) in voluntary ethanol intake (Rivera-Meza et al. 2010). Noteworthy, these studies also demonstrated the marked specificity of the intravenously administered adenoviral vector to mainly transfect the liver. Other organs investigated showed no increases in ALDH activity; namely brain, heart, kidney, lung. Spleen showed an increase in ADH activity that was only 1% of that elevated in liver. Following the administration of 1 g ethanol/kg (i.p.) arterial blood showed a brief burst in the acetaldehyde levels that subsided within 15 min. ALDH2 in peripheral organs is expected to eliminate any small elevation of arterial acetaldehyde (Guillot et al. 2009). In line with such a view, no blood acetaldehyde was found in humans who were homozygous for the active *ADH1B*2* following the administration of ethanol (Mizoi et al. 1994; Peng et al. 2007). It is of interest that subjects carrying the ADH1B*2 show a marked *protection* against esophageal and upper respiratory cancers, unlike the marked increases in such cancers in subjects carrying the ALDH2*2 (see Israel et al. 2011). No liver injury has been reported for subjects carrying the *ADH1B*2* allele. Rather, subjects carrying the *ADH1B*2* allele are protected against non-alcoholic fatty liver disease and fibrosis, independently of the level of alcohol intake (Vilar-Gomez et al. 2020).

Rivera-Meza et al. (2012) conducted a study in which a single adenoviral vector was designed to transfect both the *ADH1B*2* gene and the antisense against *ALDH2*.

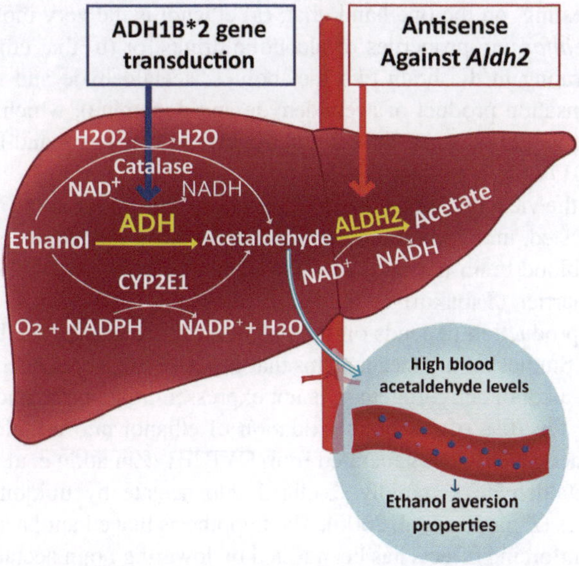

Fig. 1 Hepatic metabolism of ethanol to acetaldehyde in the liver with indication of the points where both the ADH1B * 2 transduction gene and the antisense against Aldh2 act to induce high levels of blood acetaldehyde and aversion to ethanol. Main pathways of metabolism of ethanol to acetaldehyde in the liver including the action points of both: the ADH1B * 2 transduction gene that increases the generation of acetaldehyde from ethanol and antisense against Aldh2 that inhibits the acetaldehyde degradation. Both therapies inducing increased circulating acetaldehyde levels and ethanol aversion. Abbreviations: ADH, Alcohol dehydrogenase; ALDH, Aldehyde dehydrogenase; CYP2E1, isoform 2E1 of cytochrome P450; NAD+, nicotinamide adenine dinucleotide coenzyme; NADH nicotinamide adenine dinucleotide coenzyme in its reduced form

When the adenoviral vector was intravenously delivered to rats, liver ADH activity showed a 176% increase while ALDH2 activity was reduced by 24%. The dual gene modification led to a 60% reduction in the voluntary ethanol intake of rats that had been consuming ethanol for 15 days. Upon the administration of a dose of ethanol (1 g/kg, i.p.), the rats showed arterial acetaldehyde levels that were 400% higher than those of animals administered control adenoviral vector. Venous blood acetaldehyde levels were not reported. The reader is referred to Fig. 1 which shows that both the transduction of *ALDH1B*2* gene and the antisense knockdown of ALDH2 lead to increases in blood acetaldehyde and reduce alcohol intake.

5 Brain Metabolism of Ethanol: Its Reinforcing Effects and the Role of Its Metabolic Products

Ethanol (CH3-CH2OH) is a small and relatively a non-reactive molecule, thus requiring high concentrations in the brain to directly generate effects on neurotransmission (see Deehan et al. 2013a). These characteristics have led to contrasting theories suggesting, on the one hand, that: (a) ethanol is the very molecule responsible of the *reinforcing* properties of alcoholic drinks, or (b) that ethanol acts as a prodrug generating in the brain two metabolites: acetaldehyde and salsolinol (the latter a condensation product of acetaldehyde and dopamine), which would be the molecules that would generate the reinforcing effects (see Deng and Deitrich 2008; Israel et al. 2017).

Regarding the view that ethanol acts as a prodrug, it is important to note that when ethanol is ingested, under normal conditions, systemic acetaldehyde generated does not cross the blood brain barrier due to the high activity of ALDH2 of cells of the blood brain barrier (Tabakoff et al. 1976; see Hunt 1996). Thus, in vivo brain acetaldehyde production depends on the oxidation of ethanol by the brain catalase-H_2O_2 system. Studies on the mechanisms that generate brain acetaldehyde in Wistar rats show that alcohol dehydrogenase is not expressed in the brain and that *catalase* is responsible for 70% of the brain oxidation of ethanol into acetaldehyde, while 15% of brain acetaldehyde is generated from CYP2E1 (Zimatkin et al. 2006). Brain-generated acetaldehyde is rapidly oxidized into acetate by ubiquitous aldehyde dehydrogenases (Zimatkin et al. 2006). The hypothesis that ethanol acts as a prodrug exerting its reinforcing effects has been tested by lowering brain acetaldehyde levels using the following procedures: (1) by inhibiting the brain catalase synthesis or inhibiting its enzymatic activity, (2) by activating the degradation of acetaldehyde by the brain administration of a vector with the ALDH2-coding gene or (3) by administering an agent able to trap acetaldehyde. Conversely (4) it is expected that ethanol reinforcing effects would be increased by transducing the ADH gene into the brain.

5.1 *Pharmacological Interventions Aimed at Inhibiting Brain Acetaldehyde Generation or at Lowering Its Levels*

Intracranial self-administration studies showed that P rats bred as alcohol drinkers, bar-press to self-administer either ethanol or acetaldehyde into the posterior VTA (pVTA) (Rodd et al. 2005). The question remains as to whether enough acetaldehyde is generated after moderate ethanol intake to induce its reinforcing effects during the acquisition phase of alcohol consumption. A number of studies have shown that 3-amino-1,2,4-triazole (aminotriazole), an inhibitor of catalase, reduced the acquisition of ethanol consumption in Long-Evans and UChB rats (Aragon and Amit 1992; Tampier et al. 1995) and the acquisition of operant self-administration by Wistar rats (Peana et al. 2015), in line with the view that brain acetaldehyde generated from ethanol by the action of catalase mediates both the acquisition of the voluntary ethanol consumption and the operant self-administration. However, aminotriazole also inhibited the consumption of food and of a saccharin solution, indicating that its action on ethanol intake may be nonspecific (Rotzinger et al. 1994; Tampier et al. 1995). Through an indirect study, Ledesma and Aragon (2013) showed that reducing brain hydrogen peroxide levels (required by catalase to oxidize ethanol to acetaldehyde) the ethanol-induced conditioned place preference was reduced. Further, Font et al. (2006) demonstrated that the administration of D-penicillamine, an agent that binds to acetaldehyde, significantly inhibited the voluntary intake of ethanol in rats. Again, penicillamine also shows a number of side effects (Grasedyck 1988). A recent study (Rivera-Meza et al. 2019) demonstrated that the systemic administration of a novel activator of the low-Km ALDH2 (Alda-1) dose dependently inhibited voluntary ethanol intake during the acquisition phase of ethanol consumption in UChB rats. In addition, the reader is referred to works of Quertemont et al. (2005); Deitrich (2011); Correa et al. (2012); Orrico et al. (2017) on the use of pharmacological manipulations to test the hypothesis that ethanol is a prodrug.

While from the above studies there is a general agreement that brain acetaldehyde mediates the reinforcing effects during the acquisition phase of ethanol consumption, the methodologies used are varied; in some cases employing either non-physiological concentrations and routes of administration of ethanol and acetaldehyde or non-specific inhibitors, given that it is very difficult to detect ethanol-derived acetaldehyde in the brain of living animals due to both its rapid removal and to its high reactivity (see Deng and Deitrich 2008; Peana et al. 2017). In additional studies, genetic modifications, generally considered to be more specific than pharmacological agents, were aimed at either reducing or increasing brain acetaldehyde levels were used. These are discussed below.

5.2 *Gene Modifications of Brain Acetaldehyde Generation or Its Degradation*

Two types of studies conducted with rats of the UChB line strongly conclude that the generation of acetaldehyde from ethanol in the VTA is an absolute requirement for the acquisition of ethanol consumption. In the first study, the inhibition of catalase synthesis in pVTA by administering a lentiviral vector encoding an *anti-catalase* shRNA in alcohol naive UChB rats completely inhibited (approximately 95%) the acquisition of ethanol consumption (Karahanian et al. 2011). Given that all drugs of abuse increase dopamine levels in the nucleus accumbens (Di Chiara and Bassareo 2007), it was hypothesized that the vector encoding anti-catalase shRNA could also lower nucleus accumbens dopamine levels induced by acute administration of a dose of ethanol. This was indeed demonstrated as the anti-catalase shRNA vector microinjected into pVTA completely prevented the increased dopamine level induced by an acute dose of ethanol (1 g/kg, i.p.) in the nucleus accumbens, while not affecting the dopamine release induced by amphetamine, indicating that its inhibitory effect on dopamine levels was specific for ethanol (Karahanian et al. 2011).

In a second gene therapy study, a gene encoding for the high-affinity ALDH2, which degrades acetaldehyde, was transduced into the pVTA. The administration to

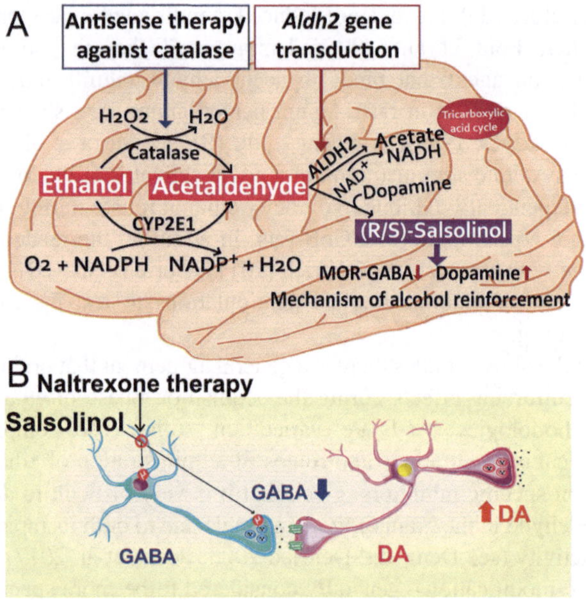

Fig. 2 Schematic representation of brain oxidation of ethanol to acetaldehyde and generation of salsolinol, indicating the points where both antisense therapy against catalase and the Aldh2 transduction gene act to inhibit the acquisition of alcohol intake, including the schematic representation of the mechanism by which naltrexone inhibits the rewarding/enhancing effect of salsolinol. (**a**) Major pathways of ethanol to acetaldehyde metabolism in the brain including the generation of salsolinol from acetaldehyde and the mechanism of its rewarding effect. The figure also includes the points where both the antisense against catalase that inhibits the generation of acetaldehyde and the transduction gene Aldh2 that increases the degradation of acetaldehyde act. (**b**) Mechanism of the rewarding effect of salsolinol and the mechanism by which naltrexone inhibits the rewarding/reinforcing effects of ethanol. *ALDH* Aldehyde dehydrogenase, *CYP2E1* isoform 2E1 of cytochrome P450, *NAD+* nicotinamide adenine dinucleotide coenzyme, *NADH* nicotinamide adenine dinucleotide coenzyme in its reduced form, *MOR* μ opioid receptor, *GABA* gamma aminobutyric acid, *DA* dopamine

naïve UChB rats of a *lentiviral* vector coding for the high-affinity human ALDH2 nearly completely abolished (90%) the acquisition of voluntary ethanol intake (Karahanian et al. 2015). The reader is referred to Fig. 2a showing both the antisense knockdown of the catalase gene and the transduction of the ALDH2 gene, both aimed at reducing VTA acetaldehyde levels and to inhibit reward/reinforcement via a reduced synthesis of salsolinol.

Given that the above studies aimed at lowering the brain acetaldehyde levels fully supported the view that brain-generated acetaldehyde is the alcohol-induced reinforcing agent, the reverse question was raised: can ethanol intake be increased by increasing the generation of acetaldehyde in the VTA? This was addressed by transducing into the pVTA a lentiviral vector encoding the hepatic ADH1B*1, which oxidizes ethanol into acetaldehyde. Hepatic ADH transduction in the VTA nearly doubled ($p < 0.001$) ethanol intake in UChB rats (Karahanian et al. 2011). Taken together, results obtained using gene specific modifications consistently indicate that ethanol-derived acetaldehyde is involved in the reinforcing properties of ethanol during the *acquisition* phase of ethanol consumption. The question remains as to whether the continuous generation of acetaldehyde in the brain is necessary to maintain a chronic ethanol consumption once a steady state intake has been achieved. This is discussed below.

5.3 *Maintenance of Chronic Ethanol Consumption: Brain Acetaldehyde* Is Not *Required to Maintain Chronic Ethanol Intake*

As described above, in alchol naïve UChB rats the early acquisition of voluntary ethanol intake was inhibited by pharmacological means or by genetic modifications aimed at reducing brain ethanol-derived acetaldehyde levels. However, in UChB rats that had chronically consumed ethanol for one to two months neither the intra-VTA administration of the anti-catalase lentiviral vector nor of the vector encoding a high-affinity ALDH2 was able to modify the chronic ethanol intake (Quintanilla et al. 2012; Karahanian et al. 2015; see Israel et al. 2017). These results were supported by studies by Peana et al. (2015) who showed that neither aminotriazole (catalase inhibitor) nor D-penicillamine (an acetaldehyde sequestering agent) was able to inhibit chronic oral operant self-administration of ethanol, although these drugs had been effective in inhibiting the acquisition of ethanol self-administration in naïve rats. Noteworthy, the latter studies were conducted in Wistar rats and not in rats selected for a high alcohol intake.

5.3.1 On the Mechanism of *Maintenance* of Chronic Ethanol Intake

Since, as described above, the mechanisms that maintain *chronic ethanol intake are not related* to the metabolism of ethanol into acetaldehyde (and thus not discussed in detail in this review), the reader is referred to literature by the authors addressing these alternative mechanisms (Israel et al. 2021; Quintanilla et al. 2020; Berríos-Cárcamo et al. 2020). Briefly, we note that while all addictive drugs act by vastly different mechanisms to release dopamine in nucleus accumbens that drive their rewarding/reinforcing effects, the mechanism of relapse of these drugs points to a hyperglutamatergic activation in response to drug-associated cues, as a causative trigger of drug-seeking behavior (Kalivas 2009). This mechanism takes precedence (over the original mechanisms) in perpetuating chronic drug relapse, as shown by the fact that relapse for virtuallly all of these drugs is markedly inhibited by toning down of the hyperglutamatergic condition in the nucleus accumbens.

Learned cues in the *chronic ethanol intake condition* are alcohol odor, and probably the alcohol taste, in heavy abusers (see Kareken et al. 2004). In rodents, it has been shown that chronic voluntary ethanol consumption (Das et al. 2015; Ding et al. 2013; Szumlinski et al. 2007) and cue-induced alcohol relapse were associated with increased glutamate levels in the nucleus accumbens (Gass et al. 2011). The finding that chronic voluntary ethanol consumption was associated both with an overflow of glutamate and a downregulation of the GLT-1 glutamate transporter (the main glutamate transporter that maintain the synaptic glutamate homeostasis) in the nucleus accumbens (Das et al. 2015) suggests (vide infra) that the drug-associated learned cues drive the perpetuation of chronic alcohol intake. This exacerbated glutamatergic activation of the nucleus accumbens would take precedence over the (catalase-mediated) release of dopamine; both of which postsynaptically activate nucleus accumbens (see Scofield et al. 2016).

Supporting this view, we have shown (Quintanilla et al. 2016) that the administration of N-acetylcysteine *does not* influence ethanol intake in rats consuming alcohol for short periods (less that 10 days), while in rats that consuming ethanol for periods over 2 months, the N-acetylcysteine greatly inhibits ethanol intake. N-acetylcysteine has been shown to tone-down the hyperglutamatergic tone induced by chronic ethanol intake (see Quintanilla et al. 2020). We have termed this differential effect of N-acetylcysteine on motivation for alcohol in the short-access versus that in the extended-access groups as an effect that occurs "beyond the first hit" (the actual title in Quintanilla et al. 2016).

5.4 *An Extended Ethanol Deprivation Allows the* Recovery *of the Reinforcing Effect of Acetaldehyde: The Alcohol Deprivation Effect (ADE)*

The alcohol deprivation effect (*ADE*) refers to a condition of heightened ethanol intake (often binge-like) seen in animals that have chronically consumed ethanol, are subsequently deprived of alcohol for a prolonged period, and are allowed re-acess to ethanol. These animals show a temporary increase in ethanol consumed in the first 60 min, extending also within the first 24 h of re-access(Sinclair and Senter 1968; Rodd et al. 2004; Spanagel and Hölter 1999; Spanagel 2009; Tampier et al. 2013; Karahanian et al. 2015). On the first day of ethanol re-access, alcohol consumed exceeds that consumed daily prior to the deprivation period. The studies described below show that such a high ethanol intake after prolonged ethanol deprivation (ADE), considered as a model of relapse-like drinking, is again dependent on brain acetaldehyde generation.

While the studies of Quintanilla et al. (2012) had shown that the anti-catalase shRNA vector did not inhibit ethanol intake in rats that had previously consumed ethanol for 60 days, an effect that was confirmed by Peana et al. (2015) using aminotriazole to inhibit catalase, the same study showed that following an ethanol deprivation period of 4 weeks the shRNA anti-catalase vector again lowered ethanol intake (Quintanilla et al. 2012). Such a return to an acetaldehyde involvement in alcohol intake was confirmed in two additional studies. Tampier et al. (2013) offered ethanol solutions (10% and 20%) to UChB rats for 67 days and following 2 weeks of alcohol deprivation they were allowed to regain access to ethanol. Results showed that the administration of the anti-catalase shRNA vector into the VTA (the first deprivation day) significantly inhibited alcohol relapse, as was demonstrated by the 60 to 80% reduction of the post-deprivation ethanol intake in these rats compared to animals injected with the control-lentiviral vector. Although this cycle of 1 week of ethanol exposure and 2 weeks of alcohol deprivation was repeated twice, the inhibition of relapse to alcohol induced by the anti-catalase shRNA vector remained significant.

An important question is whether the recovery of the reinforcing effect of acetaldehyde after deprivation and re-access to ethanol is demonstrated only by the administration of the anti-catalase vector. To answer this question, Karahanian et al. (2015) studied the effect of the intra-VTA administration of a lentiviral vector coding for the high-affinity ALDH2 on the relapse-like alcohol intake. In these studies, UChB rats that had consumed ethanol for 81 days were alcohol deprived for 7 days and subsequently were allowed ethanol re-access for 60-min. Results showed that the administration of the lentiviral vector coding for the high-affinity ALDH2 into the VTA (the first deprivation day) significantly inhibited alcohol relapse, as was demonstrated by the 75 to 80% reduction of the post-deprivation ethanol intake compared to animals injected with the control-lentiviral vector. The alcohol deprivation-reacces cycle was repeated three times of 15 days of deprivation and 7 days of re-exposure for 60 min to ethanol, with similar results. These gene transduction studies aimed at reducing brain acetaldehyde dovetail with studies of Orrico et al. (2017) showing that penicillamine, an acetaldehyde trapping agent, reduced the ADE in a dose-dependent manner in Wistar rats. As discussed below, naltrexone an opioid receptor antagonist also inhibits the high ethanol intake in the ADE condition (Cowen et al. 1999; Zhou and Kreek 2019).

Regarding the processes underlying the reactivation of acetaldehyde that led to excessive alcohol consumption during ADE, Vengeliene et al. (2014) showed the longer the deprivation period with repeated deprivation phases resulted in a stronger ADE effect. Recovery of the role of catalase-acetaldehyde (dopaminergic) in ethanol intake after prolonged alcohol deprivation (Tampier et al. 2013) suggests a partial memory loss of alcohol-associated cues and a reduction of the hyperglutamatergic state with respect to rewarding effects (dopaminergic) of ethanol generated by brain acetaldehyde. In line with this view, it has been reported that the increased glutamate concentration, found in nucleus accumbens of P rats after 2 months of chronic ethanol consumption subsided following a two-week alcohol deprivation, returning to the levels found in ethanol naive rats drinking only water (control group) (Ding et al. 2013). It is, however, noted that these animals were not exposed to the ethanol re-access condition, which likely rekindles the hyperglutamatergic state.

6 Salsolinol; An Adduct of the Condensation of Acetaldehyde and Dopamine as a Mediator of the Ethanol-Derived Acetaldehyde Effect

Acetaldehyde is a highly reactive compound that under favorable conditions, such as at physiological pH and room temperature, readily condenses with dopamine in a non-enzymatic reaction to give mainly (R/S)-salsolinol and lower amounts (about 10%) of (R/S)-isosalsolinol (Cohen and Collins 1970; King et al. 1974; Bates et al. 1986). The 85% pure *salsolinol* sold by Sigma (St. Louis, MO) for the past decades contained 10–15%of isosalsolinol (Juricic et al. 2012). The first demonstration that salsolinol can be generated from ethanol in brain tissue was provided by an ex vivo study using brain slices. In this study, Melis et al. (2015) demonstrated (1) that ethanol and acetaldehyde, but not salsolinol, failed to stimulate the activity of dopamine neurons in the pVTA slices of mice that had been administered α-methyl-p-tyrosine, an inhibitor of dopamine biosynthesis, (2) but when dopamine was applied to the slices from α-methyl-p-tyrosine-treated mice, ethanol-induced excitation of the dopaminergic neurons was restored and (3) it was possible to detect salsolinol in the medium only after the co-application of ethanol and dopamine. This effect required the ethanol oxidation into acetaldehyde by catalase, since when the H_2O_2-catalase system was inhibited by either the application of 3-amino-1,2,4-triazole to the slices or by the in vivo administration of α-lipoic acid, ethanol did not enhance dopamine cell activity (Melis et al. 2015). Recently, in an elegant in vivo study, Bassareo et al. (2021) demonstrated that brain salsolinol is generated from ethanol. These authors implanted microdialysis cannulae in both the pVTA and in nucleus accumbens shell (AcbSH) of freely moving Sprague-Dawley rats. The intragastric administration of a low dose of ethanol (1 g/kg) led to the synthesis of salsolinol in the VTA and to an increase in *dopamine* release in the AcbSH. The blockade of opiate receptors by naltrexone inhibited the release of dopamine in AcbSH, in line with studies that had suggested that the behavioral effect of salsolinol injected in the pVTA is mediated by μ-opioid receptors (Hipólito et al. 2011; Xie and Ye 2012).

By the end of the last century, several studies showed that the infusion of salsolinol into the lateral cerebral ventricle produced an increase in both preference and voluntary ethanol consumption in Sprague-Dawley rats, suggesting that the ability of salsolinol to increase ethanol consumption is centrally mediated (Myers and Melchior 1977; Duncan and Deitrich 1980). More recently, several laboratories have shown that salsolinol exhibits reinforcing properties in the absence of ethanol. These studies showed that rats developed conditioned place preference (CPP) both after systemic injections of salsolinol (10 mg/kg, i.p.), as well as after the intra-VTA administration of (30 pmol) salsolinol (Matsuzawa et al. 2000; Hipólito et al. 2011). The most compelling evidence indicating that salsolinol is reinforcing lies on the observation that the alcohol-preferring P rats bar-pressed for self-administer (R/S)-salsolinol (ca. 85% pure) into the VTA (Rodd et al. 2003). These results together with other findings by these authors indicated that ethanol, acetaldehyde, and salsolinol were intracranially self-administered into the pVTA while with different potencies. The concentrations required for the intracranial salsolinol self-administration were several orders of magnitude lower (30–100 nM) than those required for acetaldehyde (23 μM) or for ethanol (20 mM); being the reinforcing potency of salsolinol > acetaldehyde >> ethanol (Rodd et al. 2003, 2005). Other studies showed that (R/S)-salsolinol microinjected into the VTA of Wistar rats increased dopamine release in nucleus accumbens at lower concentration than that required of acetaldehyde (Deehan et al. 2013b). Taken together these results suggest that salsolinol – a secondary metabolite of ethanol – is the molecule that drives the reinforcing effect of the ethanol-derived acetaldehyde.

An added feature to note is that salsolinol has a chiral carbon and therefore the racemate can be resolved into (R)-salsolinol and (S)-salsolinol which could have a different pharmacodynamic effects. Both salsolinol stereoisomers can originate from the spontaneous condensation between dopamine and acetaldehyde. The formation of salsolinol via the non-enzymatic mechanism, following ethanol consumption, is expected to result in the formation of both enantiomers, which would not be the case through the enzymatic route. Haber et al. (1999) reported no changes in salsolinol in the striatum of alcohol-preferring rats (P rats) that had consumed ethanol for 4 weeks; however, both (R)-and (S)-salsolinol were equally increased in their adrenal grands, suggesting a non-enzymatic synthesis. It has been proposed (Naoi et al. 1996) that (R)-salsolinol could be produced enzymatically. Rojkovicova (2008) concluded that an increase in the non-enzymatic formation of (R, S)-salsolinol follows ethanol exposure. The authors showed that the ratio of R/S salsolinol present in midbrain of naïve rats was lowered following chronic ethanol intake, such that following ethanol intake the non-enzymatic synthesis of racemic (R, S)-salsolinol predominates over a stereospecific enzymatic synthesis of (R)-salsolinol.

Previous in vivo microdialysis studies shown that (*R/S*)-salsolinol reaches detectable levels (100 nM) in the neostriatum following the systemic administration of (*R/S*)-salsolinol (10 mg/kg, i.p.), isosalsolinol-free, providing evidence of the ability of salsolinol to cross the blood brain barrier (BBB) (Quintanilla et al. 2014). This result was supported by the demonstration that repeated intraperitoneal administration of (R/S)-salsolinol (10 mg/kg) elicited conditioned place preference and increased locomotor activity in a manner similar to that caused by the intra-VTA administration of repeated doses (30 pmol) of (R/S)-salsolinol in naive UChB rats (Quintanilla et al. 2014). Moreover, repeated intraperitoneal (10 mg/kg) or intra-VTA (30 pmol) administration of (R/S)-salsolinol not only sensitized rats to the locomotor stimulating effect, but also elicited a marked increases of the acquisition of ethanol intake in the UChB rats. The administration of 4 doses of salsolinol either into the VTA (30 pmol), or systemically (10 mg/kg, i.p.), every 3 days to naive UChB rats increased subsequent voluntary ethanol intake by 200 to 250%, such that the rats ingested 2.5 g of ethanol/kg in 60 min an intake which remained constant for 1 week (the duration of follow-up) after discontinuing the salsolinol administration (Quintanilla et al. 2014). Given that in this study the estimated brain half-life of salsolinol was short (30–60 min), the finding that its effect on ethanol intake lasted at least 1 week suggests that it was not the salsolinol molecule per se that perpetuated the increased ethanol consumption. The findings that salsolinol caused both a higher ethanol intake and locomotor sensitization, suggest that salsolinol elicited in these animals a greater incentive to seek ethanol (Quintanilla et al. 2014), akin to that described for amphetamine that also induces locomotor sensitization associated with a higher incentive to pursue the drug (Wise and Bozarth 1987). Importantly, intracerebral administration of salsolinol studies in UChB rats showed that the increase of ethanol intake and locomotor sensitization effects of (R/S)-salsolinol were also seen with (*R*)-salsolinol, while the (S)-salsolinol enantiomer was inactive, suggesting that in vivo the chirality of the C-1 center of (R/S)-salsolinol plays an important role in changing its affinity for transporters or receptors associated with ethanol intake (Quintanilla et al. 2015).

6.1 The Mechanism of Salsolinol Action

The action of salsolinol, as responsible for the ethanol-induced stimulation of dopamine neurons in the VTA, was blocked by pretreatment with naltrexone, a competitive antagonist of the μ-opioid receptor (*MOR*), and to a lesser extent of the κ-opioid receptor (KOR) (Niciu and Arias 2013). *Naltrexone* inhibited both the motor activation and the stimulation of the acquisition of voluntary ethanol intake induced by (R/S)-salsolinol in rats (Hipólito et al. 2010; Quintanilla et al. 2014). Xie and Ye (2012), using the patch clamp technique, showed that the ability of salsolinol to activate pVTA dopamine neurons proceeds through a mechanism that involves a decrease of GABA release onto dopamine neurons. These findings suggest that salsolinol is able to activate the (inhibitory) MOR receptors located in the VTA-GABA neuron, and as a consequence of inhibiting the inhibitory effect of GABA, the dopaminergic neurons would be activated (Xie and Ye 2012). These results were supported by a study in recombinant CHO-K1 cells that overexpressed the human μ-opioid receptor. This study showed that both (R/S)-salsolinol and its enantiomers activate the opioid receptor through the Gi-adenylate cyclase protein. However, in such in vitro study, both enantiomers showed similar efficacy, while the potency of (S)-salsolinol was greater than that of (R)-salsolinol (Berríos-Cárcamo et al. 2017). Thus, there is a discrepancy that will require resolution vis-a-vis the previous in vivo findings that showed that the (R)-salsolinol enantiomer, but not the (S)-enantiomer, generated a clear sentitization that increased voluntary ethanol intake (Quintanilla et al. 2015). The reader is referred to Fig. 2b which shows both the mechanism of salsolinol action and the mechanism by which naltrexone inhibits the rewarding effects of salsolinol.

7 Conclusions

Overall, the studies discussed here strongly suggest that: (1) ethanol is metabolized in the brain by catalase, generating acetaldehyde, which when condensed with dopamine produces salsolinol; (2) salsolinol derived from brain acetaldehyde releases dopamine and increases the acquisition of ethanol intake, an effect that is inhibited by the previous administration of naltrexone, (3) brain acetaldehyde (or salsolinol) do not influence the maintenance of chronic ethanol intake; however, (4) following a prolonged ethanol deprivation, brain acetaldehyde generation again becomes important in determining a high ethanol intake in the alcohol deprivation effect (ADE) condition, an animal model of relapse behavior; (5) the inhibitory effect of naltrexone on the high ethanol intake in the ADE condition suggests that salsolinol via an opioid receptor also contributes to the relapse-like drinking behavior, and as was mentioned above (6) a hyperglutamatergic activation of the nucleus accumbens in response to alcohol-associated cues also contributes to alcohol relapse.

Acknowledgments The authors are indebted to Dr. Lutske Tampier, Dr. Amalia Sapag, Dr. Paola Ocaranza, Dr. Eduardo Karahanian, Dr. Mario Rivera-Meza; Dr. Maria A. Juricic, Dr. Paola Morales, Dr. Ferando Ezquer; Dr. Pablo Berríos Cárcamo, Dr. Bruce Cassels, and Dr. Mario Herrera-Marschitz who contributed to many of the original studies cited here.

The studies reported here were supported by FONDECYT-Chile (grants number: 1130012, 11130241 and 1120079) and by (U.S.)NIAAA (AA-10630).

References

Adachi J, Mizoi Y, Fukunaga T, Ogawa Y, Imamichi H (1989) Comparative study on ethanol elimination and blood acetaldehyde between alcoholics and control subjects. Alcohol Clin Exp Res 13(5):601–604. https://doi.org/10.1111/j.1530-0277.1989.tb00389.x

Amit Z, Brown ZW, Rockman GE (1977) Possible involvement of acetaldehyde, norepinephrine and their tetrahydroisoquinoline derivatives in the regulation of ethanol self-administration. Drug Alcohol Depend 2(5–6):495–500. https://doi.org/10.1016/0376-8716(77)90049-7

Aragon CM, Amit Z (1992) The effect of 3-amino-1,2,4-triazole on voluntary ethanol consumption: evidence for brain catalase involvement in the mechanism of action. Neuropharmacology 31(7): 709–712. https://doi.org/10.1016/0028-3908(92)90150-n

Bassareo V, Frau R, Maccioni R, Caboni P, Manis C, Peana AT, Migheli R, Porru S, Acquas E (2021) Ethanol-dependent synthesis of Salsolinol in the posterior ventral tegmental area as key mechanism of Ethanol's action on mesolimbic dopamine. Front Neurosci 15:675061. https://doi.org/10.3389/fnins.2021.675061

Bates HA, Bagheri K, Vertino PM (1986) Effect of pH on the regioselectivity of the Pictet-Spengler reactions of 3-hydroxyphenylethylamines with formaldehyde. J Org Chem 51(15):3061–3063. https://doi.org/10.1021/jo00365a041

Bell RL, Rodd ZA, Lumeng L, Murphy JM, McBride WJ (2006) The alcohol-preferring P rat and animal models of excessive alcohol drinking. Addict Biol 11(3–4):270–288. https://doi.org/10.1111/j.1369-1600.2005.00029.x

Berríos-Cárcamo P, Quintanilla ME, Herrera-Marschitz M, Vasiliou V, Zapata-Torres G, Rivera-Meza M (2017) Racemic Salsolinol and its enantiomers act as agonists of the μ-opioid receptor by activating the Gi protein-adenylate cyclase pathway. Front Behav Neurosci 10:253. https://doi.org/10.3389/fnbeh.2016.00253

Berríos-Cárcamo P, Quezada M, Quintanilla ME, Morales P, Ezquer M, Herrera-Marschitz M, Israel Y, Ezquer F (2020) Oxidative stress and neuroinflammation as a pivot in drug abuse. A focus on the therapeutic potential of antioxidant and anti-inflammatory agents and biomolecules. Antioxidants (Basel) 9(9):830. https://doi.org/10.3390/antiox9090830

Brewer C (1984) How effective is the standard dose of disulfiram? Br J Psychiatry 144:200–202. https://doi.org/10.1192/bjp.144.2.200

Brown ZW, Amit Z, Rockman GE (1979) Intraventricular self-administration of acetaldehyde, but not ethanol, in naive laboratory rats. Psychopharmacology (Berl) 64(3):271–276. https://doi.org/10.1007/BF00427509

Cederbaum AI (2012) Alcohol metabolism. Clin Liver Dis 16(4):667–685. https://doi.org/10.1016/j.cld.2012.08.002

Cederbaum AI, Rubin E (1977) The oxidation of acetaldehyde by isolated mitochondria from various organs of the rat and hepatocellular carcinoma. Arch Biochem Biophys 179(1):46–66. https://doi.org/10.1016/0003-9861(77)90085-6

Chambers GK, Marshall SJ, Robinson GM, Maguire S, Newton-Howes J, Chong NL (2002) The genetics of alcoholism in Polynesians: alcohol and aldehyde dehydrogenase genotypes in young men. Alcohol Clin Exp Res 26(7):949–955. https://doi.org/10.1097/01.ALC.0000021145.47616.38

Chang JS, Hsiao JR, Chen CH (2017) ALDH2 polymorphism and alcohol-related cancers in Asians: a public health perspective. J Biomed Sci 24(1):19. https://doi.org/10.1186/s12929-017-0327-y

Chen YC, Peng GS, Wang MF, Tsao TP, Yin SJ (2009) Polymorphism of ethanol-metabolism genes and alcoholism: correlation of allelic variations with the pharmacokinetic and pharmacodynamic consequences. Chem Biol Interact 178(1–3):2–7. https://doi.org/10.1016/j.cbi.2008.10.029

Chen CH, Ferreira JCB, Joshi AU, Stevens MC, Li SJ, Hsu JH, Maclean R, Ferreira ND, Cervantes PR, Martinez DD, Barrientos FL, Quintanares GHR, Mochly-Rosen D (2020) Novel and prevalent non-east Asian ALDH2 variants; implications for global susceptibility to aldehydes' toxicity. EBioMedicine 55:102753. https://doi.org/10.1016/j.ebiom.2020.102753

Chick J, Gough K, Falkowski W, Kershaw P, Hore B, Mehta B, Ritson B, Ropner R, Torley D (1992) Disulfiram treatment of alcoholism. Br J Psychiatry 161:84–89. https://doi.org/10.1192/bjp.161.1.84

Christensen JK, Moller IW, Ronsted R, Angelo HR, Johansson BJ (1991) Dose-effect relationship of disulfiram in human volunteers. Pharmacol Toxicol 68(3):163–165. https://doi.org/10.1111/j.1600-0773.1991.tb01215.x

Cohen G, Collins M (1970) Alkaloids from catecholamines in adrenal tissue: possible role in alcoholism. Science 167(3926):1749–1751. https://doi.org/10.1126/science.167.3926.1749

Colombo G, Lobina C, Carai MA, Gessa GL (2006) Phenotypic characterization of genetically selected Sardinian alcohol preferring rats (sp) and non-preferring (sNP) rats. Addict Biol 11(3–4):324–338. https://doi.org/10.1111/j.1369-1600.2006.00031.x

Correa M, Salamone JD, Segovia KN, Pardo M, Longoni R, Spina L, Peana AT, Vinci S, Acquas E (2012) Piecing together the puzzle of acetaldehyde as a neuroactive agent. Neurosci Biobehav Rev 36(1):404–430. https://doi.org/10.1016/j.neubiorev.2011.07.009

Cowen MS, Rezvani AH, Jarrott B, Lawrence AJ (1999) Ethanol consumption by fawn-hooded rats following abstinence: effect of naltrexone and changes in p-opioid receptor density. Alcohol Clin Exp Res 23(6):1008–1014

Das SC, Yamamoto BK, Hristov AM, Sari Y (2015) Ceftriaxone attenuates ethanol drinking and restores extracellular glutamate concentration through normalization of GLT-1 in nucleus accumbens of male alcohol-preferring rats. Neuropharmacology 97:67–74. https://doi.org/10.1016/j.neuropharm.2015.05.009

Deehan GA Jr, Hauser SR, Wilden JA, Truitt WA, Rodd ZA (2013a) Elucidating the biological basis for the reinforcing actions of alcohol in the mesolimbic dopamine system: the role of active metabolites of alcohol. Front Behav Neurosci 7:104. https://doi.org/10.3389/fnbeh.2013.00104

Deehan GA Jr, Engleman EA, Ding ZM, McBride WJ, Rodd ZA (2013b) Microinjections of acetaldehyde or salsolinol into the posterior ventral tegmental area increase dopamine release in the nucleus accumbens shell. Alcohol Clin Exp Res 37(5):722–729. https://doi.org/10.1111/acer.12034. Epub 2012 Dec 20

Deitrich R (2011) Ethanol as a prodrug: brain metabolism of ethanol mediates its reinforcing effects – a commentary. Alcohol Clin Exp Res 35(4):581–583. https://doi.org/10.1111/j.1530-0277.2011.01454.x

Deng XS, Deitrich RA (2008) Putative role of brain acetaldehyde in ethanol addiction. Curr Drug Abuse Rev 1(1):3–8. https://doi.org/10.2174/1874473710801010003

Di Chiara G, Bassareo V (2007) Reward system and addiction: what dopamine does and doesn't do. Curr Opin Pharmacol 7(1):69–76. https://doi.org/10.1016/j.coph.2006.11.003

Ding ZM, Rodd ZA, Engleman EA, Bailey JA, Lahiri DK, McBride WJ (2013) Alcohol drinking and deprivation alter basal extracellular glutamate concentrations and clearance in the mesolimbic system of alcohol-preferring (P) rats. Addict Biol 18:297–306. https://doi.org/10.1111/adb.12018

Duncan C, Deitrich RA (1980) A critical evaluation of tetrahydroisoquinoline induced ethanol preference in rats. Pharmacol Biochem Behav 13(2):265–281. https://doi.org/10.1016/0091-3057(80)90083-0

Edenberg HJ, McClintick JN (2018) Alcohol dehydrogenases, aldehyde dehydrogenases, and alcohol use disorders: a critical review. Alcohol Clin Exp Res 42(12):2281–2297. https://doi.org/10.1111/acer.13904

Font L, Aragon CM, Miquel M (2006 Jul 15) Voluntary ethanol consumption decreases after the inactivation of central acetaldehyde by dpenicillamine. Behav Brain Res 171(1):78–86. https://doi.org/10.1016/j.bbr.2006.03.020

Gass JT, Sinclair CM, Cleva RM, Widholm JJ, Olive MF (2011) Alcohol-seeking behavior is associated with increased glutamate transmission in basolateral amygdala and nucleus accumbens as measured by glutamate-oxidase-coated biosensors. Addict Biol 16(2):215–228. https://doi.org/10.1111/j.1369-1600.2010.00262.x

Gilpin NW, Richardson HN, Cole M, Koob GF (2008) Vapor inhalation of alcohol in rats. Cur Curr Protoc Neurosci Chapter 9:Unit 9.29. https://doi.org/10.1002/0471142301.ns0929s44

Grasedyck K (1988) D-penicillamine – side effects, pathogenesis and decreasing the risks. Z Rheumatol 47(Suppl 1):17–19

Guillot A, Ren T, Jourdan T, Pawlosky RJ, Han E, Kim SJ, Zhang L, Koob GF, Gao B (2009) Targeting liver aldehyde dehydrogenase-2 prevents heavy but not moderate alcohol drinking. Proc Natl Acad Sci U S A 116(51):25974–25981. https://doi.org/10.1073/pnas.1908137116

Haber H, Dumaual N, Bare DJ, Melzig MF, McBride WF, Lumeng L, Li TK (1999) The quantitative determination of R- and S-salsolinol in the striatum and adrenal gland of rats selectively bred for disparate alcohol drinking. Addict Biol 4(2):181–189. https://doi.org/10.1080/13556219971687

Heilig M, Egli M (2006) Pharmacological treatment of alcohol dependence: target symptoms and target mechanisms. Pharmacol Ther 111(3):855–876. https://doi.org/10.1016/j.pharmthera.2006.02.001

Hellstrom E, Tottmar O (1982) Effects of aldehyde dehydrogenase inhibitors on enzymes involved in the metabolism of biogenic aldehydes in rat liver and brain. Biochem Pharmacol 31(23):3899–3905. https://doi.org/10.1016/0006-2952(82)90308-2

Hipólito L, Sanchez-Catalan MJ, Zornoza T, Polache A, Granero L (2010) Locomotor stimulant effects of acute and repeated intrategmental injections of salsolinol in rats: role of mu-opioid receptors. Psychopharmacology (Berl) 209(1):1–11. https://doi.org/10.1007/s00213-009-1751-9

Hipólito L, Martí-Prats L, Sánchez-Catalán MJ, Polache A, Granero L (2011) Induction of conditioned place preference and dopamine release by salsolinol in posterior VTA of rats: involvement of μ-opioid receptors. Neurochem Int 59(5):559–562. https://doi.org/10.1016/j.neuint.2011.04.014

Hunt WA (1996) Role of acetaldehyde in the actions of ethanol on the brain – a review. Alcohol 13(2):147–151. https://doi.org/10.1016/0741-8329(95)02026-8

Hurley TD, Edenberg HJ, Bosron WF (1990) Expression and kinetic characterization of variants of human β1β1 alcohol dehydrogenase containing substitutions at amino acid 47. J Biol Chem 265(27):16366–16372

Hurley TD, Bosron WF, Hamilton JA, Amzel LM (1991) Structure of human β1β1 alcohol dehydrogenase: catalytic effects of non-active-site substitutions. Proc Natl Acad Sci U S A 88(18):8149–8153. https://doi.org/10.1073/pnas.88.18.8149

Israel Y, Rivera-Meza M, Quintanilla ME, Sapag A, Tampier L (2011) Acetaldehyde burst protection of ADH1B*2 against alcoholism: an additional hormesis protection against esophageal cancers following alcohol consumption? Alcohol Clin Exp Res 35(5):806–810. https://doi.org/10.1111/j.1530-0277.2010.01403.x

Israel Y, Karahanian E, Ezquer F, Morales P, Ezquer M, Rivera-Meza M, Herrera-Marschitz M, Quintanilla ME (2017) Acquisition, maintenance and relapse-like alcohol drinking: lessons from the UChB rat line. Front Behav Neurosci 11:57. https://doi.org/10.3389/fnbeh.2017.00057

Israel Y, Quintanilla ME, Ezquer F, Morales P, Santapau D, Berríos-Cárcamo P, Ezquer M, Olivares B, Herrera-Marschitz M (2021) Aspirin and N-acetylcysteine co-administration markedly inhibit chronic ethanol intake and block relapse binge drinking: role of neuroinflammation-oxidative stress self-perpetuation. Addict Biol 26(1):e12853. https://doi.org/10.1111/adb.12853

Jørgensen CH, Pedersen B, Tønnesen H (2011) The efficacy of disulfiram for the treatment of alcohol use disorder. Alcohol Clin Exp Res 35(10):1749–1758. https://doi.org/10.1111/j.1530-0277.2011.01523.x

Juricic MA, Berríos-Cárcamo PA, Acevedo ML, Israel Y, Almodovar I, Cassels BK (2012) Salsolinol and isosalsolinol: condensation products of acetaldehyde and dopamine. Separation of their enantiomers in the presence of a large excess of dopamine. J Pharm Biomed Anal 63: 170–174. https://doi.org/10.1016/j.jpba.2012.02.002

Kalivas PW (2009) The glutamate homeostasis hypothesis of addiction. Nat Rev Neurosci 10:561–572. https://doi.org/10.1038/nrn2515

Karahanian E, Quintanilla ME, Tampier L, Rivera-Meza M, Bustamante D, Gonzalez-Lira V, Morales P, Herrera-Marschitz M, Israel Y (2011) Ethanol as a prodrug: brain metabolism of ethanol mediates its reinforcing effects. Alcohol Clin Exp Res 35(4):606–612. https://doi.org/10.1111/j.1530-0277.2011.01439.x

Karahanian E, Rivera-Meza M, Tampier L, Quintanilla ME, Herrera-Marschitz M, Israel Y (2015) Long-term inhibition of ethanol intake by the administration of an aldehyde dehydrogenase-2 (ALDH2) coding lentiviral vector into the ventral tegmental area of rats. Addict Biol 20(2): 336–344. https://doi.org/10.1111/adb.12130

Kareken DA, Claus ED, Sabri M, Dzemidzic M, Kosobud AEK, Radnovich AJ, Hector D, Ramchandani VA, O'Connor SJ, Lowe M, Li T-K (2004) Alcohol-related olfactory cues activate the nucleus accumbens and ventral tegmental area in high-risk drinkers: preliminary findings. Alcohol Clin Exp Res 28:550–557. https://doi.org/10.1097/01.alc.0000122764.60626.af

Kim DJ, Choi IG, Park BL, Lee BC, Ham BJ, Yoon S, Bae JS, Cheong HS, Shin HD (2008) Major genetic components underlying alcoholism in Korean population. Hum Mol Genet 17(6): 854–858. https://doi.org/10.1093/hmg/ddm357

King GS, Goodwin BL, Sandler M (1974) Isosalsolinol formation: a secondary reaction in the Pictet-Spengler condensation. J Pharm Pharmacol 26(6):476–478. https://doi.org/10.1111/j.2042-7158.1974.tb09323.x

Koob GF (2021) Drug addiction: hyperkatifeia/negative reinforcement as a framework for medications development. Pharmacol Rev 73(1):163–201. https://doi.org/10.1124/pharmrev.120.000083

Koob GF, Volkow ND (2010) Neurocircuitry of addiction. Neuropsychopharmacology 35(1):217–238. https://doi.org/10.1038/npp.2009.110

Kwon HJ, Won YS, Park O, Chang B, Duryee MJ, Thiele GE, Matsumoto A, Singh S, Abdelmegeed MA, Song BJ, Kawamoto T, Vasiliou V, Thiele GM, Gao B (2014) Aldehyde dehydrogenase 2 deficiency ameliorates alcoholic fatty liver but worsens liver inflammation and fibrosis in mice. Hepatology 60(1):146–157. https://doi.org/10.1002/hep.27036

Ledesma JC, Aragon CM (2013) Acquisition and reconditioning of ethanol-induced conditioned place preference in mice is blocked by the H_2O_2 scavenger alpha lipoic acid. Psychopharmacology (Berl) 226(4):673–685. https://doi.org/10.1007/s00213-012-2831-9

Li D, Zhao H, Gelernter J (2011) Strong association of the alcohol dehydrogenase 1B gene (ADH1B) with alcohol dependence and alcohol-induced medical diseases. Biol Psychiatry 70(6):504–512. https://doi.org/10.1016/j.biopsych.2011.02.024

Long D, Long B, Koyfman A (2017) The emergency medicine management of severe alcohol withdrawal. Am J Emerg Med 35(7):1005–1011. https://doi.org/10.1016/j.ajem.2017.02.002

Madan A, Faiman MD (1995) Characterization of diethyl dithio carbamate methylester sulfine as an intermediate in the bioactivation of disulfiram. J Pharmacol Exp Ther 272(2):775–780

Mardones J, Segovia-Riquelme N (1983) Thirty-two years of selection of rats by ethanol preference: UChA and UChB strains. Neurobehav Toxicol Teratol 5(2):171–178

Matsuzawa S, Suzuki T, Misawa M (2000) Involvement of mu-opioid receptor in the salsolinol-associated place preference in rats exposed to conditioned fear stress. Alcohol Clin Exp Res 24(3):366–372

Mays DC, Nelson AN, Fauq AH, Schriver ZH, Ververka KA (1995) S-methyl N,N diethylthiocarbamate sulfone, a potential metabolite of disulfiram and potent inhibitor of low Km mitochondrial aldehyde dehydrogenase. Biochem Pharmacol 49(5):693–700. https://doi.org/10.1016/0006-2952(94)00504-f

Meinhardt MW, Sommer WH (2015) Postdependent state in rats as a model for medication development in alcoholism. Addict Biol 20:1–21. https://doi.org/10.1111/adb.12187. Epub 2014

Melis M, Carboni E, Caboni P, Acquas E (2015) Key role of salsolinol in ethanol actions on dopamine neuronal activity of the posterior ventral tegmental area. Addict Biol 20(1):182–193. https://doi.org/10.1111/adb.12097

Mizoi Y, Yamamoto K, Ueno Y, Fukunaga T, Harada S (1994) Involvement of genetic polymorphism of alcohol and aldehyde dehydrogenases in individual variation of alcohol metabolism. Alcohol Alcohol 29(6):707–710

Myers RD, Melchior CL (1977) Differential actions on voluntary alcohol intake of tetrahydroisoquinolines or a beta-carboline infused chronically in the ventricle of the rat. Pharmacol Biochem Behav 7(4):381–392. https://doi.org/10.1016/0091-3057(77)90235-0

Naoi M, Maruyama W, Dostert P, Hashizume Y, Nakahara D, Takahashi T, Ota M (1996) Dopamine-derived endogenous 1 (R),2(N)-dimethyl-6,7-dihydroxy-1,2,3,4-tetrahydroisoquinoline, N-methyl-(R)-salsolinol, induced parkinsonism in rat: biochemical, pathological and behavioral studies. Brain Res 709(2):285–295. https://doi.org/10.1016/0006-8993(95)01325-3

Niciu MJ, Arias AJ (2013) Targeted opioid receptor antagonists in the treatment of alcohol use disorders. CNS Drugs 27(10):777–787. https://doi.org/10.1007/s40263-013-0096-4

Ocaranza P, Quintanilla ME, Tampier L, Karahanian E, Sapag A, Israel Y (2008) Gene therapy reduces ethanol intake in an animal model of alcohol dependence. Alcohol Clin Exp Res 32(1):52–57. https://doi.org/10.1111/j.1530-0277.2007.00553.x

Orrico A, Martí-Prats L, Cano-Cebrián MJ, Granero L, Polache A, Zornoza T (2017) Pre-clinical studies with D-penicillamine as a novel pharmacological strategy to treat alcoholism: updated evidence. Front Behav Neurosci 11:37. https://doi.org/10.3389/fnbeh.2017.00037

Peana AT, Porcheddu V, Bennardini F, Carta A, Rosas M, Acquas E (2015) Role of ethanol-derived acetaldehyde in operant oral self-administration of ethanol in rats. Psychopharmacology (Berl) 232(23):4269–4276. https://doi.org/10.1007/s00213-015-4049-0

Peana AT, Sánchez-Catalán MJ, Hipólito L, Rosas M, Porru S, Bennardini F, Romualdi P, Caputi FF, Candeletti S, Polache A, Granero L, Acquas E (2017) Mystic acetaldehyde: the never-ending story on alcoholism. Front Behav Neurosci 11:81. https://doi.org/10.3389/fnbeh.2017.00081

Peng GS, Chen YC, Tsao TP, Wang MF, Yin SJ (2007) Pharmacokinetic and pharmacodynamic basis for partial protection against alcoholism in Asians, heterozygous for the variant ALDH2*2 gene allele. Pharmacogenet Genomics 17(10):845–855. https://doi.org/10.1097/FPC.0b013e3282609e67

Peng GS, Chen YC, Wang MF, Lai CL, Yin SJ (2014) ALDH2*2 but not ADH1B*2 is a causative variant gene allele for Asian alcohol flushing after a low-dose challenge: correlation of the pharmacokinetic and pharmacodynamic findings. Pharmacogenet Genomics 24(12):607–617. https://doi.org/10.1097/FPC.0000000000000096

Quertemont E, Tambour S, Tirelli E (2005) The role of acetaldehyde in the neurobehavioral effects of ethanol: a comprehensive review of animal studies. Prog Neurobiol 75(4):247–274. https://doi.org/10.1016/j.pneurobio.2005.03.003

Quintanilla ME, Tampier L, Sapag A, Israel Y (2005a) Polymorphisms in the mitochondrial aldehyde dehydrogenase gene (Aldh2) determine peak blood acetaldehyde levels and voluntary ethanol consumption in rats. Pharmacogenet Genomics 15(6):427–431. https://doi.org/10.1097/01213011-200506000-00009

Quintanilla ME, Tampier L, Valle-Prieto A, Sapag A, Israel Y (2005b) Complex I regulates mutant mitochondrial aldehyde dehydrogenase activity and voluntary ethanol consumption in rats. FASEB J 19(1):36–42. https://doi.org/10.1096/fj.04-2172com

Quintanilla ME, Israel Y, Sapag A, Tampier L (2006) The UChA and UChB rat lines: metabolic and genetic differences influencing alcohol intake. Addict Biol 11(3–4):310–323. https://doi.org/10.1111/j.1369-1600.2006.00030.x

Quintanilla ME, Tampier L, Karahanian E, Rivera-Meza M, Herrera-Marschitz M, Israel Y (2012) Reward and relapse: complete gene-induced dissociation in an animal model of alcohol dependence. Alcohol Clin Exp Res 36(3):517–522. https://doi.org/10.1111/j.1530-0277.2011.01606.x

Quintanilla ME, Rivera-Meza M, Berrios-Cárcamo PA, Bustamante D, Buscaglia M, Morales P, Karahanian E, Herrera-Marschitz M, Israel Y (2014) Salsolinol, free of isosalsolinol, exerts ethanol-like motivational/sensitization effects leading to increases in ethanol intake. Alcohol 48(6):551–559. https://doi.org/10.1016/j.alcohol.2014.07.003

Quintanilla ME, Rivera-Meza M, Berríos-Cárcamo P, Cassels BK, Herrera-Marschitz M, Israel Y (2015) (R)-Salsolinol, a product of ethanol metabolism, stereospecifically induces behavioral sensitization and leads to excessive alcohol intake. Addict Biol 21(6):1063–1071. https://doi.org/10.1111/adb.12268

Quintanilla ME, Rivera-Meza M, Berríos-Cárcamo P, Salinas-Luypaert C, Herrera-Marschitz M, Israel Y (2016) Beyond the "First Hit": marked inhibition by N-acetyl cysteine of chronic ethanol intake but not of early ethanol intake. Parallel effects on ethanol-induced saccharin motivation. Alcohol Clin Exp Res 40(5):1044–1051. https://doi.org/10.1111/acer.13031

Quintanilla ME, Ezquer F, Morales P, Ezquer M, Olivares B, Santapau D, Herrera-Marschitz M, Israel Y (2020) N-acetylcysteine and acetylsalicylic acid inhibit alcohol consumption by different mechanisms: combined protection. Front Behav Neurosci 14:122. https://doi.org/10.3389/fnbeh.2020.00122

Rimondini R, Arlinde C, Sommer W, Helig M (2002) Long-lasting increase in voluntary ethanol consumption and transcriptional regulation in the rat brain afterintermittent exposure to alcohol. FASEB J 16:27–35

Rivera-Meza M, Quintanilla ME, Tampier L, Mura CV, Sapag A, Israel Y (2010) Mechanism of protection against alcoholism by an alcohol dehydrogenase polymorphism: development of an animal model. FASEB J 24(1):266–274. https://doi.org/10.1096/fj.09-132563

Rivera-Meza M, Quintanilla ME, Tampier L (2012) Reduction of ethanol consumption in alcohol-preferring rats by dual expression gene transfer. Alcohol Alcohol 47(2):102–108. https://doi.org/10.1093/alcalc/agr161

Rivera-Meza M, Vasquez D, Quintanilla ME, Lagos D, Rojas B, Herrera-Marschitz M, Israel Y (2019) Activation of mitochondrial aldehyde dehydrogenase (ALDH2) by ALDA-1 reduces both the acquisition and maintenance of ethanol intake in rats: a dual mechanism? Neuropharmacology 146:175–183. https://doi.org/10.1016/j.neuropharm.2018.12.001

Rodd ZA, Bell RL, Zhang Y, Goldstein A, Zaffaroni A, McBride WJ, Li TK (2003) Salsolinol produces reinforcing effects in the nucleus accumbens shell of alcohol-preferring (P) rats. Alcohol Clin Exp Res 27(3):440–449. https://doi.org/10.1097/01.ALC.0000056612.89957.B4

Rodd ZA, Bell RL, Sable HJK, Murphy JM, McBride WJ (2004) Recent advances in animal models of alcohol craving and relapse. Pharmacol Biochem Behav 79(3):439–450. https://doi.org/10.1016/j.pbb.2004.08.018

Rodd ZA, Bell RL, Zhang Y, Murphy JM, Goldstein A, Zaffaroni A, Li TK, McBride WJ (2005) Regional heterogeneity for the intracranial self-administration of ethanol and acetaldehyde within the ventral tegmental area of alcohol-preferring (P) rats: involvement of dopamine and serotonin. Neuropsychopharmacology 30(2):330–338. https://doi.org/10.1038/sj.npp.1300561

Rodd-Henriks ZA, Melendez RI, Zaffaroni A, Goldstein A, McBride WJ, Li TK (2002) The reinforcing effects of acetaldehyde in the posterior ventral tegmental area of alcohol-preferring rats. Pharmacol Biochem Behav 72(1–2):55–64. https://doi.org/10.1016/s0091-3057(01)00733-x

Rojkovicova T, Mechref Y, Starkey JA, Wu G, Bell RL, McBride WJ et al (2008) Quantitative chiral analysis of salsolinol in different brain regions of rats genetically predisposed to alcoholism. J Chromatogr B Analyt Technol Biomed Life Sci 863:206–214. https://doi.org/10.1016/j.jchromb.2008.01.016

Rotzinger S, Smith BR, Amit Z (1994) Catalase inhibition attenuates the acquisition of ethanol and saccharin-quinine consumption in laboratory rats. Behav Pharmacol 5(2):203–209. https://doi.org/10.1097/00008877-199404000-00012

Sapag A, Tampier L, Valle-Prieto A, Quintanilla ME, Moncada C, Israel Y (2003) Mutations in mitochondrial aldehyde dehydrogenase (ALDH2) change cofactor affinity and segregate with voluntary alcohol consumption in rats. Pharmacogenetics 13(8):509–515. https://doi.org/10.1097/00008571-200308000-00009

Scofield MD, Heinsbroek JA, Gipson CD, Kupchik YM, Spencer S, Smith AC, Roberts-Wolfe D, Kalivas PW (2016) The nucleus accumbens: mechanisms of addiction across drug classes reflect the importance of glutamate homeostasis. Pharmacol Rev 68(3):816–871. https://doi.org/10.1124/pr.116.012484

Sinclair JD, Senter RJ (1968) Development of an alcohol-deprivation effect in rats. Q J Stud Alcohol 29(4):863–867

Skinner MD, Lahmek P, Pham H, Aubin HJ (2014) Disulfiram efficacy in the treatment of alcohol dependence: a meta-analysis. PloS One 9(2):e87366. https://doi.org/10.1371/journal.pone.0087366

Sommer W, Hyytia P, Kianmaa K (2006) The alcohol-preferring AA and alcohol-avoiding rat neurobiology of the regulation of alcohol drinking. Addict Biol 11:289–309. Review. https://doi.org/10.1111/j.1369-1600.2006.00037.x

Spanagel R (2009) Alcoholism: a systems approach from molecular physiology to addictive behavior. Physiol Rev 89(2):649–705. https://doi.org/10.1152/physrev.00013.2008

Spanagel R, Hölter SM (1999) Long-term alcohol self-administration with repeated alcohol deprivation phases: an animal model of alcoholism? Alcohol Alcohol 34:231–243

Szumlinski KK, Diab ME, Friedman R, Henze LM, Lominac KD, Bowers MS (2007) Accumbens neurochemical adaptations produced by binge-like alcohol consumption. Psychopharmacology (Berl) 190(4):415–431. https://doi.org/10.1007/s00213-006-0641-7

Tabakoff B, Anderson RA, Ritzmann RF (1976) Brain acetaldehyde after ethanol administration. Biochem Pharmacol 25(11):1305–1309. https://doi.org/10.1016/0006-2952(76)90094-0

Tampier L, Quintanilla ME (2003) Involvement of brain ethanol metabolism on acute tolerance development and on ethanol consumption in alcohol-drinker (UChB) and non-drinker (UChA) rats. Addict Biol 8(3):279–286. https://doi.org/10.1080/13556210310001602185

Tampier L, Quintanilla ME, Mardones J (1995) Effects of aminotriazole on ethanol, water, and food intake and on brain catalase in UChA and UChB rats. Alcohol 12(4):341–344. https://doi.org/10.1016/0741-8329(95)00014-i

Tampier L, Quintanilla ME, Israel Y (2008) Tolerance to disulfiram induced by chronic alcohol intake in the rat. Alcohol Clin Exp Res 32(6):937–941. https://doi.org/10.1111/j.1530-0277.2008.00658.x

Tampier L, Quintanilla ME, Karahanian E, Rivera-Meza M, Herrera-Marschitz M, Israel Y (2013) The alcohol deprivation effect: marked inhibition by anticatalase gene administration into the ventral tegmental area in rats. Alcohol Clin Exp Res 37(8):1278–1285. https://doi.org/10.1111/acer.12101

Vengeliene V, Bilbao A, Spanagel R (2014) The alcohol deprivation effect model for studying relapse behavior: a comparison between rats and mice. Alcohol 48:313–320. https://doi.org/10.1016/j.alcohol.2014.03.002

Vilar-Gomez E, Sookoian S, Pirola CJ, Liang T, Gawrieh S, Cummings O, Liu W, Chalasani NP (2020) ADH1B*2 is associated with reduced severity of nonalcoholic fatty liver disease in adults, independent of alcohol consumption. Gastroenterology 159(3):929–943. https://doi.org/10.1053/j.gastro.2020.05.054

Volkow ND, Li TK (2005) Drugs and alcohol: treating and preventing abuse, addiction and their medical consequences. Pharmacol Ther 108(1):3–17. https://doi.org/10.1016/j.pharmthera.2005.06.021

Weiss F, Porrino LJ (2002) Behavioral neurobiology of alcohol addiction: recent advances and challenges. J Neurosci 22(9):3332–3337. https://doi.org/10.1523/JNEUROSCI.22-09-03332.2002

Wise RA, Bozarth MA (1987) A psychomotor stimulant theory of addiction. Psychol Rev 94(4):469–492

Xie G, He JH (2012) Salsolinol facilitates glutamatergic transmission to dopamine neurons in the posterior ventral tegmental area of rats. PLoS One 7(5):e36716. https://doi.org/10.1371/journal.pone.0036716

Yoshimura A, Kimura M, Nakayama H, Matsui T, Okudaira F, Akazawa S, Ohkawara M, Cho T, Kono Y, Hashimoto K, Kumgai M, Sahashi Y, Roh S, Higuchi S (2014) Efficacy of disulfiram for the treatment of alcohol dependence assessed with a multicenter randomized controlled trial. Alcohol Clin Exp Res 38(2):572–578. https://doi.org/10.1111/acer.12278. Epub 2013 Oct 11

Zhou Y, Kreek MJ (2019) Clinically utilized kappa-opioid receptor agonist nalfurafine combined with low-dose naltrexone prevents alcohol relapse-like drinking in male and female mice. Brain Res 1724:146410. https://doi.org/10.1016/j.brainres.2019.146410

Zimatkin SM, Pronko SP, Vasiliou V, Gonzalez FJ, Deitrich RA (2006) Enzymatic mechanisms of ethanol oxidation in the brain. Alcohol Clin Exp Res 30(9):1500–1505. https://doi.org/10.1111/j.1530-0277.2006.00181.x

Synaptic Effects Induced by Alcohol

David M. Lovinger and Marisa Roberto

Contents

D. M. Lovinger
Laboratory for Integrative Neuroscience, National Institute on Alcohol Abuse and Alcoholism (NIAAA), Rockville, MD, USA
e-mail: lovindav@mail.nih.gov

M. Roberto (✉)
Molecular Medicine Department, Scripps Research Institute, La Jolla, CA, USA
e-mail: mroberto@scripps.edu

Curr Topics Behav Neurosci (2025) 71: 217–308
https://doi.org/10.1007/7854_2022_412
Published Online: 11 February 2023

Abstract Ethanol (EtOH) has effects on numerous cellular molecular targets, and alterations in synaptic function are prominent among these effects. Acute exposure to EtOH activates or inhibits the function of proteins involved in synaptic transmission, while chronic exposure often produces opposing and/or compensatory/homeostatic effects on the expression, localization, and function of these proteins. Interactions between different neurotransmitters (e.g., neuropeptide effects on release of small molecule transmitters) can also influence both acute and chronic EtOH actions. Studies in intact animals indicate that the proteins affected by EtOH also play roles in the neural actions of the drug, including acute intoxication, tolerance, dependence, and the seeking and drinking of EtOH. The present chapter is an update of our previous Lovinger and Roberto (Curr Top Behav Neurosci 13: 31–86, 2013) chapter and reviews the literature describing these acute and chronic synaptic effects of EtOH with a focus on adult animals and their relevance for synaptic transmission, plasticity, and behavior.

Keywords Dependence · GABA · Glutamate · Intoxication · Monoamine · Neuropeptide · Neurotransmitter receptor · Postsynaptic · Presynaptic · Protein phosphorylation · Synaptic plasticity · Tolerance

1 Acute EtOH Actions

Ethanol (EtOH) produces intoxication through actions on the central nervous system (CNS) at concentrations ranging from low mM to ~100 mM (at least in non-tolerant humans and experimental animals) (Cui and Koob 2017). Several proteins involved in synaptic transmission are altered by EtOH effects within this concentration range. The target proteins include, but are not limited to, ion channels, neurotransmitter receptors, and intracellular signaling proteins (Abrahao et al. 2017; Cui et al. 2015). The first section of this article will review the literature describing the most prominent acute EtOH effects on synaptic transmission in the CNS. This review is not meant to be comprehensive, but rather to cover those effects that have been observed most consistently and that are thought to contribute to intoxication.

1.1 Ligand-Gated Ion Channels and Postsynaptic EtOH Effects

Ion channels are among the best characterized targets for acute EtOH actions (Lovinger 1997; Vengeliene et al. 2008). Ligand-gated ion channels (LGICs) are heteromeric proteins that bind extracellular neurotransmitters or intracellular messengers and transduce that binding energy into opening of an intrinsic ion pore

(Collingridge et al. 2009). Among those channels activated by extracellular neuro-transmitters, there are three classes.

1.1.1 Cys-Loop LGICs

The "cys-loop" LGICs are pentameric proteins characterized by an obligatory cysteine double bond in the n-terminal binding domain. Each subunit protein contains an extracellular ligand-binding domain, four membrane-spanning domains, and one large intracellular loop domain that also serves as a "portal" for ion permeability. This receptor class includes proteins with cation-permeable pores, the nicotinic acetylcholine (nAChR), and serotonin$_3$ (5-HT$_3$) receptors, as well as those with anion-permeable pores, the γ-aminobutyric acid$_A$ (GABA$_A$), and strychnine-sensitive glycine (GlyR) receptors. This class of receptors is distributed throughout the peripheral and central nervous systems.

Generally, acute EtOH exposure enhances the function of cys-loop LGICs (Aguayo et al. 2002; Harris 1999; Hendrickson et al. 2013; Lovinger 1997; Perkins et al. 2010; Rahman et al. 2016; Söderpalm et al. 2017), but instances of inhibition of the nAChRs and GABA$_A$Rs have been reported (Aguayo et al. 2002; Cardoso et al. 1999; Davis and De Fiebre 2006; Marszalec et al. 1994; Noori et al. 2018; Rahman et al. 2016; Roberto et al. 2003). The most common EtOH action is to potentiate channel opening in the presence of a low concentration of agonist by increasing probability of channel opening (Zhou et al. 1998) and/or increasing agonist affinity (Tonner and Miller 1995; Welsh et al. 2009). Direct EtOH binding to receptors is thought to underlie the potentiating action (Howard et al. 2014; Sauguet et al. 2013). This potentiating effect can influence both synaptic and extrasynaptic receptors (Sebe et al. 2003; Ye et al. 2001; Eggers and Berger 2004; Ziskind-Conhaim et al. 2003; Herman et al. 2016a; Herman and Roberto 2016) (Fig. 1). For example, EtOH has been shown to increase the amplitude and/or duration of GABA$_A$ and GlyR-mediated inhibitory postsynaptic currents (IPSCs) (Sebe et al. 2003; Ziskind-Conhaim et al. 2003).

Ethanol potentiation of GABA$_A$ receptor function has been extensively studied. There are 19 subunit proteins that contribute to the formation of GABA$_A$ receptors (International Union of Basic and Clinical Pharmacology, IUPHAR, database http://www.iuphar-db.org/index.jsp). Many of these subunit combinations have been examined for function and pharmacology in heterologous expression systems. To briefly summarize a large body of data, there is evidence that EtOH potentiates the function of α/β/γ-subunit-containing receptors, as well as those containing α4 or α6 along with β and δ subunits (Olsen et al. 2007; Lobo and Harris 2008; Mihic and Harris 1995; McCool et al. 2003). However, none of these findings has been uniformly replicated in all laboratories that have examined EtOH effects in heterologous systems (reviewed in Lovinger and Homanics (2007); Aguayo et al. (2002)). Using cultured and isolated neurons, several investigators have observed potentiation of GABA$_A$R function (Celentano et al. 1988; Reynolds and Prasad 1991; Aguayo 1990; Nishio and Narahashi 1990; Sapp and Yeh 1998), but this sort of

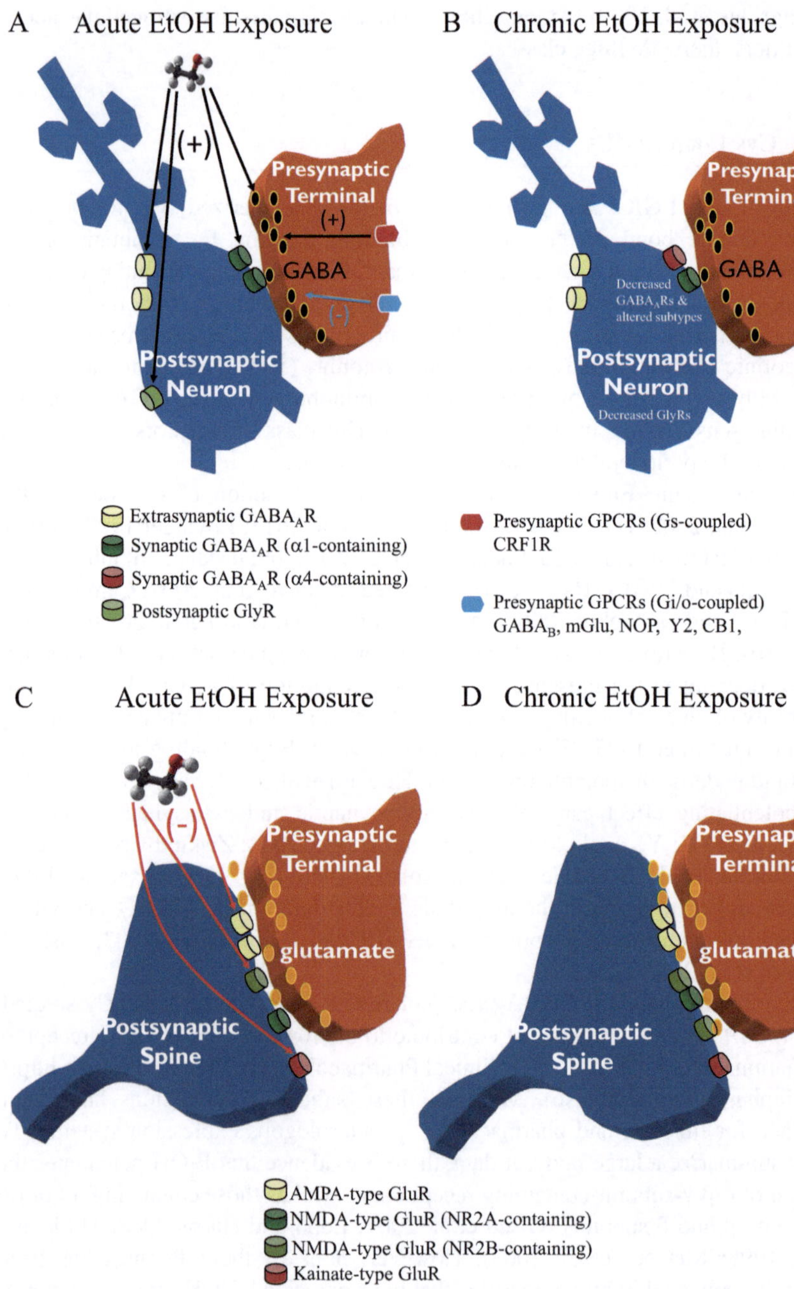

Fig. 1 Acute and chronic EtOH effects on GABAergic and glutamatergic synaptic transmission. (**a**) Schematic diagram of an inhibitory CNS synapse, including presynaptic GPCRs that modulate neurotransmitter release and postsynaptic ionotropic receptors (located both at synapses and extrasynaptically) that mediate fast synaptic transmission. The predominant presynaptic effect of acute EtOH is potentiation of GABA release (most likely by increasing the probability of vesicle fusion). This presynaptic potentiation may involve neuromodulators such as CRF, and activation of

effect has not been observed in every neuronal type examined (e.g., McCool et al. 2003; White et al. 1990; Yamashita et al. 2006). A tonic $GABA_A$-mediated current is observed in many CNS neurons and is thought to reflect the function of extrasynaptic, high-affinity GABA receptors containing the δ receptor subunit (Hanchar et al. 2005). Potentiation of this tonic current has been observed in recordings from cerebellum, hippocampus, and thalamus using the brain slice preparation (Hanchar et al. 2005; Wei et al. 2004; Glykys et al. 2007; Jia et al. 2008, although see Botta et al. (2007)). It should be noted that potentiation of GABAergic tonic current in cerebellar granule neurons does not require δ receptor subunits and involves EtOH-induced increases in interneuron firing (Diaz and Valenzuela 2016; Wadleigh and Valenzuela 2012). A recent study indicates a role for acetate-induced increases in GABA production in this EtOH action (Jin et al. 2021). Indeed, it has been suggested that EtOH inhibits GABAAR function in cerebellar granule neurons via a protein kinase C-dependent mechanism (Kaplan et al. 2013).

EtOH potentiation of $GABA_A$ receptor function appears to depend on protein phosphorylation. Messing and coworkers have shown that activity of the epsilon subunit of protein kinase C (PKC) is necessary for EtOH potentiation of γ2-subunit-containing $GABA_A$ receptors expressed heterologously in a mammalian cell line (Qi et al. 2007). This PKC action appears to involve phosphorylation of a specific serine residue on the γ2 subunit. This finding may explain data from previous studies indicating the involvement of PKC in EtOH potentiation of GABAergic transmission (Weiner et al. 1994). However, in this earlier study it was not clear if the EtOH effects on transmission involved pre- or postsynaptic mechanisms. A parallel line of investigation indicates that PKCδ is necessary for EtOH potentiation of tonic current involving δ-subunit-containing $GABA_A$Rs (Domi et al. 2021; Choi et al. 2008). It is not yet clear if acute EtOH exposure activates PKC phosphorylation of the

Fig. 1 (continued) presynaptic GPCRs and downstream signaling pathways. Postsynaptically, EtOH potentiates ionotropic $GABA_A$ and glycine receptor function. Increases in synaptic $GABA_A$R function prolong synaptic responses, while potentiation of extrasynaptic receptors increases tonic current that affects neuronal excitability. (**b**) Changes in GABAergic synapses following chronic EtOH exposure. Presynaptically, the release of GABA is decreased. Alterations in levels of neuromodulators that act on GPCRs, as well as altered function of presynaptic GPCRs may contribute to these changes. Postsynaptically, the subunit composition of $GABA_A$Rs is altered, often including increased synaptic α4-containing receptors, and fewer α1-containing synaptic receptors. Synaptic α4-containing receptors may be less sensitive to acute EtOH, promoting tolerance to synaptic effect of the drug. (**c**) Schematic diagram of a glutamatergic synapse on a dendritic spine, including postsynaptic ionotropic receptors that mediate fast synaptic transmission. The predominant effect of acute EtOH is to inhibit ionotropic glutamate receptor function, and all subclasses of these receptors are sensitive to EtOH inhibition. The most potent effects have been observed at kainate and NMDA receptor subtypes. (**d**) Changes in glutamatergic synapses following chronic EtOH exposure. Presynaptically, the release of glutamate is enhanced. Postsynaptically, NMDAR function is increased, most likely due to increased receptor density at the synapse. There is also evidence for increased numbers of NR2B-containing NMDARs, as well as evidence of increased dendritic spine volume

$GABA_AR$ or if phosphorylation on key amino acid residues is permissive for EtOH potentiation of receptor function, and this will be an interesting topic for future research.

Ethanol potentiation of glycine-activated chloride channels appears to be dependent on receptor subunit composition. Potentiation is consistently greater at receptors containing the α1 subunit (Davies et al. 2003; Mascia et al. 1996; Mihic et al. 1997), when expressed in Xenopus laevis oocytes and in neurons that express this subunit (Förstera et al. 2017; Valenzuela et al. 1998b, although see McCool et al. (2003); Yevenes et al. (2008)). Receptors containing the α2 subunit also exhibit EtOH potentiation (McCool et al. 2003; Gallegos et al. 2021), but may be less sensitive than those containing the α1 subunit (Mascia et al. 1996). Ethanol interactions with both membrane-spanning and intracellular domains within the receptor have been implicated in potentiation (Burgos et al. 2015; Mascia et al. 1996). Inclusion of the β subunit along with α2 eliminates potentiation (McCool et al. 2003). Potentiation has also been observed in neurons from brain and spinal cord, particularly in regions where the α1 subunit is expressed (Aguayo et al. 1996; Ye et al. 2001). Potentiation of the function of $GABA_A$ and glycine receptors is thought to increase inhibition of neurons. Indeed, in the prefrontal cortex, potentiation of glycine effects on GlyRs is implicated in EtOH-induced inhibition of neurons (Badanich et al. 2013). Recently, it has been shown that taurine, a glycine receptor partial agonist, modulates GABAA-mediated evoked synaptic transmission in central amygdala (CeA) of naïve rats, without affecting the acute alcohol-induced facilitation of GABAergic responses. Additionally, preapplication of the glycine receptor-specific antagonist strychnine blocked the EtOH-induced increase in GABA responses in CeA neurons from naïve rats. In CeA neurons from dependent rats, taurine no longer influenced evoked responses, but now blocked the EtOH-induced increases (Kirson et al. 2020). The relative influence of effects on synaptic versus extrasynaptic channels in producing this inhibition remains to be determined.

Acute EtOH exposure potentiates the function of $5\text{-}HT_3$ receptors that contain an intrinsic cation channel (Lovinger 1991; Machu and Harris 1994). It is yet to be determined if this action alters pre- or postsynaptic mechanisms activated by this receptor.

1.1.2 Ionotropic Glutamate Receptors

The ionotropic glutamate receptors (iGluRs) constitute the second class of neurotransmitter-activated LGICs. Three major classes of iGluRs exist, the AMPA receptors (AMPARs, gene name GRIA gives proper iGluR name, made by GluRs1-4), the NMDA receptors (NMDARs1-3, gene name GRIN), and the kainate receptors (KARs, made by GluRs5-7 and KAs1-2, gene name GRIK). These receptors are now thought to be tetrameric and each subunit contains a large n-terminal domain and an extracellular loop domain that together participate in ligand binding via a "venus fly-trap" motif (Gouaux 2004). The subunits have three membrane-spanning domains and a re-entrant pore-loop that forms the ion conduction pathway, as well as intracellular loops and a large intracellular c-terminal domain. The iGluRs are all cation-permeable, with varying ratios of Na^+, K^+, and Ca^{2+} selectivity. These receptors are present on all CNS neurons, where they mediate fast synaptic transmission and activation of intracellular signaling.

Ethanol has consistent inhibitory actions on iGluRs (although see Lu and Yeh (1999)) (Fig. 1c, d). Inhibition of NMDARs at EtOH concentrations associated with intoxication is the best characterized of these effects (Criswell et al. 2003; Dildy and Leslie 1989; Hoffman et al. 1989; Lima-Landman and Albuquerque 1989; Lovinger et al. 1989). The synaptic responses mediated by NMDARs are also reduced by EtOH (Kirson et al. 2018; Lovinger et al. 1990; Morrisett and Swartwelder 1993; Nie et al. 1994; Roberto et al. 2004b; Weitlauf and Woodward 2008; Wang et al. 2007).

Functional NMDARs always contain an obligatory NR1 subunit in combination with at least one NR2 or NR3 subunit. While EtOH inhibits all NMDAR subtypes, differences in the sensitivity to inhibition have been observed for recombinant with receptors containing different subunit compositions. The most common observation is that EtOH is less potent at receptors containing the NR1/2C composition in comparison with those containing NR1/2A or NR1/2B (Masood et al. 1994; Chu et al. 1995, but see Kuner et al. (1993); Lovinger (1995)). There are several splice variants of the NR1 subunit, and a recent comprehensive study by Woodward and coworkers showed that the NR1 splicing status, in combination with the identity of the co-assembled NR2 subunit, has small but reliable effects on EtOH sensitivity (Jin and Woodward 2006). This NR1 splice variant effect could account for the previous difference in reports of low EtOH sensitivity of NR2C-containing receptors. Receptors containing the NR3 subunit are relatively insensitive to inhibition by EtOH, but inclusion of the NR2B subunit enhances the EtOH inhibitory action on NR3-containing receptors (Jin et al. 2008). In addition, Mg^{2+} enhances EtOH inhibition of several NR1/2 and N1/2/3 receptor combinations, especially when NR2B is present (Jin et al. 2008). This finding may account for the larger effect of EtOH on NR2B containing NMDARs seen in some neuronal preparations (e.g., Fink and Göthert 1996; Lovinger 1995).

Recent studies indicate that portions of the transmembrane domains and c-terminal domain of different NMDAR subunits contribute to ethanol sensitivity of the receptor (Honse et al. 2004; Ren et al. 2003, 2007, 2012, 2013, 2017; Salous et al. 2009; Smothers et al. 2013, 2016; Smothers and Woodward 2016; Wu et al. 2019; Zhao et al. 2015, 2016).

Ethanol also inhibits the function of AMPARs, and effects can be seen at concentrations as low as 10 mM (Akinshola 2001; Akinshola et al. 2003; Dildy-Mayfield and Harris 1992; Möykkynen et al. 2003; Nieber et al. 1998; Wirkner et al. 2000). In neurons from the brain, EtOH generally shows lower potency for inhibition of AMPARs in comparison with NMDARs (Frye and Fincher 2000; Lovinger et al. 1989; Lovinger 1995). The ethanol sensitivity of recombinant AMPAR receptors is not greatly altered by changing the receptor subunit composition (Lovinger 1993), although the potency of EtOH is slightly higher for inhibition of GluR1-containing in contrast to GluR3-containing GluRs in Xenopus laevis oocytes (Akinshola 2001). In addition, recombinant AMPA receptors containing GluRs 2 and 3 exhibits slightly decreased EtOH sensitivity in comparison with those containing GluRs1, 2, and 3 or 3 alone (Akinshola et al. 2003). Recent studies suggest that this EtOH action involves increased receptor desensitization (Möykkynen et al. 2003, 2009), and thus the drug has little impact on AMPAR-mediated synaptic responses at most synapses given that desensitization does not contribute to the amplitude or time course of excitatory postsynaptic currents (EPSCs) (Lovinger et al. 1990; Ariwodola et al. 2003, but see Nie et al. (1993); Roberto et al. (2004b); Mameli et al. (2005); Zhu et al. (2007); Logrip et al. (2017); Herman et al. (2016b)). It is notable that a recent study indicates that EtOH enhances AMPAR-mediated EPSCs in the VTA via an indirect mechanism involving nicotinic ACh receptors (Engle et al. 2015).

Inhibition of KAR-mediated responses has been observed at quite low EtOH concentrations (Costa et al. 2000; Lack et al. 2008; Valenzuela et al. 1998a; Weiner et al. 1999). However, direct examination of KAR-mediated ion current has yielded mixed results, at least for the receptor constructs examined to date (Dildy-Mayfield and Harris 1992; Valenzuela et al. 1998a). Thus, it is not yet clear if EtOH inhibition of KAR function involves a direct effect on protein function or a more indirect action. Ethanol inhibition of iGluRs is generally thought to dampen neuronal excitability in many brain regions by reducing excitatory synaptic drive and inhibiting synaptic plasticity that requires iGluR activation.

1.1.3 Purinergic LGICs

The third major subtype of LGIC is the P2X purinergic receptor subclass. The P2X receptors are trimeric (Mio et al. 2005) with each subunit containing an n-terminal ligand-binding domain, two membrane-spanning domains linked by an extracellular ligand-binding domain, and a c-terminal intracellular domain of moderate length. The second membrane-spanning domain appears to serve as the lining for the ion conduction pathway. Ethanol inhibits the function of most P2X receptor subtypes, with some effects reported at concentrations associated with intoxication (Davies et al. 2002; Li et al. 1993). The P2X4 receptor appears to be the most sensitive to inhibition by EtOH, while P2X3 receptors exhibit EtOH-induced potentiation (Davies et al. 2002, 2005). At present, the physiological consequences of P2X inhibition are unclear.

1.2 G Protein-Coupled Receptors and Roles in EtOH Effects

The majority of neurotransmitter receptors are members of the G protein-coupled receptor (GPCR) superfamily. These receptors are specialized for binding a neurotransmitter, and this binding stimulates rearrangement of the protein to favor activation of intracellular signaling proteins known to bind GTP and GDP. In the GTP-bound state, the G-protein is activated. Several forms of intracellular signaling proteins are affected by activated G-proteins, including proteins that generate small molecule second messengers, as well as protein kinases and ion channels. Thus,

G-protein activation can affect neurophysiology fairly directly by altering ion channel function and can have a long-lasting influence on neuronal function by altering intracellular signaling and even gene expression.

Receptor-activated G-proteins are heterotrimeric, consisting of α, β, and γ subunits. The β and γ subunits form a tight complex, but when the G-protein is activated the α subunit affinity for the β/γ complex is reduced. The result is that two signaling elements arise from the G-protein activation and can act on different intracellular targets. The GPCRs act predominantly on three G-protein subclasses: Gi/o, Gq-like, and Gs-like (Wickman and Clapham 1995). The Gi/o G-protein class has net inhibitory effects on neuronal function, through actions of both the α and β/γ protein subunits. For example, the α subunit inhibits the enzyme adenylyl cyclase (AC) that normally generates the second messenger cAMP. The β/γ subunits activate potassium channels that inhibit neuronal activity (the so-called G protein-activated inward rectifier, GIRK, potassium channels). The β/γ subunits also inhibit the function of voltage-gated calcium channels, leading to inhibition of neurotransmitter release, and also appear to have more direct effects on vesicle fusion (Dolphin 2003; Elmslie 2003; Miller 1998; Wu and Saggau 1994). The Gq-like α subunits activate protein and lipid signaling pathways that activate ion channels that excite neurons, inhibit potassium channels, and increase neurotransmitter release. Thus, activation of the Gq subclass generally has a net excitatory effect on neuronal activity and synaptic transmission. The proximal effects of Gs-like G-protein activation are not always clear. The α subunit of these G-proteins stimulates AC/cAMP formation which can enhance synaptic transmission and inhibits some potassium channels. The effects on ion channel function of the different G-proteins are outlined in detail in previous review articles (Luo et al. 2022; Mochida 2019; Proft and Weiss 2015; Wickman and Clapham 1995).

Direct effects of acute EtOH on the function of GPCRs and G-proteins are generally weak. Furthermore, the physiological impact of these actions is not always clear. However, there are mechanisms involving these molecules that are influenced by EtOH. Studies beginning in the 1980s showed that EtOH can stimulate cAMP formation (Luthin and Tabakoff 1984; Rabin and Molinoff 1981). This may be due to direct EtOH actions on AC, but other proteins that influence GPCRs and their signaling might play roles in the neural actions of EtOH (Bjork et al. 2008, 2013; Meinhardt et al. 2022). The physiological consequences of this AC activation have long been unclear. However, recent studies indicate that acute EtOH exposure can increase neurotransmitter release (described in greater detail later in this review, Fig. 1), and activation of AC is a strong candidate to mediate these effects (Kelm et al. 2008).

In heterologous expression systems, EtOH has been shown to inhibit responses to activation of GPCRs that couple to Gq-like G-proteins. These findings mostly involve demonstrations that pharmacologically-relevant concentrations of EtOH reduce the ability of the GPCRs to activate a calcium-dependent chloride current in the Xenopus laevis oocyte preparation (Minami et al. 1997a, b, 1998). Among the GPCRs that have been examined in this context are metabotropic glutamate receptors (mGluRs), muscarinic ACh receptors, and serotonin type 2 receptors. The observation that these receptor effects are all three inhibited despite differences in the structures of the receptor molecules themselves indicates that the EtOH target site is likely downstream of the receptor itself. Indeed there is some evidence for involvement of protein kinase C, at least in the inhibition of muscarinic AChR (mAChR)-induced responses (Minami et al. 1997b).

Ethanol can also potentiate the function of GIRK-type potassium channels (Aryal et al. 2009; Kobayashi et al. 1999; Lewohl et al. 1999). This effect occurs at concentrations associated with intoxication and involves binding to a region of the channel implicated in phospholipid actions (Bodhinathan and Slesinger 2013; Glaaser and Slesinger 2017). The net effect of GIRK activation is to inhibit neuronal activity. This action of EtOH was originally observed in heterologous expression systems and in cerebellar granule neurons (Kobayashi et al. 1999; Lewohl et al. 1999), and subsequent studies have indicated similar actions in midbrain dopaminergic neurons (Federici et al. 2009). Ethanol effects on this G-protein target may contribute to intoxication. Studies by Blednov et al. (2001) indicate that loss of the GIRK2 channel subunit alters acute EtOH actions, while Tipps et al. (2016) showed enhanced ethanol conditioned place preference in mice lacking the GIRK2 subunit. The analgesic effects of ethanol are lost in mice carrying a missense mutation in GIRK2 (Kobayashi et al. 1999). Furthermore, constitutive deletion of GIRK3 in knockout (KO) mice selectively increased ethanol binge-like drinking, without affecting ethanol metabolism, sensitivity to ethanol intoxication, or continuous-access drinking (Herman et al. 2015). Notably, virally mediated expression of GIRK3 in the VTA reversed the phenotype of GIRK3 KO mice and further decreased the intake of their wild-type counterparts. In addition, GIRK3 deletion prevents ethanol-induced activation of VTA neurons and ethanol-induced release of dopamine in the nucleus accumbens (Herman et al. 2015). There is certainly a need for additional studies of how GIRK activation might contribute to intoxication.

1.3 Presynaptic Effects of EtOH

Ethanol potentiation of GABAergic synaptic inhibition is now known to result from both pre- and postsynaptic actions. As discussed in the section on LGICs, the postsynaptic effects result from potentiation of $GABA_A$/anion channels. A large literature indicates that EtOH also acts to enhance GABA release from presynaptic terminals and that this action contributes to enhanced synaptic inhibition (reviewed in Siggins et al. (2005)) (Fig. 1). Increases in fast GABAergic synaptic transmission during EtOH treatment have been observed in cerebellum, hippocampus, VTA, hypoglossal nucleus, and amygdala, both basolateral and central nuclei (Ariwodola and Weiner 2004; Ming et al. 2006; Kelm et al. 2007; Theile et al. 2008; Zhu and Lovinger 2006; Roberto et al. 2003; Sebe et al. 2003; Weiner et al. 2005; Ziskind-Conhaim et al. 2003). These studies have been carried out mostly in brain slices and isolated brain neurons. Examination of spontaneous and miniature GABAergic IPSCs allows investigators to determine if the frequency of synaptic events is altered (a likely presynaptic change) or if the amplitude is affected (likely a postsynaptic change). Such analyses have consistently shown that the frequencies of spontaneous inhibitory postsynaptic currents (sIPSC) activated by spontaneous GABA release and miniature inhibitory postsynaptic currents (mIPSC) activated by action potential-independent release of GABA quanta are increased at EtOH concentrations associated with intoxication, at least in the amygdala, cerebellum, hippocampus, and VTA (Ariwodola and Weiner 2004; Zhu and Lovinger 2006; Theile et al. 2008; Roberto et al. 2003; Kelm et al. 2007; Jimenez et al. 2019; Herman et al. 2013b; Khom et al. 2020a, b; Kirson et al. 2021; Varodayan and Harrison 2013). These effects are rapid in onset and rapidly reversible following EtOH removal from tissue.

At present, little is known about the mechanisms underlying EtOH potentiation of GABA release. The increase in mIPSC frequency suggests that the site of EtOH action is downstream of action potential generation and calcium entry into the presynaptic terminal. Experiments in the cerebellum and VTA suggest that EtOH interacts with mechanisms involved in intracellular calcium release, perhaps increasing calcium concentrations in the presynaptic terminal (Kelm et al. 2007; Theile et al. 2009). It would be helpful to know if EtOH increases calcium concentrations in the relevant population of GABAergic presynaptic terminals. However, this is difficult to determine given the small size (<1 μM diameter) of terminals and the diversity of subtypes of terminals found on any given neuron. More recently, L-type voltage-gated calcium channels (LTCCs) have been implicated in the EtOH-induced increases in CeA action potential-dependent activity (neuronal firing rates and GABA release) in naïve rats, and ethanol dependence reduces CeA LTCC membrane abundance (Varodayan et al. 2017b). Notably, nifedipine, an LTCC antagonist, prevents ethanol-induced GABA release and firing in naïve CeA, but not in dependent rats where a CRF1 antagonist (R121919) did. This switch from an LTCC- to a CRF1-based mechanism with alcohol dependence is accompanied by a shift from a role for inositol triphosphate receptor (IP3R)-mediated calcium-induced calcium release to the involvement of ryanodine receptors (RyRs) (Varodayan et al. 2017b). Furthermore, P/Q-type voltage-gated calcium channels mediate ethanol-induced CeA vesicular GABA release in a PKA and PKC-dependent manner in both naïve and dependent rats (Varodayan et al. 2017c; Cruz et al. 2011).

In fact, the role of intracellular signaling pathways in this potentiating EtOH effect has also been examined. It is well established that activation of AC or PKC potentiates transmission at synapses throughout the nervous system (see Leenders and Sheng (2005); Nguyen and Woo (2003) for review). Thus, it is logical to speculate that these signaling molecules might play a role in the acute alcohol action. Potentiation of GABA release onto cerebellar Purkinje neurons and principal neurons in the basolateral amygdala is eliminated in the presence of AC and protein kinase A (PKA) inhibitors (Kelm et al. 2008; Talani and Lovinger 2015) and is also affected by compounds targeting phospholipase C and PKC (Kelm et al. 2010). The potentiating effect of EtOH is impaired in the CeA in mice that lack PKCε (Bajo et al. 2008). Thus, PKC is implicated in both the pre- and postsynaptic effects of

EtOH at GABAergic synapses. It is notable that GABA release appears to be increased in the PKCε knockout mice prior to EtOH exposure, and thus the effect in this case may be more akin to occlusion rather than blockade of the drug action. Recently, a new class of PKCε inhibitors designed on the Rho-associated protein kinase (ROCK) inhibitor Y-27632 displayed selectivity for PKCε over other kinases and prevented ethanol-stimulated GABA release in the mouse CeA slices (Blasio et al. 2018). Nevertheless, it remains to be determined if the effects of EtOH on these signaling molecules are direct or indirect. Indeed, several studies indicate that EtOH interacts with neuromodulators such as CRF and endocannabinoids to alter GABA release (Ariwodola and Weiner 2004; Nie et al. 2004; Talani and Lovinger 2015; Roberto et al. 2010; Varodayan et al. 2015, 2016).

Inhibition of GABA transmission by acute EtOH exposure has also been observed (Blomeley et al. 2011; Wilcox et al. 2014; Patton et al. 2016). Experiments in striatal brain slices support a presynaptic mechanism of decreased GABAergic transmission onto the medium spiny projection neurons (MSNs) (Wilcox et al. 2014; Patton et al. 2016). Using an optogenetic technique in which channel rhodopsin (ChR2) was expressed in parvalbumin-containing fast-spiking striatal GABAergic interneurons (FSIs), Patton et al. (2016) found that ethanol inhibited transmission at this synapse. This inhibition involves presynaptic inhibition of GABA release due to activation of delta opiate receptors, presumably secondary to increased extracellular enkephalin.

In contrast to the effects on GABA release, the vast majority of studies indicate that acute EtOH either has no effect or inhibits release of glutamate (reviewed in Siggins et al. (2005)), although increases have been observed in some brain regions (Eggers and Berger 2004; Gioia et al. 2017; Herman et al. 2016b; Silberman et al. 2015; Xiao et al. 2009; Herman et al. 2016b). The vesicle-associated Munc 13 proteins are implicated in EtOH inhibition of glutamate release in the basolateral amygdala (Gioia et al. 2017) and neurotransmitter release in Drosophila melanogaster (Xu et al. 2018). These findings suggest differences between GABAergic and glutamatergic terminals in most brain regions that may be useful in determining what factors contribute to EtOH sensitivity of release.

1.4 Monoamines and Neurotransmitter Transport

Acute EtOH effects on neurotransmitter transport have been investigated using brain tissue and heterologous expression systems. In vivo studies indicate that EtOH increases monoamine levels in brain (reviewed in Deehan et al. (2016); Gonzales et al. (2004); LeMarquand et al. (1994); Thielen et al. (2001)), and there is also evidence for EtOH-induced increases in human ventral striatum (Aalto et al. 2015). However, most studies of neurotransmitter transporters show them to be relatively insensitive to EtOH. However, increased cell surface expression of the dopamine transporter (DAT) was observed when this protein was heterologously expressed (Mayfield et al. 2001; Maiya et al. 2002). This effect would most likely decrease striatal dopamine during acute in vivo EtOH exposure in rodents, and thus does not help to explain the findings from in vivo studies. However, there is some controversy as to whether EtOH has potent effects on dopamine uptake measured in brain tissue using voltammetric techniques (Jones et al. 2006; Mathews et al. 2006; Robinson et al. 2005; Yavich and Tiihonen 2000). The EtOH-induced increase in striatal DA levels is unperturbed in DAT knockout mice, suggesting that the drug action responsible for this effect does not involve the transporter (Mathews et al. 2006). Furthermore, studies using in vitro voltammetry and in vivo microdialysis to measure dopamine levels indicate that direct infusion of EtOH into striatum does not alter DA levels (Mathews et al. 2006; Yan 2003; Yim et al. 1998). Thus, the physiological impact of alterations in DAT function is not yet clear. Ethanol decreases DA release in striatal brain slices, albeit only at high concentrations (Budygin et al. 2001; Schilaty et al. 2014), but DAT has not been implicated in this effect. Interestingly, acute EtOH (44 mM) also decreases DA release, without impacting noradrenaline, in CeA slices of naive rats (Hedges et al. 2020).

Examination of EtOH effects on the brain serotonergic system has yielded interesting findings. In addition to potentiating 5-HT$_3$ receptor function, as mentioned in the previous section on ligand-gated ion channels, inhibition of 5-HT1c by EtOH has also been reported (Sanna et al. 1994) although it is not clear if this inhibition results from a direct effect on the receptor or on downstream signaling mechanisms. Exposure to acute EtOH also increases extracellular 5-HT levels in brain (LeMarquand et al. 1994; Thielen et al. 2001), and a recent report indicates that reduced 5-HT uptake may contribute to this effect as well as to the acute intoxicating effects of EtOH (Daws et al. 2006). A recent study showed that alcohol dependence and protracted withdrawal did not alter either 5-HT1A-mediated decrease of CeA GABA release or Htr1a expression but disrupted 5-HT2C-signaling without affecting Htr2c expression (Khom et al. 2020b). Collectively, those results provide detailed insights into modulation of CeA activity by the 5-HT system and unravel this system to chronic EtOH exposure. Thus, EtOH effects on serotonin and other monoamines require further examination.

1.5 Acetylcholine

Acute EtOH exposure has mixed effects on cholinergic synaptic transmission. As noted above, EtOH potentiates the function of some nicotinic ACh receptors, while inhibiting others (Noori et al. 2018). In addition, the Gq-coupled mAChRs are inhibited by acute EtOH (Candura et al. 1992; Kovacs et al. 1995; Larsson et al. 1995; Sanna et al. 1994; Smith 1983).

Early studies in the neuromuscular junction indicated that EtOH enhances and prolongs cholinergic synaptic transmission (Gage et al. 1975). These effects appeared to involve EtOH actions on the postjunctional (muscle) side of the synapse but were only observed at concentrations that would be near-lethal or lethal. This conclusion was supported by evidence that high concentrations of EtOH enhance responses to ACh directly applied to muscle (Bradley et al. 1980).

Mixed effects of acute EtOH on cholinergic synaptic function have been observed in different brain regions. In the striatum, the majority of ACh is provided by large, tonically active cholinergic neurons that ramify extensively and innervate many other striatal neuronal subclasses (Goldberg and Wilson 2017). Ethanol inhibits the tonic firing of these neurons, and this inhibition relieves tonic mAChR actions on striatal MSNs (Blomeley et al. 2011). The medial septum contains both cholinergic and non-cholinergic neurons. Acute EtOH application enhances the firing rate of both neuronal subtypes (Ericson et al. 1984), and the increase observed in non-cholinergic neurons is prevented by an mAChR antagonist. In the hippocampal CA1 region, acute EtOH potentiates a slow postsynaptic current mediated by mAChRs (Madamba et al. 1995) and enhances responses to applied ACh measured in vivo (Mancillas et al. 1986). These acute effects in medial septum and hippocampus appear to be due to increased ACh tone.

Ethanol has also been shown to decrease ACh release in brain slices, including studies in cortex (Carmichael and Israel 1975; Kalant and Grose 1967) and striatum (Darstein et al. 1997). In vivo studies have also shown decreased ACh levels during acute alcohol exposure in brain regions including parietal cortex and the reticular system (Erickson and Graham 1988), as well as in hippocampus (Henn et al. 1998). In contrast, ACh levels in hypothalamic slices were increased following a single in vivo exposure to alcohol (Kaneyuki et al. 1995). Increased ACh has also been observed in vivo in the striatum, brainstem, and VTA during acute exposure (Hunt and Dalton 1976; Larsson et al. 2005). It is not yet clear if the differential effects in different brain regions are due to molecular, cell type, or circuit differences in responses to EtOH. It should also be noted that these studies were performed using techniques with low temporal resolution, and thus it will be interesting to revisit alcohol effects on ACh release using newer approaches with subsecond resolution (Jing et al. 2018). More direct measurement of ACh release in brain slices would help to clarify the presynaptic effects of EtOH at cholinergic synapses in different brain regions.

1.6 EtOH and Synaptic Plasticity

Long-lasting changes in the efficacy of synaptic transmission are thought to contribute to brain development, learning and memory, and addiction (Hyman et al. 2006; Kauer and Malenka 2007). The most commonly studied forms of long-lasting synaptic plasticity are long-term potentiation (LTP), a persistent increase in synaptic transmission, and long-term depression (LTD), a persistent decrease in transmission. These types of plasticity are usually brought about by repetitive patterned activation of afferent inputs to a given postsynaptic neuron.

Effects of EtOH on LTP have been studied in different brain regions (Zorumski et al. 2014; Lovinger and Kash 2015), but the majority of information comes from studies of the Schaffer collateral inputs to the CA1 pyramidal neurons of the hippocampal formation (Blitzer et al. 1990; Morrisett and Swartzwelder 1993;

Mulkeen et al. 1987; Sinclair and Lo 1986). Acute EtOH exposure generally suppresses the induction of LTP at this and other synapses (Yin et al. 2007; Blitzer et al. 1990; Givens and McMahon 1995; Morrisett and Swartzwelder 1993; Mulkeen et al. 1987; Sinclair and Lo 1986; Wayner et al. 1993; Weitlauf et al. 2004). Effects occur at EtOH concentrations associated with intoxication, and in some studies at surprisingly low concentrations (Blitzer et al. 1990; Fujii et al. 2008). While inhibition of NMDAR function has been implicated in EtOH-induced LTP reduction (Blitzer et al. 1990; Schummers and Browning 2001), other mechanisms including enhanced GABAergic transmission, corticosterone, acetaldehyde, and neurosteroid production have also been implicated (Izumi et al. 2007, 2015; Ramachandran et al. 2015; Schummers et al. 1997; Tokuda et al. 2013). Ethanol also inhibits LTP induced by kainate receptor activation in the basolateral amygdala (Lack et al. 2008).

There is not as much information regarding EtOH effects on LTD. Two prominent subtypes of LTD can be elicited in the hippocampal CA1 region. The most widely studied form of LTD is induced by repetitive low-frequency synaptic activation and requires activation of NMDA receptors (Dudek and Bear 1992; Mulkey and Malenka 1992). In the hippocampal CA1 region LTD is enhanced by exposure to EtOH at a concentration associated with strong intoxication (Hendricson et al. 2002), although this observation has not been consistent (Izumi et al. 2005). In the nucleus accumbens (NAc), acute EtOH inhibits NMDAR-dependent LTD (Jeanes et al. 2011, 2014). Short-term in vivo exposure to EtOH prevents this LTD, and instead LTP is induced following low-frequency stimulation (Jeanes et al. 2011).

Other forms of LTD observed in hippocampus and elsewhere involve activation of mGluRs (reviewed in Lüscher and Huber (2010)). One report indicates that EtOH, at concentrations associated with severe intoxication, prevents mGluR-LTD at hippocampal synapses (Overstreet et al. 1997). At glutamatergic synapses onto cerebellar Purkinje neurons mGluR-LTD involves decreased surface expression and function of AMPARs (Ito 2001). Acute EtOH exposure inhibits this cerebellar LTD (Belmeguenai et al. 2008; Su et al. 2010), most likely due to inhibition of voltage-gated calcium channels and mGluR function. This finding is intriguing given that acute EtOH is known to impair motor coordination, and cerebellar function has been implicated in these effects. In the dorsal striatum, LTD involving these receptors also requires endocannabinoid (EC) signaling from the post- to the presynaptic neuron (retrograde EC signaling) and subsequent activation of CB1 cannabinoid receptors (Gerdeman et al. 2002). The expression of this form of LTD appears to be on the presynaptic side of the synapse. Acute EtOH increases the expression of this EC-dependent mGluR-LTD in dorsal striatum (Yin et al. 2007). It is not presently clear what mechanisms contribute to this effect of EtOH.

2 Chronic EtOH Actions

2.1 Chronic EtOH Effects on Glutamatergic Transmission and Glutamate Roles in Synaptic Plasticity

Chronic EtOH treatment in animals provides critical information relevant to central changes that take place during long-term alcohol abuse in humans (Cui et al. 2013). Persistent ethanol exposure produces both tolerance and dependence. Tolerance is manifested as a decreased behavioral response to EtOH that implies a decrease in the intoxicating effects and other responses to the drug. Therefore, higher amounts of EtOH are required to achieve the same intoxicating effects seen with acute drug administration. Ethanol dependence is generally described by symptomology elicited during and following withdrawal from EtOH (Heilig et al. 2010). These effects include anxiety, dysphoria and increased seizure susceptibility, hyperalgesia, and disruption of sleep states (Enoch 2008; Grobin et al. 1998; Kumar et al. 2009). Chronic EtOH treatment is known to induce many neuroadaptative changes in the CNS involving both glutamatergic and GABAergic synaptic transmission (reviewed in Roberto and Varodayan (2017)).

The majority of work on chronic EtOH effects on glutamatergic transmission has focused on changes in glutamate receptors, particularly in light of the sensitivity of these receptors to acute EtOH actions (see previous discussion). Chronic EtOH exposure or intake generally produces an increase in the function of NMDARs and in NMDAR-mediated glutamatergic synaptic transmission (Cebere et al. 1999; Cheng et al. 2017; Grover et al. 1998; Gulya et al. 1991; Lack et al. 2007; Ma et al. 2017; Smothers et al. 1997) (Fig. 1d), although decreases were observed in the medial prefrontal cortex (Holmes et al. 2012). Initial studies examined effects of receptor activation on neuronal calcium and nitric oxide signals either in preparations made from EtOH-exposed animals or in cultured neurons treated with ethanol in the medium (Grover et al. 1998; Gulya et al. 1991; Chandler et al. 1997; Iorio et al. 1992; Smothers et al. 1997). Exposure to EtOH for days to weeks increased NMDAR agonist-induced increases in intracellular calcium. These effects could be observed at EtOH concentrations that did not alter neuronal viability and did not affect baseline intracellular calcium levels. Furthermore, changes in responses to NMDAR activation were consistently larger than changes in the effects of activation of other ionotropic glutamate receptors (Chandler et al. 1997; Gulya et al. 1991; Smothers et al. 1997). Direct examination of ion current through the NMDAR pore has revealed effects consistent with a chronic EtOH-induced upregulation of NMDAR function (Floyd et al. 2003; Grover et al. 1998). An increase in the component of current mediated by NR2B-containing receptors has also been observed (Floyd et al. 2003; Kash et al. 2009; Roberto et al. 2004b, 2006). However, in the nucleus accumbens core an increase in synaptic receptors containing the NR2C subunit contributes to changes in glutamatergic transmission and drinking despite adverse consequences (Seif et al. 2013). Interestingly, acute EtOH inhibition of NMDARs in most brain regions is still intact or even increased after chronic in vivo exposure (Floyd et al. 2003; Roberto et al. 2004b, 2006), although a small decrease in inhibition was observed in medial septum/diagonal band neurons (Grover et al. 1998). Evidence of tolerance to EtOH inhibition during acute exposure has also been observed in hippocampal slices (Grover et al. 1994; Miyakawa et al. 1997). Overall, it appears that NMDAR function is still suppressed during intoxication even after prolonged EtOH exposure, and thus the increase in NMDAR function is likely to be dramatic after EtOH withdrawal following chronic exposure. In the mouse mPFC (Layer 5), chronic intermittent ethanol (CIE) and abstinence from CIE leads to enduring increases in synaptic glutamatergic transmission and long-term synaptic plasticity (Kroener et al. 2012). Consistent with the Kroener's report, CIE exposure (for 15 days) increased the baseline amplitude of evoked NMDA currents in layer V pyramidal neurons of mPFC of rats examined either 1 week or 4 weeks into withdrawal (Trantham-Davidson et al. 2014). Glutamatergic transmission was also enhanced in layer 2/3 mPFC of 48 h. withdrawn CIE mice compared to control mice (Pleil et al. 2015b). While this study did not separate out NMDAR- and AMPAR-mediated currents, neurons from the infralimbic of CIE mice had larger sEPSC amplitudes, indicating altered postsynaptic receptor expression/function. Also, this enhancement of glutamatergic transmission in mPFC was accompanied by a reduction in sEPSC amplitudes in the CeA of the CIE mice. A recent study investigated the concomitant alterations in basal synaptic function and neuronal excitability in the rat mPFC and dentate gyrus of the hippocampus during CIE, protracted abstinence from CIE, and re-exposure to one ethanol vapor session during protracted abstinence (Avchalumov et al. 2021a, b). Chronic ethanol consistently increased excitability of layer 2/3 pyramidal neurons in the mPFC and granule cell neurons in the DG. In the DG, this effect persisted during 21 days of abstinence. Re-exposure did not enhance excitability, suggesting resistance to vapor-induced effects. Western blotting demonstrates enhanced phosphorylation of Ca^{2+}/calmodulin-dependent protein kinase II (CaMKII) and reduced phosphorylation of NMDA receptor (N2A/2B subunits) (Avchalumov et al. 2021b; Natividad et al. 2018). One consequence of the increase in NMDAR-mediated calcium influx appears to be an increase in susceptibility to excitotoxic effects of NMDA (Chandler et al. 1993; Iorio et al. 1993), although enhanced NMDAR-mediated neuroprotection can also be observed in young cerebellar granule neurons (Pantazis et al. 1998). It has thus been postulated that excitotoxicity during EtOH withdrawal contributes to alcohol-related neuronal loss in the brain. Cortical NMDARs appear to contribute to EtOH drinking in mice (Radke et al. 2017a), and this may be related to regulation of subunit expression and receptor function after chronic intake (Radke et al. 2017b).

The mechanisms underlying the increase in NMDAR function are still under investigation, but several interesting facets of the story have already emerged. Analysis of receptor function and pharmacology, as well as examination of receptor subunit expression and location, indicate that receptors containing the NR2B subunit are the subtypes most strongly affected by chronic EtOH exposure (Carpenter-Hyland et al. 2004; Floyd et al. 2003; Kash et al. 2009; Roberto et al. 2004b) (Fig. 1d). The molecular basis of increased NR2B function is less clear. While some investigators have reported increases in NR2B mRNA expression following chronic alcohol exposure in vitro (Hu et al. 1996; Snell et al. 1996) and in vivo (Follesa and Ticku 1995; Kash et al. 2009; Roberto et al. 2006), such increases have not been observed in every brain region (Cebere et al. 1999; Floyd et al. 2003; Läck et al. 2005). Increases in NR2B, and to a lesser extent NR2A, protein expression have also been observed using immunological techniques after both in vitro and in vivo EtOH exposure (Obara et al. 2009; Snell et al. 1996; Staples et al. 2015; Avchalumov et al. 2021a, b). However, other investigators did not observe increased expression of this protein. Changes in expression of proteins that associate with NR2B may also contribute to chronic EtOH effects on transmission (Swartzwelder et al. 2016; Wills et al. 2017). Increased expression of mRNA and protein for other NR subunits and particular NR1 splice variants has been observed in some brain regions following chronic EtOH exposure (Raeder et al. 2008; Trevisan et al. 1994; Roberto et al. 2006; Winkler et al. 1999, but see Morrow et al. (1994)), but there is less evidence for increased receptor function as a result of these increases. Thus, it is not clear if increased subunit expression is the driving force behind increased receptor function, and if so, what mechanisms underlie the increase in expression or trafficking.

Changes in subcellular distribution of receptors may also contribute to altered NMDAR function following chronic EtOH exposure. In cultured hippocampal neurons, exposure to EtOH leads to increased NMDAR expression in dendritic spines, the location of glutamatergic synapses (Carpenter-Hyland et al. 2004). This increased trafficking to spines is accompanied by an increase in the contribution of NMDARs to glutamatergic transmission, but does not appear to involve increased NMDAR protein expression. The synaptic NMDARs observed following chronic EtOH exposure appear to contain the NR2B subunit. Increases in the contribution of NMDARs to glutamatergic synaptic transmission have also been observed following subacute (10 s of seconds or min) EtOH exposure, and NR2B-containing receptors also appear to contribute to these increases (Wang et al. 2007; Yaka et al. 2003). Tyrosine phosphorylation by a Fyn-like kinase has been implicated in these rapid increases in the function of NR2B-containing receptors (Wang et al. 2007), but it is yet to be determined if this mechanism plays a role in chronic EtOH effects on the receptor.

Chronic EtOH effects on AMPA and kainate receptors have been examined, with variable results. Increases in AMPA receptor subunit mRNA have been observed in hippocampus following chronic EtOH exposure (Bruckner et al. 1997). Expression of AMPAR subunit proteins was also induced by chronic exposure in primary cortical cultures (Chandler et al. 1999), while increased AMPAR binding was observed in cortical membranes from EtOH-exposed animals, and AMPA receptor binding in cortical membranes (Haugbol et al. 2005). Evidence of increased AMPAR function has also been reported following chronic EtOH exposure, as measured with intracellular calcium signals in cerebellar Purkinje neurons (Netzeband et al. 1999), and AMPA receptor-mediated synaptic responses are increased in basolateral amygdala (Lack et al. 2007). This latter effect was observed following during withdrawal but not just after the end of chronic EtOH exposure. However, other studies have reported that AMPAR expression and function are not altered following chronic EtOH exposure (e.g., Smothers et al. 1997). Chronic ethanol up-regulates neuronal activity via pentraxin (Narp) levels as well as increases in levels of the AMPAR subunits in the mouse NAcc (Ary 2012). Additionally, Marty and Spigelman (2012) reported that the amplitude and conductance of AMPAR-mediated miniature EPSCs were increased in CIE-treated rats due to an increase in a small fraction of functional postsynaptic GluA2-lacking AMPA receptors (Marty and Spigelman 2012). Similarly, CIE induced a significant increase in baseline AMPAR-mediated signaling in D1+ but not D1- MSNs in the rat NAcc (Renteria et al. 2017). The factors that underlie this variability in findings may include the type of preparation examined, the duration and pattern of EtOH exposure, and whether assays were performed just after the end of drug exposure or after withdrawal had been allowed to proceed. Increased glutamatergic transmission involving both AMPA and NMDA receptors is observed at prefrontal cortex synapses in the dorsomedial striatum following chronic alcohol consumption (Ma et al. 2017), while increased AMPAR-mediated transmission was observed in ventral hippocampus and medial prefrontal cortex (Ewin et al. 2019; Varodayan et al. 2018; Avchalumov et al. 2021b). With respect to kainate receptors, Chandler and collaborations (Chandler et al. 1999) observed no change in receptor expression in cultured cortical neurons following chronic EtOH exposure. In contrast, enhancement of both subunit protein and kainate receptor function was found in cultured hippocampal neurons (Carta et al. 2002), and chronic intermittent EtOH increased KAR-mediated synaptic transmission in basolateral amygdala (Lack et al. 2009).

Chronic alcohol has also been associated with functional upregulation of mGluR2/3 receptor signaling in the CeA and bed nucleus of the stria terminalis (BNST) (Kufahl et al. 2011), as opposed to the downregulation observed in mPFC (Meinhardt et al. 2013, 2022). Furthermore, chronic ethanol self-administration (alcohol-deprivation model) also increased sEPSC rise times indicative of compromised CeA glutamatergic receptor function (Suarez et al. 2019). Additionally, chronic intermittent ethanol treatment did not alter evoked CeA glutamate but decreased both spontaneous vesicular glutamate (mEPSCs) release and postsynaptic glutamate receptor function at rat CeA synapses (Varodayan et al. 2017a).

Chronic EtOH intake has also been shown to enhance intracellular signaling associated with mGluRs, particularly mGluR5, in the NAc (Cozzoli et al. 2009). While chronic EtOH drinking can induce increases in mGluR1 and mGluR5 protein expression in NAc and amygdala (Szumlinski et al. 2008; Obara et al. 2009), changes in mGluR5 signaling in NAc are not always associated with an increase in the protein itself (Szumlinski et al. 2008). In cultured cerebellar Purkinje neurons, exposure to EtOH for 11 days produced a decrease in mGluR-induced dendritic calcium signals (Netzeband et al. 2002). Clearly, more work is needed to determine how signaling by the many mGluR subtypes changes with long-term EtOH exposure and drinking.

Measurements of extracellular glutamate levels in brain have consistently shown increases produced by chronic EtOH exposure, especially after withdrawal or repeated cycles of withdrawal (Meinhardt et al. 2021; Dahchour and De Witte 1999, 2003; Pati et al. 2016; Rossetti and Carboni 1995; Roberto et al. 2004b).

However, reduced glutamate levels were observed following chronic ethanol drinking in mPFC (Meinhardt et al. 2021). These findings have generally been derived from measurements using in vivo microdialysis in brain. However, microdialysis measures of this type must be interpreted carefully, as both synaptic and nonsynaptic sources of glutamate contribute to the extracellular pool of this amino acid. Indeed, there is mounting evidence that changes in the cystine/glutamate exchanger generate increases in extracellular glutamate produced by some drugs of abuse (Kalivas 2009). Evidence of increased synaptic glutamate release has been observed in amygdala and hippocampus following chronic EtOH treatment (Chefer et al. 2011; Christian et al. 2013; Lack et al. 2007; Zhu and Pan 2007; Roberto et al. 2004b). Increased glutamatergic transmission onto MSNs involving presynaptic mechanisms has also been observed following chronic EtOH consumption. Amygdala inputs to dorsomedial striatum exhibit increases in glutamate release following chronic drinking (Ma et al. 2017). Presynaptic effects may be stronger at D1 receptor-expressing MSNs related to those that express D2 receptors (Cheng et al. 2017). Decreases in glutamate uptake have also been noted following chronic EtOH exposure (Melendez et al. 2005). Examination of effects of pharmacological treatments that alter extracellular glutamate levels indicates that increased glutamate in the NAc contributes to increased EtOH intake (Griffin et al. 2014), and glutamate uptake mechanisms may thus be a target for treatment of AUD (Rao et al. 2015). The mGlu2 metabotropic receptors provide feedback reduction of glutamate release, and dysfunction of these receptors appears to contribute to increased release following chronic EtOH exposure (Adermark et al. 2011a; Johnson et al. 2020; Meinhardt et al. 2013). Enhancing this feedback function may be useful in reducing excessive EtOH consumption (Griffin et al. 2014; Meinhardt et al. 2013). However, mechanisms independent of glutamate transport and group II mGluRs have also been implicated in the increase in extracellular glutamate in the NAc (Pati et al. 2016). There may be multiple factors that contribute to increased extracellular glutamate levels and increased or decreased glutamatergic transmission following chronic EtOH exposure and withdrawal.

Despite the evidence that NMDAR function and extracellular glutamate levels are increased following chronic EtOH exposure, studies of hippocampal LTP indicate that this form of synaptic plasticity is decreased under the same conditions (Drissi et al. 2019; Durand and Carlen 1984; Roberto et al. 2002; Talani et al. 2014, although see Fujii et al. (2008); Stragier et al. (2015)). Altered function of NMDA receptors containing the 2A and 2B subunits, resulting from related from changes in histone acetylation has been implicated in impaired LTP (Drissi et al. 2019). Similar results have been obtained in the amygdala (Stephens et al. 2005). In the NAc, NMDAR-dependent LTP is also impaired by repeated EtOH exposure, and this alteration is associated with sensitization to the locomotor stimulating effects of the drug as well as increased EtOH intake (Abrahao et al. 2013). In a subsequent study, loss of LTP in NAc was only observed in D2 receptor-expressing MSNs following binge drinking (Ji et al. 2017). It is not yet clear what factors underlie the decrease in LTP, but mechanisms occurring downstream of NMDAR activation in the LTP induction process may play a role. However, mice expressing ethanol-resistant NMDARs show enhanced sensitization and consumption (den Hartog et al. 2013,

2017), implicating this receptor in altered sensitization perhaps related to loss of LTP. In the NAc changes in dopaminergic transmission involving D1 receptors may play a role in LTP impairment (Ji et al. 2017). Loss of LTD in the hippocampal CA1 region has been observed in mice that are resistant to locomotor sensitization, suggesting that resilience to plasticity of glutamatergic transmission may contribute to lack of this increased response to EtOH (Coune et al. 2017). Hippocampal LTD is also impaired following two high-dose ethanol exposures, and this is associated with impaired novel object recognition (Silvestre de Ferron et al. 2015). In a recent rat study, ethanol self-administration and chronic intermittent ethanol exposure (6–7 weeks) did not alter the degree of LTP compared to naïve controls in mPFC of both females and males, and this form of LTP was dependent on both NMDA and AMPA receptors activation (Avchalumov et al. 2021b).

Disruption of mGluR-dependent hippocampal LTD has also been observed following chronic intermittent EtOH exposure (Wills et al. 2017). This change in plasticity is associated with altered expression of a number of proteins associated with the NR2B NMDAR subunit, including the ARC and Homer proteins that also interact with group I mGluRs. These proteins may thus mediate cross-talk between NMDA- and mGluR-based LTD mechanisms that are altered by EtOH and contribute to impaired plasticity.

It should be noted that LTP is enhanced following chronic EtOH exposure in some brain regions. For example, glutamatergic synapses in the prefrontal cortex show enhanced LTP in chronic EtOH-exposed mice (Kroener et al. 2012; Nimitvilai et al. 2016). Recent studies have shown that chronic ethanol drinking produces increased AMPAR function in the medial part of the dorsal striatum resembling that seen in LTP, particularly at medial PFC inputs to this striatal subregion, and synapses onto the striatal projection neurons that express D1-type dopamine receptors (Wang et al. 2012, 2015; Ma et al. 2017). A similar effect has been observed at glutamatergic synapses in the NAc and can appear after the first session of EtOH self-administration (Beckley et al. 2016). Inducing LTP and LTD in the dorsomedial striatum alters ethanol drinking (Ma et al. 2018). In the BNST, enhanced LTP of glutamatergic synapses is observed following chronic intermittent ethanol exposure (Wills et al. 2012).

Chronic EtOH exposure also alters LTD in striatal brain regions. In the NAc, chronic EtOH-induced changes in LTD vary according to neuronal subtype. In D1 receptor-expressing direct pathway MSNs, LTD appears after chronic exposure, while it is lost, and even converted to LTP in D2-expressing, indirect pathway MSNs (Jeanes et al. 2014; Renteria et al. 2017, 2018). In dorsal striatum, chronic EtOH exposure reduces or eliminates endocannabinoid-dependent LTD (Adermark et al. 2011b; Cui et al. 2011; DePoy et al. 2013). Impairment of the dampening of cortical glutamatergic inputs may contribute to enhanced activation of dorsolateral striatum and altered decision making (DePoy et al. 2013, 2015). Impairment of a form of LTD in the BNST driven by activation of alpha1 adrenergic receptors is observed following chronic EtOH exposure (McElligott et al. 2010).

In recent years it has become apparent that chronic EtOH exposure or drinking reduces presynaptic modulation by a number of G protein-coupled receptors. The affected receptors are generally those that couple to Gi/o-type G-proteins and reduce glutamate release (Ding et al. 2016; Johnson and Lovinger 2016; Johnson et al. 2020; Muñoz et al. 2018a, b; Roberto and Varodayan 2017). Activation of these receptors often results in a presynaptically-expressed form of LTD (Atwood et al. 2014). Activation of the presynaptic Gi/o-coupled mGlu2 receptor produces LTD, and mutations that lead to loss of receptor function in alcohol-preferring rats contribute to their increased EtOH consumption (Zhou et al. 2013).

2.2 Chronic EtOH and GABAergic Transmission: Postsynaptic Effects

Chronic EtOH treatment is known to induce many neuroadaptative changes in the CNS. Over the past 20 years, it has been widely demonstrated that GABAergic transmission is sensitive to EtOH in distinct brain regions and is clearly involved in ethanol tolerance and dependence (Eckardt et al. 1998; Grobin et al. 1998). Chronic EtOH exposure often results in the development of tolerance to many GABAergic effects of the drug including the anxiolytic, sedative, ataxic, and positive reinforcing effects (Kumar et al. 2004, 2009). Substantial evidence suggests that these behavioral and neural adaptations involve marked changes in the expression profile of specific $GABA_A$ receptor subunits (Grobin et al. 1998) and in the pharmacological properties of $GABA_A$ receptors (Kang et al. 1998b) (Fig. 1).

Chronic EtOH administration differentially altered the expression of distinct $GABA_A$ receptor subunit mRNAs and peptide levels in various brain regions. In the cerebral cortex, both mRNA and peptide levels for $GABA_A$ receptor α1, α2, and α3 subunits were decreased (Devaud et al. 1995, 1997). In contrast, both α4, β1, β2, β3, γ1, and γ2 subunit mRNA and peptide levels were increased (Devaud et al. 1995, 1997). These alterations in the subunit expression affect the $GABA_A$ receptor assemblage and consequently, also affect receptor function and binding. It has been reported that recombinant $GABA_A$ receptors with α4β2γ2 subunits are less sensitive to GABA and benzodiazepines compared to α1β2γ2 receptors (Whittemore et al. 1996). Therefore, these alterations may account for the decreased sensitivity to GABA in cerebral cortical synaptoneurosomes (Morrow et al. 1988) and benzodiazepines in cortical membrane vesicles (microsacs) (Buck and Harris 1990). Following chronic EtOH exposure, acute ethanol did not facilitate the GABA or muscimol-stimulated Cl^- uptake in cortex (Morrow et al. 1988) and in cerebellum (Allan and Harris 1987). Recently, Morrow and collaborators have reported in cultured rat cortical neurons two distinct populations of synaptic and extrasynaptic α4-containing $GABA_A$Rs[1,2] that are altered after chronic EtOH treatment.

In the cerebellum, chronic EtOH exposure decreased $GABA_A$ receptor α1 subunit mRNA and increased α6 subunit mRNA (Mhatre and Ticku 1992; Morrow et al. 1992). Chronic EtOH administration also decreased the polypeptide levels of the δ subunit of $GABA_A$ receptors in the rat cerebellum and hippocampus, whereas there were no changes in the δ subunit polypeptide levels in the rat cerebral cortex (Marutha Ravindran et al. 2007). Furthermore, chronic EtOH administration caused a downregulation of native δ subunit-containing $GABA_A$ receptor assemblies in the rat cerebellum as determined by [(3)H]muscimol binding to the immunoprecipitated receptor assemblies (Marutha Ravindran et al. 2007).

The alterations in $GABA_A$ receptor gene expression are regionally and temporally dependent. For example, chronic EtOH consumption produced a significant increase in the level of $GABA_A$ receptor α4 subunit peptide in the hippocampus following 40 days but not 14 days exposure (Matthews et al. 1998). The relative expression of hippocampal $GABA_A$ receptor α1, α2, α3, β(2/3), or γ2 subunits was not altered by either period of chronic EtOH exposure (Charlton et al. 1997; Matthews et al. 1998). Hippocampal α1 subunit immunoreactivity and mRNA content were also significantly reduced after 12 weeks of treatment, but not after 4 weeks of exposure. In contrast, α5 mRNA content was increased in this brain region. In marked contrast, chronic EtOH consumption for both 14 (Devaud et al. 1997) and 40 (Devaud et al. 1997; Matthews et al. 1998) days significantly increased the relative expression of cerebral cortical $GABA_A$ receptor α4 subunits and significantly decreased the relative expression of α1 subunits (Devaud et al. 1997; Matthews et al. 1998). These findings indicate that chronic EtOH consumption alters $GABA_A$ receptor gene expression in the hippocampus but in a different manner from that in either the cerebral cortex or the cerebellum (Kaplan et al. 2016, for review, see Valenzuela and Jotty (2015)). In addition, these alterations are dependent on the duration of EtOH exposure (Grobin et al. 1998).

The Olsen and Spigelman groups have developed a chronic intermittent EtOH treatment paradigm in which rats are given a 5- to 6-g/kg dose of ethanol on alternate days for 60 treatments (120 days). This chronic administration of EtOH to rats on an intermittent regimen, for 60 repeated intoxicating doses and repeated withdrawal episodes, increases levels of α4 subunit mRNA in hippocampus with no significant change in the mRNAs for the α5 subunit (Mahmoudi et al. 1997). Similarly, rats that were exposed to intermittent episodes of intoxicating EtOH and withdrawal showed increased hippocampal α4 subunit peptide expression (Cagetti et al. 2003) and alteration in the pharmacological responses of $GABA_A$ receptors to benzodiazepine agonists and inverse agonists (Cagetti et al. 2003). The mRNA levels for the γ2S and γ1 subunits were also elevated. In CA1 pyramidal slices from chronic intermittent EtOH-exposed rats, the baseline decay time of $GABA_A$R-mediated mIPSCs was decreased, and the positive GABA receptor modulation of mIPSCs was also reduced compared with control rats. However, mIPSC potentiation by the α-preferring benzodiazepine ligand bretazenil was maintained, and mIPSC potentiation by Ro15-4513 was increased (Cagetti et al. 2003; Liang et al. 2009).

In the VTA, levels of α1 subunit immunoreactivity were significantly decreased after 12 weeks but not 1–4 weeks of treatment (Charlton et al. 1997). Papadeas et al. (2001) found that in the amygdala, α1 and α4 subunit expression was significantly decreased after 2 weeks of chronic EtOH consumption. In the nucleus accumbens (NAC), α4 subunit expression was decreased, but α1 subunit expression was not altered. In the VTA, there were no changes in α1 and α4 subunit expression.

Muscimol-stimulated Cl^- uptake was enhanced in the extended amygdala, but not the NAC of EtOH-dependent rats. These results suggest that chronic EtOH exposure alters $GABA_A$ receptor expression in the amygdala and NAC and that decreased expression of α4 subunits is associated with increases in $GABA_A$ receptor function in the amygdala but not the NAC (Papadeas et al. 2001).

Alterations in subunit assembly could induce alterations in the functional properties of $GABA_A$ receptors without alterations in the total number of receptors (Devaud et al. 1995; Kumar et al. 2009; Morrow et al. 1992). The expression of $GABA_A$ receptors involves a highly regulated process of synthesis, assembly, endocytosis, and recycling or degradation. Changes in the expression and composition of various $GABA_A$ receptors could result from selective endocytosis, recycling, and/or trafficking of newly synthesized receptors to the cell surface. $GABA_A$ receptor trafficking on the cell surface following EtOH consumption is thought to contribute to the development of EtOH dependence (Kumar et al. 2004). It has been reported by Kumar et al. (2003) that chronic EtOH exposure selectively increases the internalization of α1 $GABA_A$ receptors with no change in the internalization of α4 $GABA_A$ receptors into clathrin-coated vesicles of the cerebral cortex. There is also a decrease in α1 $GABA_A$ receptors and a significant increase in α4 subunit peptide in the synaptic fraction following chronic EtOH exposure. These results suggest that the regulation of intracellular trafficking following chronic EtOH administration may alter the subtypes of $GABA_A$ receptors on the cell surface and may account for changes in the pharmacological properties of $GABA_A$ receptors (Kumar et al. 2004) (Fig. 1).

Clathrin and the adaptor complex (AP) play a crucial role in the internalization of $GABA_A$ receptors following chronic EtOH administration. Notably, in the intracellular fraction, the clathrin-α1-$GABA_A$ receptor complex is increased following chronic EtOH administration (Kumar et al. 2004). Specific $GABA_A$ receptor subunits (β2 and/or γ2) are required for recognition of the receptor by the AP-2 that precedes clathrin-dependent endocytosis (Herring et al. 2003; Kittler et al. 2008). Chronic EtOH exposure induces an increase in the expression of α4-, β2-, and β3-$GABA_A$ receptor subunits in the cerebral cortex and all of these subunits contain consensus phosphorylation sites for PKC. In contrast, α1, α2, and α3 $GABA_A$ receptor subunits are decreased in the cortex and these subunits do not contain consensus phosphorylation sites for PKC. Hence, it has been hypothesized that PKC may phosphorylate the $GABA_A$ receptor subunits and/or AP-2 following chronic EtOH administration, altering the recognition and endocytosis of $GABA_A$ receptors by blocking AP-2 binding (Macdonald 1995; Mohler et al. 1996). A single dose of EtOH also increases the internalization of $GABA_A$ receptor α4 and δ subunits (Liang et al. 2007). In rat hippocampus, chronic EtOH exposure induces a decrease in the tyrosine kinase phosphorylation of α1 subunits, an increase of β2 subunits, and no alteration in γ2 subunits (Marutha Ravindran et al. 2007).

$GABA_A$ receptor trafficking is regulated by many protein kinases, including PKC, PKA, and fyn. However, to date, the role of these protein kinases has not yet been studied in the trafficking of $GABA_A$ receptors, especially following EtOH exposure. Chronic EtOH consumption decreases association of PKCγ with α1

GABA$_A$ receptors and increases association of PKCγ with α4 GABA$_A$ receptors, accompanied by a decreased expression of the α1 subunit and an increased expression of α4 at the cell surface in cerebral cortex (Kumar et al. 2002). However, there were no alterations in the association of PKCγ with GABA$_A$ receptors in the α1 subunit expression following chronic EtOH administration in the hippocampus (Kumar et al. 2004). The increased association of PKCγ with α4 GABA$_A$ receptors may phosphorylate GABA$_A$ receptor subunits and prevent recognition of the receptor by AP-2, thus preventing its internalization. Indeed, phosphorylation of GABA$_A$ receptor subunits reduced the binding of receptors with AP- 2 and subsequent internalization (Kittler et al. 2008). Moreover, reduced PKC-dependent GABA$_A$ receptor phosphorylation increases receptor binding to the AP-2 and promotes receptor endocytosis (Terunuma et al. 2008). Chronic activation of PKA in cerebellar granule cells increases cell surface expression of GABA$_A$ receptor α1 subunit (Ives et al. 2002). Ethanol exposure alters expression and translocation of PKA (Diamond and Gordon 1994; Newton and Messing 2006) suggesting that PKA is likely also involved in the trafficking of GABA$_A$ receptors following EtOH exposure. Future studies will determine the specific role of various protein kinases in GABA$_A$ receptor trafficking following chronic EtOH administration.

Post-translational modifications such as phosphorylation and glycosylation of GABA$_A$ receptors may play a role in the development of EtOH dependence. In particular, phosphorylation of GABA$_A$ receptors has been demonstrated to modulate receptor function. In Xenopus oocytes and isolated mouse brain membrane vesicles (microsacs), PKC and PKA phosphorylation of GABA$_A$ receptors decreases receptor activation (Kellenberger et al. 1992; Krishek et al. 1994; Leidenheimer et al. 1992). Phosphorylation by CAM kinase II or tyrosine kinase enhances GABA$_A$ receptor function (Churn et al. 2002; Valenzuela et al. 1995). As discussed previously, acute EtOH induces changes in GABA$_A$ receptor function that may be dependent on phosphorylation of particular proteins. Chronic EtOH exposure might be expected to result in long-term changes in second messenger systems, including kinase activity. However, the heterogeneity of GABA$_A$ receptors expressed in vivo has precluded definitively answering this question and none of these studies have directly demonstrated that phosphorylation is involved in EtOH modulation of GABA$_A$ receptor function. The exact mechanisms involved in the alteration of GABA$_A$ receptor function following chronic EtOH exposure still remain to be determined.

From the preceding review, it is clear that the majority of early studies characterizing chronic effects of EtOH on GABAergic transmission focused mainly on postsynaptic properties and the subunit composition of the GABA$_A$ receptors themselves. Some of the disparity in the findings across laboratories on postsynaptic sites of EtOH action may reflect the differences in the chronic EtOH treatment duration and protocol, brain region examined, and methods of assessing receptor function. Most of these studies were generally in agreement that chronic EtOH exposure and withdrawal did not result in dramatic decreases in the number of GABA$_A$ receptors in most brain regions. However, many of these studies reported marked alterations in the expression of specific GABA$_A$ receptor subunits and hypothesized that those changes in the subunit composition of the $GABA_A$ receptors may account for the physiological and pharmacological alterations in GABAergic signaling associated with chronic EtOH administration (Grobin et al. 1998).

Of particular clinical importance is the development of tolerance and dependence to EtOH, and it is likely that adaptive changes in synaptic function in response to ethanol's actions on $GABA_A$ receptors play a role in this process. Indeed, it is well known that chronic EtOH treatment can lead to tolerance and physical dependence (Chandler et al. 1998) and that withdrawal following long-term EtOH consumption is associated with increased neuronal excitability (Kliethermes 2005; Weiner and Valenzuela 2006). These alterations have been hypothesized to represent, in part, a compensatory adaptation to the in vitro acute facilitatory effects of EtOH on GABAergic synapses (Siggins et al. 2005; Weiner and Valenzuela 2006). Few studies have reported the effects of long-term EtOH exposure on GABAergic synaptic transmission looking at both postsynaptic and presynaptic mechanisms using in vitro brain slice methods.

As described above, the adaptive changes in $GABA_A$ receptor expression are thought to lead to a pronounced hypofunction of GABAergic neurotransmission and possibly the development of tolerance to the in vitro acute effects of EtOH on these synapses. In the hippocampus, there is a decrease in the threshold for seizure induction by the $GABA_A$ receptor antagonist pentylenetetrazole (Kokka et al. 1993) and a decrease in $GABA_A$ receptor activity in hippocampal slices that also lasts for at least 40 days after the last EtOH dose (Cagetti et al. 2003; Kang et al. 1996; Liang et al. 2004, 2009). Using analysis of tetrodotoxin (TTX)-resistant mIPSCs recorded from CA1 pyramidal neurons of chronic EtOH-exposed and control rats, this group demonstrated a significant decrease in the amplitude and decay of these responses (Cagetti et al. 2003) possibly reflecting the observed alteration in the expression of α1 and α4 subunits. The mIPSC frequency is also slightly decreased, suggesting that chronic EtOH exposure may also be associated with a presynaptic decrease in GABA release at these synapses (see later section). Importantly, the pharmacological alterations in the properties of GABAergic synapses were consistent with the observed changes in subunit expression. For example, diazepam and the neurosteroid alfaxalone did not have any effect on mIPSCs in slices from chronic EtOH-exposed rats (Cagetti et al. 2003), possibly reflecting the loss of α1 and γ-subunits, respectively.

On the other hand, drugs with some selectivity for α4-subunits (e.g., RO 15-4513 and DMCM) showed an increased modulation of mIPSCs possibly reflecting the increase in α4 subunit expression (Kang et al. 1996, 1998a, b). Interestingly, the evoked IPSCs were still sensitive to alfaxalone (Kang et al. 1998b) suggesting differences in the populations of $GABA_A$ receptors that underlie evoked and mIPSCs. In addition, the acute effect of EtOH on evoked IPSCs was significantly increased in slices from chronic ethanol-exposed rats (Kang et al. 1998a, b). Liang et al. (2004) have also compared the effects of chronic EtOH exposure on synaptic and extrasynaptic receptor functions in CA1 neurons. These investigators found similar alterations in the synaptic mIPSCs and the tonic extrasynaptic $GABA_A$ receptor-mediated conductance associated with chronic EtOH exposure. Both mIPSCs and the tonic current show profound tolerance to α1-containing $GABA_A$ receptor selective doses of diazepam and zolpidem (Cagetti et al. 2003). As previously demonstrated (Grobin et al. 2000), chronic EtOH exposure results in a decrease in BZP-sensitive α1-subunits and an increase in BZP-insensitive α4-subunits at synaptic receptors. Thus, THIP (a high affinity and efficacy agonist of the α4-containing $GABA_A$ receptors and a partial agonist at most other $GABA_A$ receptor assemblies) activated the tonic GABA current in slices from control-untreated rats and had little effect in slices from chronic EtOH-exposed rats (Liang et al. 2004). However, THIP depressed mIPSCs in control-untreated rats but strongly increased mIPSCs in chronic EtOH-treated rats. In addition, the chronic EtOH-treated rats show a modest tolerance to the soporific effects of THIP and no change in its anxiolytic effects (Liang et al. 2004). In the last decade, significant progress has been made in understanding tonic conductance in the CeA of rodents using electrophysiology and immunohistochemistry (Herman et al. 2013a, 2016a). Two types of tonic conductance expressed in a cell-type-specific manner were also observed in rat CeA (Herman and Roberto 2016). One type is mediated by the α1-GABAA receptor subunit and is insensitive to acute ethanol exposure and the other type is mediated by the δ-GABAA receptor subunit and can be activated by increasing the ambient GABA concentration or by acute ethanol exposure. Notably, chronic ethanol exposure produces a functional switch in ongoing tonic signaling in the CeA in the specific cell populations, however there is no change in the ability of THIP and acute ethanol to further augment tonic conductance in these neurons, suggesting that these receptors are either not maximally activated or that THIP or ethanol is able to displace the ambient GABA to produce similar levels of activation as seen in naïve rats. Collectively, the presence of cell-type-specific tonic signaling in the CeA provides support for the complex mechanisms of actions of acute and chronic ethanol in inhibitory circuitry in this brain region (Herman and Roberto 2016; Herman et al. 2016a).

In the last decade, non-human primates (Cynomolgus macaques) have been a powerful model to study the effects of long-term EtOH consumption (Vivian et al. 2001). Ongoing research in the Weiner lab has provided the first evidence of neuroadaptations in the GABAergic synapses in monkey hippocampus (Weiner et al. 2005). In this paradigm of EtOH-self administration, cynomolgus macaques are trained to self-administer a 4% EtOH solution on an operant panel and then given 22 h. daily access to the ethanol solution. Control subjects were age- and sex-matched animals that had free access to food and water but were not exposed to the operant panels. The preliminary in vitro electrophysiological findings revealed a significant increase in paired-pulse facilitation (PPF) of $GABA_A$ IPSCs in dentate granule cells in slices prepared immediately following the last day of 18 months of daily EtOH drinking. Their finding is consistent with a decrease in GABA release probability (see Sect. 2.3 on presynaptic ethanol effects at GABAergic synapses) and agrees with the decrease in mIPSC frequency observed in rats following chronic intermittent EtOH exposure (Cagetti et al. 2003). Interestingly, there was lack of tolerance for both the acute facilitatory effect of EtOH and flunitrazepam on evoked $GABA_A$ IPSCs (Weiner et al. 2005). Using the same paradigm of EtOH self-administration, whole-cell patch clamp recordings on acutely dissociated amygdala neurons from ethanol-exposed cynomolgus macaques showed a decrease in the effect of flunitrazepam on the currents gated by exogenous GABA application compared with amygdala neurons from control animals (Anderson et al. 2007; Floyd et al. 2004). However, the modest inhibition of GABA-gated currents induced by acute EtOH was not affected by the chronic ethanol consumption. In addition, mRNA expression levels for the β, γ, and δ subunits in total amygdala RNA isolated from control and EtOH-drinking animals were measured. Chronic EtOH significantly reduced amygdala β1 and γ2 subunit expression. Overall, these finding demonstrate that chronic EtOH self-administration reduces the benzodiazepine sensitivity of amygdala $GABA_A$ receptors and this reduced sensitivity may reflect decreased expression of the γ subunit.

Electrophysiological studies in the monkey striatum indicate that chronic alcohol consumption decreases GABAergic synaptic transmission onto projection neurons (Cuzon Carlson et al. 2011, 2018). This effect was especially prominent in the putamen striatal subregion, and the decrease was larger in putamen of monkeys that began EtOH drinking as adolescents compared to those who started later (Cuzon Carlson et al. 2018).

Early work by Roberto et al. (2004a) assessed whether GABAergic synaptic changes occur with EtOH dependence in CeA slices. To obtain dependent rats, these investigators used an EtOH vapor inhalation method (Rogers et al. 1979). In this study, male Sprague-Dawley rats were exposed to a continuous EtOH vapor for 2–3 weeks with a targeted blood alcohol level of 150–200 mg/dL while control rats were maintained in similar chambers without EtOH vapor. On experiment days, the chronic EtOH-treated rats were maintained in the ethanol vapor chamber until preparation of the CeA slices, and recordings of GABAergic transmission were made in EtOH-free solution 2–8 h after cutting the slices (Roberto et al. 2004a). In CeA neurons from EtOH-dependent rats, both evoked IPSCs and mean baseline amplitude of mIPSCs were significantly increased compared to naïve rats, suggesting a postsynaptic effect of chronic ethanol (Roberto et al. 2004a). However, possible changes in the expression of $GABA_A$ receptor subunits were not characterized. It was also found that the baseline PPF ratio of IPSCs was significantly decreased and the mIPSC frequency was higher in neurons of EtOH-dependent rats compared to naïve rats, suggesting that GABA release was augmented in chronic ethanol-treated rats (Roberto et al. 2004a) (see later section on presynaptic change). In addition, acute EtOH (44 mM) increased IPSCs, decreased the PPF ratio of IPSCs, and increased the mIPSCs frequency to the same extent in ethanol-dependent rats and naïve rats, suggesting a lack of tolerance for the acute ethanol effects (Roberto et al. 2004a). These results have been replicated by several recent studies from the same group (Herman and Roberto 2016; Khom et al. 2020a, b; Varodayan et al. 2017c; Kirson et al. 2020, 2021; Tunstall et al. 2019) and one of the most consistent findings is the lack of tolerance for the acute potentiating effect of EtOH on GABAergic synapses in rodents after chronic ethanol exposure (up to 2 weeks of ethanol withdrawal). These studies suggest that GABAergic mechanisms may not be associated with the tolerance that is known to develop with some of the behavioral effects of EtOH (e.g., ataxia, sedation). Additional studies will be needed to determine the molecular mechanisms responsible more carefully for these adaptive changes in different brain regions and length/duration of EtOH exposure required to induce such neuroadaptations in GABAergic synapse. Moreover, these data also suggest that, as with the acute effects of EtOH, long-term exposure to ethanol results in both pre- and postsynaptic alterations and these changes may differ between brain regions (Siggins et al. 2005; Weiner and Valenzuela 2006; Roberto and Varodayan 2017).

In contrast to the rodents, in the monkey amygdala, acute ethanol application significantly increased the frequency of sIPSCs in controls, but not in abstinent drinkers, suggesting a tolerance to ethanol-enhanced GABA release in abstinent rhesus monkeys with a history of chronic ethanol self-administration and repeated abstinent drinkers (Jimenez et al. 2019). It is important to note that the loss of an acute effect of ethanol in the CeA in abstinent monkeys may be due to the extended (28-day) ethanol-abstinent protocol, which has not tested in rodent models (for review, see Roberto et al. (2020)).

2.3 *Chronic EtOH and GABAergic Transmission: Presynaptic Effects*

There are only a few studies reporting that chronic EtOH exposure can alter GABAergic transmission by effects on GABA release. Short in vitro chronic EtOH exposure (1 day) induced a transient decrease in mIPSC duration in cultured cortical neurons. Chronic EtOH exposure did not change mIPSC frequency nor did it produce a substantial cross-tolerance to a benzodiazepine in cortical neurons (Fleming et al. 2009). The results suggest that EtOH exposure in vitro has limited effects on synaptic $GABA_AR$ function and action potential-independent GABA release in cultured neurons. This group also investigated the effect of chronic EtOH exposure on GABA release in cultured hippocampal neurons (Fleming et al. 2009). These investigators found that chronic EtOH exposure did not alter mIPSC kinetics and frequencies in hippocampal neurons (Fleming et al. 2009). These results suggest that EtOH exposure in cultured cortical and hippocampal neurons may not reproduce all the effects that occur in vivo and in acute brain slices.

In fact, more results generated using in vitro brain slices show a stronger effect of EtOH on GABA release, as discussed earlier in this review (Fig. 1). In vitro brain slice preparations provide a number of highly sensitive experimental strategies that can be employed to detect presynaptic changes in transmitter release (for reviews of these approaches, see Siggins et al. (2005); Weiner and Valenzuela (2006); Roberto and Varodayan (2017)).

Studies in the hippocampus show that chronic EtOH exposure decreased long-term potentiation (LTP) by increasing the electrically stimulated (but not basal) release of tritiated GABA pre-loaded in CA1 hippocampal slices (Tremwel et al.

1994). The GABA uptake or $GABA_AR$ function was not altered, and this effect may be due to alterations in the mAChR regulation of GABA release at presynaptic terminals (Hu et al. 1999). In addition, studies using the $GABA_B$ receptor agonist baclofen to reduce release of tritiated GABA suggest that a change in $GABA_B$ autoreceptors on GABAergic terminals may also contribute to this effect of chronic EtOH exposure on LTP (Peris et al. 1997) (see later $GABA_B$ paragraph). For a general review of brain region-specific EtOH actions on the GABA system, see Criswell and Breese (2005); Siggins et al. (2005); Weiner and Valenzuela (2006). More recent studies also reported that chronic EtOH consumption induces tolerance to the impairing effects of acute ethanol treatment on induction of LTP in rat CA1 slices (Fujii et al. 2008). In CA1 slices from control rats, stable LTP was induced by tetanic stimulation, and LTP induction was blocked if the tetanus was delivered in the presence of 8.6 mM EtOH or muscimol. A decrease in the stimulation threshold for inducing LTP was found in hippocampal slices from chronic EtOH-treated rats. In addition, application of EtOH or muscimol did not affect LTP induction in these cells, suggesting that the effects of chronic ethanol exposure on LTP induction are mediated by a reduction in GABAergic inhibition in hippocampal CA1 neurons (Fujii et al. 2008).

Weiner et al. (2004) have found that voluntary EtOH drinking is associated with a significant increase in paired-pulse plasticity at GABAergic synapses in dentate gyrus neurons from the hippocampal formation of monkeys (cynomolgus macaques), consistent with a reduction in GABA release probability. In addition, a lack of tolerance to the facilitating effects of both acute EtOH and flunitrazepam on the $GABA_A$ IPSCs was reported.

In contrast, Melis et al. (2002) reported that a single EtOH exposure in vivo induces a long-lasting facilitation of GABA transmission in the VTA of ethanol-preferring C57BL/6 mice. These investigators observed that evoked $GABA_A$ IPSCs in dopaminergic neurons of EtOH-treated animals exhibited paired-pulse depression (PPD) compared with saline-treated animals, which exhibited PPF (Melis et al. 2002). An increase in frequency of mIPSCs was also observed in the EtOH-treated animals. Moreover, the $GABA_B$ receptor antagonist, CGP35348, shifted PPD to PPF, indicating that presynaptic $GABA_B$ receptor activation, likely attributable to GABA spillover, might play a role in mediating PPD in the EtOH-treated mice (see later $GABA_B$ paragraph). In a more recent study, the same group (Wanat et al. 2009) demonstrated that EtOH exposure also increased GABA release onto VTA dopamine neurons in ethanol non-preferring DBA/2 mice. However, a single EtOH exposure reduced glutamatergic transmission and LTP in VTA dopamine neurons from the ethanol non-preferring DBA strain but not ethanol-preferring C57BL/6 mice (Wanat et al. 2009). In vivo recordings in VTA indicate that acute EtOH reduces the activity of putative GABAergic neurons, while increased firing of putative dopaminergic neurons occurs on a faster time scale (Burkhardt and Adermark 2014). These findings indicate that both direct effects and indirect disinhibitory effects may contribute to EtOH-induced increases in DA release.

Additional data from Roberto and coworkers (2004a, 2010) further suggest that chronic EtOH exposure can affect CeA GABA release, perhaps via an action on

GABAergic terminals. Baseline $GABA_A$ IPSCs were significantly higher, and baseline PPF of $GABA_A$ IPSCs was significantly smaller in CeA neurons from EtOH-dependent rats compared to non-dependent rats, suggesting that evoked GABA release was augmented after chronic ethanol exposure. These investigators also reported an increase in the baseline frequency of mIPSCs in CeA neurons from EtOH-dependent rats compared to that of naïve controls. Acute superfusion of EtOH significantly enhanced $GABA_A$ IPSCs, decreased the PPF ratio of IPSCs, and increased the mIPSC frequency to the same extent in CeA slices from ethanol-dependent rats and naïve rats, suggesting a lack of tolerance to the presynaptic acute EtOH effects (Roberto et al. 2004a). In addition, these investigators estimated the interstitial GABA levels in CeA using microdialysis in freely moving rats. In agreement with the in vitro electrophysiological results, the in vivo data showed a four-fold increase of baseline dialysate GABA concentrations in CeA of EtOH-dependent rats compared to naïve rats. Moreover, local administration of EtOH by dialysis increased the dialysate GABA levels in CET rats. These findings again indicate a lack of tolerance to presynaptic acute EtOH effects on GABA release in CeA of CET rats (Roberto et al. 2004a). These studies strengthen the possibility that chronic as well as acute EtOH may alter the function of the GABAergic synapses acting at both the postsynaptic site and presynaptic terminals. As mentioned above, recent studies have also consistently replicated the increased GABA release in the CeA of rodents using the same and/or slightly different chronic ethanol exposure in rodents (Herman and Roberto 2016; Khom et al. 2020a, b; Varodayan et al. 2017c; Kirson et al. 2020; Tunstall et al. 2019). Interestingly, the data obtained in abstinent rhesus monkeys with a history of chronic ethanol self-administration and repeated abstinence agree with the rodent studies showing increased GABA release in the CeA following chronic ethanol exposure at early (2–10 h) withdrawal, and late [5–7 days (Herman et al. 2016a) and 14 days (Khom et al. 2020a, b)] withdrawal. Furthermore, a recent study showing decreased GABA transporter (GAT-3) levels and impaired GABA clearance in the CeA of alcohol-preferring rodents and in humans (Augier et al. 2018) support an elevation of GABA level, together, these data suggest that long-term exposure to EtOH causes changes at GABAergic synapses that may differ between brain regions and with the duration of chronic exposure. Further studies will be needed to more carefully determine the specific exposure durations required to elicit these changes in GABAergic synapses, the molecular mechanisms responsible for these adaptive changes, as well as their behavioral consequences with respect to withdrawal and dependence.

Evidence of decreased GABA release following chronic alcohol ingestion has also been observed in mouse striatum (Wilcox et al. 2014). In this study, mice drank alcohol in the drinking in the dark schedule that produces binge-like consumption. The frequency of action potential-independent miniature IPSCs was decreased in both dorsolateral and dorsomedial striatum in the alcohol-drinking mice.

In summary, a growing area in which action of EtOH on GABA function has been implicated is withdrawal from chronic ethanol. Withdrawal results in an increased sensitivity to induction of seizures (Allan and Harris 1987; Frye et al. 1983). Several functional and behavioral studies on benzodiazepines and other drugs with

GABAmimetic action reduced such withdrawal-related hyper-excitability (Breese et al. 2006; McCown et al. 1985; Roberto et al. 2008; Ticku and Burch 1980; Herman et al. 2016a; Khom et al. 2020a, b). Collectively, these results offer strong support for the hypothesis that at least a part of the action of EtOH was mediated by effects on neural functions associated with GABA transmission and that these effects play an important role in the maintenance of addictive drinking behavior.

The molecular basis of chronic EtOH effects on presynaptic function is just beginning to be explored, and early findings implicated changes in vesicle-associated proteins (see Das (2020) for review). In rhesus macaque monkeys, chronic alcohol consumptions alters expression of the vesicle-associated SNAP-25 protein (Alexander et al. 2018; Nimitvilai et al. 2017; Das et al. 2013; Ghosh et al. 2017; Varodayan et al. 2011; Barbier et al. (2021)). Increased expression of Munc13-1, another vesicle-associated protein has also been observed following chronic EtOH exposure in both mouse and monkeys (Alexander et al. 2018; Ghosh et al. 2017). These findings are particularly interesting as alcohol directly interacts with Munc13-1 (Das et al. 2013) and this protein has been implicated in acute EtOH effects on synaptic transmission (Gioia et al. 2017). It will be interesting to determine if chronic EtOH effects on neurotransmitter release depend on these proteins and/or other proteins involved in vesicle fusion.

2.4 $GABA_B$ Receptors and Chronic EtOH Actions

Several studies demonstrated $GABA_B$ receptor involvement in the effects of EtOH. For instance, $GABA_B$ receptor antagonists enhance the ability of acute EtOH to facilitate GABA transmission in the hippocampus (Ariwodola and Weiner 2004; Wan et al. 1996; Wu and Saggau 1994) and NAc (Nie et al. 2000). Ariwodola and Weiner (2004) suggested that the effect of EtOH to facilitate GABA transmission is limited because of GABA feedback on presynaptic $GABA_B$ receptors (Fig. 1). The presence of $GABA_B$ receptors accounted for the difference in sensitivity to EtOH influences on GABA transmission in specific subfields of the hippocampus (Weiner et al. 1997). On the other hand, $GABA_B$ receptors did not influence GABA release from neurons in the CeA (Roberto et al. 2003). Thus, the involvement of $GABA_B$ receptors on GABA release in various brain regions may not be universal, suggesting that the presence or absence of presynaptic $GABA_B$ receptors may be an important determinant for the regional specificity of ethanol to affect GABA transmission (Ariwodola and Weiner 2004).

As mentioned above, Peris et al. (1997) showed that chronic EtOH treatment, sufficient for decreasing LTP in rats, also increased 3H-GABA release from hippocampal slices in these same animals. These investigators characterized presynaptic autoreceptor modulation of 3H-GABA release in hippocampal slices from control and EtOH-dependent rats. Effects of a $GABA_B$ receptor agonist (baclofen) and antagonist [2-hydroxy (OH)-saclofen] on electrically stimulated 3H-GABA release from superfused hippocampal slices were examined. Baclofen decreased stimulated release in a dose-dependent manner and the antagonist 2-OH-saclofen increased release consistent with the presence of presynaptic $GABA_B$ autoreceptors in hippocampus. The $GABA_A$ antagonist bicuculline did not significantly modulate basal or stimulated release. Presynaptic modulation of release by baclofen and 2-OH-saclofen was decreased in animals 48 h. after withdrawal from EtOH. Using quantitative autoradiographic techniques, the density of 3H-baclofen binding sites in the hippocampus was not affected by chronic EtOH exposure, whereas the density of 3H-bicuculline binding sites was increased by 28% in EtOH-treated rats. These data may explain how chronic EtOH treatment increases presynaptic regulation of GABA release from hippocampus that may contribute to the decrease in LTP seen in rats after chronic ethanol exposure (Peris et al. 1997).

Another study assessed the impact of EtOH on postsynaptic $GABA_B$ receptors via baclofen-induced hyperpolarization of hippocampal CA1 and CA3 pyramidal neurons. These receptors activate outward K+ currents via a pertussis toxin-sensitive G-protein cascade to reduce membrane potential during the slow inhibitory postsynaptic potential and may play a role in EtOH intoxication and withdrawal excitability. In both types of pyramidal neurons, baclofen applied consecutively in increasing concentrations caused concentration-dependent hyperpolarization. There were no significant differences in resting membrane potential, input resistance, maximum baclofen-induced hyperpolarization, or EC_{50} between CA1 and CA3 neurons, although slope values were significantly smaller in the former neurons. These parameters were not significantly changed in the presence of EtOH 10–100 mM. Chronic EtOH treatment (12 days) did not shift sensitivity or maximum response to baclofen in CA1 neurons. These results suggest that $GABA_B$ receptors in this model were essentially insensitive to ethanol (Frye and Fincher 1996).

Melis et al. (2002) linked the long-lasting potentiation of GABAergic synapses on dopaminergic neurons in the VTA by systemic EtOH to an effect on presynaptic $GABA_B$ receptors. Moreover, the frequency (but not the amplitude) of mIPSCs was also significantly higher in VTA neurons of EtOH-treated animals compared to controls, further supporting an increased probability of presynaptic GABA release independent of neuronal discharge in VTA neurons treated with ethanol. Interestingly, the $GABA_B$ receptor antagonist, CGP35348, shifted PPD to PPF in EtOH-treated animals by increasing the amplitude of the second evoked $GABA_A$ IPSC and without affecting $GABA_A$ IPSC in the saline-treated animals. In addition, both the frequency and the amplitude of mIPSCs were unaffected by CGP35348 in both groups of mice. Thus, the PPD observed in the EtOH-treated mice could result from an increased probability of GABA release, which might in turn lead to activation of presynaptic $GABA_B$ receptors and decrease the second IPSC. These results further support the hypothesis that GABA levels are increased after EtOH exposure, leading to spillover onto presynaptic $GABA_B$ receptors, whose activation leads to inhibition of release (Hausser and Yung 1994; Melis et al. 2002).

In a recent study, Roberto et al. (2008) reported neuroadaptations in $GABA_B$ receptors in CeA after chronic EtOH exposure. The sensitivity of GABA IPSCs to the $GABA_B$ receptor antagonist CGP 55845A and agonist baclofen was decreased after chronic EtOH, suggesting downregulation of this system. Specifically, the $GABA_B$ receptor antagonist, CGP 55845A significantly increased the mean amplitude of evoked IPSCs in CeA from naïve rats. This increase in the IPSC amplitude was associated with a significant decrease in PPF, suggesting a tonic activation of presynaptic $GABA_B$ receptors in naïve rats. In contrast, in CeA from EtOH-dependent rats, CGP 55845A did not alter the mean-evoked IPSCs and did not affect mean PPF. Baclofen markedly depressed evoked GABA IPSC amplitudes in neurons of naïve rats, with recovery during washout. The baclofen-induced inhibition of GABA IPSCs was significantly reduced in neurons of EtOH-dependent rats. In addition, in CeA neurons from EtOH-dependent rats, baclofen-induced depression was associated with a smaller increase of the PPF ratio of GABA IPSCs compared to that in neurons of naïve rats. These data suggest that the downregulation of the $GABA_B$ system associated with EtOH dependence may explain in part the increased GABAergic tone reported in dependent rats (Roberto et al. 2008).

2.5 Glycine Receptor Roles in Chronic Alcohol Actions

In comparison with GABAergic transmission, much less is known about chronic EtOH effects on glycinergic synapses (Vengeliene et al. 2010). However, there is increasing information about how glycine receptors in the CNS contribute to alcohol-related behaviors. Using mice in which an EtOH-insensitive mutated GlyR alpha1 subunit is substituted for the wild-type receptor, investigators have shown reduced sedative responses to acute EtOH (Aguayo et al. 2014). These mice also show greater conditioned place preference for EtOH and greater EtOH intake upon first exposure to the drug (Muñoz et al. 2020). Mice carrying a similar mutation that renders the alpha2 GlyR subunit EtOH-insensitive show a similar pattern of shorter duration of EtOH-induced sedation and increased EtOH consumption (Gallegos et al. 2021). Mice lacking alpha2 GlyR subunits show reduced EtOH intake and preference and increased aversive responses to the drug while mice lacking the alpha3 subunit show increased intake and preference but a decrease in conditioned EtOH taste aversion (Blednov et al. 2015). Thus, glycinergic synaptic effects appear to have roles in acute EtOH actions and regulation of EtOH drinking. As mentioned above, agonism of the glycine receptor impacts GABAergic transmission in CeA of naïve rats, without affecting the acute alcohol-induced facilitation of GABAergic responses, and this effect is lost in neurons from alcohol-dependent rats (Kirson et al. 2020). Glycine transport in the prefrontal cortex appears to play a role in impulsivity during abstinence following chronic EtOH exposure (Irimia et al. 2017). This may involve GlyRs or glycine-sensitive NMDARs. It will be interesting to determine how chronic alcohol affects glycine release and glycine receptors.

Chronic alcohol drinking also alters the expression of a number of genes related to glycinergic transmission (Vengeliene et al. 2010). Some of these changes can be reversed by treatment with a glycine transporter antagonist that also reduced compulsive-like drinking in rat (Vengeliene et al. 2010).

2.6 *Changes in Dopaminergic Transmission Induced by Chronic Alcohol*

There are conflicting reports of chronic alcohol effects on DA release and extracellular DA (De Santis et al. 2020). Several studies indicate that increases in extracellular DA concentrations persist throughout chronic EtOH exposure and intake in self-administration paradigms and also become associated with conditions that predict drug availability (Bassareo et al. 2017; Doyon et al. 2003; Hirth et al. 2016) (Fig. 2a). Sensitization to the dopamine-increasing effects in NAc of EtOH microinjection into the VTA has also been observed following chronic EtOH consumption (Ding et al. 2016). Examination of striatal tissue from AUD patients indicates decreased DAT expression, possibly indicating a hyperdopaminergic state in these individuals (Hirth et al. 2016) (Fig. 2a). A combined analysis of extracellular

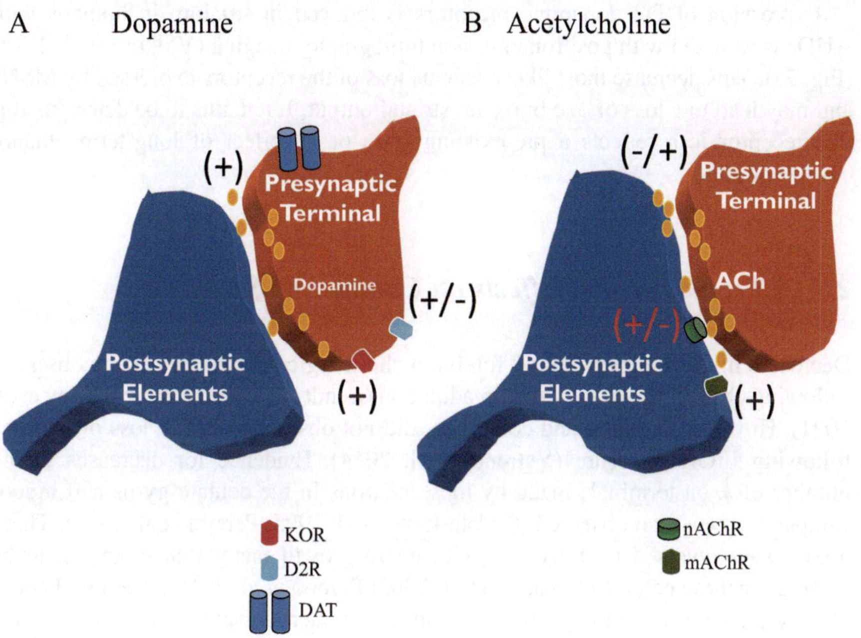

Fig. 2 Chronic EtOH effects on dopaminergic and cholinergic transmission. (**a**) Chronic EtOH exposure generally enhances DA release, although decreases have also been observed. Decreased DAT expression may contribute to hyperdopaminergic conditions following chronic exposure. Enhanced potency of kappa opiate receptor agonist inhibition of DA release indicates either greater numbers or increased sensitivity of these presynaptic receptors. The number of D2 receptors is decreased in humans with AUD, while chronic EtOH-induced changes in D2 autoreceptor function vary in different species and sexes. (**b**) Chronic EtOH exposure has mixed effects on ACh release, increasing release at some synapses, while decreasing it at others. Likewise, chronic EtOH effects on nAChR number and function are a mix of enhancement and reduction depending on receptor subtypes and cellular locus of the receptor. In contrast, chronic EtOH exposure generally increases mAChR function in the different preparations in which this has been examined

tortuosity and modeling suggested the changes in dopamine diffusion could contribute to increased availability of the neurotransmitter after chronic ethanol exposure (De Santis et al. 2020). Altogether these findings suggest that the ability of EtOH to enhance accumbal NAc either does not show tolerance with repeated exposure or undergoes adaptations that maintain high dopamine levels, and some mechanisms may even be enhanced under these circumstances.

However, other studies indicate that dopamine release is reduced following chronic EtOH consumption or forced exposure. For example, Karkhanis et al. (2016) found that acute EtOH stimulation of DA release switched to inhibition in NAc following chronic exposure. Decreased DA release in the NAc core subregion was observed in adult rats following adolescent EtOH exposure (Zandy et al. 2015). Enhanced function of kappa opiate receptors and D2 dopamine autoreceptors may contribute to the decreased DA release both in mice and monkeys (Rose et al. 2016; Siciliano et al. 2015, 2016), although decreased D2 autoreceptor function has also been observed, a change that varies across sexes (Salinas et al. 2021) (Fig. 2a).

Expression of D2 dopamine receptors is reduced in striatum in humans with AUD, as assessed with positron emission tomography imaging (Volkow et al. 2017) (Fig. 2a). This decrease most likely reflects loss of the receptors expressed by MSNs and may lead to a loss of one brake on striatal output. It remains to be determined if this receptor loss reflects a pre-existing state or an effect of long-term ethanol consumption.

2.7 *Chronic Alcohol Effects on Cholinergic Systems*

Decreases in the number of basal forebrain cholinergic neurons have been observed following chronic EtOH exposure in adult rat (Arendt et al. 1988, 1989; Smiley et al. 2021). However, Vetreno and coworkers did not observe a similar loss of neurons following EtOH exposure (Vetreno et al. 2014). Evidence for decreases in the number of axon terminals made by these neurons in the dentate gyrus and hippocampal gyri was also observed (Cadete-Leite et al. 1995; Pereira et al. 2016). These losses were reversed by treatment with nerve growth factor that is known to be trophic for these cells (Lukoyanov et al. 2003; Pereira et al. 2016). The numbers of cholinergic neurons in the pedunculopontine and laterodorsal tegmental areas were also decreased following chronic EtOH consumption and withdrawal (Pereira et al. 2020).

Mixed effects of chronic EtOH on ACh levels and release have been reported (Fig. 2b). Decreased ACh levels following chronic EtOH exposure were originally reported in several brain regions (Hunt and Dalton 1976). However, subsequent studies reported increased ACh concentration in the rat striatum 1–3 days after the cessation of a 4-day EtOH treatment (Hunt et al. 1979), and mixed results were observed in other studies examining a variety of brain regions (Parker et al. 1978; Smyth and Beck 1969). The activity of enzymes involved in ACh synthesis as well as the high-affinity choline uptake system has also been examined following chronic

EtOH exposure (Nordberg and Wahlström 1992). Activity of the synthetic enzyme choline acetyltransferase (ChAT) is increased after a few days of EtOH exposure (Ebel et al. 1979), but decreased after weeks of exposure (Smyth and Beck 1969; Pelham et al. 1980). High-affinity choline uptake is increased a few days after withdrawal from a relatively short exposure to EtOH (Hunt et al. 1979; Hunt and Majchrowicz 1979). It must be noted that the subject of ACh levels and enzyme expression/function has not be revisited with newer research approaches, and thus additional study is warranted.

Reduced preparations have also been used to examine effects of chronic EtOH exposure on ACh release. Using slices of nucleus accumbens and dorsal striatum, Nestby et al. (1997) found that 15 days of exposure to a moderate ethanol treatment enhanced electrically-evoked ACh release. No changes in ACh release from cortical or hippocampal synaptosomes were observed following chronic EtOH consumption (Sabriá et al. 2003).

Subsequent studies examined changes in ACh release in vivo following chronic ethanol exposure using microdialysis. As is the case for acute EtOH exposure, chronic EtOH in rat also generally decreases hippocampal ACh levels (Casamenti et al. 1993; Imperato et al. 1998). Decreased ChAT activity appears to be associated with the decreased release (Casamenti et al. 1993). Recovery of release was observed following 4 weeks of abstinence subsequent to 3 months of EtOH drinking, but less recovery was observed with abstinence after 6 months of drinking (Casamenti et al. 1993). In anesthetized rat, increased hippocampal ACh was observed following a 4-day EtOH exposure and subsequent withdrawal (Imperato et al. 1998). In another set of studies, mixed results were obtained within the same laboratory. Decreased hippocampal ACh was observed following 9 months of drinking in Sprague-Dawley rats, and this was correlated with impaired passive avoidance performance (Melis et al. 1996). However, these effects were not observed in the Sardinian alcohol-preferring rat (Fadda et al. 1999). Chronic exposure to EtOH can alter the initial acute drug effects. Increased ACh in the VTA during EtOH intake appears to subside with continued drinking (Larsson et al. 2005). Acute ethanol-induced increases in ACh levels in hypothalamus changed to decreases following several days of administration (Kaneyuki et al. 1995).

As is the case for acute EtOH actions on ACh and cholinergic synapses, effects of chronic EtOH exposure are mixed and depend on ACh receptor subtype and anatomical region (Fig. 2b). Radioligand-binding studies were initially employed to identify changes in receptor numbers and affinity. The most direct measures of changes in nAChRs were conducted in cell lines in which binding can be examined independent of changes in cell type or circuitry. In the PC12 neuroblastoma cell line, exposure to EtOH in the medium for 48–96 h increased the binding of epibatidine, a ligand for the $\alpha4\beta2$ subunit-containing nAChR subtype (Dohrman and Reiter 2003). Nicotine stimulation of binding was also increased by this treatment. In an M10 cell line engineered to express $\alpha4\beta2$-containing receptors, exposure to EtOH for 12–48 h had the opposite effect, decreasing epibatidine binding, but a slight increase was observed after 96 h of exposure (Dohrman and Reiter 2003). It is not clear why different effects were observed in these different cell lines, but differences in intracellular signaling are likely to be involved. The decrease in binding in the M10 cells was blocked by a protein kinase C inhibitor, but this was not tested in PC12 cells.

Mixed effects on nAChR radioligand binding have also been observed in brain tissue from animals chronically treated with EtOH in vivo. In the rat hippocampus, hypothalamus and thalamus decreased binding of nicotine was observed (Yoshida et al. 1982). However, hippocampal nicotine binding was decreased just after voluntary EtOH drinking in rat (Robles and Sabriá 2006) but increased after withdrawal from drinking (Robles and Sabriá 2008). Examination of binding of αBungarotoxin, a ligand for α7-type nAChRs, revealed differential effects of chronic EtOH drinking in the inbred Long-sleep (decreased hippocampal and increased thalamic binding) and Short-sleep mice (increased binding in cerebellum and superior colliculus) (Booker and Collins 1997). Evidence for decreased binding to α4β2-containing receptors was also observed in rhesus monkey cortex after chronic alcohol drinking (Hillmer et al. 2014).

Radioligand binding and molecular biological approaches have been used to examine chronic alcohol effects on mAChR expression. Exposing human neuroblastoma cells for several days led to an increase in mAChR-induced inositol phosphate production (Larsson et al. 1996). This effect was accompanied by increased radioligand binding that implicated M1-type mAChRs in the potentiation. A similar increase in mAChR binding was also observed following 2 days of EtOH exposure in NG108-15 neuroblastoma x glioma cells (Hu et al. 1993). In general, in vivo chronic EtOH exposure has been shown to increase mAChR binding sites in cerebral cortex, hippocampus, mammillary body, and striatum, although mixed effects have been observed in different studies (reviewed in Nordberg and Wahlström (1992); Pick et al. (1993); Pietzak et al. (1988); Rothberg et al. (1996); Tabakoff et al. (1979)). In general, the increases were largest after exposure periods of a few days and during withdrawal following longer exposure periods. The M1 mAChR is one subtype that appears to be upregulated after chronic EtOH exposure (Pietzak et al. 1988; Hoffman et al. 1986; Muller et al. 1980). Expression of the five different mAChR subtype proteins in hippocampus was also examined with immunoprecipitation, but no effect of chronic EtOH exposure was observed (Rothberg et al. 1993).

Effects of chronic EtOH exposure on the functional consequences of mAChR activation have also been examined, using both receptor-mediated inositol phosphate generation and electrophysiological changes as the functional readouts. A reduction in the ability of EtOH to inhibit mAChR-mediated stimulation of inositol triphosphate formation in mouse brain tissue and synaptosomes was observed following chronic EtOH consumption (Hoffman et al. 1986; Smith 1983). It is unclear if this tolerance is due to decreased mAChR expression or downstream signaling mechanisms, but given the general finding of increased receptor binding it is probable that decreased downstream signaling is involved. Activation of mAChRs enhances the population spike (PS) and inhibits the field excitatory postsynaptic potential (fEPSP) during extracellular field potential recordings in the hippocampal CA1 subregion. Chronic EtOH exposure reduces the population spike facilitation but does not alter fEPSP inhibition (Rothberg and Hunter 1991; Rothberg et al. 1993). Frye and coworkers also found no chronic EtOH-induced change in mAChR inhibition of hippocampal fEPSPs, as well as no change in inhibition of the afterhyperpolarization by mAChRs (Frye et al. 1995). The differential EtOH effects on these responses likely result from the fact that different mAChRs mediate the different physiological effects, with those coupled to Gi/o-type G-proteins in involved in fEPSP inhibition and Gq G-protein coupled receptors mediating the other responses.

Cholinergic neuron numbers and mAChR binding have also been examined in postmortem samples from humans with alcohol use disorder. Decreased basal forebrain cholinergic neuron numbers were observed in humans diagnosed with Korsakoff syndrome, as a consequence of prolonged alcohol drinking (Arendt et al. 1983). Radioligand binding revealed evidence of decreases in mAChRs in older AUD patients (Freund and Ballinger 1989a, b; Hellstrom-Lindahl et al. 1993; Nordberg et al. 1983; Nordberg and Wahlström 1992). It is unclear if the receptor loss is a result of the decrease in cholinergic neuron number mentioned previously, although Freund and Ballinger (1991) did not observe evidence of neurodegeneration in the brains in which they observed decreased mAChR binding sites. Activity of the ChAT enzyme is also reduced in postmortem brain samples from individuals with AUD (Antuono et al. 1980).

3 Neuropeptide Roles in Acute and Chronic Alcohol Actions

Neuropeptides are potent neuromodulators in the CNS whose actions are mediated via GPCRs. In contrast to classical neurotransmitters, neuropeptides are released in a frequency-dependent fashion and often have a longer half-life of activity after release. These factors, among others, enable neuropeptides to produce long-lasting effects on cellular functions such as excitatory and inhibitory synaptic transmission, neuronal excitability, and gene transcription (Gallagher et al. 2008). Thus, a long-lasting dysregulation of neuropeptides could have significant effects on the activity of neurons and consequently, behavior. Thus, several neuropeptideric system in different brain circuits have received a lot of attention particularly in the development of AUD (Koob and Volkow 2016; Roberto et al. 2012).

3.1 Corticotropin-Releasing Factor

Corticotropin-releasing factor (CRF) is a 41-amino acid polypeptide that has a major role in coordinating the stress response of the body by mediating hormonal, autonomic, and behavioral responses to stressors. CRF (originally called corticotropin-releasing hormone, although the International Union of Pharmacology designation is CRF) was identified through classic techniques of peptide sequencing (Vale et al. 1981). Subsequently, genes encoding three paralogs of CRF – urocortins 1, 2, and 3 (Ucn 1, Ucn 2, Ucn 3), were identified by modern molecular biological approaches. Ucn 2 and Ucn 3 are also referred to as stresscopin-related peptide and stresscopin, respectively. CRF and the urocortins have been implicated in the modulation of multiple neurobiological systems, including those that regulate feeding, anxiety and depression, hypothalamic-pituitary-adrenal (HPA) axis signaling, and EtOH consumption (Hauger et al. 2006; Heilig and Koob 2007; Ryabinin and Weitemier 2006; Smith and Vale 2006). CRF and the Ucn peptides produce their effects by binding to the G protein-coupled CRF type 1 (CRF1R) and CRF type 2 (CRF2R) receptors. CRF binds to both receptors but has greater affinity for the CRF1R (Bale and Vale 2004; Fekete and Zorrilla 2007; Hauger et al. 2006; Pioszak et al. 2008).

CRF1R and CRF2R are GPCRs that are predominantly positively linked to the activation of AC (Fig. 1), and recent reports also implicate other second messenger systems such as inositol triphosphate and PKC (Blank et al. 2003; Grammatopoulos et al. 2001). Using corticotrophins, Antoni and coworkers (2003) demonstrated a coupling of CRF1R to AC9 and AC7. The switch in coupling from AC9 to AC7 results in a more robust cAMP signal when CRF binds to the CRF1R (Antoni 2000; Antoni et al. 2003). It should be emphasized that AC7 is localized both postsynaptically (striatum, hippocampus) and presynaptically (nucleus accumbens, amygdala) (Mons et al. 1998a, b) and is anatomically positioned to receive signals from GPCRs on both dendrites and axon terminals.

Pharmacological and transgenic studies show that brain and pituitary CRF1Rs mediate many of the functional stress-like effects of the CRF system (Heinrichs and Koob 2004). CRF and the Ucn peptides have a wide distribution throughout the brain, but there are particularly high concentrations of cell bodies in the paraventricular nucleus of the hypothalamus, the basal forebrain (notably the extended amygdala), and the brainstem (Swanson et al. 1983). Ucn1 binds with equal affinity to CRF1R and CRF2R, and Ucn2 and Ucn3 are CRF2R agonists (Hauger et al. 2006; Pioszak et al. 2008). CRF and the Ucn peptides exert their behavioral and neuroendocrine actions through central hypothalamic and extrahypothalamic pathways (Hauger et al. 2006; Heilig and Koob 2007; Heinrichs and Koob 2004; Koob and Le Moal 2008).

Increasing evidence implicates CRF and its receptors in the synaptic effects of EtOH. Ethanol induces release of CRF from the hypothalamus that initiates the activation of the HPA axis (Ogilvie et al. 1998). Ethanol also modulates the extra-neuroendocrine CRF system involved in behavioral stress responses, particularly in the amygdala. Ethanol withdrawal induces an increase in CRF levels in the amygdala (Merlo Pich et al. 1995) and in the BNST (Olive et al. 2002).

The central administration of a CRF antagonist attenuates both EtOH self-administration and the anxiety-like response to stress observed during alcohol abstinence, (Valdez et al. 2002) and administration of a CRFR antagonist into the CeA reverses the anxiogenic-like effect of alcohol (Rassnick et al. 1993). Rats tested

3–5 weeks post alcohol withdrawal showed an anxiogenic-like response provoked by a mild restraint stress only in rats with a history of alcohol dependence. This stress-induced anxiogenic-like response was reversed by a competitive CRF1R antagonist (Valdez et al. 2003). The increased self-administration of alcohol observed during protracted abstinence also was blocked by a competitive CRF1R antagonist (Valdez et al. 2003). Gehlert et al. (2007) also described that a novel CRF1R antagonist, the 3-(4-Chloro-2-morpholin-4-yl-thiazol-5-yl)-8-(1-ethylpropyl)-2,6-dimethyl-imidazo[1,2-b]pyridazine (MTIP) has advantageous properties for both clinical development and in preclinical models of alcohol use disorder (AUD). MTIP dose-dependently reversed anxiogenic effects of ethanol withdrawal and blocked excessive alcohol self-administration in Wistar rats with a history of dependence (Gehlert et al. 2007). CRF also contributes to increased alcohol consumption in dependent animals, because increased EtOH self-administration is reduced by CRF1R antagonists in dependent animals but not in non-dependent animals (Funk et al. 2007; Overstreet et al. 2004) and by CRF1R deletion (Chu et al. 2007; Sillaber et al. 2002). More recently, it has been reported that chronic CRF1R antagonist treatment blocked withdrawal-induced increases in alcohol drinking by dependent rats and tempered moderate increases in alcohol consumption (Roberto et al. 2010). In addition, inactivation of the CeA CRF+ neurons prevents recruitment of this neuronal ensemble, decreases the escalation of alcohol drinking, and decreases the intensity of somatic signs of withdrawal (de Guglielmo et al. 2019). These results have led to the hypothesis that negative emotional states (including anxiety-like states) contribute to the compulsive alcohol intake associated with AUD via negative reinforcement mechanisms (Koob 2008; Zorrilla et al. 2013, 2014; Zorrilla and Koob 2010; Gilpin and Roberto 2012).

Several recent reviews (Lowery and Thiele 2010; Zorrilla et al. 2013, 2014; Zorrilla and Koob 2010; Spierling and Zorrilla 2017; Quadros et al. 2016) provide a comprehensive overview of preclinical evidence from rodent studies that suggest a promising role for CRFR antagonists in the treatment of alcohol abuse disorders. In contrast, few other reviews emphasize the preclinical results that hinder the translational of CRF pharmacology to the clinic (Pomrenze et al. 2017; Cannella et al. 2019; Agoglia et al. 2020; Agoglia and Herman 2018; Roberto et al. 2017). These reviews point to the lack of preclinical studies performed in female rodents (as most the studies have been performed in male rodents) and that would strongly suggest sex differences in the ability of CRF/CRF1-directed therapies to functionally regulate alcohol drinking in the clinical setting.

CRFR antagonists protect against excessive EtOH intake resulting from ethanol dependence without influencing ethanol intake in non-dependent animals. Similarly, CRFR antagonists block excessive binge-like ethanol drinking in non-dependent mice but do not alter ethanol intake in mice drinking moderate amounts of ethanol (Lowery and Thiele 2010). CRFR antagonists also protect against increased EtOH intake and relapse-like behaviors precipitated by exposure to a stressful event. Additionally, CRFR antagonists attenuate the negative emotional responses associated with EtOH withdrawal. The protective effects of CRFR antagonists are modulated by CRF1R. Finally, recent evidence has emerged suggesting that CRF2R agonists may also be useful for treating alcohol abuse disorders; for review, see Lowery and Thiele (2010); Spierling and Zorrilla (2017); Roberto et al. (2017).

Low CRF concentrations can influence neuronal properties in the CNS (see Aldenhoff et al. (1983); Siggins et al. (1985)). CRF decreases the slow afterhyperpolarizing potential in hippocampus (Aldenhoff et al. 1983) and CeA (Rainnie et al. 1992) and enhances R-type voltage-gated calcium channels in rat CeA neurons (Uhrig et al. 2017; Yu and Shinnick-Gallagher 1998). These and other data (Liu et al. 2004; Nie et al. 2004, 2009; Roberto et al. 2010; Ungless et al. 2003) also suggest that CRF plays an important role in regulating synaptic transmission in CNS. For example, in VTA dopamine neurons, CRF potentiates NMDA-mediated synaptic transmission via CRF_2 activation (Ungless et al. 2003), and we recently found that CRF augments GABAergic inhibitory transmission in mouse CeA neurons via CRF1 activation (Fig. 1).

3.1.1 CRF Actions in the VTA

The VTA receives CRF inputs from a number of sources including the limbic forebrain and the paraventricular nucleus of the hypothalamus (Rodaros et al. 2007). These CRF inputs form symmetric and asymmetric synapses, mostly onto dendrites, that co-release either GABA or glutamate, respectively (Tagliaferro and Morales 2008). VTA dopamine neurons express both types of CRF receptors, CRF1R and CRF2R (Ungless et al. 2003), and approximately 25% of VTA dopamine neurons express the CRF-binding protein (CRF-BP); (Wang et al. 2005; Wang and Morales 2008). CRF regulates dopamine neurons through a subtle interplay of effects at CRF1R, CRF2R, and CRF-BP. CRF increases action potential firing rate in VTA dopamine neurons via CRF1R and involves a PKC-dependent enhancement of I_h (a hyperpolarization-activated inward current) (Wanat et al. 2008). CRF enhanced the amplitude and slowed the kinetics of IPSCs following activation of D2-dopamine and $GABA_B$ receptors. This action is postsynaptic and dependent on the CRF1R. The enhancement induced by CRF was attenuated by repeated in vivo exposures to psychostimulants or restraint stress (Beckstead et al. 2009).

CRF can induce a slowly developing, but transient, potentiation of NMDAR-mediated synaptic transmission (Ungless et al. 2003). This effect involves the CRF2R and activation of the protein kinase C pathway and the requirement of CRF-BP. However, the effect of CRF is restricted to a subset of dopamine neurons expressing large I_h currents (Ungless et al. 2003).

In addition to fast, excitatory glutamate-mediated synaptic transmission, dopamine neurons also express metabotropic glutamate receptors (mGluRs) which mediate slower, inhibitory synaptic transmission (Fiorillo and Williams 1998). The rapid rise and brief duration of synaptically released glutamate in the extracellular space mediates a rapid excitation through activation of ionotropic receptors, followed by inhibition through the mGluR1 receptor (Fiorillo and Williams 1998). CRF can enhance these mGluRs via a CRF2R-PKA pathway that stimulates release of calcium from intracellular stores (Riegel and Williams 2008). The CRF modulation of VTA synaptic activity is very complex because CRF has diverse actions on dopamine neurons that are excitatory and inhibitory. Furthermore, desensitization of D2 receptors induced by dopamine or CRF on DAergic VTA neurons is associated with increased glutamatergic signaling in the VTA (Nimitvilai et al. 2014). In summary, the excitatory effects of CRF on dopamine neurons appear to affect fast events (e.g., action potential firing rate and NMDAR-mediated synaptic transmission), whereas the inhibitory effects involve slow forms of synaptic transmission. Another important aspect is that CRF1R-mediated effects do not involve interactions with the CRF-BP, whereas CRF2R-mediated effects do. Recently, the CRF-BP has been considered a potential target for its role in AUD (Haass-Koffler et al. 2016; Ketchesin et al. 2016), and its role in the escalation of alcohol drinking may involve its interaction with CRF2 (Albrechet-Souza et al. 2015; Quadros et al. 2016).

It is speculated that these effects on short-term plasticity phenomena may modulate longer-lasting forms of plasticity. For example, NMDAR activation is required for the induction of long-term potentiation in VTA dopamine neurons (Bonci and Malenka 1999; Borgland et al. 2010).

3.1.2 CRF Actions in the Central Amygdala

The CeA contains CRF receptors and abundant CRF-containing fibers (De Souza et al. 1984; Uryu et al. 1992); CRF itself is generally co-localized in CeA neurons together with GABA (Eliava et al. 2003; Asan et al. 2005). Acute EtOH augments evoked $GABA_A$ receptor-mediated inhibitory postsynaptic currents (IPSCs) by increasing GABA release in both mouse (Bajo et al. 2008; Nie et al. 2004) and rat CeA neurons (Roberto et al. 2003, 2004a, b).

CRF1Rs mediate the EtOH-induced augmentation of IPSCs in mouse CeA (Nie et al. 2004, 2009) via the PKCε signaling pathway (Bajo et al. 2008; Nie et al. 2004). Both CRF and EtOH augment evoked IPSCs in mice CeA neurons, and CRF1R (but not CRFR2) antagonists blocked both CRF and ethanol effects. In addition, CRF and EtOH augment IPSCs in wild-type and CRF2R knockout mice, but not in CRF1R knockout mice (Nie et al. 2004) or with CRF1 antagonism (Nie et al. 2009).

Electrophysiological data showed that CRF, like EtOH, also enhances GABAergic transmission in the rat CeA (Roberto et al. 2010). As in mice, CRF and EtOH actions involve presynaptic CRF1R activation at the CeA GABAergic synapses. Interestingly, the interactions between the CRF and GABAergic systems in the CeA may play an important role in alcohol reward and dependence (Roberto et al. 2010). These results suggest that the presynaptic effect of EtOH on GABA release in rodent CeA involves CRF1R and perhaps release of CRF itself. Furthermore, both CRF and EtOH decreased PPF of IPSCs in mouse and rat neurons, and the effects of both were selectively blocked by CRF1R antagonists. In addition, both EtOH and CRF increase the frequency of GABAR-mediated mIPSCs, and this effect is blocked by CRF1R antagonists (Nie et al. 2004, 2009; Roberto et al. 2010). Thus, EtOH probably enhances the release of GABA by activating CRF1R on GABAergic terminals (Nie et al. 2009; Roberto et al. 2010). Conversely, CRF1R antagonists directly increased PPF of IPSCs and decreased mIPSC frequencies, consistent with decreased GABA release, thus opposing EtOH effects. Because GABA and CRF are often co-localized in CeA neurons, the EtOH-elicited GABA release may involve release of the CRF peptide itself, perhaps even from the terminals synapsing on autoreceptors on the same cell bodies or on collaterals from other GABAergic interneurons. Thus, this example raises the possibility of involvement of other, secondary messengers in EtOH effects on GABAergic terminals.

Chronic EtOH exposure produces functional adaptation of the CRF system in CeA (Broccoli et al. 2018; Hansson et al. 2006, 2007; Sommer et al. 2008; Weiss et al. 2001). Interestingly, in CeA of dependent rats, the ability of maximal (200 nM) and a submaximal (100 nM) concentrations of CRF to augment evoked IPSCs was significantly enhanced compared to naïve CeA. A greater effect of CRF1R antagonists on basal IPSCs of dependent rats was also reported. The greater effect of CRF and CRF1R antagonists may reflect increased tonic release of endogenous CRF, constitutive CRF1R activation, increased receptor number, and/or sensitization of CRF1R in CeA of dependent rats. This is supported by increased CRF and CRF1 mRNA levels seen in the CeA of alcohol-dependent rats and by reversal of dependence-induced elevations in amygdalar GABA dialysate by a CRF1 antagonist (Roberto et al. 2010). Thus, these combined findings suggest an important EtOH–CRF interaction on GABAergic transmission in the CeA that markedly increases during development of ethanol dependence (Roberto et al. 2010).

In other studies using adult mice, one and six cycles of the drinking in the dark paradigm (DID) increases CeA CRF immunoreactivity, suggesting that the CRF system is recruited during early binge-like drinking episodes (Lowery-Gionta et al. 2012). Notably, the synaptic effects of CRF on CeA GABAergic transmission are reduced after repeated bouts of binge-like drinking (Lowery-Gionta et al. 2012).

Given the critical role of the CRF/CRF1 system and the cellular heterogeneity in the CeA, several recent studies have used a transgenic mouse line expressing the green fluorescent protein (GFP) under the Crhr1 promoter (CRF1:GFP) to readily identify neurons expressing CRF1 (CRF1+) (Justice et al. 2008; Herman et al. 2013a, 2016a) to unveil unique molecular, morphological, and functional properties that distinguish CeA CRF1+ neurons from their CRF1 non-expressing (CRF1-) neighbors. CRF1+ neurons are mainly located in the medial subdivision of the CeA and exhibit an ongoing tonic GABAergic conductance driven by action potential-dependent GABA release. In contrast, CRF1- neurons do not display tonic inhibition (Herman et al. 2013a). As described above, chronic ethanol induced functional adaptations on phasic and tonic inhibition and cell firing in CRF1+ and CRF1- CeA neurons (Herman et al. 2016a). In particular, a loss of tonic currents and a significantly higher basal firing rate were observed in CRF1+ CeA neurons projecting to the BNST of CIE vs. control mice (Herman et al. 2016a). Recent work from the Herman laboratory has shown that CRF1+ CeA neurons exhibit sex differences in sensitivity to the effects of acute alcohol, as well as CRF1 agonists and antagonists (Agoglia et al. 2020, 2021). Furthermore, chronic alcohol drinking produced neuroadaptations in CRF1+ neurons that increased the sensitivity of GABAA receptor-mediated sIPSCs to the acute effects of alcohol, CRF, and the CRF1 antagonist R121919, but these adaptations were more pronounced in male versus female mice. The CRF1 antagonism reduced voluntary alcohol drinking in both sexes and abolished sex differences in alcohol drinking. The minimal alcohol-induced changes in the female CRF1 system may be related to the elevated alcohol intake displayed by female mice and could contribute to the ineffectiveness of CRF1 antagonists in female AUD patients (Agoglia et al. 2020, 2021).

Retson and colleagues (2015) have reported similar results supporting clear sex differences in CeA CRF in rats. They found that alcohol drinking activated CeA CRF neurons and enhanced the response of these neurons to stress selectively in male but not female rats (Retson et al. 2015). Further investigation of these sex differences is necessary to clarify the contributions of CRF activity to alcohol use in both males and females.

Overall those studies have yielded significant insight into cell type-specific effects of acute and chronic alcohol in local and downstream CRF-CeA circuits. In parallel molecular studies have also assessed expression of subpopulation markers and neuropeptides, dendritic spine density and morphology, and glutamatergic transmission in CeA CRF1+ vs. CRF1- neurons (Wolfe et al. 2019). In brief, CeA CRF1+ neurons are GABAergic, but do not segregate with calbindin, calretinin, or PKCδ. Co-expression analysis using in situ hybridization revealed Crhr1 had highest co-expression with Penk and Sst and least with neuropeptide Y (NPY). Additionally, CeA CRF1+ neurons do not display differences in mature spines and accordingly no difference in basal CeA glutamate transmission. CRF application enhances overall glutamate release onto both CRF1+ and CRF1- neurons but increases postsynaptic glutamate receptor functions selectively in CRF1+ neurons (Wolfe et al. 2019).

CRF-related peptides serve as hormones and neuromodulators of the stress response and play a role in affective disorders. It has been shown that excitatory glutamatergic transmission is modulated by two endogenous CRF-related peptide ligands, CRF rat/human (r/h) and Ucn I, within the CeA and the lateral septum mediolateral nucleus (LSMLN) (Liu et al. 2004). Activation of these receptors exerts diametrically opposing actions on glutamatergic transmission in these nuclei. In the CeA, CRF(r/h) depressed excitatory glutamatergic transmission through a CRF1R-mediated postsynaptic action, whereas Ucn I facilitated synaptic responses through presynaptic and postsynaptic CRF2R-mediated mechanisms. Conversely, in the lateral septum mediolateral nucleus (LSMLN), CRF induced a CRF1R-mediated facilitation of glutamatergic transmission via postsynaptic mechanisms, whereas Ucn I depressed EPSCs by postsynaptic and presynaptic CRF2R-mediated actions. Furthermore, antagonists of these receptors also affected glutamatergic neurotransmission, indicating a tonic endogenous modulation at these synapses (Liu et al. 2004). These data show that CRF receptors in CeA and LSMLN synapses exert and maintain a significant synaptic tone and thereby regulate excitatory glutamatergic transmission. In fact, studies on CIE-induced changes in the modulation of rat glutamatergic synapses by CRF (Varodayan et al. 2017a) revealed that CRF also decreased rat CeA locally- or basolateral amygdala (BLA)-evoked glutamatergic responses. In contrast to the evoked data CRF increased mEPSC frequency similarly in naive and CIE neurons, suggesting increased vesicular glutamate release (Varodayan et al. 2017a; Herman et al. 2016b). Those studies also revealed that

CRF-induced facilitation of glutamate release is mediated by CRF1 receptors, but the mechanisms are complex and may involve both CRF1 and CRF2 receptors with opposite receptor subtype effects on glutamate release (Varodayan et al. 2017a). These rat studies agree with mouse studies showing that acute bath application of EtOH significantly increased sEPSC frequency in a concentration-dependent manner in CeA neurons, and this effect was blocked by pretreatment of co-applied CRFR1 and CRFR2 antagonists (Silberman et al. 2015).

3.1.3 CRF Actions in the Bed Nucleus of the Stria Terminalis

The BNST, a brain region associated with anxiety, has enriched expression of CRF (Ju and Han 1989) and CRFRs (Van Pett et al. 2000). A component of the extended amygdala, the BNST is anatomically well-situated to integrate stress and reward-related processing in the CNS, regulating activation of the hypothalamic-pituitary-adrenal (HPA) axis and reward circuits. The oval nucleus is a rich source of CRF neurons and terminals which may originate from local CRF neurons or from CRF neurons projecting from the CeA (Morin et al. 1999; Sakanaka et al. 1986; see also Kash et al. (2015) for review). Much evidence supports the role of CRF signaling in the BNST in general anxiety (Gafford et al. 2012; Sink et al. 2013; see also Kash et al. (2015) for review) and anxiety-like behaviors induced by ethanol withdrawal (Huang et al. 2010).

Pharmacological studies suggest that CRF signaling in the BNST is involved in anxiety (Lee and Davis 1997) and stress-induced relapse to cocaine self-administration (Erb and Stewart 1999). Moreover, a stimulus that promotes anxiogenic responses, the withdrawal of rodents from chronic EtOH exposure, produces rises in extracellular levels of CRF in the BNST (Olive et al. 2002). However, in another study, following 2 weeks of binge-like alcohol intake, adolescent rats display decreases in CRF cell number in the CeA and no changes in BNST (Karanikas et al. 2013). Interactions between CRF and GABAergic transmission in BNST were reported to play a role in regulating stress and anxiety (Kash and Winder 2006). In this study the actions of CRF on GABAergic transmission in the ventrolateral region of the BNST were examined. This region projects to both the VTA (Georges and Aston-Jones 2002; Rinker et al. 2017) and the PVN of the hypothalamus (Cullinan et al. 1993), thus providing a point of access to both reward and stress pathways. Using whole-cell recordings in a BNST slice preparation, Kash and Winder (2006) found that CRF enhances GABAergic transmission. Their pharmacological and genetic experiments suggest that CRF and urocortin CRF enhance postsynaptic responses to GABA through activation of the CRF1R. CRF1-R signaling in the BNST also enhances glutamatergic drive on neurons projecting to the VTA in a presynaptic fashion (Silberman et al. 2013). Thus, CRF can enhance both inhibitory and excitatory transmission in the BNST, albeit through distinct signaling mechanisms.

Kash et al. (2008) also showed the action of dopamine on cellular and synaptic function in the BNST using an ex vivo slice preparation. These investigators demonstrated a rapid and robust dopamine-induced enhancement of excitatory transmission in the BNST. This enhancement is activity-dependent and requires the downstream action of CRF1R, suggesting that dopamine induces CRF release through a local network mechanism. Furthermore, it was found that both in vivo and ex vivo cocaine induced a dopamine receptor and CRF1R-dependent enhancement of a form of NMDA receptor-dependent short-term potentiation in the BNST. These data highlight a direct and rapid interaction between dopamine and CRF systems that regulates excitatory transmission and plasticity in a brain region key to reinforcement and reinstatement. Because a rise in extracellular dopamine levels in the BNST is a shared consequence of multiple classes of drugs of abuse, this suggests that the CRF1R-dependent enhancement of glutamatergic transmission in this region may be a common key action of substances of abuse (Kash et al. 2008). Subsequent studies from the Kash laboratory revealed a complex interaction between CRF and NPY in the BNST in the regulation of binge alcohol drinking in both mice and monkeys (see section below and Pleil et al. (2015a)).

Francesconi et al. (2009a, b) investigated the effects of protracted withdrawal from alcohol in the juxtacapsular nucleus of the anterior division of the BNST (jcBNST). The jcBNST receives robust glutamatergic projections from the BLA, the postpiriform transition area, and the insular cortex as well as dopamine inputs from the midbrain. In turn, the jcBNST sends GABAergic projections to the medial division of the central CeA as well as other brain regions. These investigators described a form of long-term potentiation of the intrinsic excitability (LTP-IE) of neurons of the jcBNST in response to high-frequency stimulation (HFS) of the stria terminalis that was impaired during protracted withdrawal from alcohol (Francesconi et al. 2009b). Administration of the selective CRF1R antagonist (R121919), but not of the CRF2R antagonist (astressin 2B), normalized jcBNST LTP-IE in animals with a history of alcohol dependence (Francesconi et al. 2009b). In addition, repeated, but not acute, administration of CRF itself produced a decreased jcBNST LTP-IE. These investigators also showed that dopaminergic neurotransmission is required for the induction of LTP-IE of jcBNST neurons through dopamine D1 receptors (Francesconi et al. 2009b). Thus, activation of the central CRF stress system and altered dopaminergic neurotransmission during protracted withdrawal from alcohol and drugs of abuse may contribute to the disruption of LTP-IE in the jcBNST. Furthermore, the jcBNST also shows marked reductions in excitability after protracted withdrawal from CIE (Szücs et al. 2012). Overall, the impairment of this form of intrinsic neuronal plasticity in the jcBNST could result in inadequate neuronal integration and reduced inhibition of the CeA, contributing to the negative affective state that characterizes protracted abstinence in post-dependent individuals (Francesconi et al. 2009a, b).

It is important to mention that NE is another key interface in the BNST CRF with stress and chronic ethanol. Studies have examined the effects of NE on BNST CRF neuron activity and determine if these effects may be modulated by CIE exposure or a single restraint stress (Snyder et al. 2019). Stress and CIE enhance BNST CRF neuron activity via similar β-AR dependent mechanisms. Surprisingly, stress and CIE do not appear to alter NE-induced inhibition of glutamatergic inputs onto BNST CRF neurons, an effect previously shown to be α-AR dependent 2 (Fetterly et al. 2019). Together, these results indicate that stress and chronic EtOH target the activity of β-ARs on BNST CRF neurons without altering α-AR modulation of these neurons, thereby altering the α/β-AR balance within this circuitry. Thus, maintaining α/β-AR balance in BNST CRF circuits may be an important target for novel treatments for stress-related disorders and stress-induced reinstatement to alcohol seeking behaviors (Snyder et al. 2019).

3.1.4 CRF Actions in the Basolateral Amygdala

Liu et al. (2004) demonstrated that CRF and its related family of peptides act differentially at CRF1 vs. CRF2 synaptic receptors to facilitate or depress excitatory transmission in CeA and lateral septum mediolateral nucleus. Notably, the effects of CRF and its ligands occurred without any apparent direct action on membrane potential or membrane excitability, suggesting that the role of CRF at these limbic synapses is that of a "neuroregulator." The investigators suggested pre- and postsynaptic loci for CRF1 and CRF2 receptors within the glutamatergic CeA and LSMN synapses. Although both synapses exhibit a comparable pre- and postsynaptic location of CRF1 and CRF2 receptors, their functions (facilitation vs. depression of glutamatergic transmission) are opposite within each synapse (Gallagher et al. 2008). Liu et al. (2004) also demonstrated that endogenous CRF ligands induce a tonic effect on excitatory glutamatergic transmission at synapses within both of these nuclei since application of competitive, selective CRF1 or CRF2 receptor antagonists resulted in an enhancement or depression of glutamatergic EPCS. A similar tonic endogenous action of CRF ligands was not observed under control conditions in the medial prefrontal cortex (Orozco-Cabal et al. 2006). This latter result further emphasizes that CRF effects are different depending upon the CNS synapse being investigated. Most of these studies in the Gallagher group aimed to investigate the action of CRF on glutamatergic synapses in relation to cocaine administration. There is very poor data on EtOH–CRF–glutamate interaction.

Taken together these data suggest that a dysregulation of the extrahypothalamic CRF function is a major determinant of vulnerability to high alcohol intake and maintenance of alcohol and drug dependence and other aspects of AUD.

3.2 Neuropeptide Y

The inhibitory NPY peptide is produced in abundance in the hypothalamus and phylogenetically conserved across species (Allen et al. 1986). NPY is involved in regulation of food and water intake. It has recently been ascribed its prominent role in the aversive aspects of alcohol withdrawal and relapse via their actions in the CeA.

Endogenous NPY reduces anxiety via actions in the amygdala (Heilig et al. 1993; Sajdyk et al. 2002) and suppresses alcohol drinking in rats (Gilpin et al. 2003) via its actions in CeA (Gilpin et al. 2008a, b; Thorsell 2008). More specifically, NPY microinjection into the CeA exhibits an enhanced ability to suppress alcohol drinking in certain subpopulations of drinkers, including rats that are made dependent on alcohol via vapor inhalation.

NPY is generally co-localized with GABA in inhibitory interneurons. NPY mediates its actions by interacting with a family of G-protein coupled receptors (GPCRs), at least 5 of which have been cloned and designated Y1, Y2, Y4, Y5, and Y6. These receptors are widely distributed throughout the brain. NPY also has been shown to be a regulator of neuronal excitability in hippocampus, where its cellular actions have been most extensively studied (Colmers et al. 1991). In the amygdala, NPY has anxiolytic effects that are mediated via activation of Y1 receptors (Heilig et al. 1993). NPY neurons in the amygdala project to the BNST (Allen et al. 1984), which also contains Y1 receptors and Y1 and Y2 receptor mRNA. Further, the CeA receives NPYergic input from the nucleus of the solitary tract, arcuate nucleus, and the lateral septum (see Kask et al. (2002) for a review). Y1, Y2, and Y5 receptors and receptor mRNA are found in the amygdala, and each of these receptor subtypes has been implicated in anxiety (Kask et al. 2002). Y2 receptors are thought to act presynaptically as autoreceptors providing negative feedback to NPYergic nerve terminals, whereas Y1 receptors appear to act postsynaptically (Kask et al. 2002; Wolak et al. 2003).

Many in vivo studies point to the involvement of NPY in mediating some of the behavioral effects of EtOH (Caberlotto et al. 2001; Cippitelli et al. 2010; Rimondini et al. 2005). NPY KO mice show increased EtOH preference but blunted behavioral responses to ethanol, while NPY overexpressors show a lower preference and increased sensitivity to ethanol (Thiele et al. 1998). Likewise, increased NPY expression in the CeA was noted in two independent strains of alcohol-preferring rats (Hwang et al. 1999). There were increased levels of NPY in the paraventricular nucleus of the hypothalamus (PVN) and arcuate nucleus of EtOH-preferring rats and decreased NPY levels in the CeA of ethanol-preferring rats, suggesting an inverse relationship between NPY levels in the CeA and EtOH consumption. Additionally, alcohol-preferring rats show significant decreases in both cAMP-responsive element-binding protein (CREB) and NPY levels in the CeA and medial amygdala, but not the basolateral amygdala (Pandey et al. 2005). Further, virally mediated alterations in NPY levels in the CeA differentially affect EtOH consumption in rats with low and high basal levels of anxiety (Primeaux et al. 2006). Also, recent genetic and pharmacological evidence indicates that C57BL/6 J mice have low NPY levels in CeA compared to DBA/2 mice, suggesting that NPY contributes to the high EtOH consumption characteristic of C57BL/6 J mice (Hayes et al. 2005).

Electrophysiological findings suggest that NPY and EtOH have a similar profile of actions (Ehlers et al. 1998a, b, 1999). Increased sensitivity to NPY and CRF was observed in cortex and amygdala after chronic EtOH exposure, as measured by EEG activity and event-related potentials (Slawecki et al. 1999). Modulation of amygdala EEGs by NPY differs in naïve P and NP rats, suggesting that NPY has different neuromodulatory effects in these two strains (Ehlers et al. 1998a). Furthermore, NPY antagonizes the effects of CRF in the amygdala (Ehlers et al. 1998a).

At the cellular level NPY interactions with EtOH have been characterized in the CeA and other brain regions (for review, see Gilpin et al. (2015); Robinson and Thiele (2017)). Gilpin et al. (2011) found that NPY in rat CeA prevents acute alcohol-induced increases in evoked and spontaneous GABA release. Pharmacological manipulation with antagonists confirm the presynaptic site of action and suggest that NPY blocks alcohol effects via presynaptic Y2Rs. NPY also normalizes alcohol dependence-induced increases in GABA release in CeA, suggesting that chronic exposure causes neuroadaptations in NPY systems that affect inhibitory transmission. Notably, in mice, central infusion of NPY, an NPY Y1 receptor (Y1R) agonist, and a Y2R antagonist significantly blunted binge-like ethanol drinking in C57BL/6 J mice (Sparrow et al. 2012). Binge-like ethanol drinking reduced NPY and Y1R immunoreactivity in the CeA, and 24 h of ethanol withdrawal increased Y1R and Y2R immunoreactivity. Binge-like ethanol drinking also increased the ability of NPY to inhibit GABAergic transmission. Thus, binge-like ethanol drinking in C57BL/6 J mice promoted alterations of NPY signaling in the CeA (Sparrow et al. 2012), and administration of exogenous NPY compounds protected against binge-like drinking. Overall, these results in the CeA of rats and mice align with findings on NPY modulation of GABA transmission in BNST (Kash and Winder 2006) and suggest that Y2Rs function as autoreceptors regulating NPY release. NPY and CRF have opposing effects on stress and anxiety as well as on synaptic activity in BNST (Heilig et al. 1994; Kash and Winder 2006). Kash and Winder (2006) found that NPY and CRF inhibit and enhance GABAergic transmission, respectively: NPY depresses GABAergic transmission through activation of the Y2 receptors, whereas CRF and urocortin enhance GABAergic transmission through activation of CRF1 receptors. Further, NPY appears to reduce GABA release, whereas CRF enhances postsynaptic responses to GABA, suggesting potential anatomical and cellular substrates for the robust behavioral interactions between NPY and CRF in the extended amygdala. A recent study employed physiological, pharmacological, and chemogenetic approaches to identify a precise neural mechanism in the BNST underlying the interactions between NPY and CRF in the regulation of binge alcohol drinking in both mice and monkeys (Pleil et al. 2015a). The results showed that Y1R activation in the BNST suppressed binge alcohol drinking by enhancing inhibitory synaptic transmission specifically in CRF neurons via a previously unknown Gi-mediated, PKA-dependent postsynaptic mechanism. In addition, chronic alcohol drinking altered Y1R function in the BNST of both mice and monkeys, highlighting the enduring, conserved nature of this effect across mammalian species (Pleil et al. 2015a).

Chronic restraint stress also alters the NPY system (Pleil et al. 2012). Specifically, increases NPY and Y2R expression in the BNST and reduces the Y2R-mediated effect of NPY on inhibitory synaptic transmission in a stress-susceptible mouse strain (DBA/2 J), but not a stress-resilient strain (C57BL/6 J) (Pleil et al. 2012). Notably, deletion of neuropeptide Y2 receptors from GABAergic neurons in the extended amygdala differently affected affective and alcohol-drinking behaviors in male and female mice (McCall et al. 2013). Specifically, females displayed greater basal anxiety, higher levels of ethanol consumption, and faster fear conditioning than males, and that knockout mice exhibited enhanced depressive-like behavior in the forced swim test. Together, these finding support higher expression of negative affective and alcohol-drinking behaviors in females than males, and they highlight the importance of Y2R function in GABAergic systems in the expression of depressive-like behavior (McCall et al. 2013).

3.3 Orphanin FQ/Nociceptin (OFQ/N)

Nociceptin (known also as orphanin FQ) is the most recently discovered member of the endogenous opioid peptide family, albeit nearly 15 years ago. Nociceptin mediates or influences many behavioral, psychological, and neurobiological processes, including memory, anxiety, stress, and reward (Economidou et al. 2008; Martin-Fardon et al. 2010; Murphy 2010). The heptadecapeptide nociceptin is the endogenous ligand of the nociceptin opioid receptor (NOR), previously referred to as opiate receptor-like1 (ORL1). NOR is a GPCR that belongs to the opioid receptor family (Mogil et al. 1996; Mogil and Pasternak 2001). In rodents, moderate to high levels of NOR mRNA are detected in cerebral cortex, nucleus accumbens, amygdala, dorsal raphe nucleus, and hippocampus (Harrison and Grandy 2000). Nociceptin has a high structural homology with opioid peptides, especially dynorphin A (Meunier et al. 1995; Reinscheid et al. 1995), but nociceptin does not bind to MOR, DOR, or KOR (μ, δ, and κ-opioid receptors) and opioid peptides do not bind NOR (Lachowicz et al. 1995; Reinscheid et al. 1995). Nociceptin inhibits forskolin-stimulated cAMP formation (see Harrison and Grandy (2000); Hawes et al. (2000)), and protein kinase C (PKC), MAP kinases and phospholipase A2 have been linked to NOR (Fukuda et al. 1998; Hawes et al. 2000; Lou et al. 1998).

At the cellular level, nociceptin acts at NOR to augment K+ conductances in amygdalar (Meis and Pape 1998, 2001), hippocampal (Amano et al. 2000; Ikeda et al. 1997; Madamba et al. 1999; Tallent et al. 2001; Yu and Xie 1998), and thalamic neurons (Meis 2003; Meis et al. 2002), thus depressing cell excitability. Nociceptin has also been shown to decrease Ca^{2+} currents (Abdulla and Smith 1997; Calo et al. 2000; Connor et al. 1999; Henderson and McKnight 1997; Larsson et al. 2000) and to reduce the amplitude of both non-NMDA receptor-mediated excitatory postsynaptic currents (EPSCs) and IPSCs in rat lateral amygdala (Meis et al. 2002).

Roberto and Siggins (2006) found that nociceptin did not significantly alter resting membrane potential, input resistance, or spike amplitude, in accord with results reported by others in CeA (Meis and Pape 1998) and for other brain regions (Ikeda et al. 1997; Madamba et al. 1999; Tallent et al. 2001). However, nociceptin dose-dependently reduced $GABA_A$ IPSCs. This inhibition of GABAergic transmission was reversible on washout (Roberto and Siggins 2006). Nociceptin also concomitantly increased the PPF of IPSCs and decreased the frequency of mIPSCs, suggesting decreased GABA release. Thus, nociceptin decreases GABAergic transmission by reducing GABA release at CeA synapses (Roberto and Siggins 2006). Interestingly, nociceptin applied before EtOH completely prevented the ethanol-induced enhancement of GABAergic transmission in CeA opposing the enhancing action of ethanol on GABA release (Roberto and Siggins 2006). These investigators also found that the nociceptin-induced decrease of GABAergic transmission was larger in EtOH-dependent rats and might reflect neuroadaptations associated with ethanol dependence. Notably, nociceptin completely blocked the CRF-induced increase of GABA release (Cruz et al. 2012), suggesting that nociceptin antagonized the effect of CRF. Moreover, the NOP receptor antagonist [Nphe1]nociceptin(1-13)NH2 blocked the nociceptin-induced diminution of GABA but not the CRF-induced augmentation of GABA release, indicating that nociceptin modulates both ethanol and CRF effects through the NOP receptors. Nociceptin also blocked CRF-induced increases in GABAergic responses in CeA from ethanol-dependent rats (Cruz et al. 2012). Using a multidisciplinary approach, Ciccocioppo and collaborators (2014a, b) found a selective upregulation of the nociceptin and downregulation of the CRF1 receptor transcripts in the CeA and BLA after stress restraint (Ciccocioppo et al. 2014a). Notably, intra-CeA injections of nociceptin reduced anxiety-like behavior in restrained rats in the elevated plus maze. Finally, in restraint-stressed rats, baseline CeA GABAergic responses were elevated and nociceptin exerted a larger inhibition of GABA responses compared with non-restrained rats (Ciccocioppo et al. 2014a).

Nociceptin interaction on glutamatergic transmission and ethanol effects were also investigated (Kallupi et al. 2014a). Acute and chronic ethanol exposures significantly decrease glutamate transmission by both pre- and postsynaptic actions (Roberto et al. 2004b). Nociceptin diminished basal-evoked compound glutamatergic and spontaneous glutamate transmission by mainly decreasing glutamate release in the CeA of naive rats (Kallupi et al. 2014a). Nociceptin blocked the inhibition induced by acute ethanol and ethanol blocked the nociceptin-induced inhibition of glutamatergic responses in CeA neurons of naive rats. Like the GABAergic synapses, nociceptin antagonism revealed tonic inhibitory activity of NOP on CeA glutamatergic transmission only in alcohol-dependent rats. The antagonist also blocked nociceptin-induced decreases in glutamatergic responses but did not affect ethanol-induced decreases in evoked glutamate responses. Taken together, these studies implicate a potential role for the nociceptin system in regulating CeA glutamatergic and GABAergic synapses in both acute stress and alcohol dependence providing translational support for nociceptin as a "druggable" candidate system for medication development for the treatment of AUD. In support of this concept, it is important to continue to identify novel soluble non-peptidergic molecules such as nociceptin agonists (Ciccocioppo et al. 2014b; Kallupi et al. 2014b) that decrease excessive drinking and act at the cellular level in brain regions such as the amygdala that are associated with ethanol dependence.

The functional interactions of neuropeptides (CRF, NPY, nociceptin) with inhibitory and excitatory systems in the brain may play major roles in the acute reinforcement effects of EtOH. Understanding the underlying mechanisms of these interactions may offer a possible avenue for restoring "normal" function following chronic drug exposure. The neuroadaptations induced by chronic EtOH on GABAergic and glutamatergic systems may represent homeostatic or compensatory mechanisms in response to the acute ethanol actions on these systems.

4 New Approaches to Determine In Vivo Roles of Ethanol Effects on Synaptic Transmission

From the foregoing discussion it should be clear that we now know a great deal about how acute and chronic EtOH exposure alters synaptic function. However, less is known about the roles played by these synaptic effects in the in vivo physiological and behavioral effects of the drug. New genetic, optical, pharmacological, and physiological techniques allow for faster advancement in this research area.

Assessing alcohol effects on in vivo neural function has moved beyond traditional single- and multi-unit electrophysiological recordings. New systems such as the "Neuropixels" recording system allow investigators to measure the firing of 1000s of neurons in a single recording with excellent signal/noise ratios and discrimination of single neurons (Steinmetz et al. 2021). When combined with the proper analysis tools this approach has the potential to enhance our understanding of the alcohol impact of neurophysiology. Combining such recordings with genetic and pharmacological manipulations of synaptic proteins and synaptic function will allow investigators to determine how particular synapses contribute to EtOH-induced changes in neuron/circuit function and behavior.

An explosion of techniques for measurement of neuronal activity and neurotransmitter levels has taken place over the last decade. With the development of genetically-encoded fluorescent sensors for intracellular calcium, other second messengers and extracellular neurotransmitters, real-time in vivo measurements can now be made with imaging and optical fiber-based photometry (Jing et al. 2019; Liang et al. 2015; Labouesse and Patriarchi 2021; Meng et al. 2018; Siciliano and Tye 2019). Combining these approaches with behavioral analysis in awake animals is providing unprecedented analyses of how a variety of neuronal/synaptic functions are related to behavior. These techniques are already being applied to examine effects of EtOH on the function of specific afferent projections in the brain (Siciliano and Tye 2019). Studies in the coming years are sure to reveal much more detailed evidence of EtOH effects on neurotransmitter levels that can be related to drinking or other behaviors.

As mentioned earlier in this chapter, optogenetic activation of specific afferent projections has now gained widespread usage in neuroscience and alcohol research. By expressing a light-activatable opsin that induces depolarization or hyperpolarization in a specific cell type, investigators can now interrogate how EtOH and other drugs alter synaptic transmission at a given synapse. Studies investigating EtOH effects on optogenetically-activated synaptic transmission in brain slices have already been discussed. This technique is also being used to examine how altered function of specific neurons affects EtOH-related behaviors (Juarez et al. 2019). Combining optogenetic approaches with other techniques outlined in this chapter should allow investigators to determine how EtOH exposure alters afferent and synaptic function in vivo.

The development of techniques for activation of non-native LGICs and GPCRs with ligands that are normally biologically inactive has revolutionized techniques for altering neuronal and synaptic function (Campbell and Marchant 2018; Vardy et al. 2015). These approaches allow investigators to examine how activation, inhibition, and modulation of different neural cells contribute to circuit function and behavior. With regard to synaptic function, the Designer Receptor Exclusively Activated by Designer Drug (DREADD) technique is especially attractive. This technique uses genetically engineered GPCRs that can be inserted into neurons of interest and affect neuronal/synaptic function in numerous ways. For example, the DREADD variants that couple to Gi/o G-proteins inhibit neurotransmitter release (e.g., hM4Di) (Armbruster et al. 2007), as expected from other GPCRs with similar coupling. It is now possible to alter transmitter release at an identified synaptic terminal in vivo, especially when DREADD expression is combined with local injection of the designer drug receptor agonist (Gremel et al. 2016; Mahler et al. 2014; Cheng and Wang 2019). This will allow investigators to interrogate how a particular presynaptic manipulation alters EtOH-related behaviors and mimic effects of EtOH at identified synapses. Additional uses of this technique to alter pre- and postsynaptic function and interactions with EtOH will undoubtedly be used in the coming years.

5 Conclusions

In this review we have focused on acute and chronic EtOH actions on synaptic transmission. It is not possible to cover all aspects of this topic, and thus we have focused on describing the best established EtOH actions. As the review attests, EtOH affects numerous aspects of synaptic transmission both directly and indirectly, to alter brain function and behavior. Acute exposure to EtOH generally increases the function of cys-loop ligand-gated ion channels, with prominent effects of $GABA_A$ and glycine receptors. These actions increase synaptic and extrasynaptic inhibition and are thought to contribute to sedation and other aspects of intoxication. Ionotropic glutamate and P2X receptors are generally inhibited by acute EtOH exposure, with some noted exceptions. The inhibitory effect on ionotropic glutamate receptors is most prominent at NMDARs and on NMDAR-mediated synaptic responses, and this inhibitory action is thought to contribute to cognitive impairment produced by EtOH. At present the postsynaptic EtOH effects on neurotransmitter receptors appear to occur within the receptor molecules themselves, although more work is needed to elucidate the roles of posttranslation modification. On the presynaptic side, acute EtOH generally potentiates GABA release, contributing to the enhanced neuronal inhibition produced by the drug. The molecular mechanisms involved in EtOH potentiation of GABA release remain to be fully explored. Ethanol also alters other aspects of synaptic transmission involving amino acid transmitters and monoamines. The net result of the EtOH effects of transmission seems to be to dampen synaptic excitation in many brain regions and reduce most forms of synaptic plasticity (with noted exceptions).

Chronic exposure to EtOH, whether by forced administration or ingestion, generally enhances the function of NMDARs, most often those containing the NR2B subunit. Increases in glutamate release and responses to some other glutamate receptors are also observed following chronic exposure. The net effect of these increases in glutamatergic transmission appears to be a hyperexcitable CNS state during withdrawal that contributes to withdrawal symptoms and relapse. Excitotoxicity might be another result of this hyper-glutamatergic state. In general, acute EtOH effects on glutamate receptor function and glutamatergic transmission are intact even after subchronic or chronic ethanol exposure, suggesting that behavioral tolerance is not a simple function of loss of pharmacological effects at these synapses. At GABAergic synapses, chronic EtOH generally alters either the efficacy of inhibitory synaptic transmission or the types of receptors involved in transmission. Extrasynaptic $GABA_A$ receptor-mediated synaptic responses are also altered, leading to changes in tonic current in the postsynaptic neuron. The pattern of chronic EtOH effects on GABAergic transmission varies considerably across brain regions, making this subject a rich and important area for future investigation. The resultant alterations in patterns of GABAergic transmission in key brain regions may contribute to aspects of AUD including EtOH tolerance, dependence, and drug intake. More work is needed to determine the exact pattern of changes in GABAergic inhibition across brain regions, and how these changes contribute to aspects of alcohol use disorders including tolerance, dependence, and escalating intake.

The modulatory effects of neuropeptides have become subjects of intense investigation in the alcohol research field. Neuropeptides implicated in stress responses, such as CRF, appear to contribute to stress–EtOH interactions as well as drinking and relapse. Acute EtOH exposure alters the release of some neuropeptides, while others alter synaptic transmission in ways that interfere with the actions of ethanol. Chronic EtOH exposure also appears to alter neuropeptide modulatory actions. In addition to providing tools for investigation of mechanisms involved in ethanol actions, the neuropeptides may also provide new avenues for pharmacotherapies that could be used in the treatment of alcohol use disorders. Despite the great progress done and the promising results in understanding the mechanisms of action of numerous neuropeptides in well-established preclinical models of AUD, translating this knowledge to the clinical side has been ineffective. Similar issues hamper preclinical models of antidepressant activity and psychiatric domains in which neuropeptide-targeting compounds have yet to show clinical efficacy. Researchers have just begun to explore the alcohol-related actions of a few of the many neuropeptides found in brain. Thus, more work remains to fully define how peptides participate in the neural actions of alcohol.

Acknowledgments This work was supported by NIH/NIAAA grants AA021491, AA017447, AA006420, AA013498, AA027700, AA029841 to MR and ZIA000416 to DML.

References

Aalto S, Ingman K, Alakurtti K, Kaasinen V, Virkkala J, Någren K, Rinne JO, Scheinin H (2015) Intravenous ethanol increases dopamine release in the ventral striatum in humans: PET study using bolus-plus-infusion administration of [(11)C]raclopride. J Cereb Blood Flow Metab 35: 424–431

Abdulla FA, Smith PA (1997) Nociceptin inhibits T-type Ca2+ channel current in rat sensory neurons by a G-protein-independent mechanism. J Neurosci 17:8721–8728

Abrahao KP, Ariwodola OJ, Butler TR, Rau AR, Skelly MJ, Carter E, Alexander NP, McCool BA, Souza-Formigoni ML, Weiner JL (2013) Locomotor sensitization to ethanol impairs NMDA receptor-dependent synaptic plasticity in the nucleus accumbens and increases ethanol self-administration. J Neurosci 33:4834–4842

Abrahao KP, Salinas AG, Lovinger DM (2017) Alcohol and the brain: neuronal molecular targets, synapses, and circuits. Neuron 96(6):1223–1238. https://doi.org/10.1016/j.neuron.2017.10.032

Adermark L, Jonsson S, Ericson M, Söderpalm B (2011a) Chronic alcohol alters rewarded behaviors and striatal plasticity. Neuropharmacology 61(7):1160–1155

Adermark L, Jonsson S, Ericson M, Söderpalm B (2011b) Intermittent ethanol consumption depresses endocannabinoid-signaling in the dorsolateral striatum of rat. Neuropharmacology 61(7):1160–1165. https://doi.org/10.1016/j.neuropharm.2011.01.014

Agoglia AE, Herman MA (2018) The center of the emotional universe: alcohol, stress, and CRF1 amygdala circuitry. Alcohol 72:61–73

Agoglia AE, Tella J, Herman MA (2020) Sex differences in corticotropin releasing factor peptide regulation of inhibitory control and excitability in central amygdala corticotropin releasing factor receptor 1-neurons. Neuropharmacology 180:108296

Agoglia AE, Zhu M, Quadir SG, Bluitt MN, Douglass E, Hanback T, Tella J, Ying R, Hodge CW, Herman MA (2021) Sex-specific plasticity in CRF regulation of inhibitory control in central amygdala CRF1 neurons after chronic voluntary alcohol drinking. Addict Biol:e13067. https://doi.org/10.1111/adb.13067

Aguayo LG (1990) Ethanol potentiates the GABAA-activated Cl- current in mouse hippocampal and cortical neurons. Eur J Pharmacol 187(1):127–130. https://doi.org/10.1016/0014-2999(90)90349-b

Aguayo LG, Tapia JC, Pancetti FC (1996) Potentiation of the glycine-activated Cl- current by ethanol in cultured mouse spinal neurons. J Pharmacol Exp Ther 279(3):1116–1122

Aguayo LG, Peoples RW, Yeh HH, Yevenes GE (2002) GABA(A) receptors as molecular sites of ethanol action. Direct or indirect actions? Curr Top Med Chem 2(8):869–885. https://doi.org/10.2174/1568026023393426. PMID: 12171577 Review

Aguayo LG, Castro P, Mariqueo T, Muñoz B, Xiong W, Zhang L, Lovinger DM, Homanics GE (2014) Altered sedative effects of ethanol in mice with α1 glycine receptor subunits that are insensitive to Gβγ modulation. Neuropsychopharmacology 39:2538–2548

Akinshola BE (2001) Straight-chain alcohols exhibit a cutoff in potency for the inhibition of recombinant glutamate receptor subunits. Br J Pharmacol 133(5):651–658

Akinshola BE, Yasuda RP, Peoples RW, Taylor RE (2003) Ethanol sensitivity of recombinant homomeric and heteromeric AMPA receptor subunits expressed in Xenopus oocytes. Alcohol Clin Exp Res. 27(12):1876–1883

Albrechet-Souza L, Hwa LS, Han X, Zhang EY, DeBold JF, Miczek KA (2015) Corticotropin releasing factor binding protein and CRF2 receptors in the ventral tegmental area: modulation of ethanol binge drinking in C57BL/6J mice. Alcohol Clin Exp Res 39:1609–1618

Aldenhoff JB, Gruol DL, Rivier J, Vale W, Siggins GR (1983) Corticotropin releasing factor decreases postburst hyperpolarizations and excites hippocampal neurons. Science 221:875–877

Alexander NJ, Rau AR, Jimenez VA, Daunais JB, Grant KA, McCool BA (2018) SNARE complex-associated proteins in the lateral amygdala of Macaca mulatta following long-term ethanol drinking. Alcohol Clin Exp Res 42:1661–1673

Allan AM, Harris RA (1987) Acute and chronic ethanol treatments alter GABA receptor-operated chloride channels. Pharmacol Biochem Behav 27:665–670

Allen YS, Roberts GW, Bloom SR, Crow TJ, Polak JM (1984) Neuropeptide Y in the stria terminalis: evidence for an amygdalofugal projection. Brain Res 321:357–362

Allen YS, Bloom SR, Polak JM (1986) The neuropeptide Y-immunoreactive neuronal system: discovery, anatomy and involvement in neurodegenerative disease. Hum Neurobiol 5:227–234

Amano T, Matsubayashi H, Tamura Y, Takahashi T (2000) Orphanin FQ-induced outward current in rat hippocampus. Brain Res 853:269–274

Anderson NJ, Daunais JB, Friedman DP, Grant KA, McCool BA (2007) Long-term ethanol self-administration by the nonhuman primate, Macaca fascicularis, decreases the benzodiazepine sensitivity of amygdala GABA(a) receptors. Alcohol Clin Exp Res 31:1061–1070

Antoni FA (2000) Molecular diversity of cyclic AMP signalling. Front Neuroendocrinol 21:103–132

Antoni FA, Sosunov AA, Haunso A, Paterson JM, Simpson J (2003) Short-term plasticity of cyclic adenosine 3′,5′-monophosphate signaling in anterior pituitary corticotrope cells: the role of adenylyl cyclase isotypes. Mol Endocrinol 17:692–703

Antuono P, Sorbi S, Bracco L, Fusco T, Amaducci L (1980) A discrete sampling technique in senile dementia of the Alzheimer type and alcoholic dementia: study of the cholinergic system. In: Amaducci L, Davison AN, Antuono P (eds) Aging of brain and dementia. Raven Press, New York, pp 151–158

Arendt T, Bigl V, Arendt A, Tennstedt A (1983) Loss of neurons in the nucleus basalis of Meynert in Alzheimer's disease, paralysis agitans and Korsakoff's disease. Acta Neuropathol 61(2): 101–108. https://doi.org/10.1007/BF00697388

Arendt T, Hennig D, Gray JA, Marchbanks R (1988) Loss of neurons in the rat basal forebrain cholinergic projection system after prolonged intake of ethanol. Brain Res Bull 21:563–570

Arendt T, Allen Y, Marchbanks RM, Schugens MM, Sinden J, Lantos PL, Gray JA (1989) Cholinergic system and memory in the rat: effects of chronic ethanol, embryonic basal forebrain transplants and excitotoxic lesions of the cholinergic basal forebrain projection system. Neurosciene 33:435–462

Ariwodola OJ, Weiner JL (2004) Ethanol potentiation of GABAergic synaptic transmission may be self-limiting: role of presynaptic GABA(B) receptors. J Neurosci 24:10679–10686

Ariwodola OJ, Crowder TL, Grant KA, Daunais JB, Friedman DP, Weiner JL (2003) Ethanol modulation of excitatory and inhibitory synaptic transmission in rat and monkey dentate granule neurons. Alcohol Clin Exp Res 27:1632–1639

Armbruster BN, Li X, Pausch MH, Herlitze S, Roth BL (2007) Evolving the lock to fit the key to create a family of G protein-coupled receptors potently activated by an inert ligand. Proc Natl Acad Sci U S A 104:5163–5168

Aryal P, Dvir H, Choe S, Slesinger PA (2009) A discrete alcohol pocket involved in GIRK channel activation. Nat Neurosci 12(8):988–995

Ary AW, Cozzoli DK, Finn DA, Crabbe JC, Dehoff MH, Worley PF, Szumlinski KK (2012) Ethanol up-regulates nucleus accumbens neuronal activity dependent pentraxin (Narp): implications for alcohol-induced behavioral plasticity. Alcohol 46(4):377–387. https://doi.org/10.1016/j.alcohol.2011.10.003

Asan E, Yilmazer-Hanke DM, Eliava M, Hantsch M, Lesch KP, Schmitt A (2005) The corticotropin-releasing factor (CRF)-system and monoaminergic afferents in the central amygdala: investigations in different mouse strains and comparison with the rat. Neuroscience 131: 953–967

Atwood BK, Lovinger DM, Mathur BN (2014) Presynaptic long-term depression mediated by Gi/o-coupled receptors. Trends Neurosci 37(11):663–673. https://doi.org/10.1016/j.tins.2014.07.010

Augier E, Barbier E, Dulman RS, Licheri V, Augier G, Domi E, Barchiesi R, Farris S, Nätt D, Mayfield RD, Adermark L, Heilig M (2018) A molecular mechanism for choosing alcohol over an alternative reward. Science 360(6395):1321–1326. https://doi.org/10.1126/science.aao1157

Avchalumov Y, Kreisler AD, Xing N, Shayan AA, Bharadwaj T, Watson JR, Sibley B, Somkuwar SS, Trenet W, Olia S, Piña-Crespo JC, Roberto M, Mandyam CD (2021a) Sexually dimorphic prelimbic cortex mechanisms play a role in alcohol dependence: protection by endostatin. Neuropsychopharmacology. https://doi.org/10.1038/s41386-021-01075-6

Avchalumov Y, Oliver RJ, Trenet W, Heyer Osorno RE, Sibley BD, Purohit DC, Contet C, Roberto M, Woodward JJ, Mandyam CD (2021b) Chronic ethanol exposure differentially alters neuronal function in the medial prefrontal cortex and dentate gyrus. Neuropharmacology 185: 108438. https://doi.org/10.1016/j.neuropharm.2020.108438

Badanich KA, Mulholland PJ, Beckley JT, Trantham-Davidson H, Woodward JJ (2013) Ethanol reduces neuronal excitability of lateral orbito-frontal cortex neurons via a glycine receptor dependent mechanism. Neuropsychopharmacology 38:1176–1188

Bajo M, Cruz MT, Siggins GR, Messing R, Roberto M (2008) Protein kinase C epsilon mediation of CRF- and ethanol-induced GABA release in central amygdala. Proc Natl Acad Sci U S A 105:8410–8415

Bale TL, Vale WW (2004) CRF and CRF receptors: role in stress responsivity and other behaviors. Annu Rev Pharmacol Toxicol 44:525–557

Barbier E, Barchiesi R, Domi A, Chanthongdee K, Domi E, Augier G, Augier E, Xu L, Adermark L, Heilig M (2021) Downregulation of synaptotagmin 1 in the prelimbic cortex drives alcohol-associated behaviors in rats. Biol Psychiatry 89(4):398–406

Bassareo V, Cucca F, Frau R, Di Chiara G (2017) Changes in dopamine transmission in the nucleus accumbens shell and core during ethanol and sucrose self-administration. Front Behav Neurosci 11:71

Beckley JT, Laguesse S, Phamluong K, Morisot N, Wegner SA, Ron D (2016) The first alcohol drink triggers mTORC1-dependent synaptic plasticity in nucleus accumbens dopamine D1 receptor neurons. J Neurosci 36:701–713

Beckstead MJ, Gantz SC, Ford CP, Stenzel-Poore MP, Phillips PE, Mark GP, Williams JT (2009) CRF enhancement of GIRK channel-mediated transmission in dopamine neurons. Neuropsychopharmacology 34:1926–1935

Belmeguenai A, Botta P, Weber JT, Carta M, De Ruiter M, De Zeeuw CI, Valenzuela CF, Hansel C (2008) Alcohol impairs long-term depression at the cerebellar parallel fiber-Purkinje cell synapse. J Neurophysiol 100(6):3167–3174

Bjork K, Rimondini R, Hansson AC, Terasmaa A, Hyytia P, Heilig M, Sommer WH (2008) Modulation of voluntary ethanol consumption by beta-arrestin 2. FASEB J 22:2552–2560

Björk K, Tronci V, Thorsell A, Tanda G, Hirth N, Heilig M, Hansson AC, Sommer WH (2013) β-Arrestin 2 knockout mice exhibit sensitized dopamine release and increased reward in response to a low dose of alcohol. Psychopharmacology 230(3):439–449

Blank T, Nijholt I, Grammatopoulos DK, Randeva HS, Hillhouse EW, Spiess J (2003) Corticotropin-releasing factor receptors couple to multiple G-proteins to activate diverse intracellular signaling pathways in mouse hippocampus: role in neuronal excitability and associative learning. J Neurosci 23:700–707

Blasio A, Wang J, Wang D, Varodayan FP, Pomrenze MB, Miller J, Lee AM, McMahon T, Gyawali S, Wang HY, Roberto M, McHardy S, Pleiss MA, Messing RO (2018) Novel small-molecule inhibitors of protein kinase C epsilon reduce ethanol consumption in mice. Biol Psychiatry 84(3):193–201. https://doi.org/10.1016/j.biopsych.2017.10.017

Blednov YA, Stoffel M, Chang SR, Harris RA (2001) Potassium channels as targets for ethanol: studies of G-proteincoupled inwardly rectifying potassium channel 2 (GIRK2) null mutant mice. J Pharmacol Exp Ther 298(2):521–530

Blednov YA, Benavidez JM, Black M, Leiter CR, Osterndorff-Kahanek E, Harris RA (2015) Glycine receptors containing α2 or α3 subunits regulate specific ethanol-mediated behaviors. J Pharmacol Exp Ther 353:181–191

Blitzer RD, Gil O, Landau EM (1990) Long-term potentiation in rat hippocampus is inhibited by low concentrations of ethanol. Brain Res 537(1–2):203–208

Blomeley CP, Cains S, Smith R, Bracci E (2011) Ethanol affects striatal interneurons directly and projection neurons through a reduction in cholinergic tone. Neuropsychopharmacology 36(5): 1033–1046. https://doi.org/10.1038/npp.2010.241

Bodhinathan K, Slesinger PA (2013) Molecular mechanism underlying ethanol activation of G-protein-gated inwardly rectifying potassium channels. Proc Natl Acad Sci U S A 110:18309–18314

Bonci A, Malenka RC (1999) Properties and plasticity of excitatory synapses on dopaminergic and GABAergic cells in the ventral tegmental area. J Neurosci 19:3723–3730

Booker TK, Collins AC (1997) Long-term ethanol treatment elicits changes in nicotinic receptor binding in only a few brain regions. Alcohol 14:131–140

Borgland SL, Ungless MA, Bonci A (2010) Convergent actions of orexin/hypocretin and CRF on dopamine neurons: emerging players in addiction. Brain Res 1314:139–144

Botta P, Radcliffe RA, Carta M, Mameli M, Daly E, Floyd KL, Deitrich RA, Valenzuela CF (2007) Modulation of GABAA receptors in cerebellar granule neurons by ethanol: a review of genetic and electrophysiological studies. Alcohol 41(3):187–199

Bradley RJ, Peper K, Sterz R (1980) Postsynaptic effects of ethanol at the frog neuromuscular junction. Nature 284(5751):60–62. https://doi.org/10.1038/284060a0

Breese GR, Criswell HE, Carta M, Dodson PD, Hanchar HJ, Khisti RT, Mameli M, Ming Z, Morrow AL, Olsen RW, Otis TS, Parsons LH, Penland SN, Roberto M, Siggins GR, Valenzuela CF, Wallner M (2006) Basis of the gabamimetic profile of ethanol. Alcohol Clin Exp Res 30: 731–744

Broccoli L, Uhrig S, von Jonquieres G, Schönig K, Bartsch D, Justice NJ, Spanagel R, Sommer WH, Klugmann M, Hansson AC (2018) Targeted overexpression of CRH receptor subtype 1 in central amygdala neurons: effect on alcohol-seeking behavior. Psychopharmacology 235(6): 1821–1833

Bruckner MK, Rossner S, Arendt T (1997) Differential changes in the expression of AMPA receptors genes in rat brain after chronic exposure to ethanol: an in situ hybridization study. J Hirnforsch 38:369–376

Buck KJ, Harris RA (1990) Benzodiazepine agonist and inverse agonist actions on GABAA receptor-operated chloride channels. II. Chronic effects of ethanol. J Pharmacol Exp Ther 253:713–719

Budygin EA, Phillips PE, Wightman RM, Jones SR (2001) Terminal effects of ethanol on dopamine dynamics in rat nucleus accumbens: an in vitro voltammetric study. Synapse 42: 77–79

Burgos CF, Castro PA, Mariqueo T, Bunster M, Guzmán L, Aguayo LG (2015) Evidence for α-helices in the large intracellular domain mediating modulation of the α1-glycine receptor by ethanol and Gβγ. J Pharmacol Exp Ther 352:148–155

Burkhardt JM, Adermark L (2014) Locus of onset and subpopulation specificity of in vivo ethanol effect in the reciprocal ventral tegmental area-nucleus accumbens circuit. Neurochem Int 76: 122–130

Caberlotto L, Thorsell A, Rimondini R, Sommer W, Hyytia P, Heilig M (2001) Differential expression of NPY and its receptors in alcohol-preferring AA and alcohol-avoiding ANA rats. Alcohol Clin Exp Res 25:1564–1569

Cadete-Leite A, Andrade JP, Sousa N, Ma W, Ribeiro-da-Silva A (1995) Effects of chronic alcohol consumption on the cholinergic innervation of the rat hippocampal formation as revealed by choline acetyltransferase immunocytochemistry. Neuroscience 64(2):357–374. https://doi.org/10.1016/0306-4522(94)00330-8

Cagetti E, Liang J, Spigelman I, Olsen RW (2003) Withdrawal from chronic intermittent ethanol treatment changes subunit composition, reduces synaptic function, and decreases behavioral responses to positive allosteric modulators of GABAA receptors. Mol Pharmacol 63:53–64

Calo G, Bigoni R, Rizzi A, Guerrini R, Salvadori S, Regoli D (2000) Nociceptin/orphanin FQ receptor ligands. Peptides 21:935–947

Campbell EJ, Marchant NJ (2018) The use of chemogenetics in behavioural neuroscience: receptor variants, targeting approaches and caveats. Br J Pharmacol 175:994–1003

Candura SM, Manzo L, Costa LG (1992) Inhibition of muscarinic receptor- and G-protein-dependent phosphoinositide metabolism in cerebrocortical membranes from neonatal rats by ethanol. Neurotoxicology 13(1):281–288

Cannella N et al (2019) Building better strategies to develop new medications in alcohol use disorder: learning from past success and failure to shape a brighter future. Neurosci Biobehav Rev 103:384–398. https://doi.org/10.1016/j.neubiorev.2019.05.014

Cardoso RA, Brozowski SJ, Chavez-Noriega LE, Harpold M, Valenzuela CF, Harris RA (1999) Effects of ethanol on recombinant human neuronal nicotinic acetylcholine receptors expressed in Xenopus oocytes. J Pharmacol Exp Ther 289(2):774–780

Carmichael FJ, Israel Y (1975) Effects of ethanol on neurotransmitter release by rat brain cortical. J Pharmacol Exp Ther 193(3):824–834

Carpenter-Hyland EP, Woodward JJ, Chandler LJ (2004) Chronic ethanol induces synaptic but not extrasynaptic targeting of NMDA receptors. J Neurosci 24:7859–7868

Carta M, Olivera DS, Dettmer TS, Valenzuela CF (2002) Ethanol withdrawal upregulates kainate receptors in cultured rat hippocampal neurons. Neurosci Lett 327:128–132

Casamenti F, Scali C, Vannucchi MG, Bartolini L, Pepeu G (1993) Long-term ethanol consumption by rats: effect on acetylcholine release in vivo, choline acetyltransferase activity, and behavior. Neuroscience 56(2):465–471

Cebere A, Cebers G, Liljequist S (1999) Enhancement of NMDA-induced functional responses without concomitant NMDA receptor changes following chronic ethanol exposure in cerebellar granule cells. Naunyn Schmiedeberg's Arch Pharmacol 360(6):623–632

Celentano JJ, Gibbs TT, Farb DH (1988) Ethanol potentiates GABA- and glycine-induced chloride currents in chick spinal cord neurons. Brain Res 455(2):377–380

Chandler LJ, Newsom H, Sumners C, Crews F (1993) Chronic ethanol exposure potentiates NMDA excitotoxicity in cerebral cortical neurons. J Neurochem 60(4):1578–1581

Chandler LJ, Sutton G, Norwood D, Sumners C, Crews FT (1997) Chronic ethanol increases N-methyl-D-aspartate-stimulated nitric oxide formation but not receptor density in cultured cortical neurons. Mol Pharmacol 51(5):733–740

Chandler LJ, Harris RA, Crews FT (1998) Ethanol tolerance and synaptic plasticity. Trends Pharmacol Sci 19:491–495

Chandler LJ, Norwood D, Sutton G (1999) Chronic ethanol upregulates NMDA and AMPA, but not kainate receptor subunit proteins in rat primary cortical cultures. Alcohol Clin Exp Res 23: 363–370

Charlton ME, Sweetnam PM, Fitzgerald LW, Terwilliger RZ, Nestler EJ, Duman RS (1997) Chronic ethanol administration regulates the expression of GABAA receptor alpha 1 and alpha 5 subunits in the ventral tegmental area and hippocampus. J Neurochem 68:121–127

Chefer V, Meis J, Wang G, Kuzmin A, Bakalkin G, Shippenberg T (2011) Repeated exposure to moderate doses of ethanol augments hippocampal glutamate neurotransmission by increasing release. Addict Biol 16:229–237. https://doi.org/10.1111/j.1369-1600.2010.00272.x

Cheng Y, Wang J (2019) The use of chemogenetic approaches in alcohol use disorder research and treatment. Alcohol 74:39–45. https://doi.org/10.1016/j.alcohol.2018.05.012

Cheng Y, Huang CCY, Ma T, Wei X, Wang X, Lu J, Wang J (2017) Distinct synaptic strengthening of the striatal direct and indirect pathways drives alcohol consumption. Biol Psychiatry 81:918–929

Choi DS, Wei W, Deitchman JK, Kharazia VN, Lesscher HM, McMahon T, Wang D, Qi ZH, Sieghart W, Zhang C, Shokat KM, Mody I, Messing RO (2008) Protein kinase Cdelta regulates ethanol intoxication and enhancement of GABA-stimulated tonic current. J Neurosci 28(46): 11890–11899

Christian DT, Alexander NJ, Diaz MR, McCool BA (2013) Thalamic glutamatergic afferents into the rat basolateral amygdala exhibit increased presynaptic glutamate function following withdrawal from chronic intermittent ethanol. Neuropharmacology 65:134–142

Chu B, Anantharam V, Treistman SN (1995) Ethanol inhibition of recombinant heteromeric NMDA channels in the presence and absence of modulators. J Neurochem 65(1):140–148
Chu K, Koob GF, Cole M, Zorrilla EP, Roberts AJ (2007) Dependence-induced increases in ethanol self-administration in mice are blocked by the CRF1 receptor antagonist antalarmin and by CRF1 receptor knockout. Pharmacol Biochem Behav 86:813–821
Churn SB, Rana A, Lee K, Parsons JT, De Blas A, Delorenzo RJ (2002) Calcium/calmodulin-dependent kinase II phosphorylation of the GABAA receptor alpha1 subunit modulates benzodiazepine binding. J Neurochem 82:1065–1076
Ciccocioppo R, de Guglielmo G, Hansson AC, Ubaldi M, Kallupi M, Cruz MT, Oleata CS, Heilig M, Roberto M (2014a) Restraint stress alters nociceptin/orphanin FQ and CRF systems in the rat central amygdala: significance for anxiety-like behaviors. J Neurosci 34:363–372
Ciccocioppo R, Stopponi S, Economidou D, Kuriyama M, Kinoshita H, Heilig M, Roberto M, Weiss F, Teshima K (2014b) Chronic treatment with novel brain-penetrating selective NOP receptor agonist MT-7716 reduces alcohol drinking and seeking in the rat. Neuropsychopharmacology 39:2601–2610
Cippitelli A, Damadzic R, Hansson AC, Singley E, Sommer WH, Eskay R, Thorsell A, Heilig M (2010) Neuropeptide Y (NPY) suppresses yohimbine-induced reinstatement of alcohol seeking. Psychopharmacology 208:417–426
Collingridge GL, Olsen RW, Peters J, Spedding M (2009) A nomenclature for ligand-gated ion channels. Neuropharmacology 56(1):2–5
Colmers WF, Klapstein GJ, Fournier A, St-Pierre S, Treherne KA (1991) Presynaptic inhibition by neuropeptide Y in rat hippocampal slice in vitro is mediated by a Y2 receptor. Br J Pharmacol 102:41–44
Connor M, Vaughan CW, Jennings EA, Allen RG, Christie MJ (1999) Nociceptin, Phe(1)psi-nociceptin(1–13), nocistatin and prepronociceptin(154–181) effects on calcium channel currents and a potassium current in rat locus coeruleus in vitro. Br J Pharmacol 128:1779–1787
Costa ET, Soto EE, Cardoso RA, Olivera DS, Valenzuela CF (2000) Acute effects of ethanol on kainate receptors in cultured hippocampal neurons. Alcohol Clin Exp Res 24:220–225
Coune F, Silvestre de Ferron B, González-Marín MC, Antol J, Naassila M, Pierrefiche O (2017) Resistance to ethanol sensitization is associated with a loss of synaptic plasticity in the hippocampus. Synapse 71:2. https://doi.org/10.1002/syn.21899
Cozzoli DK, Goulding SP, Zhang PW, Xiao B, Hu JH, Ary AW, Obara I, Rahn A, Abou-Ziab H, Tyrrel B, Marini C, Yoneyama N, Metten P, Snelling C, Dehoff MH, Crabbe JC, Finn DA, Klugmann M, Worley PF, Szumlinski KK (2009) Binge drinking upregulates accumbens mGluR5-Homer2-PI3K signaling: functional implications for alcoholism. J Neurosci 29(27): 8655–8668
Criswell HE, Breese GR (2005) A conceptualization of integrated actions of ethanol contributing to its GABAmimetic profile: a commentary. Neuropsychopharmacology 30:1407–1425
Criswell HE, Ming Z, Griffith BL, Breese GR (2003) Comparison of effect of ethanol on N-methyl-D-aspartate- and GABA-gated currents from acutely dissociated neurons: absence of regional differences in sensitivity to ethanol. J Pharmacol Exp Ther 304:192–199
Cruz MT, Bajo M, Magnoli EM, Tabakoff B, Siggins GR, Roberto M (2011) Type 7 adenylyl cyclase is involved in the ethanol and CRF sensitivity of GABAergic synapses in mouse central amygdala. Front Neurosci 4:207
Cruz MT, Herman MA, Kallupi M, Roberto M (2012) Nociceptin/orphanin FQ blockade of corticotropin-releasing factor-induced gamma-aminobutyric acid release in central amygdala is enhanced after chronic ethanol exposure. Biol Psychiatry 71:666–676
Cui C, Koob GF (2017) Titrating tipsy targets: the neurobiology of low-dose alcohol. Trends Pharmacol Sci 38:556–568
Cui SZ, Wang SJ, Li J, Xie GQ, Zhou R, Chen L, Yuan XR (2011) Alteration of synaptic plasticity in rat dorsal striatum induced by chronic ethanol intake and withdrawal via ERK pathway. Acta Pharmacol Sin 32:175–181

Cui C, Noronha A, Morikawa H, Alvarez VA, Stuber GD, Szumlinski KK, Kash TL, Roberto M, Wilcox MV (2013 Apr) New insights on neurobiological mechanisms underlying alcohol addiction. Neuropharmacology 67:223–232. https://doi.org/10.1016/j.neuropharm.2012.09.022

Cui C, Noronha A, Warren KR, Koob GF, Sinha R, Thakkar M, Matochik J, Crews FT, Chandler LJ, Pfefferbaum A, Becker HC, Lovinger D, Everitt BJ, Egli M, Mandyam CD, Fein G, Potenza MN, Harris RA, Grant KA, Roberto M, Meyerhoff DJ, Sullivan EV (2015) Brain pathways to recovery from alcohol dependence. Alcohol 49(5):435–452. https://doi.org/10.1016/j.alcohol.2015.04.006

Cullinan WE, Herman JP, Watson SJ (1993) Ventral subicular interaction with the hypothalamic paraventricular nucleus: evidence for a relay in the bed nucleus of the stria terminalis. J Comp Neurol 332:1–20

Cuzon Carlson VC, Seabold GK, Helms CM, Garg N, Odagiri M, Rau AR, Daunais J, Alvarez VA, Lovinger DM, Grant KA (2011) Synaptic and morphological neuroadaptations in the putamen associated with long-term, relapsing alcohol drinking in primates. Neuropsychopharmacology 36(12):2513–2528. https://doi.org/10.1038/npp.2011.140

Cuzon Carlson VC, Grant KA, Lovinger DM (2018) Synaptic adaptations to chronic ethanol intake in male rhesus monkey dorsal striatum depend on age of drinking onset. Neuropharmacology 131:128–142. https://doi.org/10.1016/j.neuropharm.2017.12.010

Dahchour A, De Witte P (1999) Effect of repeated ethanol withdrawal on glutamate microdialysate in the hippocampus. Alcohol Clin Exp Res 23:1698–1703

Dahchour A, De Witte P (2003) Excitatory and inhibitory amino acid changes during repeated episodes of ethanol withdrawal: an in vivo microdialysis study. Eur J Pharmacol 459:171–178

Darstein M, Löschmann PA, Knörle R, Feuerstein TJ (1997) Strychnine-sensitive glycine receptors inducing [3H]-acetylcholine release in rat caudatoputamen: a new site of action of ethanol? Naunyn Schmiedeberg's Arch Pharmacol 356(6):738–745. https://doi.org/10.1007/pl00005112

Das J (2020) SNARE complex-associated proteins and alcohol. Alcohol Clin Exp Res 44(1):7–18. https://doi.org/10.1111/acer.14238

Das J, Xu S, Pany S, Guillory A, Shah V, Roman GW (2013) The pre-synaptic Munc13–1 binds alcohol and modulates alcohol self-administration in Drosophila. J Neurochem 126:715–726

Davies DL, Machu TK, Guo Y, Alkana RL (2002) Ethanol sensitivity in ATP-gated P2X receptors is subunit dependent. Alcohol Clin Exp Res 26(6):773–778

Davies DL, Trudell JR, Mihic SJ, Crawford DK, Alkana RL (2003) Ethanol potentiation of glycine receptors expressed in Xenopus oocytes antagonized by increased atmospheric pressure. Alcohol Clin Exp Res 27(5):743–755

Davies DL, Kochegarov AA, Kuo ST, Kulkarni AA, Woodward JJ, King BF, Alkana RL (2005) Ethanol differentially affects ATP-gated P2X(3) and P2X(4) receptor subtypes expressed in Xenopus oocytes. Neuropharmacology 49(2):243–253

Davis TJ, de Fiebre CM (2006) Alcohol's actions on neuronal nicotinic acetylcholine receptors. Alcohol Res Health 29(3):179–185

Daws LC, Montañez S, Munn JL, Owens WA, Baganz NL, Boyce-Rustay JM, Millstein RA, Wiedholz LM, Murphy DL, Holmes A (2006) Ethanol inhibits clearance of brain serotonin by a serotonin transporter-independent mechanism. J Neurosci 26(24):6431–6438

de Guglielmo G, Kallupi M, Pomrenze MB, Crawford E, Simpson S, Schweitzer P, Koob GF, Messing RO, George O (2019) Inactivation of a CRF-dependent amygdalofugal pathway reverses addiction-like behaviors in alcohol-dependent rats. Nat Commun 10(1):1238. https://doi.org/10.1038/s41467-019-09183-0

De Santis S, Cosa-Linan A, Garcia-Hernandez R, Dmytrenko L, Vargova L, Vorisek I, Stopponi S, Bach P, Kirsch P, Kiefer F, Ciccocioppo R, Sykova E, Moratal D, Sommer WH, Canals S (2020) Chronic alcohol consumption alters extracellular space geometry and transmitter diffusion in the brain. Sci Adv 6(26):eaba0154. https://doi.org/10.1126/sciadv.aba0154

De Souza EB, Perrin MH, Insel TR, Rivier J, Vale WW, Kuhar MJ (1984) Corticotropin-releasing factor receptors in rat forebrain: autoradiographic identification. Science 224:1449–1451

Deehan GA Jr, Knight CP, Waeiss RA, Engleman EA, Toalston JE, McBride WJ, Hauser SR, Rodd ZA (2016) Peripheral administration of ethanol results in a correlated increase in dopamine and serotonin within the posterior ventral tegmental area. Alcohol Alcohol 51:535–540

den Hartog CR, Beckley JT, Smothers TC, Lench DH, Holseberg ZL, Fedarovich H, Gilstrap MJ, Homanics GE, Woodward JJ (2013) Alterations in ethanol-induced behaviors and consumption in knock-in mice expressing ethanol-resistant NMDA receptors. PLoS One 8:e80541

den Hartog CR, Gilstrap M, Eaton B, Lench DH, Mulholland PJ, Homanics GE, Woodward JJ (2017) Effects of repeated ethanol exposures on NMDA receptor expression and locomotor sensitization in mice expressing ethanol resistant NMDA receptors. Front Neurosci 11:84

DePoy L, Daut R, Brigman JL, MacPherson K, Crowley N, Gunduz-Cinar O, Pickens CL, Cinar R, Saksida LM, Kunos G, Lovinger DM, Bussey TJ, Camp MC, Holmes A (2013) Chronic alcohol produces neuroadaptations to prime dorsal striatal learning. Proc Natl Acad Sci U S A 110: 14783–14788

DePoy L, Daut R, Wright T, Camp M, Crowley N, Noronha B, Lovinger D, Holmes A (2015) Chronic alcohol alters rewarded behaviors and striatal plasticity. Addict Biol 20(2):345–348. https://doi.org/10.1111/adb.12131

Devaud LL, Smith FD, Grayson DR, Morrow AL (1995) Chronic ethanol consumption differentially alters the expression of gamma-aminobutyric acidA receptor subunit mRNAs in rat cerebral cortex: competitive, quantitative reverse transcriptase-polymerase chain reaction analysis. Mol Pharmacol 48:861–868

Devaud LL, Fritschy JM, Sieghart W, Morrow AL (1997) Bidirectional alterations of GABA (A) receptor subunit peptide levels in rat cortex during chronic ethanol consumption and withdrawal. J Neurochem 69:126–130

Diamond I, Gordon AS (1994) The role of adenosine in mediating cellular and molecular responses to ethanol. EXS 71:175–183

Diaz MR, Valenzuela CF (2016) Sensitivity of GABAergic tonic currents to acute ethanol in cerebellar granule neurons is not age- or δ subunit- dependent in developing rats. Alcohol Clin Exp Res 40:83–92

Dildy JE, Leslie SW (1989) Ethanol inhibits NMDA-induced increases in intracellular Ca2+ in dissociated brain cells. Brain Res 499:383–387

Dildy-Mayfield JE, Harris RA (1992) Comparison of ethanol sensitivity of rat brain kainate, DL-alpha-amino-3-hydroxy-5-methyl-4-isoxalone proprionic acid and N-methyl-D-aspartate receptors expressed in Xenopus oocytes. J Pharmacol Exp Ther 262:487–494

Ding ZM, Ingraham CM, Rodd ZA, McBride WJ (2016) Alcohol drinking increases the dopamine-stimulating effects of ethanol and reduces D2 auto-receptor and group II metabotropic glutamate receptor function within the posterior ventral tegmental area of alcohol preferring (P) rats. Neuropharmacology 109:41–48

Dohrman DP, Reiter CK (2003) Ethanol modulates nicotine-induced upregulation of nAChRs. Brain Res 975:90–98

Dolphin AC (2003) G protein modulation of voltage-gated calcium channels. Pharmacol Rev 55 (4):607–627

Domi E, Xu L, Toivainen S, Nordeman A, Gobbo F, Venniro M, Shaham Y, Messing RO, Visser E, van den Oever MC, Holm L, Barbier E, Augier E, Heilig M (2021) A neural substrate of compulsive alcohol use. Sci Adv 7(34):eabg9045

Doyon WM, York JL, Diaz LM, Samson HH, Czachowski CL, Gonzales RA (2003) Dopamine activity in the nucleus accumbens during consummatory phases of oral ethanol self-administration. Alcohol Clin Exp Res 27(10):1573–1582. https://doi.org/10.1097/01.ALC.0000089959.66222.B8

Drissi I, Deschamps C, Fouquet G, Alary R, Peineau S, Gosset P, Sueur H, Marcq I, Debuysscher V, Naassila M, Vilpoux C, Pierrefiche O (2019) Memory and plasticity impairment after binge drinking in adolescent rat hippocampus: GluN2A/GluN2B NMDA receptor subunits imbalance through HDAC2. Addict Biol 25(3):e12760. https://doi.org/10.1111/adb.12760

Dudek SM, Bear MF (1992) Homosynaptic long-term depression in area CA1 of hippocampus and effects of N-methyl-D-aspartate receptor blockade. Proc Natl Acad Sci U S A 89(10): 4363–4367

Durand D, Carlen PL (1984) Impairment of long-term potentiation in rat hippocampus following chronic ethanol treatment. Brain Res 308(2):325–332

Ebel A, Vigran R, Mack G, Durkin T, Mandel P (1979) Cholinergic involvement in ethanol intoxication and withdrawalinduced seizure susceptibility. Psychopharmacology (Berl). 61 (3):251–254. https://doi.org/10.1007/BF00432267

Eckardt MJ, File SE, Gessa GL, Grant KA, Guerri C, Hoffman PL, Kalant H, Koob GF, Li TK, Tabakoff B (1998) Effects of moderate alcohol consumption on the central nervous system. Alcohol Clin Exp Res 22:998–1040

Economidou D, Hansson AC, Weiss F, Terasmaa A, Sommer WH, Cippitelli A, Fedeli A, Martin-Fardon R, Massi M, Ciccocioppo R, Heilig M (2008) Dysregulation of nociceptin/orphanin FQ activity in the amygdala is linked to excessive alcohol drinking in the rat. Biol Psychiatry 64: 211–218

Eggers ED, Berger AJ (2004) Mechanisms for the modulation of native glycine receptor channels by ethanol. J Neurophysiol 91(6):2685–2695

Ehlers CL, Li TK, Lumeng L, Hwang BH, Somes C, Jimenez P, Mathe AA (1998a) Neuropeptide Y levels in ethanol-naive alcohol-preferring and nonpreferring rats and in Wistar rats after ethanol exposure. Alcohol Clin Exp Res 22:1778–1782

Ehlers CL, Somes C, Cloutier D (1998b) Are some of the effects of ethanol mediated through NPY? Psychopharmacology 139:136–144

Ehlers CL, Somes C, Lumeng L, Li TK (1999) Electrophysiological response to neuropeptide Y (NPY): in alcohol-naive preferring and non-preferring rats. Pharmacol Biochem Behav 63:291–299

Eliava M, Yilmazer-Hanke D, Asan E (2003) Interrelations between monoaminergic afferents and corticotropin-releasing factor-immunoreactive neurons in the rat central amygdaloid nucleus: ultrastructural evidence for dopaminergic control of amygdaloid stress systems. Histochem Cell Biol 120:183–197

Elmslie KS (2003) Neurotransmitter modulation of neuronal calcium channels. J Bioenerg Biomembr 35(6):477–489

Engle SE, McIntosh JM, Drenan RM (2015) Nicotine and ethanol cooperate to enhance ventral tegmental area AMPA receptor function via α6-containing nicotinic receptors. Neuropharmacology 91:13–22

Enoch MA (2008) The role of GABA(A) receptors in the development of alcoholism. Pharmacol Biochem Behav 90(1):95–104

Erb S, Stewart J (1999) A role for the bed nucleus of the stria terminalis, but not the amygdala, in the effects of corticotropin-releasing factor on stress-induced reinstatement of cocaine seeking. J Neurosci 19:RC35

Erickson CK, Graham DT (1988) Alteration of cortical and reticular acetylcholine release by ethanol in vivo. Neurochem Int 12(4):447–452

Ericson M, Sama MA, Yeh HH (1984) Acute ethanol exposure elevates muscarinic tone in the septohippocampal system. Brain Res 295(1):101–112

Ewin SE, Morgan JW, Niere F, McMullen NP, Barth SH, Almonte AG, Raab-Graham KF, Weiner JL (2019) Chronic intermittent ethanol exposure selectively increases synaptic excitability in the ventral domain of the rat hippocampus. Neuroscience 398:144–157. https://doi.org/10.1016/j.neuroscience.2018.11.028

Fadda F, Cocco S, Stancampiano R, Rossetti ZL (1999) Long-term voluntary ethanol consumption affects neither spatial nor passive avoidance learning, nor hippocampal acetylcholine release in alcohol-preferring rats. Behav Brain Res 103(1):71–76

Federici M, Nisticò R, Giustizieri M, Bernardi G, Mercuri NB (2009) Ethanol enhances GABAB-mediated inhibitory postsynaptic transmission on rat midbrain dopaminergic neurons by facilitating GIRK currents. Eur J Neurosci 29(7):1369–1377

Fekete EM, Zorrilla EP (2007) Physiology, pharmacology, and therapeutic relevance of urocortins in mammals: ancient CRF paralogs. Front Neuroendocrinol 28:1–27

Fetterly TL, Basu A, Nabit BP, Awad E, Williford KM, Centanni SW, Matthews RT, Silberman Y, Winder DG (2019) alpha2A-adrenergic receptor activation decreases parabrachial nucleus excitatory drive onto BNST CRF neurons and reduces their activity in vivo. J Neurosci 39: 472–484. https://doi.org/10.1523/JNEUROSCI.1035-18.2018

Fink K, Göthert M (1996) Both ethanol and ifenprodil inhibit NMDA-evoked release of various neurotransmitters at different, yet proportional potency: potential relation to NMDA receptor subunit composition. Naunyn Schmiedeberg's Arch Pharmacol 354(3):312–319

Fiorillo CD, Williams JT (1998) Glutamate mediates an inhibitory postsynaptic potential in dopamine neurons. Nature 394:78–82

Fleming RL, Manis PB, Morrow AL (2009) The effects of acute and chronic ethanol exposure on presynaptic and postsynaptic gamma-aminobutyric acid (GABA) neurotransmission in cultured cortical and hippocampal neurons. Alcohol 43:603–618

Floyd DW, Jung KY, McCool BA (2003) Chronic ethanol ingestion facilitates N-methyl-D-aspartate receptor function and expression in rat lateral/basolateral amygdala neurons. J Pharmacol Exp Ther 307:1020–1029

Floyd DW, Friedman DP, Daunais JB, Pierre PJ, Grant KA, McCool BA (2004) Long-term ethanol self-administration by cynomolgus macaques alters the pharmacology and expression of GABAA receptors in basolateral amygdala. J Pharmacol Exp Ther 311:1071–1079

Follesa P, Ticku MK (1995) Chronic ethanol treatment differentially regulates NMDA receptor subunit mRNA expression in rat brain. Brain Res Mol Brain Res 29:99–106

Förstera B, Muñoz B, Lobo MK, Chandra R, Lovinger DM, Aguayo LG (2017) Presence of ethanol-sensitive glycine receptors in medium spiny neurons in the mouse nucleus accumbens. J Physiol 595:5285–5300

Francesconi W, Berton F, Koob GF, Sanna PP (2009a) Intrinsic neuronal plasticity in the juxtacapsular nucleus of the bed nuclei of the stria terminalis (jcBNST). Prog Neuro-Psychopharmacol Biol Psychiatry 33:1347–1355

Francesconi W, Berton F, Repunte-Canonigo V, Hagihara K, Thurbon D, Lekic D, Specio SE, Greenwell TN, Chen SA, Rice KC, Richardson HN, O'Dell LE, Zorrilla EP, Morales M, Koob GF, Sanna PP (2009b) Protracted withdrawal from alcohol and drugs of abuse impairs long-term potentiation of intrinsic excitability in the juxtacapsular bed nucleus of the stria terminalis. J Neurosci 29:5389–5401

Freund G, Ballinger WE (1991) Loss of synaptic receptors can precede morphologic changes induced by alcoholism. Alcohol Alcohol Suppl 1:385–391

Freund G, Ballinger WE Jr (1989a) Loss of muscarinic and benzodiazepine neuroreceptors from hippocampus of alcohol abusers. Alcohol 6(1):23–31. https://doi.org/10.1016/0741-8329(89)90069-4

Freund G, Ballinger WE Jr (1989b) Loss of muscarinic cholinergic receptors from the temporal cortex of alcohol abusers. Metab Brain Dis 4(2):121–141

Frye GD, Fincher A (1996) Sensitivity of postsynaptic GABAB receptors on hippocampal CA1 and CA3 pyramidal neurons to ethanol. Brain Res 735:239–248

Frye GD, Fincher A (2000) Sustained ethanol inhibition of native AMPA receptors on medial septum/diagonal band (MS/DB) neurons. Br J Pharmacol 129:87–94

Frye GD, McCown TJ, Breese GR (1983) Differential sensitivity of ethanol withdrawal signs in the rat to gamma-aminobutyric acid (GABA)mimetics: blockade of audiogenic seizures but not forelimb tremors. J Pharmacol Exp Ther 226:720–725

Frye GD, Taylor L, Grover CA, Fincher AS, Griffith WH (1995) Acute ethanol dependence or long-term ethanol treatment and abstinence do not reduce hippocampal responses to carbachol. Alcohol 2(1):29–36

Fujii S, Yamazaki Y, Sugihara T, Wakabayashi I (2008) Acute and chronic ethanol exposure differentially affect induction of hippocampal LTP. Brain Res 1211:13–21

Fukuda K, Shoda T, Morikawa H, Kato S, Mima H, Mori K (1998) Activation of phospholipase A2 by the nociceptin receptor expressed in Chinese hamster ovary cells. J Neurochem 71:2186–2192

Funk CK, Zorrilla EP, Lee MJ, Rice KC, Koob GF (2007) Corticotropin-releasing factor 1 antagonists selectively reduce ethanol self-administration in ethanol-dependent rats. Biol Psychiatry 61:78–86

Gafford GM, Guo JD, Flandreau EI, Hazra R, Rainnie DG, Ressler KJ (2012) Cell-type specific deletion of GABA(A)alpha1 in corticotropin-releasing factor-containing neurons enhances anxiety and disrupts fear extinction. Proc Natl Acad Sci U S A 109:16330–16335. https://doi.org/10.1073/pnas.1119261109

Gage PW, McBurney RN, Schneider GT (1975) Effects of some aliphatic alcohols on the conductance change caused by a quantum of acetylcholine at the toad end-plate. J Physiol 244:409–429

Gallagher JP, Orozco-Cabal LF, Liu J, Shinnick-Gallagher P (2008) Synaptic physiology of central CRH system. Eur J Pharmacol 583:215–225

Gallegos S, San Martin L, Araya A, Lovinger DM, Homanics GE, Aguayo LG (2021) Reduced sedation and increased ethanol consumption in knock-in mice expressing an ethanol insensitive alpha 2 subunit of the glycine receptor. Neuropsychopharmacology 46(3):528–536. https://doi.org/10.1038/s41386-020-0689-9

Gehlert DR, Cippitelli A, Thorsell A, Le AD, Hipskind PA, Hamdouchi C, Lu J, Hembre EJ, Cramer J, Song M, McKinzie D, Morin M, Ciccocioppo R, Heilig M (2007) 3-(4-Chloro-2-morpholin-4-yl-thiazol-5-yl)-8-(1-ethylpropyl)-2,6-dimethyl-imidazo [1,2-b]pyridazine: a novel brain-penetrant, orally available corticotropin-releasing factor receptor 1 antagonist with efficacy in animal models of alcoholism. J Neurosci 27:2718–2726

Georges F, Aston-Jones G (2002) Activation of ventral tegmental area cells by the bed nucleus of the stria terminalis: a novel excitatory amino acid input to midbrain dopamine neurons. J Neurosci 22:5173–5187

Gerdeman G, Ronesi J, Lovinger DM (2002) Postsynaptic endocannabinoid release is necessary for long-term depression in the striatum. Nat Neurosci 5(5):446–451

Ghosh A, Wooden J, Leasure J, Das J (2017) Ethanol upregulates active zone protein Munc13–1: a possible implication in presynaptic physiology and alcoholism. Alcohol Clin Ex Res 41S:24A–24A

Gilpin NW, Roberto M (2012) Neuropeptide modulation of central amygdala neuroplasticity is a key mediator of alcohol dependence. Neurosci Biobehav Rev 36:873–888

Gilpin NW, Stewart RB, Murphy JM, Li TK, Badia-Elder NE (2003) Neuropeptide Y reduces oral ethanol intake in alcohol-preferring (P) rats following a period of imposed ethanol abstinence. Alcohol Clin Exp Res 27:787–794

Gilpin NW, Misra K, Koob GF (2008a) Neuropeptide Y in the central nucleus of the amygdala suppresses dependence-induced increases in alcohol drinking. Pharmacol Biochem Behav 90:475–480

Gilpin NW, Stewart RB, Badia-Elder NE (2008b) Neuropeptide Y administration into the amygdala suppresses ethanol drinking in alcohol-preferring (P) rats following multiple deprivations. Pharmacol Biochem Behav 90:470–474

Gilpin NW, Misra K, Herman MA, Cruz MT, Koob GF, Roberto M (2011) Neuropeptide Y opposes alcohol effects on gamma-aminobutyric acid release in amygdala and blocks the transition to alcohol dependence. Biol Psychiatry 69(11):1091–1099. https://doi.org/10.1016/j.biopsych.2011.02.004

Gilpin NW, Herman MA, Roberto M (2015) The central amygdala as an integrative hub for anxiety and alcohol use disorders. Biol Psychiatry 77(10):859–869. https://doi.org/10.1016/j.biopsych.2014.09.008

Gioia DA, Alexander N, McCool BA (2017) Ethanol mediated inhibition of synaptic vesicle recycling at amygdala glutamate synapses is dependent upon Munc13–2. Front Neurosci 11:424

Givens B, McMahon K (1995) Ethanol suppresses the induction of long-term potentiation in vivo. Brain Res 688(1–2):27–33

Glaaser IW, Slesinger PA (2017) Dual activation of neuronal G proteingated inwardly rectifying potassium (GIRK) channels by cholesterol and alcohol. Sci Rep 7:4592

Glykys J, Peng Z, Chandra D, Homanics GE, Houser CR, Mody I (2007) A new naturally occurring GABA(A) receptor subunit partnership with high sensitivity to ethanol. Nat Neurosci 10(1): 40–48. https://doi.org/10.1038/nn1813

Goldberg JA, Wilson CN (2017) The cholinergic interneuron of the striatum. In: Steiner H, Tseng KY (eds) Handbook of basal ganglia structure and function, 2nd edn. Elsevier/Academic Press, Amsterdam, pp 137–155

Gonzales RA, Job MO, Doyon WM (2004) The role of mesolimbic dopamine in the development and maintenance of ethanol reinforcement. Pharmacol Ther 103(2):121–146. https://doi.org/10.1016/j.pharmthera.2004.06.002

Gouaux E (2004) Structure and function of AMPA receptors. J Physiol 554(Pt 2):249–253

Grammatopoulos DK, Randeva HS, Levine MA, Kanellopoulou KA, Hillhouse EW (2001) Rat cerebral cortex corticotropin-releasing hormone receptors: evidence for receptor coupling to multiple G-proteins. J Neurochem 76:509–519

Gremel CM, Chancey JH, Atwood BK, Luo G, Neve R, Ramakrishnan C, Deisseroth K, Lovinger DM, Costa RM (2016) Endocannabinoid modulation of orbitostriatal circuits gates habit formation. Neuron 90(6):1312–1324. https://doi.org/10.1016/j.neuron.2016.04.043

Griffin WC III, Haun HL, Hazelbaker CL, Ramachandra VS, Becker HC (2014) Increased extracellular glutamate in the nucleus accumbens promotes excessive ethanol drinking in ethanol dependent mice. Neuropsychopharmacology 39:707–717

Grobin AC, Matthews DB, Devaud LL, Morrow AL (1998) The role of GABA(A) receptors in the acute and chronic effects of ethanol. Psychopharmacology 139:2–19

Grobin AC, Fritschy JM, Morrow AL (2000) Chronic ethanol administration alters immunoreactivity for GABA(A) receptor subunits in rat cortex in a region-specific manner. Alcohol Clin Exp Res 24:1137–1144

Grover CA, Frye GD, Griffith WH (1994) Acute tolerance to ethanol inhibition of NMDA-mediated EPSPs in the CA1 region of the rat hippocampus. Brain Res 642(1–2):70–76

Grover CA, Wallace KA, Lindberg SA, Frye GD (1998) Ethanol inhibition of NMDA currents in acutely dissociated medial septum/diagonal band neurons from ethanol dependent rats. Brain Res 782:43–52

Gulya K, Grant KA, Valverius P, Hoffman PL, Tabakoff B (1991) Brain regional specificity and time-course of changes in the NMDA receptor-ionophore complex during ethanol withdrawal. Brain Res 547:129–134

Haass-Koffler CL, Henry AT, Melkus G, Simms JA, Naemmuddin M, Nielsen CK, Lasek AW, Magill M, Schwandt ML, Momenan R et al (2016) Defining the role of corticotropin releasing factor binding protein in alcohol consumption. Transl Psychiatry 6:e953

Hanchar HJ, Dodson PD, Olsen RW, Otis TS, Wallner M (2005) Alcohol-induced motor impairment caused by increased extrasynaptic GABA(A) receptor activity. Nat Neurosci 8(3):339–345

Hansson AC, Cippitelli A, Sommer WH, Fedeli A, Bjork K, Soverchia L, Terasmaa A, Massi M, Heilig M, Ciccocioppo R (2006) Variation at the rat Crhr1 locus and sensitivity to relapse into alcohol seeking induced by environmental stress. Proc Natl Acad Sci U S A 103:15236–15241

Hansson AC, Cippitelli A, Sommer WH, Ciccocioppo R, Heilig M (2007) Region-specific down-regulation of Crhr1 gene expression in alcohol-preferring msP rats following ad lib access to alcohol. Addict Biol 12:30–34

Harris RA (1999) Ethanol actions on multiple ion channels: which are important? Alcohol Clin Exp Res 23(10):1563–1570

Harrison LM, Grandy DK (2000) Opiate modulating properties of nociceptin/orphanin FQ. Peptides 21:151–172

Haugbol SR, Ebert B, Ulrichsen J (2005) Upregulation of glutamate receptor subtypes during alcohol withdrawal in rats. Alcohol Alcohol 40:89–95

Hauger RL, Risbrough V, Brauns O, Dautzenberg FM (2006) Corticotropin releasing factor (CRF) receptor signaling in the central nervous system: new molecular targets. CNS Neurol Disord Drug Targets 5:453–479

Hausser MA, Yung WH (1994) Inhibitory synaptic potentials in Guinea-pig substantia nigra dopamine neurones in vitro. J Physiol 479(Pt 3):401–422

Hawes BE, Graziano MP, Lambert DG (2000) Cellular actions of nociceptin: transduction mechanisms. Peptides 21:961–967

Hayes DM, Knapp DJ, Breese GR, Thiele TE (2005) Comparison of basal neuropeptide Y and corticotropin releasing factor levels between the high ethanol drinking C57BL/6J and low ethanol drinking DBA/2J inbred mouse strains. Alcohol Clin Exp Res 29:721–729

Hedges DM, Yorgason JT, Brundage JN, Wadsworth HA, Williams B, Steffensen SC, Roberto M (2020) Corticotropin releasing factor, but not alcohol, modulates norepinephrine release in the rat central nucleus of the amygdala. Neuropharmacology 179:108293. https://doi.org/10.1016/j.neuropharm.2020.108293

Heilig M, Koob GF (2007) A key role for corticotropin-releasing factor in alcohol dependence. Trends Neurosci 30:399–406

Heilig M, McLeod S, Brot M, Heinrichs SC, Menzaghi F, Koob GF, Britton KT (1993) Anxiolytic-like action of neuropeptide Y: mediation by Y1 receptors in amygdala, and dissociation from food intake effects. Neuropsychopharmacology 8:357–363

Heilig M, Koob GF, Ekman R, Britton KT (1994) Corticotropin-releasing factor and neuropeptide Y: role in emotional integration. Trends Neurosci 17:80–85

Heilig M, Egli M, Crabbe JC, Becker HC (2010) Acute withdrawal, protracted abstinence and negative affect in alcoholism: are they linked? Addict Biol 15(2):169–184

Heinrichs SC, Koob GF (2004) Corticotropin-releasing factor in brain: a role in activation, arousal, and affect regulation. J Pharmacol Exp Ther 311:427–440

Hellstrom-Lindahl E, Winblad B, Nordberg A (1993) Muscarinic and nicotinic receptor changes in the cortex and thalamus of brains of chronic alcoholics. Brain Res 620:42–48

Henderson G, McKnight AT (1997) The orphan opioid receptor and its endogenous ligand – nociceptin/orphanin FQ. Trends Pharmacol Sci 18:293–300

Hendrickson LM, Guildford MJ, Tapper AR (2013) Neuronal nicotinic acetylcholine receptors: common molecular substrates of nicotine and alcohol dependence. Front Psych 4:29

Hendricson AW, Miao CL, Lippmann MJ, Morrisett RA (2002) Ifenprodil and ethanol enhance NMDA receptor-dependent long-term depression. J Pharmacol Exp Ther 301(3):938–944

Henn C, Löffelholz K, Klein J (1998) Stimulatory and inhibitory effects of ethanol on hippocampal acetylcholine release. Naunyn Schmiedeberg's Arch Pharmacol 357(6):640–647

Herman MA, Roberto M (2016) Cell-type-specific tonic GABA signaling in the rat central amygdala is selectively altered by acute and chronic ethanol. Addict Biol 21:72–86

Herman MA, Contet C, Justice NJ, Vale W, Roberto M (2013a) Novel subunit-specific tonic GABA currents and differential effects of ethanol in the central amygdala of CRF receptor-1 reporter mice. J Neurosci 33:3284–3298

Herman MA, Kallupi M, Luu G, Oleata CS, Heilig M, Koob GF, Ciccocioppo R, Roberto M (2013b) Enhanced GABAergic transmission in the central nucleus of the amygdala of genetically selected Marchigian Sardinian rats: alcohol and CRF effects. Neuropharmacology 67:337–348

Herman MA, Sidhu H, Stouffer DG, Kreifeldt M, Le D, Cates-Gatto C, Munoz MB, Roberts AJ, Parsons LH, Roberto M et al (2015) GIRK3 gates activation of the mesolimbic dopaminergic pathway by ethanol. Proc Natl Acad Sci U S A 112:7091–7096

Herman MA, Contet C, Roberto M (2016a) A functional switch in tonic GABA currents alters the output of central amygdala corticotropin releasing factor receptor-1 neurons following chronic ethanol exposure. J Neurosci 36:10729–10741

Herman MA, Varodayan FP, Oleata CS, Luu G, Kirson D, Heilig M, Ciccocioppo R, Roberto M (2016b) Glutamatergic transmission in the central nucleus of the amygdala is selectively altered in Marchigian Sardinian alcohol-preferring rats: alcohol and CRF effects. Neuropharmacology 102:21–31

Herring D, Huang R, Singh M, Robinson LC, Dillon GH, Leidenheimer NJ (2003) Constitutive GABAA receptor endocytosis is dynamin-mediated and dependent on a dileucine AP2 adaptin-binding motif within the beta 2 subunit of the receptor. J Biol Chem 278:24046–24052

Hillmer AT, Tudorascu DL, Wooten DW, Lao PJ, Barnhart TE, Ahlers EO, Resch LM, Larson JA, Converse AK, Moore CF, Schneider ML, Christian BT (2014) Changes in the alpha4beta2 nicotinic acetylcholine system during chronic controlled alcohol exposure in nonhuman primates. Drug Alcohol Depend 138:216–221

Hirth N, Meinhardt MW, Noori HR, Salgado H, Torres-Ramirez O, Uhrig S, Broccoli L, Vengeliene V, Roßmanith M, Perreau-Lenz S, Köhr G, Sommer WH, Spanagel R, Hansson AC (2016) Convergent evidence from alcohol-dependent humans and rats for a hyperdopaminergic state in protracted abstinence. Proc Natl Acad Sci U S A 113:3024–3029

Hoffman PL, Moses F, Luthin GR, Tabakoff B (1986) Acute and chronic effects of ethanol on receptor-mediated phosphatidylinositol 4,5-bisphosphate breakdown in mouse brain. Mol Pharmacol 30(1):13–28

Hoffman PL, Rabe CS, Moses F, Tabakoff B (1989) N-methyl-D-aspartate receptors and ethanol: inhibition of calcium flux and cyclic GMP production. J Neurochem 52:1937–1940

Holmes A, Fitzgerald PJ, MacPherson KP, DeBrouse L, Colacicco G, Flynn SM, Masneuf S, Pleil KE, Li C, Marcinkiewcz CA, Kash TL, Gunduz-Cinar O, Camp M (2012) Chronic alcohol remodels prefrontal neurons and disrupts NMDAR-mediated fear extinction encoding. Nat Neurosci 15(10):1359–1361

Honse Y, Ren H, Lipsky RH, Peoples RW (2004) Sites in the fourth membrane-associated domain regulate alcohol sensitivity of the NMDA receptor. Neuropharmacology 46(5):647–654. https://doi.org/10.1016/j.neuropharm.2003.11.006

Howard RJ, Trudell JR, Harris RA (2014) Seeking structural specificity: direct modulation of pentameric ligand-gated ion channels by alcohols and general anesthetics. Pharmacol Rev 66: 396–412

Hu G, Querimit LA, Downing LA, Charness ME (1993) Ethanol differentially increases 72-adrenergic and muscarinic acetylcholine receptor gene expression in NG108-15 cells. J Biol Chem 268:23441–23447

Hu XJ, Follesa P, Ticku MK (1996) Chronic ethanol treatment produces a selective upregulation of the NMDA receptor subunit gene expression in mammalian cultured cortical neurons. Brain Res Mol Brain Res 36(2):211–218

Hu M, Walker DW, Vickroy TW, Peris J (1999) Chronic ethanol exposure increases 3H-GABA release in rat hippocampus by presynaptic muscarinic receptor modulation. Alcohol Clin Exp Res 23:1587–1595

Huang MM, Overstreet DH, Knapp DJ, Angel R, Wills TA, Navarro M, Rivier J, Vale W, Breese GR (2010) Corticotropin-releasing factor (CRF) sensitization of ethanol withdrawal-induced anxiety-like behavior is brain site specific and mediated by CRF-1 receptors: relation to stress-induced sensitization. J Pharmacol Exp Ther 332:298–307. https://doi.org/10.1124/jpet.109.159186

Hunt WA, Dalton TK (1976) Regional brain acetylcholine levels in rats acutely treated with ethanol or rendered ethanol-dependent. Brain Res 109(3):628–631. https://doi.org/10.1016/0006-8993(76)90043-3

Hunt WA, Majchrowicz E (1979) Regional differences in high-affinity choline uptake in brain after acute and chronic treatment with ethanol. Drug Alcohol Depend 4(3–4):245–248. https://doi.org/10.1016/0376-8716(79)90005-x

Hunt WA, Majchrowicz E, Dalton TK (1979) Alterations in high-affinity choline uptake in brain after acute and chronic ethanol treatment. J Pharmacol Exp Ther 210:259–263

Hwang BH, Zhang JK, Ehlers CL, Lumeng L, Li TK (1999) Innate differences of neuropeptide Y (NPY) in hypothalamic nuclei and central nucleus of the amygdala between selectively bred rats with high and low alcohol preference. Alcohol Clin Exp Res 23:1023–1030

Hyman SE, Malenka RC, Nestler EJ (2006) Neural mechanisms of addiction: the role of reward-related learning and memory. Annu Rev Neurosci 29:565–598

Ikeda K, Kobayashi K, Kobayashi T, Ichikawa T, Kumanishi T, Kishida H, Yano R, Manabe T (1997) Functional coupling of the nociceptin/orphanin FQ receptor with the G- protein-activated K+ (GIRK) channel. Brain Res Mol Brain Res 45:117–126

Imperato A, Dazzi L, Carta G, Colombo G, Biggio G (1998) Rapid increase in basal acetylcholine release in the hippocampus of freely moving rats induced by withdrawal from long-term ethanol intoxication. Brain Res 784(1–2):347–350

Iorio KR, Reinlib L, Tabakoff B, Hoffman PL (1992) Chronic exposure of cerebellar granule cells to ethanol results in increased N-methyl-D-aspartate receptor function. Mol Pharmacol 41(6): 1142–1148

Iorio KR, Tabakoff B, Hoffman PL (1993) Glutamate-induced neurotoxicity is increased in cerebellar granule cells exposed chronically to ethanol. Eur J Pharmacol 248(2):209–212

Irimia C, Buczynski MW, Natividad LA, Laredo SA, Avalos N, Parsons LH (2017) Dysregulated glycine signaling contributes to increased impulsivity during protracted alcohol abstinence. J Neurosci 37:1853–1861

Ito M (2001) Cerebellar long-term depression: characterization, signal transduction, and functional roles. Physiol Rev 81(3):1143–1195

Ives JH, Drewery DL, Thompson CL (2002) Differential cell surface expression of GABAA receptor alpha1, alpha6, beta2 and beta3 subunits in cultured mouse cerebellar granule cells influence of cAMP-activated signalling. J Neurochem 80:317–327

Izumi Y, Nagashima K, Murayama K, Zorumski CF (2005) Acute effects of ethanol on hippocampal long-term potentiation and long-term depression are mediated by different mechanisms. Neuroscience 136(2):509–517

Izumi Y, Murayama K, Tokuda K, Krishnan K, Covey DF, Zorumski CF (2007) GABAergic neurosteroids mediate the effects of ethanol on long-term potentiation in rat hippocampal slices. Eur J Neurosci 26:1881–1888

Izumi Y, O'Dell KA, Zorumski CF (2015) Corticosterone enhances the potency of ethanol against hippocampal long-term potentiation via local neurosteroid synthesis. Front Cell Neurosci 9:254

Jeanes ZM, Buske TR, Morrisett RA (2011) *In vivo* chronic intermittent ethanol exposure reverses the polarity of synaptic plasticity in the nucleus accumbens shell. J Pharmacol Exp Ther 336(1): 155–164

Jeanes ZM, Buske TR, Morrisett RA (2014) Cell type-specific synaptic encoding of ethanol exposure in the nucleus accumbens shell. Neuroscience 277:184–195

Ji X, Saha S, Kolpakova J, Guildford M, Tapper AR, Martin GE (2017) Dopamine receptors differentially control binge alcohol drinking-mediated synaptic plasticity of the core nucleus accumbens direct and indirect pathways. J Neurosci 37(22):5463–5474. https://doi.org/10.1523/JNEUROSCI.3845-16

Jia F, Chandra D, Homanics GE, Harrison NL (2008) Ethanol modulates synaptic and extrasynaptic GABAA receptors in the thalamus. J Pharmacol Exp Ther 326(2):475–482

Jimenez VA, Herman MA, Cuzon Carlson VC, Walter NA, Grant KA, Roberto M (2019) Synaptic adaptations in the central amygdala and hypothalamic paraventricular nucleus associated with protracted ethanol abstinence in male rhesus monkeys. Neuropsychopharmacology 44(5): 982–993. https://doi.org/10.1038/s41386-018-0290-7

Jin C, Woodward JJ (2006) Effects of 8 different NR1 splice variants on the ethanol inhibition of recombinant NMDA receptors. Alcohol Clin Exp Res 30(4):673–679

Jin C, Smothers CT, Woodward JJ (2008) Enhanced ethanol inhibition of recombinant N-methyl-D-aspartate receptors by magnesium: role of NR3A subunits. Alcohol Clin Exp Res 32(6): 1059–1066

Jin S, Cao Q, Yang F, Zhu H, Xu S, Chen Q, Wang Z, Lin Y, Cinar R, Pawlosky RJ, Zhang Y, Xiong W, Gao B, Koob GF, Lovinger DM, Zhang L (2021) Brain ethanol metabolism by astrocytic ALDH2 drives the behavioural effects of ethanol intoxication. Nat Metab 3(3): 337–351. https://doi.org/10.1038/s42255-021-00357-z

Jing M, Zhang P, Wang G, Feng J, Mesik L, Zeng J, Jiang H, Wang S, Looby JC, Guagliardo NA, Langma LW, Lu J, Zuo Y, Talmage DA, Role LW, Barrett PQ, Zhang LI, Luo M, Song Y, Zhu JJ, Li Y (2018) A genetically encoded fluorescent acetylcholine indicator for in vitro and in vivo studies. Nat Biotechnol 36(8):726–737. https://doi.org/10.1038/nbt.4184

Jing M, Zhang Y, Wang H, Li Y (2019) G-protein-coupled receptor-based sensors for imaging neurochemicals with high sensitivity and specificity. J Neurochem 151:279–288

Johnson KA, Lovinger DM (2016) Presynaptic G protein-coupled receptors: gatekeepers of addiction? Front Cell Neurosci 10:264. https://doi.org/10.3389/fncel.2016.00264

Johnson KA, Liput DJ, Homanics GE, Lovinger DM (2020) Age-dependent impairment of metabotropic glutamate receptor 2-dependent long-term depression in the mouse striatum by chronic ethanol exposure. Alcohol 82:11–21. https://doi.org/10.1016/j.alcohol.2019.06.003

Jones SR, Mathews TA, Budygin EA (2006) Effect of moderate ethanol dose on dopamine uptake in rat nucleus accumbens in vivo. Synapse 60(3):251–255

Ju G, Han ZS (1989) Coexistence of corticotropin releasing factor and neurotensin within oval nucleus neurons in the bed nuclei of the stria terminalis in the rat. Neurosci Lett 99:246–250

Juarez B, Liu Y, Zhang L, Han MH (2019) Optogenetic investigation of neural mechanisms for alcohol-use disorder. Alcohol 74:29–38. https://doi.org/10.1016/j.alcohol.2018.05.005

Justice NJ, Yuan ZF, Sawchenko PE, Vale W (2008) Type 1 corticotropin-releasing factor receptor expression reported in BAC transgenic mice: implications for reconciling ligand-receptor mismatch in the central corticotropin-releasing factor system. J Comp Neurol. 511(4):479–496. https://doi.org/10.1002/cne.21848

Kalant H, Grose W (1967) Effects of ethanol and pentobarbital on release of acetylcholine from cerebral cortex slices. J Pharmacol Exp Ther 158(3):386–393

Kalivas PW (2009) The glutamate homeostasis hypothesis of addiction. Nat Rev Neurosci 10(8): 561–572

Kallupi M, Varodayan FP, Oleata CS, Correia D, Luu G, Roberto M (2014a) Nociceptin/orphanin FQ decreases glutamate transmission and blocks ethanol-induced effects in the central amygdala of naive and ethanol-dependent rats. Neuropsychopharmacology 39:1081–1092

Kallupi M, Oleata CS, Luu G, Teshima K, Ciccocioppo R, Roberto M (2014b) MT-7716, a novel selective nonpeptidergic NOP receptor agonist, effectively blocks ethanol-induced increase in GABAergic transmission in the rat central amygdala. Front Integr Neurosci 8:18. https://doi.org/10.3389/fnint.2014.00018

Kaneyuki T, Morimasa T, Shohmori T (1995) Neurotransmitter interactions in the striatum and hypothalamus of mice after single and repeated ethanol treatment. Acta Med Okayama 49(1): 13–17

Kang M, Spigelman I, Sapp DW, Olsen RW (1996) Persistent reduction of GABA(a) receptor-mediated inhibition in rat hippocampus after chronic intermittent ethanol treatment. Brain Res 709:221–228

Kang J, Jiang L, Goldman SA, Nedergaard M (1998a) Astrocyte-mediated potentiation of inhibitory synaptic transmission. Nat Neurosci 1:683–692

Kang MH, Spigelman I, Olsen RW (1998b) Alteration in the sensitivity of GABA(a) receptors to allosteric modulatory drugs in rat hippocampus after chronic intermittent ethanol treatment. Alcohol Clin Exp Res 22:2165–2173

Kaplan JS, Mohr C, Rossi DJ (2013) Opposite actions of alcohol on tonic GABA(A) receptor currents mediated by nNOS and PKC activity. Nat Neurosci 16:1783–1793

Kaplan JS, Nipper MA, Richardson BD, Jensen J, Helms M, Finn DA, Rossi DJ (2016) Pharmacologically counteracting a phenotypic difference in cerebellar GABAA receptor response to alcohol prevents excessive alcohol consumption in a high alcohol-consuming rodent genotype. J Neurosci 36:9019–9025

Karanikas CA, Lu Y-L, Richardson HN (2013) Adolescent drinking targets corticotropin-releasing factor peptide-labeled cells in the central amygdala of male and female rats. Neuroscience 249: 98–105. https://doi.org/10.1016/j.neuroscience.2013.04.024

Karkhanis AN, Huggins KN, Rose JH, Jones SR (2016) Switch from excitatory to inhibitory actions of ethanol on dopamine levels after chronic exposure: role of kappa opioid receptors. Neuropharmacology 110(Pt A):190–197

Kash TL, Winder DG (2006) Neuropeptide Y and corticotropin-releasing factor bi-directionally modulate inhibitory synaptic transmission in the bed nucleus of the stria terminalis. Neuropharmacology 51:1013–1022

Kash TL, Nobis WP, Matthews RT, Winder DG (2008) Dopamine enhances fast excitatory synaptic transmission in the extended amygdala by a CRF-R1-dependent process. J Neurosci 28:13856–13865

Kash TL, Baucum AJ II, Conrad KL, Colbran RJ, Winder DG (2009) Alcohol exposure alters NMDAR function in the bed nucleus of the stria terminalis. Neuropsychopharmacology 34(11): 2420–2429

Kash TL, Pleil KE, Marcinkiewcz CA, Lowery-Gionta EG, Crowley N, Mazzone C, Sugam JJ, Hardaway A, ZA ME (2015) Neuropeptide regulation of signaling and behavior in the BNST. Mol Cells 38:1–13. https://doi.org/10.14348/molcells.2015.2261

Kask A, Harro J, von Horsten S, Redrobe JP, Dumont Y, Quirion R (2002) The neurocircuitry and receptor subtypes mediating anxiolytic-like effects of neuropeptide Y. Neurosci Biobehav Rev 26:259–283

Kauer JA, Malenka RC (2007) Synaptic plasticity and addiction. Nat Rev Neurosci 8(11):844–858

Kellenberger S, Malherbe P, Sigel E (1992) Function of the alpha 1 beta 2 gamma 2S gamma-aminobutyric acid type A receptor is modulated by protein kinase C via multiple phosphorylation sites. J Biol Chem 267:25660–25663

Kelm MK, Criswell HE, Breese GR (2007) Calcium release from presynaptic internal stores is required for ethanol to increase spontaneous gamma-aminobutyric acid release onto cerebellum Purkinje neurons. J Pharmacol Exp Ther 323(1):356–364

Kelm MK, Criswell HE, Breese GR (2008) The role of protein kinase A in the ethanol-induced increase in spontaneous GABA release onto cerebellar Purkinje neurons. J Neurophysiol 100(6):3417–3428

Kelm MK, Weinberg RJ, Criswell HE, Breese GR (2010) The PLC/IP 3 R/PKC pathway is required for ethanol-enhanced GABA release. Neuropharmacology 58(7):1179–1186

Ketchesin KD, Stinnett GS, Seasholtz AF (2016) Binge drinking decreases corticotropin-releasing factor-binding protein expression in the medial prefrontal cortex of mice. Alcohol Clin Exp Res 40:1641–1650

Khom S, Steinkellner T, Hnasko TS, Roberto M (2020a) Alcohol dependence potentiates substance P/neurokinin-1 receptor signaling in the rat central nucleus of amygdala. Sci Adv 6(12): eaaz1050. https://doi.org/10.1126/sciadv.aaz1050

Khom S, Wolfe SA, Patel RR, Kirson D, Hedges DM, Varodayan FP, Bajo M, Roberto M (2020b) Alcohol dependence and withdrawal impair serotonergic regulation of GABA transmission in the rat central nucleus of the amygdala. J Neurosci 40(36):6842–6853. https://doi.org/10.1523/JNEUROSCI.0733-20.2020

Kirson D, Oleata CS, Parsons LH, Ciccocioppo R, Roberto M (2018) CB1 and ethanol effects on glutamatergic transmission in the central amygdala of male and female msP and Wistar rats. Addict Biol 23(2):676–688. https://doi.org/10.1111/adb.12525

Kirson D, Oleata CS, Roberto M (2020) Taurine suppression of central amygdala GABAergic inhibitory signaling via glycine receptors is disrupted in alcohol dependence. Alcohol Clin Exp Res 44(2):445–454. https://doi.org/10.1111/acer.14252

Kirson D, Khom S, Rodriguez L, Wolfe SA, Varodayan FP, Gandhi PJ, Patel RR, Vlkolinsky R, Bajo M, Roberto M (2021) Sex differences in acute alcohol sensitivity of naïve and alcohol dependent central amygdala GABA synapses. Alcohol Alcohol:agab034. https://doi.org/10.1093/alcalc/agab034

Kittler JT, Chen G, Kukhtina V, Vahedi-Faridi A, Gu Z, Tretter V, Smith KR, McAinsh K, Arancibia-Carcamo IL, Saenger W, Haucke V, Yan Z, Moss SJ (2008) Regulation of synaptic inhibition by phospho-dependent binding of the AP2 complex to a YECL motif in the GABAA receptor gamma2 subunit. Proc Natl Acad Sci U S A 105:3616–3621

Kliethermes CL (2005) Anxiety-like behaviors following chronic ethanol exposure. Neurosci Biobehav Rev 28:837–850

Kobayashi T, Ikeda K, Kojima H, Niki H, Yano R, Yoshioka T, Kumanishi T (1999) Ethanol opens G-protein activated inwardly rectifying K^+ channels. Nat Neurosci 2:1091–1097

Kokka N, Sapp DW, Taylor AM, Olsen RW (1993) The kindling model of alcohol dependence: similar persistent reduction in seizure threshold to pentylenetetrazol in animals receiving chronic ethanol or chronic pentylenetetrazol. Alcohol Clin Exp Res 17:525–531

Koob GF (2008) A role for brain stress systems in addiction. Neuron 59:11–34

Koob GF, Le Moal M (2008) Addiction and the brain antireward system. Annu Rev Psychol 59:29–53

Koob GF, Volkow ND (2016) Neurobiology of addiction: a neurocircuitry analysis. Lancet Psychiatry 3:760–777

Kovacs KA, Kavanagh TJ, Costa LG (1995) Ethanol inhibits muscarinic receptor-stimulated phosphoinositide metabolism and calcium mobilization in rat primary cortical cultures. Neurochem Res 20(8):939–949

Krishek BJ, Xie X, Blackstone C, Huganir RL, Moss SJ, Smart TG (1994) Regulation of GABAA receptor function by protein kinase C phosphorylation. Neuron 12:1081–1095

Kroener S, Mulholland PJ, New NN, Gass JT, Becker HC, Chandler LJ (2012) Chronic alcohol exposure alters behavioral and synaptic plasticity of the rodent prefrontal cortex. PLoS One 7: e37541

Kufahl PR, Martin-Fardon R, Weiss F (2011) Enhanced sensitivity to attenuation of conditioned reinstatement by the mGluR 2/3 agonist LY379268 and increased functional activity of mGluR 2/3 in rats with a history of ethanol dependence. Neuropsychopharmacology 36:2762–2773

Kumar S, Sieghart W, Morrow AL (2002) Association of protein kinase C with GABA(A) receptors containing alpha1 and alpha4 subunits in the cerebral cortex: selective effects of chronic ethanol consumption. J Neurochem 82:110–117

Kumar S, Kralic JE, O'Buckley TK, Grobin AC, Morrow AL (2003) Chronic ethanol consumption enhances internalization of alpha1 subunit-containing GABAA receptors in cerebral cortex. J Neurochem 86:700–708

Kumar S, Fleming RL, Morrow AL (2004) Ethanol regulation of gamma-aminobutyric acid A receptors: genomic and nongenomic mechanisms. Pharmacol Ther 101:211–226

Kumar S, Porcu P, Werner DF, Matthews DB, Diaz-Granados JL, Helfand RS, Morrow AL (2009) The role of GABA(A) receptors in the acute and chronic effects of ethanol: a decade of progress. Psychopharmacology 205:529–564

Kuner T, Schoepfer R, Korpi ER (1993) Ethanol inhibits glutamate-induced currents in heteromeric NMDA receptor subtypes. Neuroreport 5(3):297–300. https://doi.org/10.1097/00001756-199312000-00029

Labouesse MA, Patriarchi P (2021) A versatile GPCR toolkit to track in vivo neuromodulation: not a one-size-fits-all sensor. Neuropsychopharmacology 46:2043–2047

Lachowicz JE, Shen Y, Monsma FJ Jr, Sibley DR (1995) Molecular cloning of a novel G protein-coupled receptor related to the opiate receptor family. J Neurochem 64:34–40

Läck AK, Floyd DW, McCool BA (2005) Chronic ethanol ingestion modulates proanxiety factors expressed in rat central amygdala. Alcohol 36(2):83–90

Lack AK, Diaz MR, Chappell A, DuBois DW, McCool BA (2007) Chronic ethanol and withdrawal differentially modulate pre- and postsynaptic function at glutamatergic synapses in rat basolateral amygdala. J Neurophysiol 98:3185–3196

Lack AK, Ariwodola OJ, Chappell AM, Weiner JL, McCool BA (2008) Ethanol inhibition of kainate receptor-mediated excitatory neurotransmission in the rat basolateral nucleus of the amygdala. Neuropharmacology 55:661–668

Lack AK, Christian DT, Diaz MR, McCool BA (2009) Chronic ethanol and withdrawal effects on kainate receptor-mediated excitatory neurotransmission in the rat basolateral amygdala. Alcohol 43:25–33

Larsson C, Simonsson P, Hoek JB, Alling C (1995) Ethanol inhibits the peak of muscarinic receptor-stimulated formation of inositol 1,4,5-trisphosphate in neuroblastoma SH-SY5Y cells. Biochem Pharmacol 50(5):647–654

Larsson C, Simonsson P, Caron M, Alling C (1996) Long-term exposure to ethanol increases the number and function of muscarinic M1 receptors in human neuroblastoma cells. J Pharmacol Exp Ther 278(1):313–319

Larsson KP, Olsen UB, Hansen AJ (2000) Nociceptin is a potent inhibitor of N-type Ca(2+) channels in rat sympathetic ganglion neurons. Neurosci Lett 296:121–124

Larsson A, Edstrom L, Svensson L, Soderpalm B, Engel JA (2005) Voluntary ethanol intake increases extracellular acetylcholine levels in the ventral tegmental area in the rat. Alcohol Alcohol 40(5):349–358. https://doi.org/10.1093/alcalc/agh180

Lee Y, Davis M (1997) Role of the hippocampus, the bed nucleus of the stria terminalis, and the amygdala in the excitatory effect of corticotropin-releasing hormone on the acoustic startle reflex. J Neurosci 17:6434–6446

Leenders AG, Sheng ZH (2005) Modulation of neurotransmitter release by the second messenger-activated protein kinases: implications for presynaptic plasticity. Pharmacol Ther 105(1):69–84

Leidenheimer NJ, McQuilkin SJ, Hahner LD, Whiting P, Harris RA (1992) Activation of protein kinase C selectively inhibits the gamma-aminobutyric acidA receptor: role of desensitization. Mol Pharmacol 41:1116–1123

LeMarquand D, Pihl RO, Benkelfat C (1994) Serotonin and alcohol intake, abuse, and dependence: findings of animal studies. Biol Psychiatry 36:395–421

Lewohl JM, Wilson WR, Mayfield RD, Brozowski SJ, Morrisett RA, Harris RA (1999) G protein-coupled inwardly rectifying potassium channels are targets of alcohol action. Nat Neurosci 2: 1084–1090

Li C, Aguayo L, Peoples RW, Weight FF (1993) Ethanol inhibits a neuronal ATP-gated ion channel. Mol Pharmacol 44(4):871–875

Liang J, Cagetti E, Olsen RW, Spigelman I (2004) Altered pharmacology of synaptic and extrasynaptic GABAA receptors on CA1 hippocampal neurons is consistent with subunit changes in a model of alcohol withdrawal and dependence. J Pharmacol Exp Ther 310(3): 1234–1245

Liang J, Suryanarayanan A, Abriam A, Snyder B, Olsen RW, Spigelman I (2007) Mechanisms of reversible GABAA receptor plasticity after ethanol intoxication. J Neurosci 27:12367–12377

Liang J, Spigelman I, Olsen RW (2009) Tolerance to sedative/hypnotic actions of GABAergic drugs correlates with tolerance to potentiation of extrasynaptic tonic currents of alcohol-dependent rats. J Neurophysiol 102:224–233

Liang R, Broussard GJ, Tian L (2015) Imaging chemical neurotransmission with genetically encoded fluorescent sensors. ACS Chem Neurosci 6:84–93

Lima-Landman MTR, Albuquerque EX (1989) Ethanol potentiates and blocks NMDA-activated single-channel currents in rat hippocampal pyramidal cells. FEBS Lett 247:61–67

Liu J, Yu B, Neugebauer V, Grigoriadis DE, Rivier J, Vale WW, Shinnick-Gallagher P, Gallagher JP (2004) Corticotropin-releasing factor and Urocortin I modulate excitatory glutamatergic synaptic transmission. J Neurosci 24:4020–4029

Lobo IA, Harris RA (2008) GABA(A) receptors and alcohol. Pharmacol Biochem Behav 90(1): 90–94

Logrip ML, Oleata C, Roberto M (2017) Sex differences in responses of the basolateral-central amygdala circuit to alcohol, corticosterone and their interaction. Neuropharmacology 114:123–134. https://doi.org/10.1016/j.neuropharm.2016.11.021

Lou LG, Zhang Z, Ma L, Pei G (1998) Nociceptin/orphanin FQ activates mitogen-activated protein kinase in Chinese hamster ovary cells expressing opioid receptor-like receptor. J Neurochem 70: 1316–1322

Lovinger DM (1991) Ethanol potentiates 5-HT$_3$ receptor-mediated ion current in NCB-20 neuroblastoma cells. Neurosci Lett 122:54–56

Lovinger DM (1993) High ethanol sensitivity of recombinant AMPA-type glutamate receptors expressed in mammalian cells. Neurosci Lett 159:83–87

Lovinger DM (1995) Developmental decrease in ethanol inhibition of N-methyl-D-aspartate receptors in rat neocortical neurons: relation to the actions of ifenprodil. J Pharmacol Exp Ther 274:164–172

Lovinger DM (1997) Alcohols and neurotransmitter gated ion channels: past, present and future. Naunyn Schmiedeberg's Arch Pharmacol 356:267–282

Lovinger DM, Homanics GE (2007) Tonic for what ails us? High affinity GABAA receptors and alcohol. Alcohol 41(3):139–143

Lovinger DM, Kash TL (2015) Mechanisms of neuroplasticity and ethanol's effects on plasticity in the striatum and bed nucleus of the stria terminalis. Alcohol Res 37:109–124

Lovinger DM, Roberto M (2013) Synaptic effects induced by alcohol. Curr Top Behav Neurosci 13:31–86

Lovinger DM, White G, Weight FF (1989) Ethanol inhibits NMDA-activated ion current in hippocampal neurons. Science 243:1721–1724

Lovinger DM, White G, Weight FF (1990) NMDA receptor-mediated synaptic excitation selectively inhibited by ethanol in hippocampal slice from adult rat. J Neurosci 10:1372–1379

Lowery EG, Thiele TE (2010) Pre-clinical evidence that corticotropin-releasing factor (CRF) receptor antagonists are promising targets for pharmacological treatment of alcoholism. CNS Neurol Disord Drug Targets 9:77–86

Lowery-Gionta EG, Navarro M, Li C, Pleil KE, Rinker JA, Cox BR, Sprow GM, Kash TL, Thiele TE (2012) Corticotropin releasing factor signaling in the central amygdala is recruited during binge-like ethanol consumption in C57BL/6J mice. J Neurosci 32:3405–3413. https://doi.org/10.1523/JNEUROSCI.6256-11.2012

Lu SM, Yeh HH (1999) Ethanol modulates AMPA-induced current responses of primary somatosensory cortical neurons. Neurochem Int 35(2):175–183

Lukoyanov NV, Pereira PA, Paula-Barbosa MM, Cadete-Leite A (2003) Nerve growth factor improves spatial learning and restores hippocampal cholinergic fibers in rats withdrawn from chronic treatment with ethanol. Exp Brain Res 148(1):88–94. https://doi.org/10.1007/s00221-002-1290-7

Luo H, Fernandez M, de Velasco E, Wickman K (2022) Neuronal G protein-gated K+ channels. Am J Physiol Cell Physiol 323(2):C439–C460. https://doi.org/10.1152/ajpcell.00102.2022

Lüscher C, Huber KM (2010) Group 1 mGluR-dependent synaptic long-term depression: mechanisms and implications for circuitry and disease. Neuron 65(4):445–459

Luthin GR, Tabakoff B (1984) Activation of adenylate cyclase by alcohols requires the nucleotide-binding protein. J Pharmacol Exp Ther 228(3):579–887

Ma T, Barbee B, Wang X, Wang J (2017) Alcohol induces input-specific aberrant synaptic plasticity in the rat dorsomedial striatum. Neuropharmacology 123:46–54

Ma T, Cheng Y, Roltsch Hellard E, Wang X, Lu J, Gao X, Huang CCY, Wei XY, Ji JY, Wang J (2018) Bidirectional and long-lasting control of alcohol-seeking behavior by corticostriatal LTP and LTD. Nat Neurosci 21(3):373–383. https://doi.org/10.1038/s41593-018-0081-9

Macdonald RL (1995) Ethanol, gamma-aminobutyrate type A receptors, and protein kinase C phosphorylation. Proc Natl Acad Sci U S A 92:3633–3635

Machu TK, Harris RA (1994) Alcohols and anesthetics enhance the function of 5-hydroxytryptamine3 receptors expressed in Xenopus laevis oocytes. J Pharmacol Exp Ther 271(2):898–905

Madamba SG, Hsu M, Schweitzer P, Siggins GR (1995) Ethanol enhances muscarinic cholinergic neurotransmission in rat hippocampus in vitro. Brain Res 685(1–2):21–32

Madamba SG, Schweitzer P, Siggins GR (1999) Nociceptin augments K(+) currents in hippocampal CA1 neurons by both ORL-1 and opiate receptor mechanisms. J Neurophysiol 82:1776–1785

Mahler SV, Vazey EM, Beckley JT, Keistler CR, McGlinchey EM, Kaufling J, Wilson SP, Deisseroth K, Woodward JJ, Aston-Jones G (2014) Designer receptors show role for ventral pallidum input to ventral tegmental area in cocaine seeking. Nat Neurosci 17:577–585

Mahmoudi M, Kang MH, Tillakaratne N, Tobin AJ, Olsen RW (1997) Chronic intermittent ethanol treatment in rats increases GABA(A) receptor alpha4-subunit expression: possible relevance to alcohol dependence. J Neurochem 68:2485–2492

Maiya R, Buck KJ, Harris RA, Mayfield RD (2002) Ethanol-sensitive sites on the human dopamine transporter. J Biol Chem 277(34):30724–30729

Mameli M, Zamudio PA, Carta M, Valenzuela CF (2005) Developmentally regulated actions of alcohol on hippocampal glutamatergic transmission. J Neurosci 25:8027–8036

Mancillas JR, Siggins GR, Bloom FE (1986) Systemic ethanol: selective enhancement of responses to acetylcholine and somatostatin in hippocampus. Science 231(4734):161–163

Marszalec W, Kurata Y, Hamilton BJ, Carter DB, Narahashi T (1994) Selective effects of alcohols on gamma-aminobutyric acid A receptor subunits expressed in human embryonic kidney cells. J Pharmacol Exp Ther 269(1):157–163

Martin-Fardon R, Zorrilla EP, Ciccocioppo R, Weiss F (2010) Role of innate and drug-induced dysregulation of brain stress and arousal systems in addiction: focus on corticotropin-releasing factor, nociceptin/orphanin FQ, and orexin/hypocretin. Brain Res 1314:145–161

Marty VN, Spigelman I (2012) Long-lasting alterations in membrane properties, k(+) currents, and glutamatergic synaptic currents of nucleus accumbens medium spiny neurons in a rat model of alcohol dependence. Front Neurosci 6:86. https://doi.org/10.3389/fnins.2012.00086

Marutha Ravindran CR, Mehta AK, Ticku MK (2007) Effect of chronic administration of ethanol on the regulation of the delta-subunit of GABA(A) receptors in the rat brain. Brain Res 1174: 47–52

Mascia MP, Mihic SJ, Valenzuela CF, Schofield PR, Harris RA (1996) A single amino acid determines differences in ethanol actions on strychnine-sensitive glycine receptors. Mol Pharmacol 50(2):402–406

Masood K, Wu C, Brauneis U, Weight FF (1994) Differential ethanol sensitivity of recombinant N-methyl-D-aspartate receptor subunits. Mol Pharmacol 45:324–329

Mathews TA, John CE, Lapa GB, Budygin EA, Jones SR (2006) No role of the dopamine transporter in acute ethanol effects on striatal dopamine dynamics. Synapse 60(4):288–294

Matthews DB, Devaud LL, Fritschy JM, Sieghart W, Morrow AL (1998) Differential regulation of GABA(A) receptor gene expression by ethanol in the rat hippocampus versus cerebral cortex. J Neurochem 70:1160–1166

Mayfield RD, Maiya R, Keller D, Zahniser NR (2001) Ethanol potentiates the function of the human dopamine transporter expressed in Xenopus oocytes. J Neurochem 79(5):1070–1079

McCall NM, Sprow GM, Delpire E, Thiele TE, Kash TL, Pleil KE (2013) Effects of sex and deletion of neuropeptide Y2 receptors from GABAergic neurons on affective and alcohol drinking behaviors in mice. Front Integr Neurosci 7:100. https://doi.org/10.3389/fnint.2013.00100

McCool BA, Frye GD, Pulido MD, Botting SK (2003) Effects of chronic ethanol consumption on rat GABA(A) and strychnine-sensitive glycine receptors expressed by lateral/basolateral amygdala neurons. Brain Res 963(1–2):165–177

McCown TJ, Frye GD, Breese GR (1985) Evidence for site specific ethanol actions in the CNS. Alcohol Drug Res 6:423–429

McElligott ZA, Klug JR, Nobis WP, Patel S, Grueter BA, Kash TL, Winder DG (2010) Distinct forms of Gq-receptor-dependent plasticity of excitatory transmission in the BNST are differentially affected by stress. Proc Natl Acad Sci USA 107(5):2271–2276

Meinhardt MW, Hansson AC, Perreau-Lenz S, Bauder-Wenz C, Stählin O, Heilig M, Harper C, Drescher KU, Spanagel R, Sommer WH (2013) Rescue of infralimbic mGluR2 deficit restores control over drug-seeking behavior in alcohol dependence. J Neurosci 33:2794–2806

Meinhardt MW, Pfarr S, Fouquet G, Rohleder C, Meinhardt ML, Barroso-Flores J, Hoffmann R, Jeanblanc J, Paul E, Wagner K, Hansson AC, Köhr G, Meier N, von Bohlen Und Halbach O,

Bell RL, Endepols H, Neumaier B, Schönig K, Bartsch D, Naassila M, Spanagel R, Sommer WH (2021) Psilocybin targets a common molecular mechanism for cognitive impairment and increased craving in alcoholism. Sci Adv 7(47):eabh2399

Meinhardt MW, Giannone F, Hirth N, Bartsch D, Spampinato SM, Kelsch W, Spanagel R, Sommer WH, Hansson AC (2022) Disrupted circadian expression of β-arrestin 2 affects reward-related μ-opioid receptor function in alcohol dependence. J Neurochem 160(4):454–468

Meis S (2003) Nociceptin/orphanin FQ: actions within the brain. Neuroscientist 9:158–168

Meis S, Pape HC (1998) Postsynaptic mechanisms underlying responsiveness of amygdaloid neurons to nociceptin/orphanin FQ. J Neurosci 18:8133–8144

Meis S, Pape HC (2001) Control of glutamate and GABA release by nociceptin/orphanin FQ in the rat lateral amygdala. J Physiol 532:701–712

Meis S, Munsch T, Pape HC (2002) Antioscillatory effects of nociceptin/orphanin FQ in synaptic networks of the rat thalamus. J Neurosci 22:718–727

Melendez RI, Hicks MP, Cagle SS, Kalivas PW (2005) Ethanol exposure decreases glutamate uptake in the nucleus accumbens. Alcohol Clin Exp Res 29(3):326–333

Melis F, Stancampiano R, Imperato A, Carta G, Fadda F (1996) Chronic ethanol consumption in rats: correlation between memory performance and hippocampal acetylcholine release in vivo. Neuroscience 74(1):155–159

Melis M, Camarini R, Ungless MA, Bonci A (2002) Long-lasting potentiation of GABAergic synapses in dopamine neurons after a single in vivo ethanol exposure. J Neurosci 22:2074–2082

Meng C, Zhou J, Papaneri A, Peddada T, Xu K, Cui G (2018) Spectrally resolved fiber photometry for multi-component analysis of brain circuits. Neuron 98:707–717.e4. https://doi.org/10.1016/j.neuron.2018.04.012

Merlo Pich E, Lorang M, Yeganeh M, Rodriguez de Fonseca F, Raber J, Koob GF, Weiss F (1995) Increase of extracellular corticotropin-releasing factor-like immunoreactivity levels in the amygdala of awake rats during restraint stress and ethanol withdrawal as measured by microdialysis. J Neurosci 15:5439–5447

Meunier JC, Mollereau C, Toll L, Suaudeau C, Moisand C, Alvinerie P, Butour JL, Guillemot JC, Ferrara P, Monsarrat B, Marzagull H, Vassart G, Parmentier M, Costentin J (1995) Isolation and structure of the endogenous agonist of opioid receptor- like ORL1 receptor. Nature 377:532–535

Mhatre MC, Ticku MK (1992) Chronic ethanol administration alters gamma-aminobutyric acidA receptor gene expression. Mol Pharmacol 42:415–422

Mihic JS, Harris RA (1995) Alcohol actions at the GABAA receptor/choloride complex. In: Deitrich R, Erwin VG (eds) Pharmacological effects of ethanol on the nervous system. CRC Press, Boca Raton, pp 51–72

Mihic SJ, Ye Q, Wick MJ, Koltchine VV, Krasowski MD, Finn SE, Mascia MP, Valenzuela CF, Hanson KK, Greenblatt EP, Harris RA, Harrison NL (1997) Sites of alcohol and volatile anaesthetic action on GABA(a) and glycine receptors. Nature 389(6649):385–389

Miller RJ (1998) Presynaptic receptors. Ann Rev Pharmacool Toxicol 38:201–207

Minami K, Minami M, Harris RA (1997a) Inhibition of 5-hydroxytryptamine type 2A receptor-induced currents by n-alcohols and anesthetics. J Pharmacol Exp Ther 281(3):1136–1143. [PubMed] [Google Scholar]

Minami K, Vanderah TW, Minami M, Harris RA (1997b) Inhibitory effects of anesthetics and ethanol on muscarinic receptors expressed in Xenopus oocytes. Eur J Pharmacol. 339(2–3):237–244

Minami K, Gereau RW, Minami M, Heinemann SF, Harris RA (1998) Effects of ethanol and anesthetics on type 1 and 5 metabotropic glutamate receptors expressed in Xenopus laevis oocytes. Mol Pharmacol. 53(1):148–156

Ming Z, Criswell HE, Yu G, Breese GR (2006) Competing presynaptic and postsynaptic effects of ethanol on cerebellar purkinje neurons. Alcohol Clin Exp Res 30(8):1400–1407

Mio K, Kubo Y, Ogura T, Yamamoto T, Sato C (2005) Visualization of the trimeric P2X$_2$ receptor with a crown-capped extracellular domain. Biochem Biophys Res Commun 337:998–1005

Miyakawa T, Yagi T, Kitazawa H, Yasuda M, Kawai N, Tsuboi K, Niki H (1997) Fyn-kinase as a determinant of ethanol sensitivity: relation to NMDA-receptor function. Science 278(5338): 698–701

Mochida S (2019) Presynaptic calcium channels. Int J Mol Sci 20(9):2217. https://doi.org/10.3390/ijms20092217

Mogil JS, Pasternak GW (2001) The molecular and behavioral pharmacology of the orphanin FQ/nociceptin peptide and receptor family. Pharmacol Rev 53:381–415

Mogil JS, Grisel JE, Zhangs G, Belknap JK, Grandy DK (1996) Functional antagonism of mu-, delta- and kappa-opioid antinociception by orphanin FQ. Neurosci Lett 214:131–134

Mohler H, Fritschy JM, Luscher B, Rudolph U, Benson J, Benke D (1996) The GABAA receptors. From subunits to diverse functions. Ion Channels 4:89–113

Mons N, Decorte L, Jaffard R, Cooper DM (1998a) Ca2+−sensitive adenylyl cyclases, key integrators of cellular signalling. Life Sci 62:1647–1652

Mons N, Yoshimura M, Ikeda H, Hoffman PL, Tabakoff B (1998b) Immunological assessment of the distribution of type VII adenylyl cyclase in brain. Brain Res 788:251–261

Morin SM, Ling N, Liu XJ, Kahl SD, Gehlert DR (1999) Differential distribution of urocortin- and corticotropin-releasing factor-like immunoreactivities in the rat brain. Neuroscience 92:281–291. https://doi.org/10.1016/s0306-4522(98)00732-5

Morrisett RA, Swartzwelder HS (1993) Attenuation of hippocampal long-term potentiation by ethanol: a patch-clamp analysis of glutamatergic and GABAergic mechanisms. J Neurosci 13(5):2264–2272. https://doi.org/10.1523/JNEUROSCI.13-05-02264.1993

Morrisett RA, Swartzwelder HS (1993) Attenuation of hippocampal long-term potentiation by ethanol: a patch-clamp analysis of glutamatergic and GABAergic mechanisms. J Neurosci 13 (5):2264–2272. https://doi.org/10.1523/JNEUROSCI.13-05-02264

Morrow AL, Suzdak PD, Karanian JW, Paul SM (1988) Chronic ethanol administration alters gamma-aminobutyric acid, pentobarbital and ethanol-mediated 36Cl- uptake in cerebral cortical synaptoneurosomes. J Pharmacol Exp Ther 246:158–164

Morrow AL, Herbert JS, Montpied P (1992) Differential effects of chronic ethanol administration on GABA(A) receptor alpha1 and alpha6 subunit mRNA levels in rat cerebellum. Mol Cell Neurosci 3:251–258

Morrow AL, Devaud LL, Bucci D, Smith FD (1994) GABAA and NMDA receptor subunit mRNA expression in ethanol dependent rats. Alcohol Alcohol Suppl 2:89–95

Möykkynen T, Korpi ER, Lovinger DM (2003) Ethanol inhibits alpha-amino-3-hydyroxy-5-methyl-4-isoxazolepropionic acid (AMPA) receptor function in central nervous system neurons by stabilizing desensitization. J Pharmacol Exp Ther 306(2):546–555

Möykkynen TP, Coleman SK, Keinänen K, Lovinger DM, Korpi ER (2009) Ethanol increases desensitization of recombinant GluR-D AMPA receptor and TARP combinations. Alcohol 43 (4):277–284

Mulkeen D, Anwyl R, Rowan MJ (1987) Enhancement of long-term potentiation by the calcium channel agonist Bayer K8644 in CA1 of the rat hippocampus in vitro. Neurosci Lett 80:351–355

Mulkey RM, Malenka RC (1992) Mechanisms underlying induction of homosynaptic long-term depression in area CA1 of the hippocampus. Neuron 9(5):967–975

Muller P, Britton RS, Seeman P (1980) The effects of long-term ethanol on brain receptors for dopamine, acetylcholine, serotonin and noradrenaline. Eur J Pharmacol 65(1):31–37

Muñoz B, Fritz BM, Yin F, Atwood BK (2018a) Alcohol exposure disrupts mu opioid receptor-mediated long-term depression at insular cortex inputs to dorsolateral striatum. Nat Commun 9(1):1318. https://doi.org/10.1038/s41467-018-03683-1

Muñoz B, Yevenes GE, Förstera B, Lovinger DM, Aguayo LG (2018b) Presence of inhibitory glycinergic transmission in medium spiny neurons in the nucleus accumbens. Front Mol Neurosci 11:228. https://doi.org/10.3389/fnmol.2018.00228

Muñoz B, Gallegos S, Peters C, Murath P, Lovinger DM, Homanics GE, Aguayo LG (2020) Influence of nonsynaptic alpha1 glycine receptors on ethanol consumption and place preference. Addict Biol 25(2):e12726. https://doi.org/10.1111/adb.12726

Murphy NP (2010) The nociceptin/orphanin FQ system as a target for treating alcoholism. CNS Neurol Disord Drug Targets 9:87–93

Natividad LA, Steinman MQ, Laredo SA, Irimia C, Polis IY, Lintz R, Buczynski MW, Martin-Fardon R, Roberto M, Parsons LH (2018) Phosphorylation of calcium/calmodulin-dependent protein kinase II in the rat dorsal medial prefrontal cortex is associated with alcohol-induced cognitive inflexibility. Addict Biol 23(5):1117–1129. https://doi.org/10.1111/adb.12568

Nestby P, Vanderschuren LJ, De Vries TJ, Hogenboom F, Wardeh G, Mulder AH, Schoffelmeer AN (1997) Ethanol, like psychostimulants and morphine, causes long-lasting hyperreactivity of dopamine and acetylcholine neurons of rat nucleus accumbens: possible role in behavioural sensitization. Psychopharmacology 133(1):69–76

Netzeband JG, Trotter C, Caguioa JN, Gruol DL (1999) Chronic ethanol exposure enhances AMPA-elicited Ca2+ signals in the somatic and dendritic regions of cerebellar Purkinje neurons. Neurochem Int 35:163–174

Netzeband JG, Schneeloch JR, Trotter C, Caguioa-Aquino JN, Gruol DL (2002) Chronic ethanol treatment and withdrawal alter ACPD-evoked calcium signals in developing Purkinje neurons. Alcohol Clin Exp Res 26(3):386–393

Newton PM, Messing RO (2006) Intracellular signaling pathways that regulate behavioral responses to ethanol. Pharmacol Ther 109:227–237

Nguyen PV, Woo NH (2003) Regulation of hippocampal synaptic plasticity by cyclic AMP-dependent protein kinases. Prog Neurobiol 71(6):401–437

Nie Z, Yuan X, Madamba SG, Siggins GR (1993) Ethanol decreases glutamatergic synaptic transmission in rat nucleus accumbens in vitro: naloxone reversal. J Pharmacol Exp Ther 266: 1705–1712

Nie Z, Madamba SG, Siggins GR (1994) Ethanol inhibits glutamatergic neurotransmission in nucleus accumbens neurons by multiple mechanisms. J Pharmacol Exp Ther 271:1566–1573

Nie Z, Madamba SG, Siggins GR (2000) Ethanol enhances gamma-aminobutyric acid responses in a subpopulation of nucleus accumbens neurons: role of metabotropic glutamate receptors. J Pharmacol Exp Ther 293:654–661

Nie Z, Schweitzer P, Roberts AJ, Madamba SG, Moore SD, Siggins GR (2004) Ethanol augments GABAergic transmission in the central amygdala via CRF1 receptors. Science 303:1512–1514

Nie Z, Zorrilla EP, Madamba SG, Rice KC, Roberto M, Siggins GR (2009) Presynaptic CRF1 receptors mediate the ethanol enhancement of GABAergic transmission in the mouse central amygdala. ScientificWorldJournal 9:68–85

Nieber K, Poelchen W, Sieler D, Illes P (1998) Inhibition by ethanol of excitatory amino acid receptors in rat locus coeruleus neurons in vitro. Naunyn Schmiedeberg's Arch Pharmacol 357: 299–308

Nimitvilai S, Herman M, You C, Arora DS, McElvain MA, Roberto M, Brodie MS (2014) Dopamine D2 receptor desensitization by dopamine or corticotropin releasing factor in ventral tegmental area neurons is associated with increased glutamate release. Neuropharmacology 82: 28–40. https://doi.org/10.1016/j.neuropharm.2014.03.006

Nimitvilai S, Lopez MF, Mulholland PJ, Woodward JJ (2016) Chronic intermittent ethanol exposure enhances the excitability and synaptic plasticity of lateral orbitofrontal cortex neurons and induces a tolerance to the acute inhibitory actions of ethanol. Neuropsychopharmacology 41:1112–1127

Nimitvilai S, Uys JD, Woodward JJ, Randall PK, Ball LE, Williams RW, Jones BC, Lu L, Grant KA, Mulholland PJ (2017) Orbitofrontal neuroadaptations and cross-species synaptic biomarkers in heavy-drinking macaques. J Neurosci 37:3646–3660

Nishio M, Narahashi T (1990) Ethanol enhancement of GABA-activated chloride current in rat dorsal root ganglion neurons. Brain Res 518(1–2):283–286

Noori HR, Mücksch C, Vengeliene V, Schönig K, Takahashi TT, Mukhtasimova N, Bagher Oskouei M, Mosqueira M, Bartsch D, Fink R, Urbassek HM, Spanagel R, Sine SM (2018) Alcohol reduces muscle fatigue through atomistic interactions with nicotinic receptors. Commun Biol 1:159

Nordberg A, Wahlström GJ (1992) Cholinergic mechanisms in physical dependence on barbiturates, ethanol and benzodiazepines. Neural Transm Gen Sect 88(3):199–221. https://doi.org/10.1007/BF01244733

Nordberg A, Larsson C, Perdahl E, Winblad B (1983) Changes in cholinergic activity in human hippocampus following chronic alcohol abuse. Pharmacol Biochem Behav Suppl 1:397–400. https://doi.org/10.1016/0091-3057(83)90206-x

Obara I, Bell RL, Goulding SP, Reyes CM, Larson LA, Ary AW, Truitt WA, Szumlinski KK (2009) Differential effects of chronic ethanol consumption and withdrawal on homer/glutamate receptor expression in subregions of the accumbens and amygdala of P rats. Alcohol Clin Exp Res 33(11):1924–1934

Ogilvie KM, Lee S, Rivier C (1998) Divergence in the expression of molecular markers of neuronal activation in the parvocellular paraventricular nucleus of the hypothalamus evoked by alcohol administration via different routes. J Neurosci 18:4344–4352

Olive MF, Koenig HN, Nannini MA, Hodge CW (2002) Elevated extracellular CRF levels in the bed nucleus of the stria terminalis during ethanol withdrawal and reduction by subsequent ethanol intake. Pharmacol Biochem Behav 72:213–220

Olsen RW, Hanchar HJ, Meera P, Wallner M (2007) GABAA receptor subtypes: the "one glass of wine" receptors. Alcohol 41(3):201–209

Orozco-Cabal L, Pollandt S, Liu J, Shinnick-Gallagher P, Gallagher JP (2006) Regulation of synaptic transmission by CRF receptors. Rev Neurosci 17:279–307

Overstreet LS, Pasternak JF, Colley PA, Slater NT, Trommer BL (1997) Metabotropic glutamate receptor mediated long-term depression in developing hippocampus. Neuropharmacology 36(6):831–844

Overstreet DH, Knapp DJ, Breese GR (2004) Modulation of multiple ethanol withdrawal-induced anxiety-like behavior by CRF and CRF1 receptors. Pharmacol Biochem Behav 77:405–413

Pandey SC, Zhang H, Roy A, Xu T (2005) Deficits in amygdaloid cAMP-responsive element-binding protein signaling play a role in genetic predisposition to anxiety and alcoholism. J Clin Invest 115:2762–2773

Pantazis NJ, West JR, Dai D (1998) The nitric oxide-cyclic GMP pathway plays an essential role in both promoting cell survival of cerebellar granule cells in culture and protecting the cells against ethanol neurotoxicity. J Neurochem 70(5):1826–1838

Papadeas S, Grobin AC, Morrow AL (2001) Chronic ethanol consumption differentially alters GABA(A) receptor alpha1 and alpha4 subunit peptide expression and GABA(A) receptor-mediated 36 Cl(−) uptake in mesocorticolimbic regions of rat brain. Alcohol Clin Exp Res 25:1270–1275

Parker TH, Roberts RK, Henderson GI, Hoyumpa AM, Schmidt DE, Schenker S (1978) The effects of ethanol on cerebral regional acetylcholine concentration and utilization. Proc Soc Exp Biol Med 159(2):270–275. https://doi.org/10.3181/00379727-159-40330

Pati D, Kelly K, Stennett B, Frazier CJ, Knackstedt LA (2016) Alcohol consumption increases basal extra-cellular glutamate in the nucleus accumbens core of Sprague-Dawley rats without increasing spontaneous glutamate release. Eur J Neurosci 44:1896–1905

Patton MH, Roberts BM, Lovinger DM, Mathur BN (2016) Ethanol disinhibits dorsolateral striatal medium spiny neurons through activation of a presynaptic delta opioid receptor. Neuropsychopharmacology 41:1831–1840

Pelham RW, Marquis JK, Kugelmann K, Munsat TL (1980) Prolonged ethanol consumption produces persistent alterations of cholinergic function in rat brain. Alcohol Clin Exp Res 4(3): 282–287

Pereira PA, Rocha JP, Cardoso A, Vilela M, Sousa S, Madeira MD (2016) Effects of chronic alcohol consumption, withdrawal and nerve growth factor on neuropeptide Y expression and cholinergic innervation of the rat dentate hilus. Neurotoxicology 54:153–160. https://doi.org/10.1016/j.neuro.2016.04.007

Pereira PA, Gonçalves E, Silva A, Millner T, Madeira MD (2020) Effects of chronic alcohol consumption and withdrawal on the cholinergic neurons of the pedunculopontine and laterodorsal tegmental nuclei of the rat: an unbiased stereological study. Neurotoxicology 76: 58–66. https://doi.org/10.1016/j.neuro.2019.10.005

Peris J, Eppler B, Hu M, Walker DW, Hunter BE, Mason K, Anderson KJ (1997) Effects of chronic ethanol exposure on GABA receptors and GABAB receptor modulation of 3H-GABA release in the hippocampus. Alcohol Clin Exp Res 21:1047–1052

Perkins DI, Trudell JR, Crawford DK, Alkana RL, Davies DL (2010) Molecular targets and mechanisms for ethanol action in glycine receptors. Pharmacol Ther 127(1):53–65

Pick CG, Cooperman M, Trombka D, Rogel-Fuchs Y, Yanai J (1993) Hippocampal cholinergic alterations and related behavioral deficits after early exposure to ethanol. Int J Dev Neurosci 11: 379–385

Pietzak ER, Wilce PA, Shanley BC (1988) The effect of chronic ethanol consumption on muscarinic receptors in rat brain. Neurochem Int 12(4):447–452. https://doi.org/10.1016/0197-0186(88)90027-7

Pioszak AA, Parker NR, Suino-Powell K, Xu HE (2008) Molecular recognition of corticotropin-releasing factor by its G-protein-coupled receptor CRFR1. J Biol Chem 283:32900–32912

Pleil KE, Lopez A, McCall N, Jijon AM, Bravo JP, Kash TL (2012) Chronic stress alters neuropeptide Y signaling in the bed nucleus of the stria terminalis in DBA/2J but not C57BL/6J mice. Neuropharmacology 62(4):1777–1786. https://doi.org/10.1016/j.neuropharm.2011.12.002

Pleil KE, Rinker JA, Lowery-Gionta EG, Mazzone CM, McCall NM, Kendra AM, Olson DP, Lowell BB, Grant KA, Thiele TE, Kash TL (2015a) NPY signaling inhibits extended amygdala CRF neurons to suppress binge alcohol drinking. Nat Neurosci 18(4):545–552. https://doi.org/10.1038/nn.3972

Pleil KE, Lowery-Gionta EG, Crowley NA, Li C, Marcinkiewcz CA, Rose JH, McCall NM, Maldonado-Devincci AM, Morrow AL, Jones SR, Kash TL (2015b) Effects of chronic ethanol exposure on neuronal function in the prefrontal cortex and extended amygdala. Neuropharmacology 99:735–749. https://doi.org/10.1016/j.neuropharm.2015.06.017

Pomrenze MB, Fetterly TL, Winder DG, Messing RO (2017) The corticotropin releasing factor receptor 1 in alcohol use disorder: still a valid drug target? Alcohol Clin Exp Res 41:1986–1999. https://doi.org/10.1111/acer.13507

Primeaux SD, Wilson SP, Bray GA, York DA, Wilson MA (2006) Overexpression of neuropeptide Y in the central nucleus of the amygdala decreases ethanol self-administration in "anxious" rats. Alcohol Clin Exp Res 30:791–801

Proft J, Weiss N (2015) G protein regulation of neuronal calcium channels: back to the future. Mol Pharmacol 87(6):890–906. https://doi.org/10.1124/mol.114.096008

Qi ZH, Song M, Wallace MJ, Wang D, Newton PM, McMahon T, Chou WH, Zhang C, Shokat KM, Messing RO (2007) Protein kinase C epsilon regulates gamma-aminobutyrate type A receptor sensitivity to ethanol and benzodiazepines through phosphorylation of gamma2 subunits. J Biol Chem 282(45):33052–33063

Quadros IM, Macedo GC, Domingues LP, Favoretto CA (2016) An update on CRF mechanisms underlying alcohol use disorders and dependence. Front Endocrinol 7:134

Rabin RA, Molinoff PB (1981) Activation of adenylate cyclase by ethanol in mouse striatal tissue. J Pharmacol Exp Ther 216(1):129–134

Radke AK, Jury NJ, Delpire E, Nakazawa K, Holmes A (2017a) Reduced ethanol drinking following selective cortical interneuron deletion of the GluN2B NMDA receptors subunit. Alcohol 58:47–51

Radke AK, Jury NJ, Kocharian A, Marcinkiewcz CA, Lowery-Gionta EG, Pleil KE, McElligott ZA, McKlveen JM, Kash TL, Holmes A (2017b) Chronic EtOH effects on putative measures of compulsive behavior in mice. Addict Biol 22:423–434

Raeder H, Holter SM, Hartmann AM, Spanagel R, Hans-Moller J, Rujescu D (2008) Expression of N-methyl-d-aspartate (NMDA) receptor subunits and splice variants in an animal model of long-term voluntary alcohol self-administration. Drug Alcohol Depend 96(1-2):16–21. https://doi.org/10.1016/j.drugalcdep.2007.12.013

Rahman S, Engleman EA, Bell RL (2016) Recent advances in nicotinic receptor signaling in alcohol abuse and alcoholism. Prog Mol Biol Transl Sci 137:183–201

Rainnie DG, Fernhout BJ, Shinnick-Gallagher P (1992) Differential actions of corticotropin releasing factor on basolateral and central amygdaloid neurones, in vitro. J Pharmacol Exp Ther 263:846–858

Ramachandran B, Ahmed S, Zafar N, Dean C (2015) Ethanol inhibits long-term potentiation in hippocampal CA1 neurons, irrespective of lamina and stimulus strength, through neurosteroidogenesis. Hippocampus 25:106–118

Rao PS, Bell RL, Engleman EA, Sari Y (2015) Targeting glutamate uptake to treat alcohol use disorders. Front Neurosci 9:144

Rassnick S, Heinrichs SC, Britton KT, Koob GF (1993) Microinjection of a corticotropin-releasing factor antagonist into the central nucleus of the amygdala reverses anxiogenic-like effects of ethanol withdrawal. Brain Res 605:25–32

Reinscheid RK, Nothacker HP, Bourson A, Ardati A, Henningsen RA, Bunzow JR, Grandy DK, Langen H, Monsma FJ Jr, Civelli O (1995) Orphanin FQ: a neuropeptide that activates an opioidlike G protein- coupled receptor. Science 270:792–794

Ren H, Honse Y, Karp BJ, Lipsky RH, Peoples RW (2003) A site in the fourth membrane-associated domain of the N-methyl-D-aspartate receptor regulates desensitization and ion channel gating. J Biol Chem 278(1):276–283. https://doi.org/10.1074/jbc.M209486200

Ren H, Salous AK, Paul JM, Lipsky RH, Peoples RW (2007) Mutations at F637 in the NMDA receptor NR2A subunit M3 domain influence agonist potency, ion channel gating and alcohol action. Br J Pharmacol 151(6):749–757. https://doi.org/10.1038/sj.bjp.0707254

Ren H, Zhao Y, Dwyer DS, Peoples RW (2012) Interactions among positions in the third and fourth membrane-associated domains at the intersubunit interface of the N-methyl-D-aspartate receptor forming sites of alcohol action. J Biol Chem 287(33):27302–27312. https://doi.org/10.1074/jbc.M111.338921

Ren H, Zhao Y, Wu M, Peoples RW (2013) A novel alcohol-sensitive position in the N-methyl-D-aspartate receptor GluN2A subunit M3 domain regulates agonist affinity and ion channel gating. Mol Pharmacol 84(4):501–510. https://doi.org/10.1124/mol.113.085993

Ren H, Zhao Y, Wu M, Dwyer DS, Peoples RW (2017) Two adjacent phenylalanines in the NMDA receptor GluN2A subunit M3 domain interactively regulate alcohol sensitivity and ion channel gating. Neuropharmacology 114:20–33. https://doi.org/10.1016/j.neuropharm.2016.11.013

Renteria R, Maier EY, Buske TR, Morrisett RA (2017) Selective alterations of NMDAR function and plasticity in D1 and D2 medium spiny neurons in the nucleus accumbens shell following chronic intermittent ethanol exposure. Neuropharmacology 112:164–171

Renteria R, Buske TR, Morrisett RA (2018) Long-term subregion-specific encoding of enhanced ethanol intake by D1DR medium spiny neurons of the nucleus accumbens. Addict Biol 23(2): 689–698. https://doi.org/10.1111/adb.12526

Retson TA, Hoek JB, Sterling RC, Van Bockstaele EJ (2015) Amygdalar neuronal plasticity and the interactions of alcohol, sex, and stress. Brain Struct Funct 220(6):3211–3232

Reynolds JN, Prasad A (1991) Ethanol enhances GABAA receptor-activated chloride currents in chick cerebral cortical neurons. Brain Res 564(1):138–142

Riegel AC, Williams JT (2008) CRF facilitates calcium release from intracellular stores in midbrain dopamine neurons. Neuron 57:559–570

Rimondini R, Thorsell A, Heilig M (2005) Suppression of ethanol self-administration by the neuropeptide Y (NPY) Y2 receptor antagonist BIIE0246: evidence for sensitization in rats with a history of dependence. Neurosci Lett 375:129–133

Rinker JA, Marshall SA, Mazzone CM, Lowery-Gionta EG, Gulati V, Pleil KE, Kash TL, Navarro M, Thiele TE (2017) Extended amygdala to ventral tegmental area corticotropin-releasing factor circuit controls binge ethanol intake. Biol Psychiatry 81:930–940

Roberto M, Siggins GR (2006) Nociceptin/orphanin FQ presynaptically decreases GABAergic transmission and blocks the ethanol-induced increase of GABA release in central amygdala. Proc Natl Acad Sci U S A 103:9715–9720

Roberto M, Varodayan FP (2017) Synaptic targets: chronic alcohol actions. Neuropharmacology 122:85–99

Roberto M, Nelson TE, Ur CL, Gruol DL (2002) Long-term potentiation in the rat hippocampus is reversibly depressed by chronic intermittent ethanol exposure. J Neurophysiol 87:2385–2397

Roberto M, Madamba SG, Moore SD, Tallent MK, Siggins GR (2003) Ethanol increases GABAergic transmission at both pre- and postsynaptic sites in rat central amygdala neurons. Proc Natl Acad Sci U S A 100:2053–2058

Roberto M, Madamba SG, Stouffer DG, Parsons LH, Siggins GR (2004a) Increased GABA release in the central amygdala of ethanol-dependent rats. J Neurosci 24:10159–10166

Roberto M, Schweitzer P, Madamba SG, Stouffer DG, Parsons LH, Siggins GR (2004b) Acute and chronic ethanol alter glutamatergic transmission in rat central amygdala: an in vitro and in vivo analysis. J Neurosci 24:1594–1603

Roberto M, Bajo M, Crawford E, Madamba SG, Siggins GR (2006) Chronic ethanol exposure and protracted abstinence alter NMDA receptors in central amygdala. Neuropsychopharmacology 31:988–996

Roberto M, Gilpin NW, O'Dell LE, Cruz MT, Morse AC, Siggins GR, Koob GF (2008) Cellular and behavioral interactions of gabapentin with alcohol dependence. J Neurosci 28:5762–5771

Roberto M, Cruz MT, Gilpin NW, Sabino V, Schweitzer P, Bajo M, Cottone P, Madamba SG, Stouffer DG, Zorrilla EP, Koob GF, Siggins GR, Parsons LH (2010) Corticotropin releasing factor-induced amygdala gamma-aminobutyric acid release plays a key role in alcohol dependence. Biol Psychiatry 67(9):831–839

Roberto M, Gilpin NW, Siggins GR (2012) The central amygdala and alcohol: role of GABA, glutamate and neuropeptides. In: Pierce C, Kenny PJ (eds) Addiction, vol 1. Cold Spring Harbor Laboratory Press, p 2

Roberto M, Spierling SR, Kirson D, Zorrilla EP (2017) Corticotropin-releasing factor (CRF) and addictive behaviors. Int Rev Neurobiol 136:5–51

Roberto M, Kirson D, Khom S (2020) The role of the central amygdala in alcohol dependence. In: Pierce C, Kenny PJ (eds) Addiction. Cold Spring Harbor Laboratory Press. https://doi.org/10.1101/cshperspect.a039339

Robinson SL, Thiele TE (2017) The role of neuropeptide Y (NPY) in alcohol and drug abuse disorders. Int Rev Neurobiol 136:177–197. https://doi.org/10.1016/bs.irn.2017.06.005

Robinson DL, Volz TJ, Schenk JO, Wightman RM (2005) Acute ethanol decreases dopamine transporter velocity in rat striatum: in vivo and in vitro electrochemical measurements. Alcohol Clin Exp Res 29:746–755

Robles N, Sabriá J (2006) Ethanol consumption produces changes in behavior and on hippocampal alpha7 and alpha4beta2 nicotinic receptors. J Mol Neurosci 30(1–2):119–120

Robles N, Sabriá J (2008) Effects of moderate chronic ethanol consumption on hippocampal nicotinic receptors and associative learning. Neurobiol Learn Mem 89:497–503

Rodaros D, Caruana DA, Amir S, Stewart J (2007) Corticotropin-releasing factor projections from limbic forebrain and paraventricular nucleus of the hypothalamus to the region of the ventral tegmental area. Neuroscience 150:8–13

Rogers J, Wiener SG, Bloom FE (1979) Long-term ethanol administration methods for rats: advantages of inhalation over intubation or liquid diets. Behav Neural Biol 27:466–486

Rose JH, Karkhanis AN, Chen R, Gioia D, Lopez MF, Becker HC, McCool BA, Jones SR (2016) Supersensitive kappa opioid receptors promotes ethanol withdrawal-related behaviors and reduce dopamine signaling in the nucleus accumbens. Int J Neuropsychopharmacol 19:pyv127

Rossetti ZL, Carboni S (1995) Ethanol withdrawal is associated with increased extracellular glutamate in the rat striatum. Eur J Pharmacol 283:177–183

Rothberg BS, Hunter BE (1991) Chronic ethanol treatment differentially affects muscarinic receptor responses in rat hippocampus. Neurosci Lett 132(2):243–246

Rothberg BS, Yasuda RP, Satkus SA, Wolfe BB, Hunter BE (1993) Effects of chronic ethanol on cholinergic actions in rat hippocampus: electrophysiological studies and quantification of m1-m5 muscarinic receptor subtypes. Brain Res 631(2):227–234

Rothberg BS, Hunter BE, Walker DW, Anderson JF, Anderson KJ (1996) Long-term effects of chronic ethanol on muscarinic receptor binding in rat brain. Alcohol Clin Exp Res 20(9): 1613–1617

Ryabinin AE, Weitemier AZ (2006) The urocortin 1 neurocircuit: ethanol-sensitivity and potential involvement in alcohol consumption. Brain Res Rev 52:368–380

Sabriá J, Torres D, Pastó M, Peralba JM, Allali-Hassani A, Parés X (2003) Release of neurotransmitters from rat brain nerve terminals after chronic ethanol ingestion: differential effects in cortex and hippocampus. Addict Biol 8(3):287–294. https://doi.org/10.1080/13556210310001602194

Sajdyk TJ, Schober DA, Gehlert DR (2002) Neuropeptide Y receptor subtypes in the basolateral nucleus of the amygdala modulate anxiogenic responses in rats. Neuropharmacology 43:1165–1172

Sakanaka M, Shibasaki T, Lederis K (1986) Distribution and efferent projections of corticotropin-releasing factor-like immunoreactivity in the rat amygdaloid complex. Brain Res 382:213–238

Salinas AG, Mateo Y, Carlson VCC, Stinnett GS, Luo G, Seasholtz AF, Grant KA, Lovinger DM (2021) Long-term alcohol consumption alters dorsal striatal dopamine release and regulation by D2 dopamine receptors in rhesus macaques. Neuropsychopharmacology 46(8):1432–1441. https://doi.org/10.1038/s41386-020-00938-8

Salous AK, Ren H, Lamb KA, Hu XQ, Lipsky RH, Peoples RW (2009) Differential actions of ethanol and trichloroethanol at sites in the M3 and M4 domains of the NMDA receptor GluN2A (NR2A) subunit. Br J Pharmacol 158(5):1395–1404. https://doi.org/10.1111/j.1476-5381.2009.00397.x

Sanna E, Dildy-Mayfield JE, Harris RA (1994) Ethanol inhibits the function of 5-hydroxytryptamine type 1c and muscarinic M1 G protein-linked receptors in Xenopus oocytes expressing brain mRNA: role of protein kinase C. Mol Pharmacol 45(5):1004–1012

Sapp DW, Yeh HH (1998) Ethanol-GABAA receptor interactions: a comparison between cell lines and cerebellar Purkinje cells. J Pharmacol Exp Ther 284(2):768–776

Sauguet L, Howard RJ, Malherbe L, Lee US, Corringer PJ, Harris RA, Delarue M (2013) Structural basis for potentiation by alcohols and anaesthetics in a ligand-gated ion channel. Nat Commun 4:1697

Schilaty ND, Hedges DM, Jang EY, Folsom RJ, Yorgason JT, McIntosh JM, Steffensen SC (2014) Acute ethanol inhibits dopamine release in the nucleus accumbens via α6 nicotinic acetylcholine receptors. J Pharmacol Exp Ther 349:559–567

Schummers J, Browning MD (2001) Evidence for a role for GABA(A) and NMDA receptors in ethanol inhibition of long-term potentiation. Brain Res Mol Brain Res 94(1–2):9–14. https://doi.org/10.1016/s0169-328x(01)00161-9

Schummers J, Bentz S, Browning MD (1997) Ethanol's inhibition of LTP may not be mediated solely via direct effects on the NMDA receptor. Alcohol Clin Exp Res 21(3):404–408. https://doi.org/10.1111/j.1530-0277.1997.tb03783.x

Sebe JY, Eggers ED, Berger AJ (2003) Differential effects of ethanol on GABA(A) and glycine receptor-mediated synaptic currents in brain stem motoneurons. J Neurophysiol 90(2):870–875

Seif T, Chang SJ, Simms JA, Gibb SL, Dadgar J, Chen BT, Harvey BK, Ron D, Messing RO, Bonci A, Hopf FW (2013) Cortical activation of accumbens hyperpolarization-active NMDARs mediates aversion-resistant alcohol intake. Nat Neurosci 16:1094–1100

Siciliano CA, Tye KM (2019) Leveraging calcium imaging to illuminate circuit dysfunction in addiction. Alcohol 74:47–63. https://doi.org/10.1016/j.alcohol.2018.05.013

Siciliano CA, Calipari ES, Cuzon Carlson VC, Helms CM, Lovinger DM, Grant KA, Jones SR (2015) Voluntary ethanol intake predicts κ-opioid receptor supersensitivity and regionally distinct dopaminergic adaptations in macaques. J Neurosci 35:5959–5968

Siciliano CA, Calipari ES, Yorgason JT, Mateo Y, Helms CM, Lovinger DM, Grant KA, Jones SR (2016) Chronic ethanol self-administration in macaques shifts dopamine feedback inhibition to predominantly D2 receptors in nucleus accumbens core. Drug Alcohol Depend 158:159–163. https://doi.org/10.1016/j.drugalcdep.2015.10.031

Siggins GR, Gruol D, Aldenhoff J, Pittman Q (1985) Electrophysiological actions of corticotropin-releasing factor in the central nervous system. Fed Proc 44:237–242

Siggins GR, Roberto M, Nie Z (2005) The tipsy terminal: presynaptic effects of ethanol. Pharmacol Ther 107:80–98

Silberman Y, Matthews RT, Winder DG (2013) A corticotropin releasing factor pathway for ethanol regulation of the ventral tegmental area in the bed nucleus of the stria terminalis. J Neurosci 33(3):950–960

Silberman Y, Fetterly TL, Awad EK, Milano EJ, Usdin TB, Winder DG (2015) Ethanol produces corticotropin-releasing factor receptor-dependent enhancement of spontaneous glutamatergic transmission in the mouse central amygdala. Alcohol Clin Exp Res 39:2154–2162

Sillaber I, Rammes G, Zimmermann S, Mahal B, Zieglgansberger W, Wurst W, Holsboer F, Spanagel R (2002) Enhanced and delayed stress-induced alcohol drinking in mice lacking functional CRH1 receptors. Science 296:931–933

Silvestre de Ferron B, Bennouar KE, Kervern M, Alaux-Cantin S, Robert A, Rabiant K, Antol J, Naassila M, Pierrefiche O (2015) Two binges of ethanol a day keep the memory away in adolescent rats: key role for GLUN2B subunit. Int J Neuropsychopharmacol 19:pyv087

Sinclair JG, Lo GF (1986) Ethanol blocks tetanic and calcium-induced long-term potentiation in the hippocampal slice. Gen Pharmacol 17:231–233

Sink KS, Chung A, Ressler KJ, Davis M, Walker DL (2013) Anxiogenic effects of CGRP within the BNST may be mediated by CRF acting at BNST CRFR1 receptors. Behav Brain Res 243: 286–293. https://doi.org/10.1016/j.bbr.2013.01.024

Slawecki CJ, Somes C, Ehlers CL (1999) Effects of chronic ethanol exposure on neurophysiological responses to corticotropin-releasing factor and neuropeptide Y. Alcohol Alcohol 34:289–299

Smiley JF, Bleiwas C, Canals-Baker S, Williams SZ, Sears R, Teixeira CM, Wilson DA, Saito M (2021) Neonatal ethanol causes profound reduction of cholinergic cell number in the basal forebrain of adult animals. Alcohol 97:1–11. https://doi.org/10.1016/j.alcohol.2021.08.005

Smith TL (1983) Influence of chronic ethanol consumption on muscarinic cholinergic receptors and their linkage to phospholipid metabolism in mouse synaptosomes. Neuropharmacology 22(5): 661–663

Smith SM, Vale WW (2006) The role of the hypothalamic-pituitary-adrenal axis in neuroendocrine responses to stress. Dialogues Clin Neurosci 8:383–395

Smothers CT, Woodward JJ (2016) Differential effects of TM4 tryptophan mutations on inhibition of N-methyl-d-aspartate receptors by ethanol and toluene. Alcohol 56:15–19

Smothers CT, Mrotek JJ, Lovinger DM (1997) Chronic ethanol exposure leads to a selective enhancement of N-methyl-D-aspartate receptor function in cultured hippocampal neurons. J Pharmacol Exp Ther 283(3):1214–1222

Smothers CT, Jin C, Woodward JJ (2013) Deletion of the N-terminal domain alters the ethanol inhibition of N-methyl-D-aspartate receptors in a subunit-dependent manner. Alcohol Clin Exp Res 37:1882–1890

Smothers CT, Szumlinski KK, Worley PF, Woodward JJ (2016) Altered NMDA receptor function in primary cultures of hippocampal neurons from mice lacking the Homer2 gene. Synapse. 70 (1):33–39. https://doi.org/10.1002/syn.21869

Smyth RD, Beck H (1969) The effect of time and concentration of ethanol administration on brain acetylcholine metabolism. Arch Int Pharmacodyn Ther 182:295–299

Snell LD, Nunley KR, Lickteig RL, Browning MD, Tabakoff B, Hoffman PL (1996) Regional and subunit specific changes in NMDA receptor mRNA and immunoreactivity in mouse brain following chronic ethanol ingestion. Brain Res Mol Brain Res 40:71–78

Snyder AE, Salimando GJ, Winder DG, Silberman Y (2019) Chronic intermittent ethanol and acute stress similarly modulate BNST CRF neuron activity via noradrenergic signaling. Alcohol Clin Exp Res 43(8):1695–1701. https://doi.org/10.1111/acer.14118

Söderpalm B, Lidö HH, Ericson M (2017) The glycine receptor: a functionally important primary brain target of ethanol. Alcohol Clin Exp Res 41:1816–1830

Sommer WH, Rimondini R, Hansson AC, Hipskind PA, Gehlert DR, Barr CS, Heilig MA (2008) Upregulation of voluntary alcohol intake, behavioral sensitivity to stress, and amygdala crhr1 expression following a history of dependence. Biol Psychiatry 63:139–145

Sparrow AM, Lowery-Gionta EG, Pleil KE, Li C, Sprow GM, Cox BR, Rinker JA, Jijon AM, Navarro M, Kash TL, Thiele TE (2012) Central neuropeptide Y modulates binge-like ethanol drinking in C57BL/6J mice via Y1 and Y2 receptors. Neuropsychopharmacology 37(6): 1409–1421. https://doi.org/10.1038/npp.2011.327

Spierling SR, Zorrilla EP (2017) Don't stress about CRF: assessing the translational failures of CRF1antagonists. Psychopharmacology (Berl). 234(9–10):1467–1481. https://doi.org/10.1007/s00213-017-4556-2. Epub 2017 Mar 7

Staples MC, Kim A, Mandyam CD (2015) Dendritic remodeling of hippocampal neurons is associated with altered NMDA receptor expression in alcohol dependent rats. Mol Cell Neurosci 65:153–162. https://doi.org/10.1016/j.mcn.2015.03.008

Steinmetz NA, Aydin C, Lebedeva A, Okun M, Pachitariu M, Bauza M, Beau M, Bhagat J, Böhm C, Broux M, Chen S, Colonell J, Gardner RJ, Karsh B, Kloosterman F, Kostadinov D, Mora-Lopez C, O'Callaghan J, Park J, Putzeys J, Sauerbrei B, van Daal RJJ, Vollan AZ, Wang S, Welkenhuysen M, Ye Z, Dudman JT, Dutta B, Hantman AW, Harris KD, Lee AK, Moser EI, O'Keefe J, Renart A, Svoboda K, Häusser M, Haesler S, Carandini M, Harris TD (2021) Neuropixels 2.0: a miniaturized high-density probe for stable, long-term brain recordings. Science 372(6539):eabf4588. https://doi.org/10.1126/science.abf4588

Stephens DN, Ripley TL, Borlikova G, Schubert M, Albrecht D, Hogarth L, Duka T (2005) Repeated ethanol exposure and withdrawal impairs human fear conditioning and depresses long-term potentiation in rat amygdala and hippocampus. Biol Psychiatry 58:392–400

Stragier E, Martin V, Davenas E, Poilbout C, Mongeau R, Corradetti R, Lanfumey L (2015) Brain plasticity and cognitive functions after ethanol consumption in C57BL/6J mice. Transl Psychiatry 5:e696. https://doi.org/10.1038/tp.2015.183

Su LD, Sun CL, Shen Y (2010) Ethanol acutely modulates mGluR1-dependent long-term depression in cerebellum. Alcohol Clin Exp Res 34(7):1140–1145

Suarez J, Khom S, Alen F, Natividad LA, Varodayan FP, Patel RR, Kirson D, Arco R, Ballesta A, Bajo M et al (2019) Cessation of fluoxetine treatment increases alcohol seeking during relapse and dysregulates endocannabinoid and glutamatergic signaling in the central amygdala. Addict Biol 25(5):e12813

Swanson LW, Sawchenko PE, Rivier J, Vale WW (1983) Organization of ovine corticotropin-releasing factor immunoreactive cells and fibers in the rat brain: an immunohistochemical study. Neuroendocrinology 36:165–186

Swartzwelder HS, Risher ML, Miller KM, Colbran RJ, Winder DG, Wills TA (2016) Changes in the adult GluN2B associated proteome following adolescent intermittent ethanol exposure. PLoS One 11:e0155951. https://doi.org/10.1371/journal.pone.0155951

Szücs A, Berton F, Sanna PP, Francesconi W (2012) Excitability of jcBNST neurons is reduced in alcohol-dependent animals during protracted alcohol withdrawal. PLoS One 7:e42313. https://doi.org/10.1371/journal.pone.0042313

Szumlinski KK, Ary AW, Lominac KD, Klugmann M, Kippin TE (2008) Accumbens Homer2 over-expression facilitates alcohol-induced neuroplasticity in C57BL/6J mice. Neuropsychopharmacology 33:1365–1378

Tabakoff B, Munoz-Marcus M, Fields JZ (1979) Chronic ethanol feeding produces an increase in muscarinic cholinergic receptors in mouse brain. Life Sci 25(26):2173–2180

Tagliaferro P, Morales M (2008) Synapses between corticotropin-releasing factor-containing axon terminals and dopaminergic neurons in the ventral tegmental area are predominantly glutamatergic. J Comp Neurol 506:616–626

Talani G, Lovinger DM (2015) Interactions between ethanol and the endocannabinoid system at GABAergic synapses on basolateral amygdala principal neurons. Alcohol 49:781–794

Talani G, Licheri V, Masala N, Follesa P, Mostallino MC, Biggio G, Sanna E (2014) Increased voluntary ethanol consumption and changes in hippocampal synaptic plasticity in isolated C57BL/6J mice. Neurochem Res 39(6):997–1004

Tallent MK, Madamba SG, Siggins GR (2001) Nociceptin reduces epileptiform events in CA3 hippocampus via presynaptic and postsynaptic mechanisms. J Neurosci 21:6940–6948

Terunuma M, Xu J, Vithlani M, Sieghart W, Kittler J, Pangalos M, Haydon PG, Coulter DA, Moss SJ (2008) Deficits in phosphorylation of GABA(A) receptors by intimately associated protein kinase C activity underlie compromised synaptic inhibition during status epilepticus. J Neurosci 28:376–384

Theile JW, Morikawa H, Gonzales RA, Morrisett RA (2008) Ethanol enhances GABAergic transmission onto dopamine neurons in the ventral tegmental area of the rat. Alcohol Clin Exp Res 32(6):1040–1048

Theile JW, Morikawa H, Gonzales RA, Morrisett RA (2009) Role of 5-hydroxytryptamine2C receptors in Ca2+−dependent ethanol potentiation of GABA release onto ventral tegmental area dopamine neurons. J Pharmacol Exp Ther 329(2):625–633

Thiele TE, Marsh DJ, Ste Marie L, Bernstein IL, Palmiter RD (1998) Ethanol consumption and resistance are inversely related to neuropeptide Y levels. Nature 396:366–369

Thielen RJ, Morzorati SL, McBride WJ (2001) Effects of ethanol on the dorsal raphe nucleus and its projections to the caudate putamen. Alcohol 23:131–139

Thorsell A (2008) Central neuropeptide Y in anxiety- and stress-related behavior and in ethanol intake. Ann N Y Acad Sci 1148:136–140

Ticku MK, Burch T (1980) Alterations in gamma-aminobutyric acid receptor sensitivity following acute and chronic ethanol treatments. J Neurochem 34:417–423

Tipps ME, Raybuck JD, Kozell LB, Lattal KM, Buck KJ (2016) G protein-gated inwardly rectifying potassium channel subunit 3 knock-out mice show enhanced ethanol reward. Alcohol Clin Exp Res 40:857–864

Tokuda K, Izumi Y, Zorumski CF (2013) Locally-generated acetaldehyde contributes to the effects of ethanol on neurosteroids and LTP in the hippocampus. Neurol Clin Neurosci 1:138–147

Tonner PH, Miller KW (1995) Molecular sites of general anaesthetic action on acetylcholine receptors. Eur J Anaesthesiol 12(1):21–30

Trantham-Davidson H, Burnett EJ, Gass JT, Lopez MF, Mulholland PJ, Centanni SW, Floresco SB, Chandler LJ (2014) Chronic alcohol disrupts dopamine receptor activity and the cognitive function of the medial prefrontal cortex. J Neurosci 34(10):3706–3718. https://doi.org/10.1523/JNEUROSCI.0623-13.2014

Tremwel MF, Hunter BE, Peris J (1994) Chronic ethanol exposure enhances [3H]GABA release and does not affect GABAA receptor mediated 36Cl uptake. Synapse 17:149–154

Trevisan L, Fitzgerald LW, Brose N, Gasic GP, Heinemann SF, Duman RS, Nestler EJ (1994) Chronic ingestion of ethanol up-regulates NMDAR1 receptor subunit immunoreactivity in rat hippocampus. J Neurochem 62(4):1635–1638

Tunstall BJ, Kirson D, Zallar LJ, McConnell SA, Vendruscolo JCM, Ho CP, Oleata CS, Khom S, Manning M, Lee MR et al (2019) Oxytocin blocks enhanced motivation for alcohol in alcohol dependence and blocks alcohol effects on GABAergic transmission in the central amygdala. PLoS Biol 17:e2006421

Uhrig S, Vandael D, Marcantoni A, Dedic N, Bilbao A, Vogt MA, Hirth N, Broccoli L, Bernardi RE, Schönig K, Gass P, Bartsch D, Spanagel R, Deussing JM, Sommer WH, Carbone E, Hansson AC (2017) Differential roles for L-type calcium channel subtypes in alcohol dependence. Neuropsychopharmacology 42(5):1058–1069

Ungless MA, Singh V, Crowder TL, Yaka R, Ron D, Bonci A (2003) Corticotropin-releasing factor requires CRF binding protein to potentiate NMDA receptors via CRF receptor 2 in dopamine neurons. Neuron 39:401–407

Uryu K, Okumura T, Shibasaki T, Sakanaka M (1992) Fine structure and possible origins of nerve fibers with corticotropin-releasing factor-like immunoreactivity in the rat central amygdaloid nucleus. Brain Res 577:175–179

Valdez GR, Roberts AJ, Chan K, Davis H, Brennan M, Zorrilla EP, Koob GF (2002) Increased ethanol self-administration and anxiety-like behavior during acute ethanol withdrawal and protracted abstinence: regulation by corticotropin-releasing factor. Alcohol Clin Exp Res 26: 1494–1501

Valdez GR, Zorrilla EP, Roberts AJ, Koob GF (2003) Antagonism of corticotropin-releasing factor attenuates the enhanced responsiveness to stress observed during protracted ethanol abstinence. Alcohol 29:55–60

Vale W, Spiess J, Rivier C, Rivier J (1981) Characterization of a 41-residue ovine hypothalamic peptide that stimulates secretion of corticotropin and beta-endorphin. Science 213:1394–1397

Valenzuela CF, Jotty K (2015) Mini-review: effects of ethanol on Gabaa receptor-mediated neurotransmission in the cerebellar cortex-recent advances. Cerebellum 14:438–446

Valenzuela CF, Machu TK, McKernan RM, Whiting P, VanRenterghem BB, McManaman JL, Brozowski SJ, Smith GB, Olsen RW, Harris RA (1995) Tyrosine kinase phosphorylation of GABAA receptors. Brain Res Mol Brain Res 31:165–172

Valenzuela CF, Bhave S, Hoffman P, Harris RA (1998a) Acute effects of ethanol on pharmacologically isolated kainate receptors in cerebellar granule neurons: comparison with NMDA and AMPA receptors. J Neurochem 71:1777–1780

Valenzuela CF, Cardoso RA, Wick MJ, Weiner JL, Dunwiddie TV, Harris RA (1998b) Effects of ethanol on recombinant glycine receptors expressed in mammalian cell lines. Alcohol Clin Exp Res 22(5):1132–1136

Van Pett K, Viau V, Bittencourt JC, Chan RK, Li HY, Arias C, Prins GS, Perrin M, Vale W, Sawchenko PE (2000) Distribution of mRNAs encoding CRF receptors in brain and pituitary of rat and mouse. J Comp Neurol 428:191–212

Vardy E, Robinson JE, Li C, RHJ O, JF DB, Giguere PM, Sassano FM, Huang XP, Zhu H, Urban DJ, White KL, Rittiner JE, Crowley NA, Pleil KE, Mazzone CM, Mosier PD, Song J, Kash TL, Malanga CJ, Krashes MJ, Roth BL (2015) A new DREADD facilitates the multiplexed chemogenetic interrogation of behavior. Neuron 86:936–946

Varodayan FP, Harrison NL (2013) HSF1 transcriptional activity mediates alcohol induction of Vamp2 expression and GABA release. Front Integr Neurosci 7:89

Varodayan FP, Pignataro L, Harrison NL (2011) Alcohol induces synaptotagmin 1 expression in neurons via activation of heat shock factor 1. Neuroscience 193:63–71

Varodayan F, Soni N, Bajo M, Luu G, Madamba S, Schweitzer P, Parsons LH, Roberto M (2015) Chronic ethanol exposure decreases cannabinoid CB(1) receptor function at the GABAergic synapses in rat central amygdala. Addict Biol. https://doi.org/10.1111/adb.12256

Varodayan F, Soni N, Bajo M, Luu G, Madamba S, Schweitzer P, Roberto M (2016) Chronic ethanol exposure alters cannabinoid CB(1) receptor function at the GABAergic synapses in rat basolateral amygdala. Addict Biol. https://doi.org/10.1111/adb.12369

Varodayan FP, Correia D, Kirson D, Khom S, Oleata CS, Luu G, Schweitzer P, Roberto M (2017a) CRF modulates glutamate transmission in the central amygdala of naive and ethanol-dependent rats. Neuropharmacology 125:418–428

Varodayan FP, de Guglielmo G, Logrip ML, George O, Roberto M (2017b) Alcohol dependence disrupts amygdalar L-type voltage-gated calcium channel mechanisms. J Neurosci 37(17): 4593–4603. https://doi.org/10.1523/JNEUROSCI.3721-16.2017

Varodayan FP, Logrip ML, Roberto M (2017c) P/Q-type voltage-gated calcium channels mediate the ethanol and CRF sensitivity of central amygdala GABAergic synapses. Neuropharmacology 125:197–206

Varodayan FP, Sidhu H, Kreifeldt M, Roberto M, Contet C (2018) Morphological and functional evidence of increased excitatory signaling in the prelimbic cortex during ethanol withdrawal. Neuropharmacology 133:470–480

Vengeliene V, Bilbao A, Molander A, Spanagel R (2008) Neuropharmacology of alcohol addiction. Br J Pharmacol 154(2):299–315

Vengeliene V, Leonardi-Essmann F, Sommer WH, Marston HM, Spanagel R (2010) Glycine transporter-1 blockade leads to persistently reduced relapse-like alcohol drinking in rats. Biol Psychiatry 68(8):704–711

Vetreno RP, Broadwater M, Liu W, Spear LP, Crews FT (2014) Adolescent, but not adult, binge ethanol exposure leads to persistent global reductions of choline acetyltransferase expressing neurons in brain. PLoS One 9(11):e113421. https://doi.org/10.1371/journal.pone.0113421

Vivian JA, Green HL, Young JE, Majerksy LS, Thomas BW, Shively CA, Tobin JR, Nader MA, Grant KA (2001) Induction and maintenance of ethanol self-administration in cynomolgus monkeys (Macaca fascicularis): long-term characterization of sex and individual differences. Alcohol Clin Exp Res 25:1087–1097

Volkow ND, Wiers CE, Shokri-Kojori E, Tomasi D, Wang GJ, Baler R (2017) Neurochemical and metabolic effects of acute and chronic alcohol in the human brain: studies with positron emission tomography. Neuropharmacology 122:175–188

Wadleigh A, Valenzuela CF (2012) Ethanol increases GABAergic transmission and excitability in cerebellar molecular layer interneurons from GAD67-GFP knock-in mice. Alcohol Alcohol 47(1):1–8. https://doi.org/10.1093/alcalc/agr147

Wan FJ, Berton F, Madamba SG, Francesconi W, Siggins GR (1996) Low ethanol concentrations enhance GABAergic inhibitory postsynaptic potentials in hippocampal pyramidal neurons only after block of GABAB receptors. Proc Natl Acad Sci U S A 93:5049–5054

Wanat MJ, Hopf FW, Stuber GD, Phillips PE, Bonci A (2008) Corticotropin-releasing factor increases mouse ventral tegmental area dopamine neuron firing through a protein kinase C-dependent enhancement of Ih. J Physiol 586:2157–2170

Wanat MJ, Sparta DR, Hopf FW, Bowers MS, Melis M, Bonci A (2009) Strain specific synaptic modifications on ventral tegmental area dopamine neurons after ethanol exposure. Biol Psychiatry 65:646–653

Wang HL, Morales M (2008) Corticotropin-releasing factor binding protein within the ventral tegmental area is expressed in a subset of dopaminergic neurons. J Comp Neurol 509:302–318

Wang B, Shaham Y, Zitzman D, Azari S, Wise RA, You ZB (2005) Cocaine experience establishes control of midbrain glutamate and dopamine by corticotropin-releasing factor: a role in stress-induced relapse to drug seeking. J Neurosci 25:5389–5396

Wang J, Carnicella S, Phamluong K, Jeanblanc J, Ronesi JA, Chaudhri N, Janak PH, Lovinger DM, Ron D (2007) Ethanol induces long-term facilitation of NR2B-NMDA receptor activity in the dorsal striatum: implications for alcohol drinking behavior. J Neurosci 27(13):3593–3602

Wang J, Ben Hamida S, Darcq E, Zhu W, Gibb SL, Lanfranco MF, Carnicella S, Ron D (2012) Ethanol-mediated facilitation of AMPA receptor function in the dorsomedial striatum: implications for alcohol drinking behavior. J Neurosci 32:15124–15132

Wang J, Cheng Y, Wang X, Roltsch Hellard E, Ma T, Gil H, Ben Hamida S, Ron D (2015) Alcohol elicits functional and structural plasticity selectively in dopamine D1 receptor-expressing neurons of the dorsomedial striatum. J Neurosci 35:11634–11643

Wayner MJ, Armstrong DL, Polan-Curtain JL, Denny JB (1993) Ethanol and diazepam inhibition of hippocampal LTP is mediated by angiotensin II and AT1 receptors. Peptides 14(3):441–444

Wei W, Faria LC, Mody I (2004) Low ethanol concentrations selectively augment the tonic inhibition mediated by delta subunit-containing GABAA receptors in hippocampal neurons. J Neurosci 24(38):8379–8382

Weiner JL, Valenzuela CF (2006) Ethanol modulation of GABAergic transmission: the view from the slice. Pharmacol Ther 111(3):533–554

Weiner JL, Zhang L, Carlen PL (1994) Potentiation of GABAA-mediated synaptic current by ethanol in hippocampal CA1 neurons: possible role of protein kinase C. J Pharmacol Exp Ther 268(3):1388–1395

Weiner JL, Gu C, Dunwiddie TV (1997) Differential ethanol sensitivity of subpopulations of GABAA synapses onto rat hippocampal CA1 pyramidal neurons. J Neurophysiol 77:1306–1312

Weiner JL, Dunwiddie TV, Valenzuela CF (1999) Ethanol inhibition of synaptically evoked kainate responses in rat hippocampal CA3 pyramidal neurons. Mol Pharmacol 56:85–90

Weiner JL, Ariwodola OJ, Bates WH, Davenport AT, Daunais JB, Grant KA, Friedman DP (2004) The effect of long-term voluntary ethanol consumption on GABAergic and glutamatergic synaptic transmission in the monkey CNS. Alcohol Clin Exp Res 28

Weiner JL, Ariwodola OJ, Bates WH, Bryant V, Silberman Y, Daunais JB, Grant KA, Friedman DP (2005) Presynaptic mechanisms underlying ethanol actions at GABAergic synapses in rat and monkey hippocampus. Alcohol Clin Exp Res 29(5):187A

Weiss F, Ciccocioppo R, Parsons LH, Katner S, Liu X, Zorrilla EP, Valdez GR, Ben-Shahar O, Angeletti S, Richter RR (2001) Compulsive drug-seeking behavior and relapse. Neuroadaptation, stress, and conditioning factors. Ann N Y Acad Sci 937:1–26

Weitlauf C, Woodward JJ (2008) Ethanol selectively attenuates NMDAR-mediated synaptic transmission in the prefrontal cortex. Alcohol Clin Exp Res 32(4):690–698

Weitlauf C, Egli RE, Grueter BA, Winder DG (2004) High-frequency stimulation induces ethanol-sensitive long-term potentiation at glutamatergic synapses in the dorsolateral bed nucleus of the stria terminalis. J Neurosci 24(25):5741–5747

Welsh BT, Goldstein BE, Mihic SJ (2009) Single-channel analysis of ethanol enhancement of glycine receptor function. J Pharmacol Exp Ther 330(1):198–205

White G, Lovinger DM, Weight FF (1990) Ethanol inhibits NMDA-activated current but does not alter GABA-activated current in an isolated adult mammalian neuron. Brain Res 507(2): 332–336

Whittemore ER, Yang W, Drewe JA, Woodward RM (1996) Pharmacology of the human gamma-aminobutyric acidA receptor alpha 4 subunit expressed in Xenopus laevis oocytes. Mol Pharmacol 50:1364–1375

Wickman K, Clapham DE (1995) Ion channel regulation by G proteins. Physiol Rev 75(4): 865–885. https://doi.org/10.1152/physrev.1995.75.4.865

Wilcox MV, Cuzon Carlson VC, Sherazee N, Sprow GM, Bock R, Thiele TE, Lovinger DM, Alvarez VA (2014) Repeated binge-like ethanol drinking alters ethanol drinking patterns and depresses striatal GABAergic transmission. Neuropsychopharmacology 39:579–594

Wills TA, Klug JR, Silberman Y, Baucum AJ, Weitlauf C, Colbran RJ, Delpire E, Winder DG (2012) GluN2B subunit deletion reveals key role in acute and chronic ethanol sensitivity of glutamate synapses in bed nucleus of the stria terminalis. Proc Natl Acad Sci U S A 109(5): E278–E287

Wills TA, Baucum AJ II, Holleran KM, Chen Y, Pasek JG, Delpire E, Tabb DL, Colbran RJ, Winder DG (2017) Chronic intermittent alcohol disrupts the GluN2B-associated proteome and specifically regulates group I mGlu receptor-dependent long-term depression. Addict Biol 22: 275–290

Winkler A, Mahal B, Kiianmaa K, Zieglgänsberger W, Spanagel R (1999) Effects of chronic alcohol consumption on the expression of different NR1 splice variants in the brain of AA and ANA lines of rats. Brain Res Mol Brain Res 72(2):166–175

Wirkner K, Eberts C, Poelchen W, Allgaier C, Illes P (2000) Mechanism of inhibition by ethanol of NMDA and AMPA receptor channel functions in cultured rat cortical neurons. Naunyn Schmiedebergs Arch Pharmacol 362(6):568–776

Wolak ML, DeJoseph MR, Cator AD, Mokashi AS, Brownfield MS, Urban JH (2003) Comparative distribution of neuropeptide Y Y1 and Y5 receptors in the rat brain by using immunohistochemistry. J Comp Neurol 464:285–311

Wolfe SA, Sidhu H, Patel RR, Kreifeldt M, D'Ambrosio SR, Contet C, Roberto M (2019) Molecular, morphological, and functional characterization of corticotropin-releasing factor receptor 1-expressing neurons in the central nucleus of the amygdala. eNeuro 6(3). https://doi.org/10.1523/ENEURO.0087-19.2019

Wu LG, Saggau P (1994) Presynaptic calcium is increased during normal synaptic transmission and paired-pulse facilitation, but not in long-term potentiation in area CA1 of hippocampus. J Neurosci 14:645–654

Wu M, Katti P, Zhao Y, Peoples RW (2019) Positions in the N-methyl-D-aspartate receptor GluN2C subunit M3 and M4 domains regulate alcohol sensitivity and receptor kinetics. Alcohol Clin Exp Res 43(6):1180–1190. https://doi.org/10.1111/acer.14042

Xiao C, Shao XM, Olive MF, Griffin WC III, Li KY, Krnjević K, Zhou C, Ye JH (2009) Ethanol facilitates glutamatergic transmission to dopamine neurons in the ventral tegmental area. Neuropsychopharmacology 34(2):307–318

Xu S, Pany S, Benny K, Tarique K, Al-Hatem O, Gajewski K, Leasure JL, Das J, Roman G (2018) Ethanol regulates presynaptic activity and sedation through presynaptic Unc13 proteins in Drosophila. eNeuro 5

Yaka R, Tang KC, Camarini R, Janak PH, Ron D (2003) Fyn kinase and NR2B-containing NMDA receptors regulate acute ethanol sensitivity but not ethanol intake or conditioned reward. Alcohol Clin Exp Res 27:1736–1742

Yamashita M, Marszalec W, Yeh JZ, Narahashi T (2006) Effects of ethanol on tonic GABA currents in cerebellar granule cells and mammalian cells recombinantly expressing GABA (A) receptors. J Pharmacol Exp Ther 319(1):431–438

Yan QS (2003) Involvement of non-exocytotic mechanisms in ethanol-induced in vivo dopamine release: comparisons with cocaine. Eur J Pharmacol. 477(1):37–44

Yavich L, Tiihonen J (2000) Ethanol modulates evoked dopaminerelease in mouse nucleus accumbens: dependence on social stress and dose. Eur J Pharmacol 401:365–373

Ye JH, Tao L, Ren J, Schaefer R, Krnjevic K, Liu PL, Schiller DA, McArdle JJ (2001) Ethanol potentiation of glycine-induced responses in dissociated neurons of rat ventral tegmental area. J Pharmacol Exp Ther 296(1):77–83

Yevenes GE, Moraga-Cid G, Peoples RW, Schmalzing G, Aguayo LG (2008) A selective G betagamma-linked intracellular mechanism for modulation of a ligand-gated ion channel by ethanol. Proc Natl Acad Sci U S A 105(51):20523–20528

Yin HH, Park BS, Adermark L, Lovinger DM (2007) Ethanol reverses the direction of long-term synaptic plasticity in the dorsomedial striatum. Eur J Neurosci 25(11):3226–3232

Yim HJ, Schallert T, Randall PK, Gonzales RA (1998) Comparison of local and systemic ethanol effects on extracellular dopamine concentration in rat nucleus accumbens by microdialysis. Alcohol Clin Exp Res 22(2):367–374

Yoshida K, Engel J, Liljequist S (1982) The effect of chronic ethanol administration of high affinity 3H-nicotinic binding in rat brain. Naunyn Schmiedeberg's Arch Pharmacol 321(1):74–76. https://doi.org/10.1007/BF00586353

Yu B, Shinnick-Gallagher P (1998) Corticotropin-releasing factor increases dihydropyridine- and neurotoxin-resistant calcium currents in neurons of the central amygdala. J Pharmacol Exp Ther 284:170–179

Yu TP, Xie CW (1998) Orphanin FQ/nociceptin inhibits synaptic transmission and long-term potentiation in rat dentate gyrus through postsynaptic mechanisms. J Neurophysiol 80:1277–1284

Zandy SL, Matthews DB, Tokunaga S, Miller AD, Blaha CD, Mittleman G (2015) Reduced dopamine release in the nucleus accumbens core of adult rats following adolescent binge alcohol exposure: age and dose-dependent analysis. Psychopharmacology 232:777–784

Zhao Y, Ren H, Dwyer DS, Peoples RW (2015) Different sites of alcohol action in the NMDA receptor GluN2A and GluN2B subunits. Neuropharmacology 97:240–250. https://doi.org/10.1016/j.neuropharm.2015.05.018

Zhao Y, Ren H, Peoples RW (2016) Intersubunit interactions at putative sites of ethanol action in the M3 and M4 domains of the NMDA receptor GluN1 and GluN2B subunits. Br J Pharmacol 173(12):1950–1965. https://doi.org/10.1111/bph.13487

Zhou Q, Verdoorn TA, Lovinger DM (1998) Alcohols potentiate the function of 5-HT3 receptor-channels on NCB-20 neuroblastoma cells by favouring and stabilizing the open channel state. J Physiol 507(2):335–352

Zhou Z, Karlsson C, Liang T, Xiong W, Kimura M, Tapocik JD, Yuan Q, Barbier E, Feng A, Flanigan M, Augier E, Enoch M-A, Hodgkinson DA, Shen P-H, Lovinger DM, Edenberg HJ,

Heilig M, Goldman D (2013) Loss of metabotropic glutamate receptor 2 escalates alcohol consumption. Proc Natl Acad Sci U S A 110:16963–11696

Zhu PJ, Lovinger DM (2006) Ethanol potentiates GABAergic synaptic transmission in a postsynaptic neuron/synaptic Bouton preparation from basolateral amygdala. J Neurophysiol 96(1): 433–441

Zhu W, Bie B, Pan ZZ (2007) Involvement of non-NMDA glutamate receptors in central amygdala in synaptic actions of ethanol and ethanol-induced reward behavior. J Neurosci 27:289–298

Ziskind-Conhaim L, Gao BX, Hinckley C (2003) Ethanol dual modulatory actions on spontaneous postsynaptic currents in spinal motoneurons. J Neurophysiol 89:806–813

Zorrilla EP, Koob GF (2010) Progress in corticotropin-releasing factor-1 antagonist development. Drug Discov Today 15:371–383

Zorrilla EP, Heilig M, de Wit H, Shaham Y (2013) Behavioral, biological, and chemical perspectives on targeting CRF(1) receptor antagonists to treat alcoholism. Drug Alcohol Depend 128: 175–186

Zorrilla EP, Logrip ML, Koob GF (2014) Corticotropin releasing factor: a key role in the neurobiology of addiction. Front Neuroendocrinol 35:234–244

Zorumski CF, Mennerick S, Izumi Y (2014) Acute and chronic effects of ethanol on learning-related synaptic plasticity. Alcohol 48:1–17

Epigenetic Dysregulation in Alcohol-Associated Behaviors: Preclinical and Clinical Evidence

Esi Domi, Riccardo Barchiesi, and Estelle Barbier

Contents

Abstract Alcohol use disorder (AUD) is characterized by loss of control over intake and drinking despite harmful consequences. At a molecular level, AUD is associated with long-term neuroadaptations in key brain regions that are involved in

E. Domi
Department of Biomedical and Clinical Sciences, Center for Social and Affective Neuroscience, Linköping University, Linköping, Sweden

School of Pharmacy, Pharmacology Unit, Center for Neuroscience, University of Camerino, Camerino, Italy

R. Barchiesi
Department of Neuroscience, Waggoner Center for Alcohol and Alcohol Addiction Research, University of Texas at Austin, Austin, TX, USA

E. Barbier (✉)
Department of Biomedical and Clinical Sciences, Center for Social and Affective Neuroscience, Linköping University, Linköping, Sweden
e-mail: estelle.barbier@liu.se

Curr Topics Behav Neurosci (2025) 71: 309–346
https://doi.org/10.1007/7854_2022_410
Published Online: 31 January 2023

reward processing and decision-making. Over the last decades, a great effort has been made to understand the neurobiological basis underlying AUD. Epigenetic mechanisms have emerged as an important mechanism in the regulation of long-term alcohol-induced gene expression changes. Here, we review the literature supporting a role for epigenetic processes in AUD. We particularly focused on the three most studied epigenetic mechanisms: DNA methylation, Histone modification and non-coding RNAs. Clinical studies indicate an association between AUD and DNA methylation both at the gene and global levels. Using behavioral paradigms that mimic some of the characteristics of AUD, preclinical studies demonstrate that changes in epigenetic mechanisms can functionally impact alcohol-associated behaviors. While many studies support a therapeutic potential for targeting epigenetic enzymes, more research is needed to fully understand their role in AUD. Identification of brain circuits underlying alcohol-associated behaviors has made major advances in recent years. However, there are very few studies that investigate how epigenetic mechanisms can affect these circuits or impact the neuronal ensembles that promote alcohol-associated behaviors. Studies that focus on the role of circuit-specific and cell-specific epigenetic changes for clinically relevant alcohol behaviors may provide new insights on the functional role of epigenetic processes in AUD.

Keywords Alcohol use disorder · DNA methylation · Epigenetic mechanisms · Histone modifications · Non-coding RNAs

Abbreviations

3′UTR	Three prime untranslated region
ACSS2	Acyl-CoA synthetase short-chain family member 2
AMG	Amygdala complex
AUD	Alcohol use disorder
AUDIT	Alcohol use disorder identification test
BAC	Blood alcohol concentration
BDNF	Brain-derived neurotrophic factor
BDNF-AS	Brain-derived neurotrophic factor anti sense
CHIP	Chromatin immunoprecipitation
CHIP-seq	Chromatin immunoprecipitation sequencing
CIE	Chronic intermittent exposure
circRNA	Circular RNA
CPP	Conditioned place preference
CUT&RUN	Cleavage under target and release using nuclease
CUT&TAG	Cleavage under targets and tagmentation
DA	Dopamine
DID	Drinking in the dark
DNMT	DNA methyltransferase

DRD2	Dopamine receptor D2 gene
EWAS	Epigenome-wide association studies
FDA	Food and Drug Administration
fMRI	Functional magnetic resonance imaging
GABA	Gamma-aminobutyric acid
GADD45B	Growth arrest and DNA damage-inducible beta
GAS5	Growth arrest–specific five gene
HDAC	Histone deacetylase
I.C.V.	Intracerebroventricular
I.P.	Intraperitoneal injection
KD	Knockdown
KDM6B	Lysine(K)–specific demethylase 6B
lncRNAs	Long non-coding RNAs
MBD-seq	Methyl binding protein followed by next-generation sequencing
MeDIP-seq	Methylated DNA immunoprecipitation followed by next-generation sequencing
miRNA	MicroRNA
mPFC	Medial prefrontal cortex
NAc	Nucleus accumbens
NP	Non preferring
Nr2b	N-methyl-D-aspartate receptor subunit 2B
P	Preferring
Pdyn	Prodynorphin
PFC	Prefrontal cortex
piRNAs	Piwi-interacting RNAs
Pnoc	Prepronociceptin
PR C2	Polycomb repressive complex
Prdm2	PR domain containing 2
RRBS	Reduced representation bisulfite
SAHA	Suberanilohydroxamic acid
SAM	S-adenosyl-methionine
siRNA	Small interfering RNA
Syt/SYT	Synaptotagmin 1 (gene/protein)
TSA	Trichostatin A
WGBS	Whole genome bisulfite sequencing

1 Introduction

Alcohol use accounts for 5.1% of the global disease burden and 5.3% of all deaths, making it a major public health problem (Collaborators 2018). The disease burden of alcohol is mainly associated with alcohol use disorder (AUD), a diagnostic category which in its moderate and severe forms can be equated with alcohol addiction. AUD is a complex psychiatric disorder characterized by loss of control over intake, excessive use despite negative consequences ("compulsive use") and choice of alcohol over natural rewards. These behavioral symptoms are thought to reflect the emergence of persistent neuroadaptations in key brain structures that exert control over motivated behavior (Carvalho et al. 2019; Heilig et al. 2019, 2021).

During early, recreational stages, alcohol use is primarily thought to be consumed for the resulting pleasurable effects (Boileau et al. 2003; Gilman et al. 2008; Ramchandani et al. 2011b). The positively reinforcing or "rewarding" effects of alcohol are thought to occur through the activation of the brain reward system, which results in release of dopamine (DA) from terminals of mesolimbic DA neurons in the nucleus accumbens (NAc) (Spanagel 2009b). AUD is characterized by persistence of excessive alcohol intake (Begleiter 1975) and progression into more severe stages is associated with a transition into a relief-driven (negatively reinforced) alcohol seeking and intake. Negatively reinforcing properties of alcohol develop through persistent changes in brain function. It is thought that these neuroadaptations result in a progressive recruitment of brain systems that promote stress response (Gilpin et al. 2015; Gilpin and Roberto 2012).

Among regular alcohol users, only a minority (approximately 15%) develop AUD (Anthony et al. 1994). Known risk factors include genetics, amount of alcohol consumed, early-life trauma, stress and pattern of drinking such as binge drinking (Cheng et al. 2004). The genetic basis is currently believed to be responsible for more than 50% of the individual risk for becoming alcoholic (Dick and Bierut 2006) where specific gene changes might relate to an increased susceptibility for developing the disorder (Bierut et al. 2010; Edenberg and Foroud 2006; Ramchandani et al. 2011a; Schuckit 2009). Remarkably, genome-wide studies failed to identify genetic variants that could account for more than 0.1% of AUD heritability (Heath et al. 2011). Studies using animal model of alcohol-associated behaviors suggest that epigenetic inheritance may explain part of the "missing" heritability of AUD. Evidence indicates that epigenetic modifications can be passed down to the next generation and therefore promote AUD (See review (Finegersh et al. 2015). However, due to the large complexity of the mechanisms implicated in epigenetic inheritance, we believe that this topic requires a review on its own and we will therefore not discuss it in here. For a critical analysis of this topic see (Chastain and Sarkar 2017; Rompala and Homanics 2019).

Prolonged heavy drinking also contributes to the development of AUD per se. Prolonged exposure of the brain to cycles of alcohol intoxication and withdrawal induces persistent neuroadaptations and gene expression changes that in turn promote alcohol seeking, -taking and -relapse (Heilig et al. 2010; Heilig and Koob 2007; Meinhardt and Sommer 2015). These persistent neuroadaptations are thought to be in part mediated through epigenetic reprogramming of the transcriptome in key brain regions (Barbier et al. 2015, 2017; Heilig et al. 2017; Tapocik et al. 2014). After a brief description of the epigenetic mechanisms, including DNA methylation, histone modifications and non-coding RNAs, we will review the genetic, behavioral, and pharmacological evidence suggesting a key role of the epigenetic mechanisms in the development of AUD.

1.1 *Epigenetic Mechanisms*

Epigenetic mechanisms regulate gene expression via modulation of the chromatin conformation. Chromatin is formed by negatively charged DNA wrapped around positively charged histone protein octamers that contain two copies of histone proteins H2A, H2B, H3, and H4. The chromatin structure plays a key role in regulating gene expression. A tight chromatin conformation (i.e., heterochromatin) promotes gene silencing, while a loose chromatin conformation (i.e., euchromatin) promotes gene expression by allowing the binding of the transcription machinery (i.e., transcription factors, polymerase etc.) to the functional parts of the genome. Epigenetic modifications regulate chromatin conformation and gene expression through three mechanisms: DNA methylation, histone modifications and non-coding RNAs (Lyko 2018; Mattick and Makunin 2006; Szyf 2009) (Fig. 1).

1.1.1 DNA Methylation

DNA methylation refers to the addition of methyl groups at the C5 position of the cytosine to form 5-methylcytosine. Generally, DNA methylation represses gene expression by blocking the binding of transcription factors to the gene promoter region or by recruiting other proteins such as MeCp2 or methyl CpG-binding proteins (Goll and Bestor 2005). However, more recent findings showed high DNA methylation levels in the promoter and coding regions of actively transcribed genes, suggesting a more complex role of DNA methylation on gene expression (Suzuki and Bird 2008; Weber et al. 2007). Several studies also indicate that DNA methylation can modulate gene splicing (Lev Maor et al. 2015). DNA methylation is catalyzed by enzymes called DNA methyltransferases (DNMTs). These enzymes

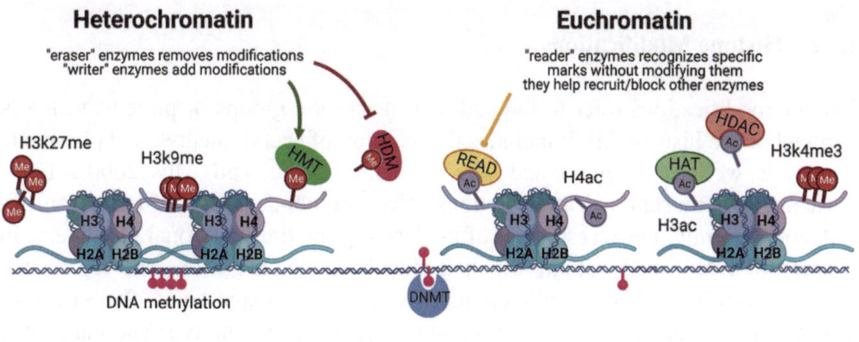

Fig. 1 Epigenetic mechanisms regulate the availability of genes to the transcriptional machinery. In the heterochromatin state (left), DNA is tightly packed around the nucleosomes and is unavailable to the transcriptional machinery. In the euchromatin state (right), genes are instead available for transcription. Ac acetyl group, DNMT DNA methyltransferase, H histone, HAT histone acetyltransferase, HDAC histone deacetylase, HDM histone demethylase, HMT histone methyl transferase, K lysine, Me methyl group

can either add new methyl groups to unmethylated DNA (i.e., DNMT3a and DNMT3b) or copy the DNA methylation pattern from the parental DNA strand onto the newly synthesized daughter strand. DNA methylation is essential during development as it controls genomic imprinting and X-chromosome inactivation and is also important for regulating gene expression in mature neurons (Li and Zhang 2014).

Over the last decades, several methods have been developed to analyze genome-wide DNA methylation patterns (Feng and Lou 2019). Part of these methods is based on antibodies that target DNA methylation. These techniques include methylated DNA immunoprecipitation followed by next-generation sequencing (MeDIP-seq (Down et al. 2008) and methyl binding protein followed by next-generation sequencing (MBD-seq (Aberg et al. 2018). MeDIP-seq implicates immunoprecipitation with methylated cytosine antibodies and can therefore identify lower CpG density (<5 CpG per 100 bp) compared to other techniques. In opposite, MBD-seq, which involved antibodies that bind methyl CpG-binding proteins, is biased towards CpG-dense regions. The other category of genome-wide DNA methods relies on bisulfite conversion of cytosine residues to uracil/thymine residues followed by next-generation sequencing. This includes, for instance, whole genome bisulfite sequencing (WGBS) and reduced representation bisulfite (RRBS (Gu et al. 2011). Both methods permit the analysis of DNA methylation levels of single cytosines. The use of restriction enzymes, combined with bisulfite treatment in the RRBS method increases the sequencing coverage of CpG-dense regions and allows for a better-read depth compared to WGBS. However, due to its bias towards CpG-dense regions, RRBS covers only part of the genome. It is important to note that the methods currently used to assess DNA methylation do not allow DNA methylation analysis of the entire genome. Instead the specifics of each methods determine the genomic regions analyzed and therefore provide different sets of information (Beck et al. 2022).

1.1.2 Histone Modifications

Histone modifications refer to the addition of specific groups or proteins to the N terminal of the histone tail. It includes the addition of acetyl, methyl, and phosphate groups as well as ubiquitin and sumo proteins (Strahl and Allis 2000). These covalent post-translational modifications affect the condensation of the chromatin and, consequently, the accessibility of the DNA to the transcriptional machinery. In contrast to DNA methylation, which is mostly associated with gene repression, histone modifications can bidirectionally affect gene expression. The effect of histone modifications on gene expression is determined by the type and number of functional groups that bind the histone tail (e.g., acetyl vs. methyl groups; mono vs. di- or tri-methylation). Histone acetylation is associated with gene activation, whereas histone methylation can either be associated with gene expression or gene repression (Strahl and Allis 2000). For instance, histone acetylation and histone H3 methylation on lysine 4, 36 and 79 (H3K4me; H3K36me and H3K79me) contribute to the formation of an open chromatin conformation that promotes gene expression. In opposite, histone H3 methylation on lysine 9 (H3K9me) and histone H4 on lysine 20 (H4K20me) are associated with gene repression (Allis et al. 2007). Covalent histone modifications are regulated by epigenetic enzymes that can either add (i.e., writers) or remove (i.e., erasers) the functional groups from the histone tail. Another class of epigenetic enzymes called "readers," such as the bromodomain family, recognizes specific histone modifications and recruits other proteins to regulate gene expression (Szyf 2009). Accurate measurements of normal and altered histone marks are important to understand their role in regulating gene expression and cell phenotype. Several methods have therefore been developed to study histone modifications. Over the past years, chromatin immunoprecipitation (CHIP) followed by deep sequencing (ChIP-seq) has been one of the most used approaches to map genome-wide patterns of histone modifications (Park 2009). ChIP consists of cell crosslinking with formaldehyde followed by fragmentation and solubilization of the chromatin. In this approach, an antibody targeting the protein of interest (e.g., histone mark) is added to immunoprecipitate the target protein together with bound DNA, which is then purified and sequenced using next-generation sequencing. ChIP-seq presents the advantage of base-pair resolution mapping. However, the crosslinking step can lead to false positive binding sites (Meyer and Liu 2014). The recently developed methods CUT&RUN (Cleavage Under Target and Release Using Nuclease) and CUT&TAG (Cleavage Under Targets and Tagmentation) can map specific interactions between protein and DNA with better efficiency than ChIP (Skene and Henikoff 2017). In contrast to ChIP, where the antibody binds target proteins after DNA fragmentation, the CUT&RUN and CUT&TAG approaches are based on antibody binding to the target proteins in intact cells. These methods use protein A and protein G to fix the enzymatic domains to antibody-bound chromatin for in situ cleavage of DNA-bound target protein. Compared to the ChIP, both approaches allow histone modifications mapping using low cell/nuclei input and are therefore compatible with single-cell technologies.

1.1.3 Non-coding RNAs

Gene expression is also regulated by non-coding RNAs, which are RNA molecules that are not translated into proteins. Non-coding RNAs are divided into two classes: long non-coding RNAs (>200 nucleotides; lncRNAs) and short non-coding RNAs (<30 nucleotides). Recent advances in lncRNAs research have demonstrated their role in regulating gene expression by modulating mRNA stability and modulating the accessibility of chromatin (Dykes and Emanueli 2017). LncRNAs can act as guide molecules for chromatin remodeling complex such as polycomb repressive complex 2 (PRC2; (Tsai et al. 2010). LncRNAs can also directly regulate gene expression by interacting with methyl CpG-binding domain protein 1 or transcription factors, to block transcription initiation, or modulate transcription elongation (Dykes and Emanueli 2017; Monnier et al. 2013). Short non-coding RNAs, which include piwi-interacting RNAs (piRNAs), small interfering RNAs (siRNA) and micro RNAs (miRNA), have also been shown to regulate gene expression (Bartel 2004; Siomi et al. 2011). The role of miRNAs in brain functions has been the most investigated amongst short non-coding RNAs. miRNAs repress gene expression by directly interacting with partially complementary target sites located in the 3′ untranslated region of target mRNA (3′UTR). This targets the transcript for degradation, effectively downregulating its expression (Finnegan and Pasquinelli 2013).

Genome-wide profiles of non-coding RNAs can be assessed using RNA sequencing. In small RNA-seq, size-selected RNA is ligated to adapters and amplified through reverse transcription and PCR before sequencing on a next-generation sequencing platform (Mehta 2014). Protocols for small RNA-seq are now available for very low amounts of input RNA, which allows for studying small RNAs at a single-cell level. This can be a powerful method for understanding biological processes (Hagemann-Jensen et al. 2018). Besides sequencing, microarray-based methods are also available, as well as targeted approaches using different PCR technologies (Chen et al. 2011; Forero et al. 2019).

Overall, epigenetic systems have an ability to translate environmental stimuli into gene expression changes and cellular adaptations. Their ability to simultaneously regulate multiple transcripts also offers a mechanism by which, for example, alcohol consumption can induce the broad changes in gene expression that are often observed in humans and rodents alike (Farris and Mayfield 2014; Kisby et al. 2021; Tapocik et al. 2013). Several studies have now demonstrated the role of epigenetic modifications following alcohol exposure, and these modifications may be important for the behavioral symptoms observed in AUD.

2 Epigenetic Mechanisms in Human Studies of AUD

Epigenetic processes have been proposed to be a core mechanism in promoting the long-term neuroadaptations that are associated with AUD. So far, comparing methylation profiles in peripheral blood cells between cases and healthy controls has been the most common way to investigate epigenetic mechanisms in alcohol use (Wedemeyer et al. 2020). One of the largest epigenome-wide association studies (EWAS) of alcohol consumption included over 13,000 subjects and found 144 differentially methylated CpG sites between non-drinkers and current heavy drinkers in whole blood, suggesting that DNA methylation could be used as a biomarker and diagnostic tool for heavy drinking. When performing an ancestry-stratified meta-analysis, this increased to 328 and 165 differentially methylated CpG sites in European-ancestry and African-ancestry samples, respectively. In the European-ancestry subset, they further identified specific CpG sites in Gamma-aminobutyric acid (GABA) A and B receptor subunit genes that correlated with the amount of alcohol exposure, as well as with the expression levels of several genes involved in immune signaling (Liu et al. 2018), supporting the evidence that GABA and GABA receptors are also important in the crosstalk between the CNS and the immune system (Jin et al. 2013). Another recent study by Liang et al., similarly investigated the possibility to use EWAS and DNA methylation as an objective biomarker of hazardous alcohol drinking, rather than self-reported drinking. Using machine learning, they found that the DNA methylation signature on phosphatidyl ethanol but not on self-reported alcohol consumption predicted hazardous alcohol drinking in two distinct populations (Liang et al. 2021). More studies have investigated gene methylation patterns as possible biomarkers. For example, one double-blind study examined whether methylation patterns in the OPRM1 gene had any correlation with the outcome of treatment with the opioid antagonist naltrexone for alcohol dependence; but the authors did not find any such relationship on relapse (Lin et al. 2020). Overall, these and other studies do suggest that methylation patterns may become useful clinical biomarkers, as it may provide a more objective measure of recent alcohol consumption.

Attempts have also been made to connect peripheral methylation profiles to brain activation patterns in relevant behaviors. In an EWAS with 18 monozygotic twin pairs, discordant for AUD, Ruggeri et al. identified an association between AUD and increased methylation levels within the promoter region of the gene PPM1G, an important gene for cell stress response. The finding was validated in a larger cohort of 499 adolescents, where they saw that methylation of PPM1G was associated with "impulsivity." When examining the PPM1G methylation with performance in a functional magnetic resonance imaging (fMRI) stop signal task, which assesses behavioral (motor) inhibition, they found a positive correlation in the same cohort (Ruggeri et al. 2015). In another study, Bidwell et al. used fMRI to test 383 heavy drinkers for associations between DA receptor D2 gene (DRD2) methylation and AUD. They found a positive correlation between methylation levels of DRD2 and signal levels in several brain reward regions in response to alcohol cues, as well as a positive correlation with clinical severity (using the AUD identification test: AUDIT) (Bidwell et al. 2019). Similar to Bidwell and colleagues, Hagerty et al. recently showed that average DRD2 methylation was negatively correlated with altered functional connectivity in executive control networks (Hagerty et al. 2020). Together, these studies suggest that methylation patterns are associated with behaviors and brain region-specific activity that are relevant to the development and maintenance of AUD.

Longitudinal studies that capture more than a single snapshot in time are also important to understand the dynamic mechanisms that underpin pathology as well as the effects of interventions. In 2017, Brückmann et al. compared the methylation patterns in 24 patients and 23 healthy controls before and after a 3-week alcohol treatment program. Here, they found 59 differentially methylated CpG sites between patients and healthy controls before entering treatment, and they found 48 CpG sites between patients before and after treatment. Interestingly, they found that the mean global DNA methylation returned to similar levels of healthy controls after the treatment (Bruckmann et al. 2017). In another study by Witt et al., large changes in methylation patterns were found between 99 male patients and 95 healthy controls at the time of both acute withdrawal and after 14 days of protracted withdrawal (Witt et al. 2020). Specifically, pathways related to immune function were identified, corroborating other studies implicating the immune system in alcohol withdrawal and consumption (Coleman and Crews 2018). Similar to what Brückmann and colleagues saw, Witt et al. also found that differences between patients and healthy controls were less pronounced after withdrawal. This suggests that methylation levels may revert to baseline levels after prolonged abstinence, and that, in extension, alcohol and alcohol withdrawal may directly influence DNA methylation patterns. Genes that do not return to baseline methylation levels may indicate genes and pathways that play a role in alcohol addiction itself or the vulnerability to become addicted. However, the important differences in patients compared to healthy controls may also simply not be visible until challenged with alcohol, cues, or stressors. In 69 patients that were followed for 12 months, no difference in methylation levels between patients who relapsed and those who abstained during the period was found (Friedel et al. 2020), indicating that the relationship between alcohol consumption and DNA methylation may indeed be dynamic in nature.

In clinical studies, DNA methylation changes are mostly investigated in whole- or cells found in peripheral blood. Tissues from other important organs, such as brain and liver, are more difficult to obtain and standardize. While there is evidence for some concordance between the methylation patterns of peripheral, blood, and brain tissue, with correlation coefficients estimated to range between 0.33 and 0.40 (Edgar et al. 2017; Hagerty et al. 2016), there is also an evident need to investigate the epigenome in the brain; where the behavioral changes symptomatic to addictions are produced.

One of the first EWAS in human postmortem brain tissue was performed by Wang and colleagues. They investigated the methylome in the prefrontal cortex (PFC; Brodmann Area 9) of 16 male and 7 female pairs of AUD patients and healthy controls. They observed no differences in methylation between cases and controls in female subjects, but a total of 1,812 differentially methylated CpG sites were found to be associated with AUD case/control status in males, indicating a possible effect of sex. Biological processes enriched for genes in these two modules included transcriptional regulation and neural development (Wang et al. 2016). More recently, Zillich and colleagues performed a large EWAS comparing five different brain regions between cases with severe AUD and healthy controls. One of their main findings was a dysregulation of immune-related pathways in the ventral striatum, in line with other studies finding an effect of alcohol on the methylation patterns of immune-related genes. They did, however, not identify any epigenome-wide significant CpG sites in the Broadman Area 9, anterior cingulate cortex, or putamen (Zillich et al. 2021). In another investigation of the methylome in tissue from Brodmann Area 9, Meng and colleagues identified a differentially methylated region, DMR-DLGAP2, associated with alcohol dependence. Methylation at the DMR-DLGAP2 regulated expression of DLGAP2 in vitro, a membrane-associated kinase, e.g., relevant for the organization of synapses (Meng et al. 2021). *Dlgap2* deficient mice were further shown to drink less alcohol, by extension in animal models, indicating a functional role of an epigenetic mark found in human postmortem brain tissue.

The notion that AUD is associated with a progressive recruitment of stress systems in the brain has also been observed at an epigenetic level. In 2019, Gatta et al. investigated DNA methylation in the PFC (Brodmann Area 10) of 25 pairs of AUD cases and controls. While the authors identified a high number of methylation sites that were nominally significant, their analysis highlighted processes containing genes that related to stress adaptation, including the glucocorticoid receptor NR3C1. They found that chronic alcohol consumption was associated with increased methylation levels of NR3C1, and decreased expression of the encoded protein. They also observed changes in the expression level of several other stress-responsive genes in the PFC of AUD cases, such as the well-known CRF, POMC, and FKBP5 (Gatta et al. 2021). Moreover, in one of the larger EWAS conducted so far, the evidence for a dysregulated glucocorticoid signaling in AUD was corroborated. Here, AUD was associated with altered DNA methylation in genes associated with glucocorticoid signaling and immune-related responses. The EWAS was conducted in several independent cohorts and results were replicated across multiple tissues (including PFC and AMG postmortem brain tissue). A probe that was consistently significant across cohorts was the lncRNA growth arrest-specific five gene (GAS5), which has been implicated in the regulation of glucocorticoid receptors. The authors further showed, using neuro imaging (fMRI) and fear conditioning paradigms, that methylation levels in GAS5 network-related probes were associated with stress phenotypes (Lohoff et al. 2021). Together, these studies provide evidence in support of a role for epigenetic mechanisms not only in the initial response to alcohol, but also for the progression of AUD.

Evidence has also suggested a role of other lncRNAs in AUD, where lifetime consumption of alcohol has been correlated with a network of coordinately expressed lncRNAs in the human brain (Farris et al. 2015a, b). Following alcohol withdrawal, expression of the lncRNA MALAT1 was significantly increased in multiple brain-regions of human alcoholics (Kryger et al. 2012). Another lncRNA implicated in AUD is the well-described and evolutionarily conserved brain-derived neurotrophic factor (BDNF) anti-sense lncRNA (BDNF-AS) (Modarresi et al. 2012), a gene that has been shown significantly upregulated in human amygdala (AMG) during the early onset of alcohol abuse (Bohnsack et al. 2019).

Among the non-coding RNAs, microRNAs are currently the most well studied (Farris and Mayfield 2021). MicroRNA expression levels, and their downstream targets, can be altered by drugs of abuse (Smith and Kenny 2018). For example, microarray studies of the superior prefrontal gyrus from human postmortem brain tissue have shown an up-regulation of 48 miRNAs in cases with chronic alcohol abuse compared to healthy controls (Lewohl et al. 2011). A similar list of alcohol-induced miRNA expression changes has also been reported for human neuroblastoma cells (Yadav et al. 2011).

Postmortem brain tissue brings us one step closer to understanding how alcohol-induced epigenetic changes may promote alcohol-related behaviors, and evidence is accumulating in support of an epigenetic regulation in the AUD psychopathology. Several challenges remain, with cell type heterogeneity, and variations in tissue collection. While efforts and sample sizes are growing, sample sizes of 50–100 patients are likely not always enough to achieve epigenome-wide significance, and information may still be lost. Many candidate gene association studies may similarly require replication. To date, most studies have also focused on DNA methylation and, to a lesser extent, non-coding RNAs. These are most likely only a part of the epigenetic mechanisms involved in the regulation of AUD, where histone modifications in particular may also play an important role (Hamilton and Nestler 2019). Remarkably, histone modifications are very understudied in postmortem brain tissue of AUD cases. As we will show below, evidence from preclinical studies demonstrates important histone modifications in alcohol-associated behaviors. Preclinical studies also show that several drugs targeting HDACS are able to mitigate alcohol-related behaviors, suggesting an important role of these epigenetic processes and the need to study alcohol-associated histone modifications in humans.

In clinical studies, it is also difficult to control for the environmental factors, or personal experiences that may affect the expression of epigenetic marks and gene expression changes. The major limitation from human findings is that they are observational and descriptive in nature, and a causal relationship is difficult to establish. Therefore, the use of animal models to study AUD becomes crucial to advance our understanding of the epigenetic mechanisms that underly different stages and symptoms of the disorder. Before introducing the role of epigenetic processes in preclinical AUD models, in the next section, we provide a brief summary of the current animal models that mimic relevant features of alcohol-related behaviors.

3 Animal Models of AUD

Animal models provide means to investigate candidates identified in human tissue as well as to discover novel mechanisms that might translate into humans. The use of clinical-based animal models and sophisticated tools have identified key neural substrates of alcohol reward, seeking, alcohol choice over a natural reward, and compulsive alcohol taking and seeking behavior (Augier et al. 2018; Domi et al. 2021b; Giuliano et al. 2018; Seif et al. 2013; Siciliano et al. 2019). The characterization of different lines of rats and mice genetically predisposed to alcohol drinking has further helped to elucidate several aspects of the neurobiology underpinning AUD (Bell et al. 2006; McBride et al. 2014).

In here, we provide an overview of current models used for studying epigenetics mechanisms in AUD. We also discuss recent animal models that consider the role of individual differences in the vulnerability to develop alcohol-related behaviors in which the role of epigenetic mechanisms has not been investigated.

3.1 Alcohol Taking and Escalation of Intake

In rodents, alcohol taking is commonly assessed using volitional drinking models including free-choice home cage alcohol consumption and operant self-administration procedures where alcohol drinking may be preceded by a nose-poke or a lever-press taking response.

3.1.1 Home Cage Free-Choice Alcohol Consumption

Richter & Campbell were the first to report that laboratory rats voluntarily drink alcohol. They showed that rats allocate their drinking between a water bottle and a bottle containing a diluted alcohol solution, which originated the two-bottle preference test (Richter and Campbell 1940). The free-choice method, using one or more bottles of alcohol solutions at different alcohol concentrations is useful to estimate the voluntary and spontaneous intake as the animal is not forced to drink. Significant escalation patterns of drinking that lead to pharmacologically relevant blood alcohol concentrations (BACs) are obtained when alcohol is offered intermittently instead of continuously (Carnicella et al. 2014; Cippitelli et al. 2012; Domi et al. 2018; Simms et al. 2008; Wise 1973).

Escalation of alcohol drinking can also be modeled in mice using intermittent free-choice drinking procedures (Griffin 2014; Hwa et al. 2011). In contrast to rats, binge patterns in mice are rather obtained using the drinking in the dark (DID) procedure which takes advantage of the most active circadian period by providing a limited 2-h alcohol access presented at 3 h into their dark cycle (Domi et al. 2020; Rhodes et al. 2005; Thiele and Navarro 2014).

Contrary to the free choice drinking procedures, including the liquid diet administration (Rogers et al. 1979), alcohol vapor inhalation leads more readily to physical dependence. Exposure to alcohol vapor induces high BACs (~200 mg%) and the insurgence of the withdrawal symptoms that peak between 12 and 24 h after cessation of alcohol inhalation (Macey et al. 1996; Rimondini et al. 2002; Roberts et al. 1996; Sommer et al. 2008). Prolonged chronic intermittent exposure (CIE) and alcohol vapor procedures induce long-lasting neuroadaptations that reflect in phenotypic traits of AUD and that could be driven by epigenetic mechanisms (Meinhardt and Sommer 2015).

3.1.2 Operant Alcohol Self-Administration

Contrary to the free-choice home cage alcohol consumption, alcohol self-administration procedures allow measurements of a constellation of addiction-like behaviors such as motivation for drinking measured using the progressive ratio paradigm (Hodos 1961), seeking behavior (Domi et al. 2021a) and volitional choice procedures (Augier et al. 2018; Pfarr et al. 2018).

While voluntary drinking using free-choice bottle procedures mainly capture the consummatory aspects of alcohol taking, operant self-administration paradigms, besides the consummatory response, evaluate appetitive and motivational components of alcohol taking. A combination of dependence models in the operant self-administration paradigms constitutes a valid tool to investigate the neurobiology of AUD.

3.2 *Compulsive Alcohol Taking*

Continued alcohol drinking despite adverse legal, health, economic, and societal consequences is a central hallmark of AUD (Association 2013; Luscher et al. 2020; Spanagel 2009a; Tiffany and Conklin 2000). Compulsive alcohol taking and seeking, defined by resistance to adverse and deleterious consequences, represents a major challenge when attempting to treat AUD (Koob and Volkow 2010). As in humans, rodents have been shown to develop persistent drug or alcohol consumption where intake persists despite overt pairing with aversive consequences. Most preclinical alcohol studies have assessed aversion-resistant alcohol taking, by investigating the persistence of the animals to drink alcohol despite adulteration with the bitter tastant, quinine (Wolffgramm 1991; Wolffgramm and Heyne 1995) or lithium chloride (Dickinson et al. 2002). Another way to model compulsive drinking in rodents is persistent alcohol responding despite the presence of an electric shock (Augier et al. 2018; Domi et al. 2021b; Halladay et al. 2020; Seif et al. 2013).

3.3 *Relapse to Alcohol Seeking*

A main challenge in the treatment of AUD is to prevent relapse after patients achieve abstinence. In patients suffering from AUD, places or contexts previously associated with alcohol use, re-exposure to a small amount of alcohol or stressful conditions can provoke relapse during abstinence (Brownell et al. 1986; Hendershot et al. 2011; O'Brien et al. 1992; Wikler 1973). As in humans, studies in animals have confirmed that reinstatement of responding for alcohol under extinction conditions (i.e., in the absence of the reinforcer) can be induced by discriminative and discrete alcohol-associated cues (Sinclair et al. 2012), contextual cues (Chaudhri et al. 2009; Hamlin et al. 2009) and stress (Domi et al. 2021a; Martin-Fardon and Weiss 2013). Less common studies have explored alcohol-seeking behavior using the conditioned place preference (CPP) task in both mice and rats (Mueller and Stewart 2000). After receiving an alcohol injection, subjects are immediately confined to a specific environment. Subsequently, in a drug-free state, subjects can either explore the alcohol-associated context or a second, familiar context. The degree of preference to which the animals seek and spend time for a previously alcohol-paired context represents an index of the strength for alcohol seeking.

A robust alcohol reinstatement is commonly produced as well by physical (electric foot shock) (Le and Shaham 2002; Le et al. 2002) or pharmacological stressors such as the anxiogenic drug yohimbine (Cippitelli et al. 2010; Le et al. 2005).

3.4 *Animal Models Assessing Individual Differences in the Vulnerability to Develop AUD*

Only a small portion of individuals who drink will develop AUD (Grant et al. 2015), suggesting that research to discover novel treatments should consider individual differences in vulnerability for clinically relevant behaviors (Deroche-Gamonet et al. 2004; Domi et al. 2019; Heilig et al. 2019; Jadhav et al. 2017; Piazza and Deroche-Gamonet 2013). Similar to the clinical situation, only a fraction of rats or mice that take alcohol will display enhanced alcohol motivation and compulsive alcohol taking and seeking behaviors (Augier et al. 2018; Domi et al. 2021b; Siciliano et al. 2019). Preclinical research efforts that focus on this subset of vulnerable animals might provide translational biomarkers and prevention strategies in the AUD field (Heilig et al. 2016).

3.5 *Genetic Animal Models of AUD*

The development of valid animal models of AUD is challenging. Outbred rat strains have the advantage to mimic human heterogeneity; however, they do not readily consume alcohol to reach pharmacologically significant blood alcohol concentrations to promote alcohol dependence. To overcome these limitations, rats that showed high alcohol intake and preference have been selectively bred across the years, to develop alcohol-preferring rat lines (Barkley-Levenson and Crabbe 2014; Bell et al. 2016; Borruto et al. 2021; Ciccocioppo 2013; Colombo et al. 2006; Crabbe et al. 2010). These lines have been instrumental in expanding our understanding of the neurobiology of AUD and enabling pharmacological studies to screen medications to treat alcohol-related behaviors (Bell et al. 2016). However, it is important to note that genetic rodent models display innate altered expression of several genes which can promote the expression of traits not related to alcohol, limiting their generalizability in AUD research.

In summary, over the past years, researchers have developed several animal models of AUD that allow measurements of a constellation of addiction-like traits. In addition to the studies based on the effects of acute and chronic alcohol injections, such models provide valuable tools to study the neurobiology and the molecular mechanisms such as epigenetic processes underlying relevant alcohol-associated behaviors.

4 Epigenetic Mechanisms in Animal Models of Alcohol-Associated Behaviors

Here, we will introduce a synopsis of epigenetic modifications observed in preclinic research using some of the aforementioned animal models of AUD. We also incorporate the findings from pharmacological and viral vector approaches, providing evidence for a functional role of the epigenetic enzymes on alcohol-associated behaviors.

4.1 DNA Methylation and Alcohol-Associated Behaviors

Early studies have suggested that the levels of methyl groups that are contained in a diet can modulate alcohol consumption. For instance, Williams et al. observed an increased alcohol consumption in rats exposed to a diet deficient in methyl donors (e.g., folates, choline). In opposite, rats with methyl-enriched diets showed decreased alcohol intake (Williams et al. 1949). In line with this study, treatment with the methyl donor S-adenosyl-methionine (SAM) prevented alcohol escalation following CIE in mice (Qiang et al. 2014). Preclinical work using animal models of alcohol-associated behaviors indicated that alcohol can in turn modulate DNA methylation levels. For instance, we showed that protracted abstinence after chronic intermittent alcohol exposure increased DNA methylation in the dorsomedial PFC (Barbier et al. 2015). Additionally, a study using male rhesus macaques observed similar findings and further found differential DNA methylation levels relative to drinking patterns (Cervera-Juanes et al. 2017). Rhesus macaques that were categorized as "low drinkers" showed decreased overall DNA methylation in the NAc compared to alcohol-naïve rhesus macaques. In contrast, "high drinker" rhesus macaques presented an overall increase in DNA methylation levels in the NAc when compared to alcohol-naïve controls. Substantial evidence also reported alcohol-induced changes in the expression level of enzymes that modulate DNA methylation. *Dnmt1* expression was found upregulated in the NAc of mice exposed to an intermittent two-bottle choice paradigm compared to control mice (Warnault et al. 2013). We observed similar findings using a rat model of chronic intermittent vapor exposure. In this study, alcohol CIE resulted in increased *Dnmt1* expression in the dorsomedial PFC, 3 weeks after protracted abstinence (Barbier et al. 2015). Increased expression of *Dnmt1* was associated with decreased expression of genes related to synaptic neurotransmitter release. These changes were abolished by intracerebroventricular (i.c.v) infusion of the DNMT1 inhibitor RG108, suggesting a regulatory role of alcohol-induced DNA hypermethylation.

Mice show differential drinking pattern as a function of their strain. Particularly, C57BL/6J mice are known to drink high amounts of alcohol, whereas DBA/2J mice show low level of alcohol consumption. Using this characteristic, researchers have investigated DNMTs expression profile in these two strains. No changes were observed in *Dnmt1*, *Dnmt3a*, or *Dnmt3b* levels in C57 mice when compared to DBA/2J mice (Gavin et al. 2016). However, the expression level of the Growth Arrest and DNA Damage-Inducible Beta (GADD45B), a protein known to promote DNA demethylation was found downregulated in C57 mice compared to DBA/2J mice (Ma et al. 2009). Consistent with a role of GADD45B in alcohol consumption, mice lacking GADD45B displayed higher alcohol consumption compared to wild-type mice in both two-bottle free choice and DID paradigm (Gavin et al. 2016).

At the gene level, substantial evidence indicates gene-specific DNA methylation changes induced by alcohol exposure. Several studies reported a decreased DNA methylation at the N-methyl-D-aspartate receptor subunit 2B (*Nr2b*) gene promoter in the PFC of mice exposed to CIE compared to control mice (Marutha Ravindran and Ticku 2004; Qiang et al. 2014). DNA methylation at the promoter region of *Nr2b* was dependent on the pattern of alcohol exposure as acute alcohol treatment did not result in *Nr2b* DNA methylation changes (Marutha Ravindran and Ticku 2004). Using a rat model of chronic intermittent alcohol exposure, we found increased DNA methylation at the exon1 of synaptotagmin 2 (*Syt2*) in the dorsomedial PFC (Barbier et al. 2015). DNA hypermethylation at exon1 was associated with decreased expression of *Syt2*. Additionally, i.c.v. infusion of the DNMT inhibitor RG108 rescued alcohol-induced *Syt2* downregulation together with the expression of another synaptotagmin: *Syt1*, suggesting a role of DNA methylation in *Syt1* and *Syt2* silencing. Knockdown of *Syt1* or *Syt2* in the dorsomedial PFC increased compulsive alcohol intake as shown by increased tolerance to quinine adulteration of the alcohol solution (Barbier et al. 2015, 2021). Altogether these findings point towards a role of DNA methylation on alcohol-associated behaviors. Consequently, targeting the enzymes that modulate DNA methylation (i.e., DNMT inhibitors) may present a good therapeutic strategy for treating AUD. Accordingly, several studies have put the focus on determining the role of DNMT inhibitors on alcohol-associated behaviors. For instance, treatment with the non-specific DNMT inhibitor 5-azacytidine prior CIE, potentiated alcohol intake (Qiang et al. 2014). In contrast, other studies reported an opposite effect of the DNMT inhibitor on alcohol consumption. Warnault and colleagues, observed a decreased alcohol intake in mice after treatment with 5-azacytidine (Warnault et al. 2013). We also found a decreased alcohol self-administration in alcohol post-dependent rats after chronic infusion of the DNMT1 inhibitor RG108 (Barbier et al. 2015). The differential effect of the DNMT inhibitor observed by these studies may be explained by the different experimental procedures used. Specifically, mice were administered with 5-azacytidine during chronic intermittent alcohol exposure, whereas in Warnault and our studies, the DNMT inhibitor was injected after 8 weeks of intermittent alcohol intake and 7 weeks of alcohol exposure, respectively.

4.2 *Histone Modifications and Alcohol-Associated Behaviors*

A large number of studies implicate histone modifications in alcohol-associated behaviors. Because most of these studies focused on histone acetylation and methylation, this chapter will mainly report the findings on the role of these two epigenetic marks.

4.2.1 Histone Acetylation

High amounts of alcohol consumption in rodents have been associated with low levels of histone acetylation (Pandey et al. 2008; Warnault et al. 2013). Specifically, decreased histone H4 acetylation was found in the NAc of mice with a history of excessive alcohol intake when compared to alcohol-naïve mice (Warnault et al. 2013). Lower acetylation of H3K9 was also observed in the AMG of alcohol-preferring (P) compared to their alcohol non-preferring counterparts (NP) (Moonat et al. 2013). Similar results were found after 24 h of alcohol deprivation following chronic alcohol intake. In this study, the anxiolytic effects of alcohol withdrawal were associated with a significant decrease of global H3 and H4 acetylation together with increased histone deacetylase (HDAC) activity (Pandey et al. 2008). In contrast to these findings, several studies observed increased rather than decreased levels of histone acetylation. For instance, Simon O'Brian et al. found increased levels of histone H3K9 acetylation in the PFC and AMG of alcohol dependent compared to non-dependent rats after acute withdrawal from ethanol vapor (Simon-O'Brien et al. 2015). Increased histone acetylation was also observed after acute alcohol exposure. One single injection of ethanol (2 g/kg; i.p.) resulted in a significant increase of H3K9 and H4K8 acetylation in the AMG (Sakharkar et al. 2012). Similarly, 10 days of alcohol self-administration (2 h) resulted in histone hyperacetylation together with a decreased expression of HDAC4 and HDAC5 mRNA levels in the NAc (Griffin et al. 2017). The discrepancies found in the histone acetylation profile and HDAC activity may be due to different patterns of alcohol exposure. In line with this hypothesis, Pandey et al. showed that the effect of alcohol withdrawal on histone acetylation depends on the duration of alcohol exposure. For instance, 24 h withdrawal after acute alcohol exposure resulted in an increased global acetylation, whereas 24 h withdrawal after chronic alcohol exposure led to a decreased histone acetylation and increased HDAC activity (Pandey et al. 2008).

Over the past years, researchers have identified several genes with altered levels of histone acetylation following alcohol exposure. Strong evidence supports the role of epigenetic mechanisms in alcohol-induced *Bdnf* expression changes. Notably, lower expression of *Bdnf* in the AMG of P rats was associated with higher H3K9 acetylation on the promoter region of *Bdnf* exon IV compared to NP rats (Moonat et al. 2013). In another study, treatment with the HDAC inhibitor trichostatin A prevented *Bdnf* downregulation following alcohol withdrawal (You et al. 2014). Similar results were found in the hippocampus of mice exposed to the free alcohol choice paradigm. Chronic alcohol consumption resulted in an increased expression of *Bdnf* exons II, III, and VI associated with an enrichment in histone H3 acetylation on the *Bdnf* promoter VI in the hippocampus (Stragier et al. 2015).

The neuronal activity marker activity-regulated cytoskeleton-associated protein gene (*Arc*), which is one of the downstream targets of BDNF has also been shown to be modulated by histone acetylation. Alcohol-preferring P rats exhibit lower levels of *Arc* mRNA and H3K9 acetylation in absence of alcohol exposure. In contrast, P rats showed increased levels of *Arc* mRNA and H3K9 acetylation on *Arc* promoter when exposed to alcohol (Moonat et al. 2011, 2013). Knockdown of *Hdac2* in the AMG with siRNA, increased histone acetylation of both *Bdnf* and *Arc* and resulted in increased levels of BDNF and ARC, suggesting a regulation by histone acetylation (Moonat et al. 2013). Additionally, several studies proposed that alcohol-induced expression changes of *Nr2b* may in part be regulated by epigenetic mechanisms. Using an in vitro mouse model of alcohol CIE followed by withdrawal, Qiang et al., showed that CIE upregulated the expression of *Nr2b* and that it was associated with an increase in H3K9 acetylation on the 5′ regulatory region of *Nr2b* gene (Qiang et al. 2011). Similar findings were observed in rats exposed to chronic alcohol. Both *Nr2b* mRNA expression and histone H3K9 acetylation of the *Nr2b* gene were found upregulated during withdrawal (Li et al. 2019). Together these results strongly suggest a key role of histone acetylation on alcohol-induced neuroadaptations and point towards a potential benefit to inhibiting enzymes that regulate histone acetylation for preventing relapse to alcohol seeking.

Several drugs targeting HDACs have been tested in animal models of alcohol-associated behaviors. To date, four HDAC inhibitors (i.e., Vorinostat, Romidepsin, Panobinostat, and Belinostat) have been approved by the Food and Drug Administration (FDA), reinforcing the possible clinical impact. Vorinostat also known as suberanilohydroxamic acid (SAHA) inhibits both class I and II HDAC activities by binding to the pocket of the catalytic site (Finnin et al. 1999). Vorinostat is used to treat cutaneous T-cell lymphoma when the disease persists or deteriorates (McClure et al. 2018). Vorinostat has also been shown to reduce alcohol consumption in rodents. Systemic injection of Vorinostat (50–100 mg/kg) dose-dependently decreased alcohol intake in a mouse DID model and the dose of 50 mg/kg reduced alcohol self-administration in rats (Warnault et al. 2013). Intraperitoneal injection of Vorinostat in rats exposed to chronic alcohol exposure also prevented withdrawal-induced hyperalgesia (Pradhan et al. 2019), suggesting that Vorinostat does not only affect alcohol consumption but also other alcohol-related behaviors. Similar to Vorinostat, trichostatin A (TSA) inhibits both HDAC class I and II (Chuang et al. 2009; Kim et al. 2019). Administration of TSA after chronic alcohol exposure in rats prevented the downregulation of *Gabra1* expression as well as reduced alcohol consumption (Bohnsack et al. 2018). Consistent with these findings, intraperitoneal injection of TSA (2 mg/kg; once daily; 3 days) reduced both anxiety-like behavior and alcohol intake in the alcohol-preferring P rats but not in the non-preferring NP rats (Sakharkar et al. 2014). In line with the pharmacological effect of the HDAC inhibitors on alcohol consumption, studies have also demonstrated that the HDAC inhibitor class I and II sodium butyrate can reduce alcohol intake. Treatment with sodium butyrate decreased excessive alcohol self-administration in rats with a history of alcohol dependence but had no effect in non-dependent animals (Simon-O'Brien et al. 2015). HDAC inhibitors specific to HDAC class I have also been shown to reduce alcohol consumption. I.c.v. injection of entinostat, also known as MS-275 (500 μM), decreased motivation to consume alcohol and cue-induced relapse of alcohol seeking in heavy drinking rats (Jeanblanc et al. 2015). Similarly, knockdown of the HDAC class 1 *Hdac2* decreased voluntary drinking (Moonat et al. 2013). These findings suggest a role of HDACs class I enzymes in reducing alcohol consumption and therefore provide a rationale for developing compounds that can target this specific class of HDACs.

To date, researchers have mostly been focusing on the enzymes that directly modulate histone acetylation. However, a recent study suggests that alcohol-induced histone acetylation in the brain may not only be driven by HDAC/ histone acetyltransferase activity but may also partly come from alcohol metabolism. In the liver, the breakdown of alcohol results in an increased peripheral level of acetate that is incorporated into acetyl-coenzyme A, a substrate for histone acetylation (Moghe et al. 2011). Using in vivo stable-isotope labeling, a study demonstrated that a portion of acetate-derived from the metabolism of alcohol contributes to histone acetylation in the brain. Knockdown of the acyl-CoA synthetase short-chain family member 2 (ACSS2), which generates the acetyl donor acetyl-CoA from acetate, decreased expression of alcohol-mediated CPP, suggesting a role of ACSS2 in alcohol-associative memory (Mews et al. 2019). Pharmacological inhibition of ACSS2 may therefore be a promising approach for intervention in alcohol use disorders.

4.2.2 Histone Methylation

In comparison to DNA methylation and histone acetylation, histone methylation has been understudied in the AUD field. Nonetheless, increasing evidence has suggested an important role of histone methylation for alcohol-associated behaviors. Like DNA methylation and histone acetylation, methylation levels on specific histone marks have been found to be regulated by alcohol exposure. For instance, a genome-wide analysis from the PFC of mice with a history of 3 weeks alcohol vapor exposure indicated a global increase in the repressive mark H3K27me3 together with a decreased expression of the activating mark H3K4me3 when measured after 21 days of protracted abstinence (Gavin et al. 2018). These epigenetic modifications were associated with long-term gene expression changes. Five genes (i.e., *Pard3*, *Plagl1*, *Calu*, *Ezr*, and *Dgkb*) showed differential expression levels that matched the predicted expression based on their histone methylation profile. In contrast, another study found a global increase in H3K4me3 in the mouse cerebral cortex after acute alcohol administration (3 g/kg; i.p.) (Finegersh and Homanics 2014). The discrepancy between these two studies may result from differential alcohol exposure. In the former study, the levels of H3K4me3 were evaluated after 21 days of protracted abstinence from chronic alcohol exposure, whereas in the latter study, levels of H3k4me3 were measured after acute alcohol exposure.

Researchers have also identified alcohol-induced changes in histone methylation at specific gene promoters. For example, acute alcohol exposure (1.5 g/kg, oral administration) resulted in a decreased expression of the repressive mark H3K27me3 and an increased expression of the activating mark H3K9ac on the prepronociceptin (*Pnoc*) and prodynorphin (*Pdyn*) promoters (D'Addario et al. 2013). This effect was associated with an enhanced expression of both *Pnoc* and *Pdyn*, suggesting that acute alcohol exposure increased their expression through epigenetic regulation.

Another study showed an increased expression of H3K4me3 and H3K27me2 at different promoter regions of *Bdnf* following chronic alcohol consumption (Stragier et al. 2015). These changes were correlated with an overexpression of *Bdnf* after chronic alcohol exposure, indicating a regulation of *Bdnf* expression via histone methylation. Similar findings were observed for the glutamatergic receptor subunit NR2B and the GABAergic receptor subunit GABA-Aα5. Chronic intermittent alcohol exposure led to H3K9me2 and H3K9me3 downregulation at the *Nr2b* gene promoter (Qiang et al. 2011) and H3K4me3 upregulation at the *Gaba-aα5* gene promoter region in the rat PFC (Zeng et al. 2018). In line with H3K4me3 enrichment (activating mark) at the *Gaba-aα5* gene promoter region, *Gaba-aα5* was upregulated in the PFC of rats exposed to chronic alcohol exposure compared to control rats. Interestingly, this study also found increased expression of *Gaba-aα5* in rat offspring of alcohol-exposed rats compared to the offspring of alcohol naïve animals. This suggests that alcohol-induced epigenetic changes may be passed on to the next generation, therefore increasing their vulnerability to develop AUD. Alcohol drinking during adolescence has also been shown to modulate histone methylation. For example, the activating mark H3K4me2 was found to be upregulated in the promoter region of cFos, Cdk5, and FosB of the PFC of adolescent rats that received chronic intermittent injections of alcohol compared to saline controls (3 g/kg; i.p.) (Pascual et al. 2012).

To date, many enzymes that regulate histone methylation processes have been identified (Allis et al. 2007). However, only a few studies have investigated the role of these enzymes in alcohol-associated behaviors. Kyzar et al. observed a decreased expression of the lysine-specific demethylase 1, (*Lsd1)* and *Lsd1 + 8a* in the AMG of adult rats previously exposed to chronic alcohol during their adolescence (Kyzar et al. 2017). LSD1 specifically removed methyl marks from di-methylated or mono-methylated H3K4 (Shi et al. 2004) and H3K9 (Cloos et al. 2008). In line with the functional role of LSD1, chronic intermittent alcohol exposure during adolescence resulted in an increased expression of H3K9me2 in the AMG. Additionally, the authors found that an acute challenge of alcohol during adulthood restored the levels of *Lsd1* mRNA and reduced the anxiety-like behavior that is characteristic of adult rats exposed to alcohol during adolescence. Together these results suggest a potential role of LSD1 in alcohol-induced anxiety-like behavior and might represent a common target in the comorbidity between AUD and anxiety disorders (Kyzar et al. 2017). Another epigenetic enzyme, lysine(K)–specific demethylase 6B (KDM6B) has been found to be upregulated in the PFC and NAc of rats with a history of alcohol dependence as well as in human alcoholic brains (Johnstone et al. 2021). KDM6B specifically demethylates di- or tri-methylated H3K27 and has been shown to play an important role in modulating the inflammatory responses (Shentu et al. 2021). Consistent with a role of KDM6B in these processes, alcohol-induced increased expression of *Kdm6B* correlated with a significant enrichment in H3K27me3 at genes in the IL6 signaling pathway. This suggests a potential role of KDM6B in the dysregulation of the inflammatory system often observed in AUD (Szabo and Lippai 2014). We also found a significant downregulation of the histone methyltransferase PR domain containing 2 (PRDM2) in the prelimbic cortex of rats exposed to intermittent alcohol vapor exposure (Barbier et al. 2017). PRDM2 preferentially catalyzes the addition of a methyl group at H3K9 (Congdon et al. 2014). Knockdown of *Prdm2* in the prelimbic cortex of non-dependent rats, led to the development of dependent-like characteristics including increased alcohol self-administration, increased tolerance to quinine adulteration and increased stress-induced reinstatement, suggesting a functional role of PRDM2 in alcohol-associated behaviors (Barbier et al. 2017).

4.3 Non-coding RNAs

Studies have also shown that alcohol can modify expression levels of non-coding RNAs. For example, mice exposed to chronic alcohol produced changes in circular RNA (circRNA) profiles in liver and heart tissue (Dou et al. 2020; Meng et al. 2019; Yang et al. 2018). CircRNAs can regulate miRNAs (Memczak et al. 2013) and are characterized by a circular structure that renders them resilient to RNA degradation and thus very stable (Chen and Yang 2015; Chien et al. 2020). This makes circRNAs interesting as putative biomarkers for AUD. Such attempts have indeed been made, by e.g., Liu et al. who found a correlation between serum hsa_circ_0004771 and the severity of alcohol dependence (Liu et al. 2021). Some functional investigations have also been conducted, such as by Lu and colleagues who demonstrated a proinflammatory role of circ_1639 in murine alcoholic liver disease (Lu et al. 2019). However, the role of circRNAs in alcohol-induced pathogenesis remains largely unclear (Chien et al. 2020).

In contrast to circRNAs, numerous studies have investigated the role of miRNAs in alcohol-related behaviors, resulting in the discovery of several miRNAs and miRNA networks that may be of importance to AUD in both rodents (Nunez et al. 2013) and humans (Lim et al. 2021). In a model of adolescent binge drinking, Kyzar et al., showed that miR-137, which targets LSD1, was upregulated in the AMG of adult rats. The infusion of an antagomir targeting miR-137 in the central AMG was sufficient to rescue aberrant alcohol- and anxiety characteristics, suggesting a functional role of miR-137 in these behaviors (Kyzar et al. 2019). Further, Darcq et al. showed that miR-30a-5p in the PFC plays a role in the transition from moderate to excessive alcohol consumption in mice, and that overexpressing this miRNA in the mPFC was sufficient to escalate alcohol consumption in mice (Darcq et al. 2015). Tapocik et al. also showed that the miRNA-206 has similar effects in the mPFC of rats (Tapocik et al. 2014). Both miRNAs of these latter studies shared a common target: BDNF. This demonstrates that alcohol-induced Bdnf expression changes are not only regulated by DNA methylation and histone modification but also by non-coding RNAs.

More broadly, this suggests that epigenetic mechanisms often work in concert to regulate gene expression and, in extension, behavior. Pharmacological and viral approaches that target epigenetic mechanisms in animal models of alcohol-associated behaviors are summarized in Table 1.

Table 1 Pharmacological and viral approaches that target epigenetic mechanisms in AUD animal models

Drug/gene	Epigenetic mechanism	Result	Paradigm	Route	Species	Reference
SAM	Methyl donor	Decreased preference and drinking	2BC after vapor CIE Exposure	i.p	Male C57BL/6J mice	Qiang et al. (2014)
RG108	DNMT1 inhibitor	Decreased drinking	Alcohol self-administration after vapor CIE	i.c.v; mPFC	Male Wistar rats	Barbier et al. (2015)
GADD45B	Promotes DNA demethylation	Increased drinking	2BC		Male Gadd45b KO mice	Gavin et al. (2016)
			DID			
Decreased expression Syt1/Syt2	Promotes DNA methylation	Increased compulsive drinking	Quinine adulteration of alcohol self-administration	KD in the dmPFC	Male Wistar rats	Barbier et al. (2021)
5-azacytidine	Non specific DNMT inhibitor	Increased drinking	2BC after vapor CIE Exposure	i.c.v	Male C57BL/6J mice	Qiang et al. (2014)
		Decreased drinking	I2BC	i.p	Male C57BL/6J mice	Warnault et al. (2013)
Vorinostat	HDAC I-II inhibitor	Decreased drinking	Alcohol self-administration	i.p	Male Long Evans rats	
		Decreased drinking	DID	i.p	Male C57BL/6J mice	
Trichostatin A	HDAC I-II inhibitor	Decreased drinking	I2BC	i.p	Male Sprague Dawley rats	Bohnsack et al. (2018)
		Decreased drinking	DID	i.p	Male C57BL/6J mice	Warnault et al. (2013
		Decreased drinking	2BC	i.p	Male P and NP rats	Sakharkar et al. (2014)
		Increased drinking	2BC after vapor CIE	i.p	Male C57BL/6J mice	Qiang et al. (2014)
				i.c.v		

(continued)

Table 1 (continued)

Drug/gene	Epigenetic mechanism	Result	Paradigm	Route	Species	Reference
Sodium butyrate (NAB)	HDAC I-II inhibitor	Decreased drinking	Alcohol self-administration after vapor CIE	i.p	Male Wistar rats	Simon-O'Brien et al. (2015)
				i.c.v		
		Decreased drinking	I2BC	i.p	Male Long-Evans rats	
		Decreased relapse	ADE	i.p	Male Sprague-Dawley rats	
Entinostat	HDAC I inhibitor	Decreased drinking, motivation and seeking	Alcohol self-administration; PR; Cue-induced relapse	i.c.v	Male Long Evans rats	Jeanblanc et al. (2015)
Decreased expression *Hdac2*	Increased histone acetylation	Decreased drinking	2BC	KD in the central amygdala	Male P rats	Moonat et al. (2013)
ACSS2	Acetyl donor	Decreased alcohol reward	CPP	KD in the dorsal hippocampus	Male C57BL/6J mice	Mews et al. (2019)
Prdm2 knockdown	Histone Methyltransferase	Increased drinking	Alcohol self-administration	KD in the PL	Male Wistar rats	Barbier et al. (2017)
		Increased compulsive drinking	Quinine adulteration			
		Increased seeking	Stress induced relapse			
miR-30a-5p	microRNA	Increased drinking	I2BC	mPFC	Male C57BL/6J mice	Darcq et al. (2015)
miRNA-206	microRNA	Increased drinking	Alcohol self-administration	mPFC	Male Wistar rats	Tapocik et al. (2014)

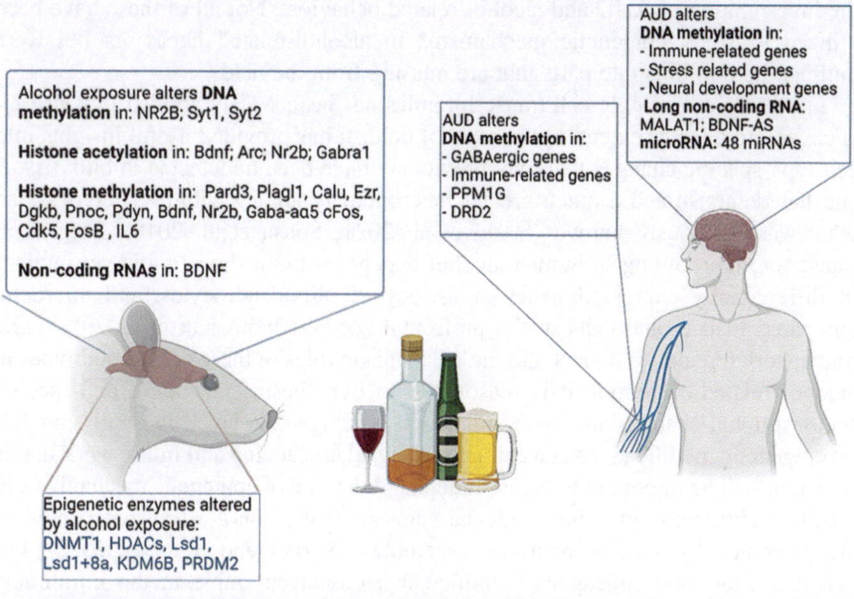

Fig. 2 Epigenetic mechanisms and gene expression modifications in AUD: preclinical and clinical evidence

5 Concluding Remarks and Future Considerations

A considerable effort has been made to understand the epigenetic underpinnings of AUD. Altogether, the literature reviewed in here, supports the role of epigenetic mechanisms in alcohol-associated behaviors (Fig. 2). Of importance for the clinic, studies using pharmacological, or vector KD approaches provide a rationale for utilizing pharmacotherapies that target epigenetic enzymes to treat AUD (Table 1). However, the existing FDA-approved drugs that inhibit epigenetic enzymes generate considerable side effects and cannot readily be applied for psychiatric diseases. A better understanding of the epigenetic mechanisms and their function in different stages of alcohol-related behaviors is central for the development of novel epigenetic-based therapeutics for AUD. To date, findings on epigenetic mechanisms in AUD rely mainly on studies that use a rather small set of alcohol-associated behavioral paradigms (e.g., alcohol drinking and operant alcohol self-administration). This is a major limitation for a translational validity of the findings, which unfortunately do not capture the broad constellation of maladaptive behaviors characteristic of the disorder. There is thus a need to go beyond the simple drinking procedures and investigate the role of epigenetic mechanisms in alcohol-associated behaviors that better model clinical features of human AUD (e.g., relapse, persistence to drink despite negative consequences or aberrant choice behavior (Goltseker et al. 2019). In this chapter, we gave a review of several preclinical models used in the investigation of AUD and alcohol-related behaviors. Not all of these have been utilized to study epigenetic mechanisms in alcohol-related behaviors but were outlined here to illustrate parts that are missing from the field.

In recent years, single-cell transcriptomics has been widely applied in neuroscience, and to a lesser extent, in the alcohol field. It has provided useful insights into cell type-specific changes that may previously have been undetected in bulk tissue, and has demonstrated a much greater heterogeneity among cellular subtypes than what was previously known (Hasel et al. 2021; Sousa et al. 2018). Single-cell transcriptome profiling of human alcohol-dependent brain showed a larger number of differentially expressed genes in astrocytes, oligodendrocytes, and microglia compared to neuronal cells in the prefrontal cortex (Brenner et al. 2020). While much work remains to understand the mechanistic roles of these cellular subtypes in alcohol related-behaviors, it is reasonable to hypothesize that these cell-specific transcriptional changes are also regulated at an epigenetic level. Presently, studies on epigenetic modifications at a cell-specific level are lacking and future work in this direction will be important to better understand the role of epigenetic mechanisms in AUD. Additionally, growing evidence suggests that alcohol-associated behaviors are promoted by specific neuronal ensembles (Korber and Sommer 2022). For instance, Pfarr and colleagues identified a neuronal ensemble in the infralimbic cortex that was activated during presentation of alcohol-associated cues. Inactivation of these neurons resulted in increased cue-induced reinstatement, indicating an inhibitory control from these neurons over the seeking response (Pfarr et al. 2015). Another study demonstrated that individual differences in vulnerability for compulsive alcohol self-administration were in part driven by an ensemble of protein kinase Cδ-positive (PKCδ+) inhibitory neurons in the lateral subdivision of the central AMG (Domi et al. 2021b). These findings highlight the importance of investigating the role of epigenetic processes in alcohol-associated neuronal ensembles rather than bulk tissues. Several technologies including CUT&RUN, CUT&TAG, DNA bisulfite sequencing and RBSS can now be combined with single-cell methodologies and therefore allow the study of the epigenetic landscape at a cellular level (Ahn et al. 2021). Because epigenetic modifications are closely linked with gene expression, combining these techniques with a method that assesses chromatin accessibility (i.e., ATAC-sequencing) may provide us with a more extensive knowledge on how epigenetic processes regulate gene expression changes in AUD.

Another gap in the field of alcohol and epigenetics lies in the lack of studies using females in human postmortem tissue-studies (Wedemeyer et al. 2020), as well as in animal models of alcohol-associated behaviors. All the preclinical studies cited in this chapter were conducted in male animals. AUD has a higher prevalence in men than in women, and males account for more of the total alcohol-related harm than females, including deaths (White et al. 2015). However, the gap in the prevalence of AUD between men and women has decreased over time, reinforcing the need to study AUD in females. Additionally, women have a higher co-occurrence of anxiety disorders (Kessler et al. 1997), and are more likely to cite stressful life experiences and negative emotions as reasons for substance use and relapse. This suggests that women may be more prone to rely on alcohol to reduce distress of negative affective states (Erol and Karpyak 2015; Smith and Randall 2012). Thus, genetic and epigenetic mechanisms that promote AUD in females may differ from males. Moreover, epigenetic studies that consider the individual differences in the vulnerability to develop AUD are warranted.

In summary, a large body of evidence indicates a role of epigenetic processes on alcohol-associated behaviors. However, more work is needed to fully understand the intricate relationship between epigenetic modifications and AUD. Specifically, the application of state-of the-art next-generation sequencing approaches may provide useful insights into how epigenetic mechanisms can impact the neuronal ensembles that promote alcohol-associated behaviors.

References

Aberg KA, Chan RF, Xie L, Shabalin AA, van den Oord E (2018) Methyl-CpG-binding domain sequencing: MBD-seq. Methods Mol Biol 1708:171–189. https://doi.org/10.1007/978-1-4939-7481-8_10

Ahn J, Heo S, Lee J, Bang D (2021) Introduction to single-cell DNA methylation profiling methods. Biomol Ther 11:1013. https://doi.org/10.3390/biom11071013

Allis CD, Berger SL, Cote J, Dent S, Jenuwien T, Kouzarides T, Pillus L, Reinberg D, Shi Y, Shiekhattar R et al (2007) New nomenclature for chromatin-modifying enzymes. Cell 131:633–636. https://doi.org/10.1016/j.cell.2007.10.039

Anthony JC, Warner LA, Kessler RC (1994) Comparative epidemiology of dependence on tobacco, alcohol, controlled substances, and inhalants: basic findings from the National Comorbidity Survey. Exp Clin Psychopharmacol 2:244–268. https://doi.org/10.1037/1064-1297.2.3.244

Association AP (2013) Diagnostic and statistical manual of mental disorders (DSM-5®). American Psychiatric Pub

Augier E, Barbier E, Dulman RS, Licheri V, Augier G, Domi E, Barchiesi R, Farris S, Natt D, Mayfield RD et al (2018) A molecular mechanism for choosing alcohol over an alternative reward. Science 360:1321–1326. https://doi.org/10.1126/science.aao1157

Barbier E, Tapocik JD, Juergens N, Pitcairn C, Borich A, Schank JR, Sun H, Schuebel K, Zhou Z, Yuan Q et al (2015) DNA methylation in the medial prefrontal cortex regulates alcohol-induced behavior and plasticity. J Neurosci 35:6153–6164. https://doi.org/10.1523/JNEUROSCI.4571-14.2015

Barbier E, Johnstone AL, Khomtchouk BB, Tapocik JD, Pitcairn C, Rehman F, Augier E, Borich A, Schank JR, Rienas CA et al (2017) Dependence-induced increase of alcohol self-administration and compulsive drinking mediated by the histone methyltransferase PRDM2. Mol Psychiatry 22:1746–1758. https://doi.org/10.1038/mp.2016.131

Barbier E, Barchiesi R, Domi A, Chanthongdee K, Domi E, Augier G, Augier E, Xu L, Adermark L, Heilig M (2021) Downregulation of synaptotagmin 1 in the prelimbic cortex drives alcohol-associated behaviors in rats. Biol Psychiatry 89:398–406. https://doi.org/10.1016/j.biopsych.2020.08.027

Barkley-Levenson AM, Crabbe JC (2014) High drinking in the dark mice: a genetic model of drinking to intoxication. Alcohol 48:217–223. https://doi.org/10.1016/j.alcohol.2013.10.007

Bartel DP (2004) MicroRNAs: genomics, biogenesis, mechanism, and function. Cell 116:281–297. https://doi.org/10.1016/s0092-8674(04)00045-5

Beck D, Ben Maamar M, Skinner MK (2022) Genome-wide CpG density and DNA methylation analysis method (MeDIP, RRBS, and WGBS) comparisons. Epigenetics 17:518–530. https://doi.org/10.1080/15592294.2021.1924970

Begleiter H (1975) Ethanol consumption subsequent to physical dependence. Adv Exp Med Biol 59:373–378. https://doi.org/10.1007/978-1-4757-0632-1_26

Bell RL, Rodd ZA, Lumeng L, Murphy JM, McBride WJ (2006) The alcohol-preferring P rat and animal models of excessive alcohol drinking. Addict Biol 11:270–288. https://doi.org/10.1111/j.1369-1600.2005.00029.x

Bell RL, Hauser S, Rodd ZA, Liang T, Sari Y, McClintick J, Rahman S, Engleman EA (2016) A genetic animal model of alcoholism for screening medications to treat addiction. Int Rev Neurobiol 126:179–261. https://doi.org/10.1016/bs.irn.2016.02.017

Bidwell LC, Karoly HC, Thayer RE, Claus ED, Bryan AD, Weiland BJ, YorkWilliams S, Hutchison KE (2019) DRD2 promoter methylation and measures of alcohol reward: functional activation of reward circuits and clinical severity. Addict Biol 24:539–548. https://doi.org/10.1111/adb.12614

Bierut LJ, Agrawal A, Bucholz KK, Doheny KF, Laurie C, Pugh E, Fisher S, Fox L, Howells W, Bertelsen S et al (2010) A genome-wide association study of alcohol dependence. Proc Natl Acad Sci U S A 107:5082–5087. https://doi.org/10.1073/pnas.0911109107

Bohnsack JP, Hughes BA, O'Buckley TK, Edokpolor K, Besheer J, Morrow AL (2018) Histone deacetylases mediate GABAA receptor expression, physiology, and behavioral maladaptations in rat models of alcohol dependence. Neuropsychopharmacology 43:1518–1529. https://doi.org/10.1038/s41386-018-0034-8

Bohnsack JP, Teppen T, Kyzar EJ, Dzitoyeva S, Pandey SC (2019) The lncRNA BDNF-AS is an epigenetic regulator in the human amygdala in early onset alcohol use disorders. Transl Psychiatry 9:34. https://doi.org/10.1038/s41398-019-0367-z

Boileau I, Assaad JM, Pihl RO, Benkelfat C, Leyton M, Diksic M, Tremblay RE, Dagher A (2003) Alcohol promotes dopamine release in the human nucleus accumbens. Synapse 49:226–231. https://doi.org/10.1002/syn.10226

Borruto AM, Stopponi S, Li H, Weiss F, Roberto M, Ciccocioppo R (2021) Genetically selected alcohol-preferring msP rats to study alcohol use disorder: anything lost in translation? Neuropharmacology 186:108446. https://doi.org/10.1016/j.neuropharm.2020.108446

Brenner E, Tiwari GR, Kapoor M, Liu Y, Brock A, Mayfield RD (2020) Single cell transcriptome profiling of the human alcohol-dependent brain. Hum Mol Genet 29:1144–1153. https://doi.org/10.1093/hmg/ddaa038

Brownell KD, Marlatt GA, Lichtenstein E, Wilson GT (1986) Understanding and preventing relapse. Am Psychol 41:765–782. https://doi.org/10.1037//0003-066x.41.7.765

Bruckmann C, Islam SA, MacIsaac JL, Morin AM, Karle KN, Di Santo A, Wust R, Lang I, Batra A, Kobor MS, Nieratschker V (2017) DNA methylation signatures of chronic alcohol dependence in purified CD3(+) T-cells of patients undergoing alcohol treatment. Sci Rep 7:6605. https://doi.org/10.1038/s41598-017-06847-z

Carnicella S, Ron D, Barak S (2014) Intermittent ethanol access schedule in rats as a preclinical model of alcohol abuse. Alcohol 48:243–252. https://doi.org/10.1016/j.alcohol.2014.01.006

Carvalho AF, Heilig M, Perez A, Probst C, Rehm J (2019) Alcohol use disorders. Lancet 394:781–792. https://doi.org/10.1016/S0140-6736(19)31775-1

Cervera-Juanes R, Wilhelm LJ, Park B, Grant KA, Ferguson B (2017) Genome-wide analysis of the nucleus accumbens identifies DNA methylation signals differentiating low/binge from heavy alcohol drinking. Alcohol 60:103–113. https://doi.org/10.1016/j.alcohol.2016.11.003

Chastain LG, Sarkar DK (2017) Alcohol effects on the epigenome in the germline: role in the inheritance of alcohol-related pathology. Alcohol 60:53–66. https://doi.org/10.1016/j.alcohol.2016.12.007

Chaudhri N, Sahuque LL, Janak PH (2009) Ethanol seeking triggered by environmental context is attenuated by blocking dopamine D1 receptors in the nucleus accumbens core and shell in rats. Psychopharmacology 207:303–314. https://doi.org/10.1007/s00213-009-1657-6

Chen LL, Yang L (2015) Regulation of circRNA biogenesis. RNA Biol 12:381–388. https://doi.org/10.1080/15476286.2015.1020271

Chen C, Tan R, Wong L, Fekete R, Halsey J (2011) Quantitation of microRNAs by real-time RT-qPCR. Methods Mol Biol 687:113–134. https://doi.org/10.1007/978-1-60761-944-4_8

Cheng AT, Gau SF, Chen TH, Chang JC, Chang YT (2004) A 4-year longitudinal study on risk factors for alcoholism. Arch Gen Psychiatry 61:184–191. https://doi.org/10.1001/archpsyc.61.2.184

Chien Y, Tsai PH, Lai YH, Lu KH, Liu CY, Lin HF, Huang CS, Wu WW, Wang CY (2020) CircularRNA as novel biomarkers in liver diseases. J Chin Med Assoc 83:15–17. https://doi.org/10.1097/JCMA.0000000000000230

Chuang DM, Leng Y, Marinova Z, Kim HJ, Chiu CT (2009) Multiple roles of HDAC inhibition in neurodegenerative conditions. Trends Neurosci 32:591–601. https://doi.org/10.1016/j.tins.2009.06.002

Ciccocioppo R (2013) Genetically selected alcohol preferring rats to model human alcoholism. Curr Top Behav Neurosci 13:251–269. https://doi.org/10.1007/7854_2012_199

Cippitelli A, Damadzic R, Hansson AC, Singley E, Sommer WH, Eskay R, Thorsell A, Heilig M (2010) Neuropeptide Y (NPY) suppresses yohimbine-induced reinstatement of alcohol seeking. Psychopharmacology 208:417–426. https://doi.org/10.1007/s00213-009-1741-y

Cippitelli A, Damadzic R, Singley E, Thorsell A, Ciccocioppo R, Eskay RL, Heilig M (2012) Pharmacological blockade of corticotropin-releasing hormone receptor 1 (CRH1R) reduces voluntary consumption of high alcohol concentrations in non-dependent Wistar rats. Pharmacol Biochem Behav 100:522–529. https://doi.org/10.1016/j.pbb.2011.10.016

Cloos PA, Christensen J, Agger K, Helin K (2008) Erasing the methyl mark: histone demethylases at the center of cellular differentiation and disease. Genes Dev 22:1115–1140. https://doi.org/10.1101/gad.1652908

Coleman LG Jr, Crews FT (2018) Innate immune signaling and alcohol use disorders. Handb Exp Pharmacol 248:369–396. https://doi.org/10.1007/164_2018_92

Collaborators GBDA (2018) Alcohol use and burden for 195 countries and territories, 1990-2016: a systematic analysis for the Global Burden of Disease Study 2016. Lancet 392:1015–1035. https://doi.org/10.1016/S0140-6736(18)31310-2

Colombo G, Lobina C, Carai MA, Gessa GL (2006) Phenotypic characterization of genetically selected Sardinian alcohol-preferring (sP) and -non-preferring (sNP) rats. Addict Biol 11:324–338. https://doi.org/10.1111/j.1369-1600.2006.00031.x

Congdon LM, Sims JK, Tuzon CT, Rice JC (2014) The PR-Set7 binding domain of Riz1 is required for the H4K20me1-H3K9me1 trans-tail ‘histone code’ and Riz1 tumor suppressor function. Nucleic Acids Res 42:3580–3589. https://doi.org/10.1093/nar/gkt1377

Crabbe JC, Phillips TJ, Belknap JK (2010) The complexity of alcohol drinking: studies in rodent genetic models. Behav Genet 40:737–750. https://doi.org/10.1007/s10519-010-9371-z

D’Addario C, Caputi FF, Ekstrom TJ, Di Benedetto M, Maccarrone M, Romualdi P, Candeletti S (2013) Ethanol induces epigenetic modulation of prodynorphin and pronociceptin gene expression in the rat amygdala complex. J Mol Neurosci 49:312–319. https://doi.org/10.1007/s12031-012-9829-y

Darcq E, Warnault V, Phamluong K, Besserer GM, Liu F, Ron D (2015) MicroRNA-30a-5p in the prefrontal cortex controls the transition from moderate to excessive alcohol consumption. Mol Psychiatry 20:1219–1231. https://doi.org/10.1038/mp.2014.120

Deroche-Gamonet V, Belin D, Piazza PV (2004) Evidence for addiction-like behavior in the rat. Science 305:1014–1017. https://doi.org/10.1126/science.1099020

Dick DM, Bierut LJ (2006) The genetics of alcohol dependence. Curr Psychiatry Rep 8:151–157. https://doi.org/10.1007/s11920-006-0015-1

Dickinson A, Wood N, Smith JW (2002) Alcohol seeking by rats: action or habit? Q J Exp Psychol B 55:331–348. https://doi.org/10.1080/0272499024400016

Domi E, Barbier E, Augier E, Augier G, Gehlert D, Barchiesi R, Thorsell A, Holm L, Heilig M (2018) Preclinical evaluation of the kappa-opioid receptor antagonist CERC-501 as a candidate therapeutic for alcohol use disorders. Neuropsychopharmacology 43:1805–1812. https://doi.org/10.1038/s41386-018-0015-y

Domi A, Stopponi S, Domi E, Ciccocioppo R, Cannella N (2019) Sub-dimensions of alcohol use disorder in alcohol preferring and non-preferring rats, a comparative study. Front Behav Neurosci 13:3. https://doi.org/10.3389/fnbeh.2019.00003

Domi E, Domi A, Ubaldi M, Somaini L, Demopulos G, Gaitanaris G, Ciccocioppo R (2020) Further evidence for the involvement of the PPARgamma system on alcohol intake and sensitivity in rodents. Psychopharmacology 237:2983–2992. https://doi.org/10.1007/s00213-020-05586-w

Domi E, Domi A, Adermark L, Heilig M, Augier E (2021a) Neurobiology of alcohol seeking behavior. J Neurochem 157:1585–1614. https://doi.org/10.1111/jnc.15343

Domi E, Xu L, Toivainen S, Nordeman A, Gobbo F, Venniro M, Shaham Y, Messing RO, Visser E, van den Oever MC et al (2021b) A neural substrate of compulsive alcohol use. Sci Adv 7: eabg9045. https://doi.org/10.1126/sciadv.abg9045

Dou X, Feng L, Ying N, Ding Q, Song Q, Jiang F, Wang C, Li S (2020) RNA sequencing reveals a comprehensive circular RNA expression profile in a mouse model of alcoholic liver disease. Alcohol Clin Exp Res 44:415–422. https://doi.org/10.1111/acer.14265

Down TA, Rakyan VK, Turner DJ, Flicek P, Li H, Kulesha E, Graf S, Johnson N, Herrero J, Tomazou EM et al (2008) A Bayesian deconvolution strategy for immunoprecipitation-based DNA methylome analysis. Nat Biotechnol 26:779–785. https://doi.org/10.1038/nbt1414

Dykes IM, Emanueli C (2017) Transcriptional and post-transcriptional gene regulation by long non-coding RNA. Genom Proteom Bioinform 15:177–186. https://doi.org/10.1016/j.gpb.2016.12.005

Edenberg HJ, Foroud T (2006) The genetics of alcoholism: identifying specific genes through family studies. Addict Biol 11:386–396. https://doi.org/10.1111/j.1369-1600.2006.00035.x

Edgar RD, Jones MJ, Meaney MJ, Turecki G, Kobor MS (2017) BECon: a tool for interpreting DNA methylation findings from blood in the context of brain. Transl Psychiatry 7:e1187. https://doi.org/10.1038/tp.2017.171

Erol A, Karpyak VM (2015) Sex and gender-related differences in alcohol use and its consequences: contemporary knowledge and future research considerations. Drug Alcohol Depend 156:1–13. https://doi.org/10.1016/j.drugalcdep.2015.08.023

Farris SP, Mayfield RD (2014) RNA-Seq reveals novel transcriptional reorganization in human alcoholic brain. Int Rev Neurobiol 116:275–300. https://doi.org/10.1016/B978-0-12-801105-8.00011-4

Farris SP, Mayfield RD (2021) Epigenetic and non-coding regulation of alcohol abuse and addiction. Int Rev Neurobiol 156:63–86. https://doi.org/10.1016/bs.irn.2020.08.006

Farris SP, Arasappan D, Hunicke-Smith S, Harris RA, Mayfield RD (2015a) Transcriptome organization for chronic alcohol abuse in human brain. Mol Psychiatry 20:1438–1447. https://doi.org/10.1038/mp.2014.159

Farris SP, Harris RA, Ponomarev I (2015b) Epigenetic modulation of brain gene networks for cocaine and alcohol abuse. Front Neurosci 9:176. https://doi.org/10.3389/fnins.2015.00176

Feng L, Lou J (2019) DNA methylation analysis. Methods Mol Biol 1894:181–227. https://doi.org/10.1007/978-1-4939-8916-4_12

Finegersh A, Homanics GE (2014) Acute ethanol alters multiple histone modifications at model gene promoters in the cerebral cortex. Alcohol Clin Exp Res 38:1865–1873. https://doi.org/10.1111/acer.12465

Finegersh A, Rompala GR, Martin DI, Homanics GE (2015) Drinking beyond a lifetime: new and emerging insights into paternal alcohol exposure on subsequent generations. Alcohol 49:461–470. https://doi.org/10.1016/j.alcohol.2015.02.008

Finnegan EF, Pasquinelli AE (2013) MicroRNA biogenesis: regulating the regulators. Crit Rev Biochem Mol Biol 48:51–68. https://doi.org/10.3109/10409238.2012.738643

Finnin MS, Donigian JR, Cohen A, Richon VM, Rifkind RA, Marks PA, Breslow R, Pavletich NP (1999) Structures of a histone deacetylase homologue bound to the TSA and SAHA inhibitors. Nature 401:188–193. https://doi.org/10.1038/43710

Forero DA, Gonzalez-Giraldo Y, Castro-Vega LJ, Barreto GE (2019) qPCR-based methods for expression analysis of miRNAs. BioTechniques 67:192–199. https://doi.org/10.2144/btn-2019-0065

Friedel E, Walter H, Veer IM, Zimmermann US, Heinz A, Frieling H, Zindler T (2020) Impact of long-term alcohol consumption and relapse on genome-wide DNA methylation changes in alcohol-dependent subjects: a longitudinal study. Alcohol Clin Exp Res 44:1356–1365. https://doi.org/10.1111/acer.14354

Gatta E, Grayson DR, Auta J, Saudagar V, Dong E, Chen Y, Krishnan HR, Drnevich J, Pandey SC, Guidotti A (2021) Genome-wide methylation in alcohol use disorder subjects: implications for an epigenetic regulation of the cortico-limbic glucocorticoid receptors (NR3C1). Mol Psychiatry 26:1029–1041. https://doi.org/10.1038/s41380-019-0449-6

Gavin DP, Kusumo H, Zhang H, Guidotti A, Pandey SC (2016) Role of growth arrest and DNA damage-inducible, beta in alcohol-drinking behaviors. Alcohol Clin Exp Res 40:263–272. https://doi.org/10.1111/acer.12965

Gavin DP, Hashimoto JG, Lazar NH, Carbone L, Crabbe JC, Guizzetti M (2018) Stable histone methylation changes at proteoglycan network genes following ethanol exposure. Front Genet 9: 346. https://doi.org/10.3389/fgene.2018.00346

Gilman JM, Ramchandani VA, Davis MB, Bjork JM, Hommer DW (2008) Why we like to drink: a functional magnetic resonance imaging study of the rewarding and anxiolytic effects of alcohol. J Neurosci 28:4583–4591. https://doi.org/10.1523/JNEUROSCI.0086-08.2008

Gilpin NW, Roberto M (2012) Neuropeptide modulation of central amygdala neuroplasticity is a key mediator of alcohol dependence. Neurosci Biobehav Rev 36:873–888. https://doi.org/10.1016/j.neubiorev.2011.11.002

Gilpin NW, Herman MA, Roberto M (2015) The central amygdala as an integrative hub for anxiety and alcohol use disorders. Biol Psychiatry 77:859–869. https://doi.org/10.1016/j.biopsych.2014.09.008

Giuliano C, Pena-Oliver Y, Goodlett CR, Cardinal RN, Robbins TW, Bullmore ET, Belin D, Everitt BJ (2018) Evidence for a long-lasting compulsive alcohol seeking phenotype in rats. Neuropsychopharmacology 43:728–738. https://doi.org/10.1038/npp.2017.105

Goll MG, Bestor TH (2005) Eukaryotic cytosine methyltransferases. Annu Rev Biochem 74:481–514. https://doi.org/10.1146/annurev.biochem.74.010904.153721

Goltseker K, Hopf FW, Barak S (2019) Advances in behavioral animal models of alcohol use disorder. Alcohol 74:73–82. https://doi.org/10.1016/j.alcohol.2018.05.014

Grant BF, Goldstein RB, Saha TD, Chou SP, Jung J, Zhang H, Pickering RP, Ruan WJ, Smith SM, Huang B, Hasin DS (2015) Epidemiology of DSM-5 alcohol use disorder: results from the national epidemiologic survey on alcohol and related conditions III. JAMA Psychiat 72:757–766. https://doi.org/10.1001/jamapsychiatry.2015.0584

Griffin WC 3rd (2014) Alcohol dependence and free-choice drinking in mice. Alcohol 48:287–293. https://doi.org/10.1016/j.alcohol.2013.11.006

Griffin EA Jr, Melas PA, Zhou R, Li Y, Mercado P, Kempadoo KA, Stephenson S, Colnaghi L, Taylor K, Hu MC et al (2017) Prior alcohol use enhances vulnerability to compulsive cocaine self-administration by promoting degradation of HDAC4 and HDAC5. Sci Adv 3:e1701682. https://doi.org/10.1126/sciadv.1701682

Gu H, Smith ZD, Bock C, Boyle P, Gnirke A, Meissner A (2011) Preparation of reduced representation bisulfite sequencing libraries for genome-scale DNA methylation profiling. Nat Protoc 6:468–481. https://doi.org/10.1038/nprot.2010.190

Hagemann-Jensen M, Abdullayev I, Sandberg R, Faridani OR (2018) Small-seq for single-cell small-RNA sequencing. Nat Protoc 13:2407–2424. https://doi.org/10.1038/s41596-018-0049-y

Hagerty SL, Bidwell LC, Harlaar N, Hutchison KE (2016) An exploratory association study of alcohol use disorder and DNA methylation. Alcohol Clin Exp Res 40:1633–1640. https://doi.org/10.1111/acer.13138

Hagerty SL, YorkWilliams SL, Bidwell LC, Weiland BJ, Sabbineni A, Blaine SK, Bryan AD, Hutchison KE (2020) DRD2 methylation is associated with executive control network connectivity and severity of alcohol problems among a sample of polysubstance users. Addict Biol 25:e12684. https://doi.org/10.1111/adb.12684

Halladay LR, Kocharian A, Piantadosi PT, Authement ME, Lieberman AG, Spitz NA, Coden K, Glover LR, Costa VD, Alvarez VA, Holmes A (2020) Prefrontal regulation of punished ethanol self-administration. Biol Psychiatry 87:967–978. https://doi.org/10.1016/j.biopsych.2019.10.030

Hamilton PJ, Nestler EJ (2019) Epigenetics and addiction. Curr Opin Neurobiol 59:128–136. https://doi.org/10.1016/j.conb.2019.05.005

Hamlin AS, Clemens KJ, Choi EA, McNally GP (2009) Paraventricular thalamus mediates context-induced reinstatement (renewal) of extinguished reward seeking. Eur J Neurosci 29:802–812. https://doi.org/10.1111/j.1460-9568.2009.06623.x

Hasel P, Rose IVL, Sadick JS, Kim RD, Liddelow SA (2021) Neuroinflammatory astrocyte subtypes in the mouse brain. Nat Neurosci 24:1475–1487. https://doi.org/10.1038/s41593-021-00905-6

Heath AC, Whitfield JB, Martin NG, Pergadia ML, Goate AM, Lind PA, McEvoy BP, Schrage AJ, Grant JD, Chou YL et al (2011) A quantitative-trait genome-wide association study of alcoholism risk in the community: findings and implications. Biol Psychiatry 70:513–518. https://doi.org/10.1016/j.biopsych.2011.02.028

Heilig M, Koob GF (2007) A key role for corticotropin-releasing factor in alcohol dependence. Trends Neurosci 30:399–406. https://doi.org/10.1016/j.tins.2007.06.006

Heilig M, Egli M, Crabbe JC, Becker HC (2010) Acute withdrawal, protracted abstinence and negative affect in alcoholism: are they linked? Addict Biol 15:169–184. https://doi.org/10.1111/j.1369-1600.2009.00194.x

Heilig M, Sommer WH, Spanagel R (2016) The need for treatment responsive translational biomarkers in alcoholism research. Curr Top Behav Neurosci 28:151–171. https://doi.org/10.1007/7854_2015_5006

Heilig M, Barbier E, Johnstone AL, Tapocik J, Meinhardt MW, Pfarr S, Wahlestedt C, Sommer WH (2017) Reprogramming of mPFC transcriptome and function in alcohol dependence. Genes Brain Behav 16:86–100. https://doi.org/10.1111/gbb.12344

Heilig M, Augier E, Pfarr S, Sommer WH (2019) Developing neuroscience-based treatments for alcohol addiction: a matter of choice? Transl Psychiatry 9:255. https://doi.org/10.1038/s41398-019-0591-6

Heilig M, MacKillop J, Martinez D, Rehm J, Leggio L, Vanderschuren L (2021) Addiction as a brain disease revised: why it still matters, and the need for consilience. Neuropsychopharmacology 46(10):1715–1723. https://doi.org/10.1038/s41386-020-00950-y

Hendershot CS, Witkiewitz K, George WH, Marlatt GA (2011) Relapse prevention for addictive behaviors. Subst Abuse Treat Prev Policy 6:17. https://doi.org/10.1186/1747-597X-6-17

Hodos W (1961) Progressive ratio as a measure of reward strength. Science 134:943–944. https://doi.org/10.1126/science.134.3483.943

Hwa LS, Chu A, Levinson SA, Kayyali TM, DeBold JF, Miczek KA (2011) Persistent escalation of alcohol drinking in C57BL/6J mice with intermittent access to 20% ethanol. Alcohol Clin Exp Res 35:1938–1947. https://doi.org/10.1111/j.1530-0277.2011.01545.x

Jadhav KS, Magistretti PJ, Halfon O, Augsburger M, Boutrel B (2017) A preclinical model for identifying rats at risk of alcohol use disorder. Sci Rep 7:9454. https://doi.org/10.1038/s41598-017-09801-1

Jeanblanc J, Lemoine S, Jeanblanc V, Alaux-Cantin S, Naassila M (2015) The class I-specific HDAC inhibitor MS-275 decreases motivation to consume alcohol and relapse in heavy drinking rats. Int J Neuropsychopharmacol 18:pyv029. https://doi.org/10.1093/ijnp/pyv029

Jin Z, Mendu SK, Birnir B (2013) GABA is an effective immunomodulatory molecule. Amino Acids 45:87–94. https://doi.org/10.1007/s00726-011-1193-7

Johnstone AL, Andrade NS, Barbier E, Khomtchouk BB, Rienas CA, Lowe K, Van Booven DJ, Domi E, Esanov R, Vilca S et al (2021) Dysregulation of the histone demethylase KDM6B in alcohol dependence is associated with epigenetic regulation of inflammatory signaling pathways. Addict Biol 26:e12816. https://doi.org/10.1111/adb.12816

Kessler RC, Crum RM, Warner LA, Nelson CB, Schulenberg J, Anthony JC (1997) Lifetime co-occurrence of DSM-III-R alcohol abuse and dependence with other psychiatric disorders in the National Comorbidity Survey. Arch Gen Psychiatry 54:313–321. https://doi.org/10.1001/archpsyc.1997.01830160031005

Kim M, Park C, Jung J, Yeo SG (2019) The histone deacetylase class I, II inhibitor trichostatin A delays peripheral neurodegeneration. J Mol Histol 50:167–178. https://doi.org/10.1007/s10735-019-09815-1

Kisby BR, Farris SP, McManus MM, Varodayan FP, Roberto M, Harris RA, Ponomarev I (2021) Alcohol dependence in rats is associated with global changes in gene expression in the central amygdala. Brain Sci 11:1149. https://doi.org/10.3390/brainsci11091149

Koob GF, Volkow ND (2010) Neurocircuitry of addiction. Neuropsychopharmacology 35:217–238. https://doi.org/10.1038/npp.2009.110

Korber C, Sommer WH (2022) From ensembles to meta-ensembles: specific reward encoding by correlated network activity. Front Behav Neurosci 16:977474. https://doi.org/10.3389/fnbeh.2022.977474

Kryger R, Fan L, Wilce PA, Jaquet V (2012) MALAT-1, a non protein-coding RNA is upregulated in the cerebellum, hippocampus and brain stem of human alcoholics. Alcohol 46:629–634. https://doi.org/10.1016/j.alcohol.2012.04.002

Kyzar EJ, Zhang H, Sakharkar AJ, Pandey SC (2017) Adolescent alcohol exposure alters lysine demethylase 1 (LSD1) expression and histone methylation in the amygdala during adulthood. Addict Biol 22:1191–1204. https://doi.org/10.1111/adb.12404

Kyzar EJ, Bohnsack JP, Zhang H, Pandey SC (2019) MicroRNA-137 drives epigenetic reprogramming in the adult amygdala and behavioral changes after adolescent alcohol exposure. eNeuro 6:PMC6917896. https://doi.org/10.1523/ENEURO.0401-19.2019

Le A, Shaham Y (2002) Neurobiology of relapse to alcohol in rats. Pharmacol Ther 94:137–156. https://doi.org/10.1016/s0163-7258(02)00200-0

Le AD, Harding S, Juzytsch W, Fletcher PJ, Shaham Y (2002) The role of corticotropin-releasing factor in the median raphe nucleus in relapse to alcohol. J Neurosci 22:7844–7849

Le AD, Harding S, Juzytsch W, Funk D, Shaham Y (2005) Role of alpha-2 adrenoceptors in stress-induced reinstatement of alcohol seeking and alcohol self-administration in rats. Psychopharmacology 179:366–373. https://doi.org/10.1007/s00213-004-2036-y

Lev Maor G, Yearim A, Ast G (2015) The alternative role of DNA methylation in splicing regulation. Trends Genet 31:274–280. https://doi.org/10.1016/j.tig.2015.03.002

Lewohl JM, Nunez YO, Dodd PR, Tiwari GR, Harris RA, Mayfield RD (2011) Up-regulation of microRNAs in brain of human alcoholics. Alcohol Clin Exp Res 35:1928–1937. https://doi.org/10.1111/j.1530-0277.2011.01544.x

Li E, Zhang Y (2014) DNA methylation in mammals. Cold Spring Harb Perspect Biol 6:a019133. https://doi.org/10.1101/cshperspect.a019133

Li D, Zhang Y, Zhang Y, Wang Q, Miao Q, Xu Y, Soares JC, Zhang X, Zhang R (2019) Correlation between the epigenetic modification of histone H3K9 acetylation of NR2B gene promoter in rat hippocampus and ethanol withdrawal syndrome. Mol Biol Rep 46:2867–2875. https://doi.org/10.1007/s11033-019-04733-7

Liang X, Justice AC, So-Armah K, Krystal JH, Sinha R, Xu K (2021) DNA methylation signature on phosphatidylethanol, not on self-reported alcohol consumption, predicts hazardous alcohol consumption in two distinct populations. Mol Psychiatry 26:2238–2253. https://doi.org/10.1038/s41380-020-0668-x

Lim Y, Beane-Ebel JE, Tanaka Y, Ning B, Husted CR, Henderson DC, Xiang Y, Park IH, Farrer LA, Zhang H (2021) Exploration of alcohol use disorder-associated brain miRNA-mRNA regulatory networks. Transl Psychiatry 11:504. https://doi.org/10.1038/s41398-021-01635-w

Lin Y, Kranzler HR, Farrer LA, Xu H, Henderson DC, Zhang H (2020) An analysis of the effect of mu-opioid receptor gene (OPRM1) promoter region DNA methylation on the response of naltrexone treatment of alcohol dependence. Pharmacogenomics J 20:672–680. https://doi.org/10.1038/s41397-020-0158-1

Liu C, Marioni RE, Hedman AK, Pfeiffer L, Tsai PC, Reynolds LM, Just AC, Duan Q, Boer CG, Tanaka T et al (2018) A DNA methylation biomarker of alcohol consumption. Mol Psychiatry 23:422–433. https://doi.org/10.1038/mp.2016.192

Liu Y, Li J, Bu H, Wang H, Zhang Y, Shen Q, Li M, Lu Z, Rong X, Zheng D, Peng Y (2021) Circular RNA expression alteration identifies a novel circulating biomarker in serum exosomal for detection of alcohol dependence. Addict Biol 26:e13031. https://doi.org/10.1111/adb.13031

Lohoff FW, Roy A, Jung J, Longley M, Rosoff DB, Luo A, O'Connell E, Sorcher JL, Sun H, Schwandt M et al (2021) Epigenome-wide association study and multi-tissue replication of individuals with alcohol use disorder: evidence for abnormal glucocorticoid signaling pathway gene regulation. Mol Psychiatry 26:2224–2237. https://doi.org/10.1038/s41380-020-0734-4

Lu X, Liu Y, Xuan W, Ye J, Yao H, Huang C, Li J (2019) Circ_1639 induces cells inflammation responses by sponging miR-122 and regulating TNFRSF13C expression in alcoholic liver disease. Toxicol Lett 314:89–97. https://doi.org/10.1016/j.toxlet.2019.07.021

Luscher C, Robbins TW, Everitt BJ (2020) The transition to compulsion in addiction. Nat Rev Neurosci 21:247–263. https://doi.org/10.1038/s41583-020-0289-z

Lyko F (2018) The DNA methyltransferase family: a versatile toolkit for epigenetic regulation. Nat Rev Genet 19:81–92. https://doi.org/10.1038/nrg.2017.80

Ma DK, Jang MH, Guo JU, Kitabatake Y, Chang ML, Pow-Anpongkul N, Flavell RA, Lu B, Ming GL, Song H (2009) Neuronal activity-induced Gadd45b promotes epigenetic DNA demethylation and adult neurogenesis. Science 323:1074–1077. https://doi.org/10.1126/science.1166859

Macey DJ, Schulteis G, Heinrichs SC, Koob GF (1996) Time-dependent quantifiable withdrawal from ethanol in the rat: effect of method of dependence induction. Alcohol 13:163–170. https://doi.org/10.1016/0741-8329(95)02030-6

Martin-Fardon R, Weiss F (2013) Modeling relapse in animals. Curr Top Behav Neurosci 13:403–432. https://doi.org/10.1007/7854_2012_202

Marutha Ravindran CR, Ticku MK (2004) Changes in methylation pattern of NMDA receptor NR2B gene in cortical neurons after chronic ethanol treatment in mice. Brain Res Mol Brain Res 121:19–27. https://doi.org/10.1016/j.molbrainres.2003.10.025

Mattick JS, Makunin IV (2006) Non-coding RNA. Hum Mol Genet 15 Spec No 1:R17–R29. https://doi.org/10.1093/hmg/ddl046

McBride WJ, Rodd ZA, Bell RL, Lumeng L, Li TK (2014) The alcohol-preferring (P) and high-alcohol-drinking (HAD) rats--animal models of alcoholism. Alcohol 48:209–215. https://doi.org/10.1016/j.alcohol.2013.09.044

McClure JJ, Li X, Chou CJ (2018) Advances and challenges of HDAC inhibitors in cancer therapeutics. Adv Cancer Res 138:183–211. https://doi.org/10.1016/bs.acr.2018.02.006

Mehta JP (2014) Sequencing small RNA: introduction and data analysis fundamentals. Methods Mol Biol 1182:93–103. https://doi.org/10.1007/978-1-4939-1062-5_9

Meinhardt MW, Sommer WH (2015) Postdependent state in rats as a model for medication development in alcoholism. Addict Biol 20:1–21. https://doi.org/10.1111/adb.12187

Memczak S, Jens M, Elefsinioti A, Torti F, Krueger J, Rybak A, Maier L, Mackowiak SD, Gregersen LH, Munschauer M et al (2013) Circular RNAs are a large class of animal RNAs with regulatory potency. Nature 495:333–338. https://doi.org/10.1038/nature11928

Meng H, Wang L, You H, Huang C, Li J (2019) Circular RNA expression profile of liver tissues in an EtOH-induced mouse model of alcoholic hepatitis. Eur J Pharmacol 862:172642. https://doi.org/10.1016/j.ejphar.2019.172642

Meng W, Sjoholm LK, Kononenko O, Tay N, Zhang D, Sarkisyan D, Geske JR, Ing A, Qiu W, Watanabe H et al (2021) Genotype-dependent epigenetic regulation of DLGAP2 in alcohol use and dependence. Mol Psychiatry 26:4367–4382. https://doi.org/10.1038/s41380-019-0588-9

Mews P, Egervari G, Nativio R, Sidoli S, Donahue G, Lombroso SI, Alexander DC, Riesche SL, Heller EA, Nestler EJ et al (2019) Alcohol metabolism contributes to brain histone acetylation. Nature 574:717–721. https://doi.org/10.1038/s41586-019-1700-7

Meyer CA, Liu XS (2014) Identifying and mitigating bias in next-generation sequencing methods for chromatin biology. Nat Rev Genet 15:709–721. https://doi.org/10.1038/nrg3788

Modarresi F, Faghihi MA, Lopez-Toledano MA, Fatemi RP, Magistri M, Brothers SP, van der Brug MP, Wahlestedt C (2012) Inhibition of natural antisense transcripts in vivo results in gene-specific transcriptional upregulation. Nat Biotechnol 30:453–459. https://doi.org/10.1038/nbt.2158

Moghe A, Joshi-Barve S, Ghare S, Gobejishvili L, Kirpich I, McClain CJ, Barve S (2011) Histone modifications and alcohol-induced liver disease: are altered nutrients the missing link? World J Gastroenterol 17:2465–2472. https://doi.org/10.3748/wjg.v17.i20.2465

Monnier P, Martinet C, Pontis J, Stancheva I, Ait-Si-Ali S, Dandolo L (2013) H19 lncRNA controls gene expression of the Imprinted Gene Network by recruiting MBD1. Proc Natl Acad Sci U S A 110:20693–20698. https://doi.org/10.1073/pnas.1310201110

Moonat S, Sakharkar AJ, Zhang H, Pandey SC (2011) The role of amygdaloid brain-derived neurotrophic factor, activity-regulated cytoskeleton-associated protein and dendritic spines in anxiety and alcoholism. Addict Biol 16:238–250. https://doi.org/10.1111/j.1369-1600.2010.00275.x

Moonat S, Sakharkar AJ, Zhang H, Tang L, Pandey SC (2013) Aberrant histone deacetylase2-mediated histone modifications and synaptic plasticity in the amygdala predisposes to anxiety and alcoholism. Biol Psychiatry 73:763–773. https://doi.org/10.1016/j.biopsych.2013.01.012

Mueller D, Stewart J (2000) Cocaine-induced conditioned place preference: reinstatement by priming injections of cocaine after extinction. Behav Brain Res 115:39–47. https://doi.org/10.1016/s0166-4328(00)00239-4

Nunez YO, Truitt JM, Gorini G, Ponomareva ON, Blednov YA, Harris RA, Mayfield RD (2013) Positively correlated miRNA-mRNA regulatory networks in mouse frontal cortex during early stages of alcohol dependence. BMC Genomics 14:725. https://doi.org/10.1186/1471-2164-14-725

O'Brien CP, Childress AR, McLellan AT, Ehrman R (1992) A learning model of addiction. Res Publ Assoc Res Nerv Ment Dis 70:157–177

Pandey SC, Ugale R, Zhang H, Tang L, Prakash A (2008) Brain chromatin remodeling: a novel mechanism of alcoholism. J Neurosci 28:3729–3737. https://doi.org/10.1523/JNEUROSCI.5731-07.2008

Park PJ (2009) ChIP-seq: advantages and challenges of a maturing technology. Nat Rev Genet 10:669–680. https://doi.org/10.1038/nrg2641

Pascual M, Do Couto BR, Alfonso-Loeches S, Aguilar MA, Rodriguez-Arias M, Guerri C (2012) Changes in histone acetylation in the prefrontal cortex of ethanol-exposed adolescent rats are associated with ethanol-induced place conditioning. Neuropharmacology 62:2309–2319. https://doi.org/10.1016/j.neuropharm.2012.01.011

Pfarr S, Meinhardt MW, Klee ML, Hansson AC, Vengeliene V, Schonig K, Bartsch D, Hope BT, Spanagel R, Sommer WH (2015) Losing control: excessive alcohol seeking after selective inactivation of cue-responsive neurons in the infralimbic cortex. J Neurosci 35:10750–10761. https://doi.org/10.1523/JNEUROSCI.0684-15.2015

Pfarr S, Schaaf L, Reinert JK, Paul E, Herrmannsdorfer F, Rossmanith M, Kuner T, Hansson AC, Spanagel R, Korber C, Sommer WH (2018) Choice for drug or natural reward engages largely overlapping neuronal ensembles in the infralimbic prefrontal cortex. J Neurosci 38:3507–3519. https://doi.org/10.1523/JNEUROSCI.0026-18.2018

Piazza PV, Deroche-Gamonet V (2013) A multistep general theory of transition to addiction. Psychopharmacology 229:387–413. https://doi.org/10.1007/s00213-013-3224-4

Pradhan AA, Tipton AF, Zhang H, Akbari A, Pandey SC (2019) Effect of histone deacetylase inhibitor on ethanol withdrawal-induced hyperalgesia in rats. Int J Neuropsychopharmacol 22:523–527. https://doi.org/10.1093/ijnp/pyz031

Qiang M, Denny A, Lieu M, Carreon S, Li J (2011) Histone H3K9 modifications are a local chromatin event involved in ethanol-induced neuroadaptation of the NR2B gene. Epigenetics 6:1095–1104. https://doi.org/10.4161/epi.6.9.16924

Qiang M, Li JG, Denny AD, Yao JM, Lieu M, Zhang K, Carreon S (2014) Epigenetic mechanisms are involved in the regulation of ethanol consumption in mice. Int J Neuropsychopharmacol 18: pyu072. https://doi.org/10.1093/ijnp/pyu072

Ramchandani VA, Umhau J, Pavon FJ, Ruiz-Velasco V, Margas W, Sun H, Damadzic R, Eskay R, Schoor M, Thorsell A et al (2011a) A genetic determinant of the striatal dopamine response to alcohol in men. Mol Psychiatry 16:809–817. https://doi.org/10.1038/mp.2010.56

Ramchandani VA, Umhau J, Pavon FJ, Ruiz-Velasco V, Margas W, Sun H, Damadzic R, Eskay R, Schoor M, Thorsell A et al (2011b) A genetic determinant of the striatal dopamine response to alcohol in men. Mol Psychiatry 16:809–817. https://doi.org/10.1038/mp.2010.56.

Rhodes JS, Best K, Belknap JK, Finn DA, Crabbe JC (2005) Evaluation of a simple model of ethanol drinking to intoxication in C57BL/6J mice. Physiol Behav 84:53–63. https://doi.org/10.1016/j.physbeh.2004.10.007

Richter CP, Campbell KH (1940) Alcohol taste thresholds and concentrations of solution preferred by rats. Science 91:507–508. https://doi.org/10.1126/science.91.2369.507

Rimondini R, Arlinde C, Sommer W, Heilig M (2002) Long-lasting increase in voluntary ethanol consumption and transcriptional regulation in the rat brain after intermittent exposure to alcohol. FASEB J 16:27–35. https://doi.org/10.1096/fj.01-0593com

Roberts AJ, Cole M, Koob GF (1996) Intra-amygdala muscimol decreases operant ethanol self-administration in dependent rats. Alcohol Clin Exp Res 20:1289–1298. https://doi.org/10.1111/j.1530-0277.1996.tb01125.x

Rogers J, Wiener SG, Bloom FE (1979) Long-term ethanol administration methods for rats: advantages of inhalation over intubation or liquid diets. Behav Neural Biol 27:466–486. https://doi.org/10.1016/s0163-1047(79)92061-2

Rompala GR, Homanics GE (2019) Intergenerational effects of alcohol: a review of paternal preconception ethanol exposure studies and epigenetic mechanisms in the male germline. Alcohol Clin Exp Res 43:1032–1045. https://doi.org/10.1111/acer.14029

Ruggeri B, Nymberg C, Vuoksimaa E, Lourdusamy A, Wong CP, Carvalho FM, Jia T, Cattrell A, Macare C, Banaschewski T et al (2015) Association of protein phosphatase PPM1G with alcohol use disorder and brain activity during behavioral control in a genome-wide methylation analysis. Am J Psychiatry 172:543–552. https://doi.org/10.1176/appi.ajp.2014.14030382

Sakharkar AJ, Zhang H, Tang L, Shi G, Pandey SC (2012) Histone deacetylases (HDAC)-induced histone modifications in the amygdala: a role in rapid tolerance to the anxiolytic effects of ethanol. Alcohol Clin Exp Res 36:61–71. https://doi.org/10.1111/j.1530-0277.2011.01581.x

Sakharkar AJ, Zhang H, Tang L, Baxstrom K, Shi G, Moonat S, Pandey SC (2014) Effects of histone deacetylase inhibitors on amygdaloid histone acetylation and neuropeptide Y expression: a role in anxiety-like and alcohol-drinking behaviours. Int J Neuropsychopharmacol 17: 1207–1220. https://doi.org/10.1017/S1461145714000054

Schuckit MA (2009) An overview of genetic influences in alcoholism. J Subst Abus Treat 36:S5–S14

Seif T, Chang SJ, Simms JA, Gibb SL, Dadgar J, Chen BT, Harvey BK, Ron D, Messing RO, Bonci A, Hopf FW (2013) Cortical activation of accumbens hyperpolarization-active NMDARs mediates aversion-resistant alcohol intake. Nat Neurosci 16:1094–1100. https://doi.org/10.1038/nn.3445

Shentu Y, Tian Q, Yang J, Liu X, Han Y, Yang D, Zhang N, Fan X, Wang P, Ma J et al (2021) Upregulation of KDM6B contributes to lipopolysaccharide-induced anxiety-like behavior via modulation of VGLL4 in mice. Behav Brain Res 408:113305. https://doi.org/10.1016/j.bbr.2021.113305

Shi Y, Lan F, Matson C, Mulligan P, Whetstine JR, Cole PA, Casero RA, Shi Y (2004) Histone demethylation mediated by the nuclear amine oxidase homolog LSD1. Cell 119:941–953. https://doi.org/10.1016/j.cell.2004.12.012

Siciliano CA, Noamany H, Chang CJ, Brown AR, Chen X, Leible D, Lee JJ, Wang J, Vernon AN, Vander Weele CM et al (2019) A cortical-brainstem circuit predicts and governs compulsive alcohol drinking. Science 366:1008–1012. https://doi.org/10.1126/science.aay1186

Simms JA, Steensland P, Medina B, Abernathy KE, Chandler LJ, Wise R, Bartlett SE (2008) Intermittent access to 20% ethanol induces high ethanol consumption in Long-Evans and Wistar rats. Alcohol Clin Exp Res 32:1816–1823. https://doi.org/10.1111/j.1530-0277.2008.00753.x

Simon-O'Brien E, Alaux-Cantin S, Warnault V, Buttolo R, Naassila M, Vilpoux C (2015) The histone deacetylase inhibitor sodium butyrate decreases excessive ethanol intake in dependent animals. Addict Biol 20:676–689. https://doi.org/10.1111/adb.12161

Sinclair CM, Cleva RM, Hood LE, Olive MF, Gass JT (2012) mGluR5 receptors in the basolateral amygdala and nucleus accumbens regulate cue-induced reinstatement of ethanol-seeking behavior. Pharmacol Biochem Behav 101:329–335. https://doi.org/10.1016/j.pbb.2012.01.014

Siomi MC, Sato K, Pezic D, Aravin AA (2011) PIWI-interacting small RNAs: the vanguard of genome defence. Nat Rev Mol Cell Biol 12:246–258. https://doi.org/10.1038/nrm3089

Skene PJ, Henikoff S (2017) An efficient targeted nuclease strategy for high-resolution mapping of DNA binding sites. elife 6:e21856. https://doi.org/10.7554/eLife.21856

Smith ACW, Kenny PJ (2018) MicroRNAs regulate synaptic plasticity underlying drug addiction. Genes Brain Behav 17:e12424. https://doi.org/10.1111/gbb.12424

Smith JP, Randall CL (2012) Anxiety and alcohol use disorders: comorbidity and treatment considerations. Alcohol Res 34:414–431

Sommer WH, Rimondini R, Hansson AC, Hipskind PA, Gehlert DR, Barr CS, Heilig MA (2008) Upregulation of voluntary alcohol intake, behavioral sensitivity to stress, and amygdala crhr1 expression following a history of dependence. Biol Psychiatry 63:139–145. https://doi.org/10.1016/j.biopsych.2007.01.010

Sousa C, Golebiewska A, Poovathingal SK, Kaoma T, Pires-Afonso Y, Martina S, Coowar D, Azuaje F, Skupin A, Balling R et al (2018) Single-cell transcriptomics reveals distinct inflammation-induced microglia signatures. EMBO Rep 19:e46171. https://doi.org/10.15252/embr.201846171

Spanagel R (2009a) Alcoholism: a systems approach from molecular physiology to addictive behavior. Physiol Rev 89:649–705. https://doi.org/10.1152/physrev.00013.2008

Spanagel R (2009b) Alcoholism: a systems approach from molecular physiology to addictive behavior. Physiol Rev 89:649–705. https://doi.org/10.1152/physrev.00013.2008.

Stragier E, Massart R, Salery M, Hamon M, Geny D, Martin V, Boulle F, Lanfumey L (2015) Ethanol-induced epigenetic regulations at the Bdnf gene in C57BL/6J mice. Mol Psychiatry 20: 405–412. https://doi.org/10.1038/mp.2014.38

Strahl BD, Allis CD (2000) The language of covalent histone modifications. Nature 403:41–45. https://doi.org/10.1038/47412

Suzuki MM, Bird A (2008) DNA methylation landscapes: provocative insights from epigenomics. Nat Rev Genet 9:465–476. https://doi.org/10.1038/nrg2341

Szabo G, Lippai D (2014) Converging actions of alcohol on liver and brain immune signaling. Int Rev Neurobiol 118:359–380. https://doi.org/10.1016/B978-0-12-801284-0.00011-7

Szyf M (2009) Epigenetics, DNA methylation, and chromatin modifying drugs. Annu Rev Pharmacol Toxicol 49:243–263. https://doi.org/10.1146/annurev-pharmtox-061008-103102

Tapocik JD, Solomon M, Flanigan M, Meinhardt M, Barbier E, Schank JR, Schwandt M, Sommer WH, Heilig M (2013) Coordinated dysregulation of mRNAs and microRNAs in the rat medial prefrontal cortex following a history of alcohol dependence. Pharmacogenomics J 13:286–296. https://doi.org/10.1038/tpj.2012.17

Tapocik JD, Barbier E, Flanigan M, Solomon M, Pincus A, Pilling A, Sun H, Schank JR, King C, Heilig M (2014) microRNA-206 in rat medial prefrontal cortex regulates BDNF expression and alcohol drinking. J Neurosci 34:4581–4588. https://doi.org/10.1523/JNEUROSCI.0445-14.2014

Thiele TE, Navarro M (2014) "Drinking in the dark" (DID) procedures: a model of binge-like ethanol drinking in non-dependent mice. Alcohol 48:235–241. https://doi.org/10.1016/j.alcohol.2013.08.005

Tiffany ST, Conklin CA (2000) A cognitive processing model of alcohol craving and compulsive alcohol use. Addiction 95:145–153. https://doi.org/10.1080/09652140050111717

Tsai MC, Manor O, Wan Y, Mosammaparast N, Wang JK, Lan F, Shi Y, Segal E, Chang HY (2010) Long noncoding RNA as modular scaffold of histone modification complexes. Science 329:689–693. https://doi.org/10.1126/science.1192002

Wang F, Xu H, Zhao H, Gelernter J, Zhang H (2016) DNA co-methylation modules in postmortem prefrontal cortex tissues of European Australians with alcohol use disorders. Sci Rep 6:19430. https://doi.org/10.1038/srep19430

Warnault V, Darcq E, Levine A, Barak S, Ron D (2013) Chromatin remodeling--a novel strategy to control excessive alcohol drinking. Transl Psychiatry 3:e231. https://doi.org/10.1038/tp.2013.4

Weber M, Hellmann I, Stadler MB, Ramos L, Paabo S, Rebhan M, Schubeler D (2007) Distribution, silencing potential and evolutionary impact of promoter DNA methylation in the human genome. Nat Genet 39:457–466. https://doi.org/10.1038/ng1990

Wedemeyer F, Kaminski JA, Zillich L, Hall ASM, Friedel E, Witt SH (2020) Prospects of genetics and epigenetics of alcohol use disorder. Curr Addict Rep 7:446–452. https://doi.org/10.1007/s40429-020-00331-x

White A, Castle IJ, Chen CM, Shirley M, Roach D, Hingson R (2015) Converging patterns of alcohol use and related outcomes among females and males in the United States, 2002 to 2012. Alcohol Clin Exp Res 39:1712–1726. https://doi.org/10.1111/acer.12815

Wikler A (1973) Dynamics of drug dependence. Implications of a conditioning theory for research and treatment. Arch Gen Psychiatry 28:611–616. https://doi.org/10.1001/archpsyc.1973.01750350005001

Williams RJ, Berry LJ, Beerstecher E (1949) Individual metabolic patterns, alcoholism, genetotrophic diseases. Proc Natl Acad Sci U S A 35:265–271. https://doi.org/10.1073/pnas.35.6.265

Wise RA (1973) Voluntary ethanol intake in rats following exposure to ethanol on various schedules. Psychopharmacologia 29:203–210. https://doi.org/10.1007/BF00414034

Witt SH, Frank J, Frischknecht U, Treutlein J, Streit F, Foo JC, Sirignano L, Dukal H, Degenhardt F, Koopmann A et al (2020) Acute alcohol withdrawal and recovery in men lead to profound changes in DNA methylation profiles: a longitudinal clinical study. Addiction 115:2034–2044. https://doi.org/10.1111/add.15020

Wolffgramm J (1991) An ethopharmacological approach to the development of drug addiction. Neurosci Biobehav Rev 15:515–519. https://doi.org/10.1016/s0149-7634(05)80142-3

Wolffgramm J, Heyne A (1995) From controlled drug intake to loss of control: the irreversible development of drug addiction in the rat. Behav Brain Res 70:77–94. https://doi.org/10.1016/0166-4328(95)00131-c

Yadav S, Pandey A, Shukla A, Talwelkar SS, Kumar A, Pant AB, Parmar D (2011) miR-497 and miR-302b regulate ethanol-induced neuronal cell death through BCL2 protein and cyclin D2. J Biol Chem 286:37347–37357. https://doi.org/10.1074/jbc.M111.235531

Yang Y, Chen H, Ding N, Wang S, Duan Z, Birnbaum Y, Ye Y, Qian J (2018) Expression profiling of circular RNAs and micrornas in heart tissue of mice with alcoholic cardiomyopathy. Cell Physiol Biochem 46:2284–2296. https://doi.org/10.1159/000489596

You C, Zhang H, Sakharkar AJ, Teppen T, Pandey SC (2014) Reversal of deficits in dendritic spines, BDNF and Arc expression in the amygdala during alcohol dependence by HDAC inhibitor treatment. Int J Neuropsychopharmacol 17:313–322. https://doi.org/10.1017/S1461145713001144

Zeng K, Xie A, Zhang X, Zhong B, Liu X, Hao W (2018) Chronic alcohol treatment-induced GABA-Aalpha5 histone H3K4 trimethylation upregulation leads to increased GABA-Aalpha5 expression and susceptibility to alcohol addiction in the offspring of Wistar rats. Front Psych 9:468. https://doi.org/10.3389/fpsyt.2018.00468

Zillich L, Frank J, Streit F, Friske MM, Foo JC, Sirignano L, Heilmann-Heimbach S, Dukal H, Degenhardt F, Hoffmann P et al (2021) Epigenome-wide association study of alcohol use disorder in five brain regions. Neuropsychopharmacology 47(4):832–839. https://doi.org/10.1038/s41386-021-01228-7

The Relationship Between Oxytocin and Alcohol Dependence

Jonas Schimmer, Ryan Patwell, Stephanie Küppers, and Valery Grinevich

Contents

Abstract The hypothalamic neuropeptide oxytocin (OT) is well known for its prosocial, anxiolytic, and ameliorating effects on various psychiatric conditions, including alcohol use disorder (AUD). In this chapter, we will first introduce the basic neurophysiology of the OT system and its interaction with other neuromodulatory and neurotransmitter systems in the brain. Next, we provide an overview over the current state of research examining the effects of acute and chronic alcohol exposure on the OT system as well as the effects of OT system manipulation on alcohol-related behaviors in rodents and humans. In rodent models of AUD, OT has been repeatedly shown to reduce ethanol consumption, particularly in models of acute alcohol exposure. In humans however, the results of OT administration on alcohol-related behaviors are promising but not yet conclusive. Therefore, we further discuss several physiological and methodological limitations to the effective application of OT in the clinic and how they may be mitigated by the

Jonas Schimmer and Ryan Patwell contributed equally to this work.

J. Schimmer, R. Patwell, S. Küppers, and V. Grinevich (✉)
Department of Neuropeptide Research in Psychiatry, Medical Faculty Mannheim, Central Institute of Mental Health, University of Heidelberg, Mannheim, Germany
e-mail: jonas.schimmer@zi-mannheim.de; ryan.patwell@zi-mannheim.de; stephanie.kueppers@zi-mannheim.de; valery.grinevich@zi-mannheim.de

Curr Topics Behav Neurosci (2025) 71: 347–378
https://doi.org/10.1007/7854_2023_444
Published Online: 12 September 2023

application of synthetic OT receptor (OTR) agonists. Finally, we discuss the potential efficacy of cutting-edge pharmacology and gene therapies designed to specifically enhance endogenous OT release and thereby rescue deficient expression of OT in the brains of patients with severe forms of AUD and other incurable mental disorders.

Keywords Alcohol use disorder · Dopamine · Oxytocin · Oxytocin receptor · Stress axis · Vasopressin

Abbreviations

5HTR	5-Hydroxytryptamine (serotonin) receptor
AAV	Adeno-associated virus
AN	Accessory nuclei
AUD	Alcohol use disorders
BBB	Blood-brain barrier
BNST	Bed nucleus of the stria terminalis
CeA	Central amygdala
CIE	Chronic intermittent ethanol vapor exposure
CNS	Central nervous system
CPP	Conditioned place preference
Cpu	Caudate putamen
CRH	Corticotropin-releasing hormone
CSF	Cerebrospinal fluid
DA	Dopamine
GABA	Gamma-aminobutyric acid
HPA	Hypothalamic–pituitary–adrenocortical
i.p.	Intraperitoneal
i.v.	Intravenous
ICV	Intracerebroventricular
IN	Intranasal
LS	Lateral septum
magnOT	Magnocellular oxytocin
MC4R	Melanocortin receptor 4
MpoA	Medial preoptic area
NA	Noradrenaline
NAc	Nucleus accumbens
OT	Oxytocin
OTR	Oxytocin receptor
parvOT	Parvocellular oxytocin
PFC	Prefrontal cortex
PP	Posterior pituitary
PVN	Paraventricular nucleus

s.c.	Subcutaneous
SON	Supraoptic nucleus
ST	Serotonin
VMH	Ventromedial hypothalamic nucleus
NVP	Arginine vasopressin
VPa	Ventral pallidum
VTA	Ventral tegmental area
αMSH	Alpha melanocyte-stimulating hormone

1 Introduction: Basic Mechanisms of the Oxytocin System

Oxytocin (OT) is a neuropeptide composed of nine amino acids. It was originally considered to be a classical neurohormone after the identification of its first physiological effects at the beginning of the twentieth century, where injection of pituitary extract in several animal species was found to induce uterine contractions during parturition (Dale 1906). Inspired by these effects, the molecule was given the name "oxytocin," a Greek word that translates as "quick labor." Nearly half a century later in 1953, the team of Vincent du Vigneaud produced OT in the lab after uncovering its amino acid sequence (Cys-Tyr-Ile-Gln-Asn-Cys-Pro-Leu-GlyNH_2; Fig. 1a), thereby making OT the first ever synthesized polypeptide (Duvigneaud et al. 1954). OT is present in all mammals and its homologs can be found in several different invertebrate phyla (Theofanopoulou et al. 2021). The first appearance of OT or its precursor has been estimated to occur more than 700 million years ago (Beets et al. 2013; Grinevich et al. 2016a, b).

Despite the long existence of the OT system, there is currently only one known version of the mammalian OT receptor (OTR), which is expressed throughout the central nervous system and some peripheral tissues. This receptor is a member of the rhodopsin-type G-protein coupled receptor family which has seven transmembrane domains composed of three intracellular and three extracellular loops (Gimpl and Fahrenholz 2001). Depending on the brain region and local OT concentration, the proximal C-terminus of the OTR couples to $G_{\alpha q11}$ or Gi/Go and thus leads to changes in the activity of phospholipase C and protein kinase C (Busnelli and Chini 2018). The expression of OTR has been shown to be influenced by many mechanisms including epigenetic modifications, ligand availability, and estrogen-receptor activation. Thus, OTR binding can be precisely regulated and differs between reproductive, social, and developmental stages as well as states of acute or chronic stress (Grinevich and Neumann 2021). The precise mapping of OTR distribution in numerous brain regions has been achieved using a variety of techniques such as transgenic animal models, receptor autoradiography, and in situ hybridization (Lee et al. 2009). OTRs are most abundantly expressed in the ventral hippocampus and layer CA2 of the general hippocampal formation, the bed nucleus of the stria terminalis (BNST), ventromedial hypothalamic nucleus (VMH), and the

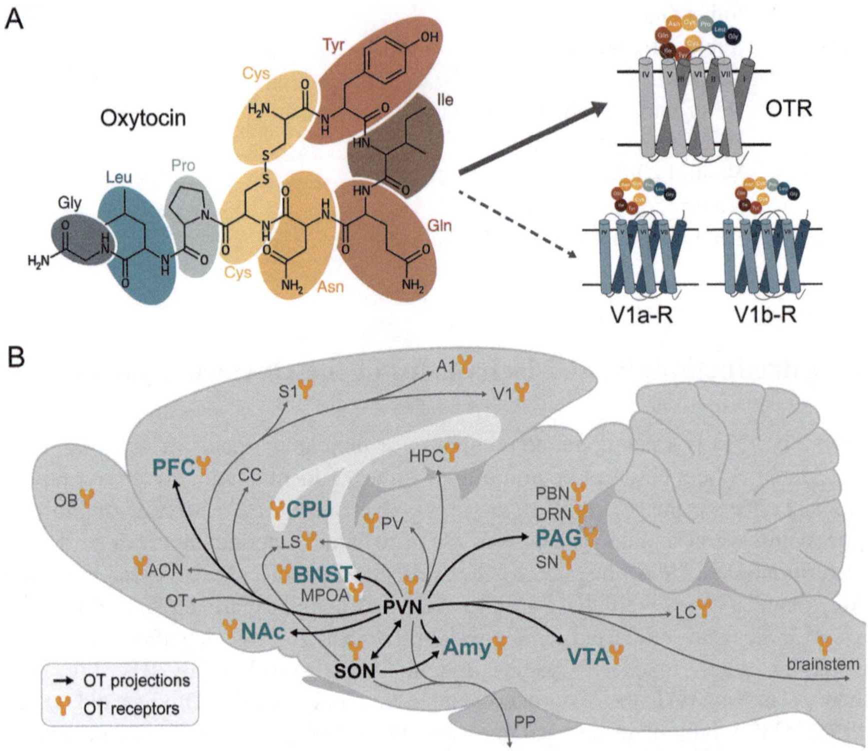

Fig. 1 Central oxytocin system overview. (**a**) OT is a highly conserved peptide hormone that acts primarily through the GPCR, OTR, resulting in differential signaling cascades depending on which Gα subunit it is coupled to. At high enough concentrations, OT can also bind and activate two receptors of the closely related Vasopressin system. (**b**) OT is produced primarily in the paraventricular (PVN) and supraoptic (SON) nuclei of the hypothalamus, where local microcircuits between magnOT and parvOT cells control the endogenous release of central and peripheral OT. OT neurons originating in the PVN project widely throughout the brain enabling the central OT system to exert influence on several OTR-expressing regions implicated in alcohol addiction-related behaviors and phenotypes

central amygdala (CeA) (Young and Song 2020; Burton et al. 2020). OTRs are also highly expressed in brain regions involved in the regulation of emotion, social, and reward related behaviors, as well as addiction (Fig. 1b). These regions include the brainstem, ventral pallidum (VPa), nucleus accumbens (NAc), caudate putamen (Cpu), medial preoptic area (MpoA), lateral septum (LS), ventral tegmental area (VTA), and prefrontal cortex (PFC) (Freeman and Young 2016). Sexual dimorphisms of OTR binding have been observed in some of these regions in rodents, namely the BNST, NAc, MPoA, and the hippocampus, in which OTR mRNA expression in males is generally higher than in females (Dumais and Veenema 2016). However, this sex difference is not consistent between species and a plethora of behavioral and environmental factors, such as social isolation and aggression, early life stress and gonadal hormones can alter OTR expression (Carter 2017b). Indeed, recent evidence suggests that OTR expression between individual humans is highly variable even in the same brain regions (Quintana et al. 2019).

The introduction of genetically modified rodents in recent years has greatly facilitated the mapping of OTR expression to specific neuronal subtypes across several brain areas, many of which are known to be involved in addiction (Grinevich and Neumann 2021; Nakajima et al. 2014; Peris et al. 2017). For example, OTR expression has been found in glutamatergic, GABAergic, and dopaminergic neurons in the VTA (Peris et al. 2017), interneurons in cortical areas (Li et al. 2016), interneurons and astrocytes in the NAc (Dolen et al. 2013), and corticotropin-releasing hormone (CRH) expressing neurons in the BNST (Dabrowska et al. 2011). These areas are all known to play a role in the modulation of processes such as reward, reinforcement, drug relapse and withdrawal. Taken together, the precise regional distribution and cell-type specific expression of OTRs in these brain areas suggests the central OT system likely plays a complex role in addiction-related processes, thereby making it a potential therapeutic target for the disease. However, despite the known distribution of OTRs throughout the brain, the complex neuronal circuitry that leads to their activation is still under heavy investigation.

The rat brain contains a total of ~7,600 neurons expressing OT, whereas the approximate number in humans is estimated to be over 50,000 cells (Althammer and Grinevich 2017). In both cases, these neurons are found almost exclusively in the hypothalamus, specifically within the paraventricular nucleus (PVN), supraoptic nucleus (SON), and smaller accessory nuclei (AN) (Knobloch and Grinevich 2014; Jurek and Neumann 2018; Swanson and Sawchenko 1983). Traditionally, OT neurons are divided into two different types: magnocellular (magnOT) and parvocellular (parvOT), which can be distinguished by their size, shape, projection targets, location, electrophysiological properties, and level of OT expression.

The majority of OT neurons are of the magnOT type, which are large cells with a diameter of about 30 μm. The large, round nuclei of magnOTs contain one or two nucleoli and only marginal amounts of Nissl substance (Armstrong et al. 1980). They are found in both of the primary OT producing subnuclei of the hypothalamus: the PVN and SON. The various peripheral actions of OT are mediated by bundles of magnOT neurons that directly project to the posterior lobe of the pituitary gland, which together form the hypothalamic–neurohypophyseal tract. Secretion of OT into the general blood circulation is activated by calcium-dependent depolarization of magnOT axonal terminals, which subsequently triggers the exocytosis of OT-rich dense core vesicles through fenestrated capillaries (Brown et al. 2013).

MagnOT cells also play a major role in the central actions of OT, the extent of which is still under investigation. One minor, less understood pathway recently observed in mammals involves the direct release of OT by magnOT neurons into the cerebrospinal fluid (CSF) via the third ventricle, which may be important for regulating the activity of the many OTR-expressing brain regions that are positioned adjacent to ventricles (Knobloch and Grinevich 2014). However, magnOT neurons primarily exert most of their influence by innervating a variety of different structures where they form axon collaterals (Son et al. 2022; Zhang et al. 2021). Many forebrain regions have been shown to receive efferents from magnOT neurons, such as the NAc (Ross et al. 2009), BNST (Stoop 2012), hippocampus (Smith et al. 2016), CeA (Knobloch et al. 2012), LS (Menon et al. 2018), and the anterior olfactory nucleus (Oettl et al. 2016). OT exerts a variety of specific behavioral effects in these areas including the reduction of stress and fear responsiveness via its action in the CeA (Ebner et al. 2005; Knobloch et al. 2012; Hasan et al. 2019) and prosocial behavior via its actions in the LS (Lukas et al. 2013) and other brain regions (Grinevich and Neumann 2021). The impact of OT specifically in the context of addiction-related behaviors is discussed in detail below.

In contrast to magnOT neurons, parvOT neurons have a diameter of ~15 μm and are exclusively located in the PVN (Althammer and Grinevich 2017). ParvOT neurons do not project axons to the posterior lobe of the pituitary gland and thus cannot directly secrete OT into the periphery via the blood. In contrast, parvOT neurons have been shown to form synapses onto magnOT neurons which allows them to indirectly modulate OT release into the general blood circulation (Eliava et al. 2016). Additionally, parvOT neurons can affect other physiological functions such as cardiovascular control and breathing (Mack et al. 2002), pain (Eliava et al. 2016), fear (Hasan et al. 2019), and somatosensory processing (Tang et al. 2022) via their projections to the spinal cord, brain stem, and midbrain (Sawchenko and Swanson 1982). Given these observations, it is likely that this relatively smaller population of parvOT neurons is playing a much more important role as a regulator of OT circuitry than has been appreciated thus far. The full extent of which will be determined by current and future research efforts.

1.1 Neurobiological Interactions of OT with Addiction-Related Components

1.1.1 Mesolimbic Dopamine Pathway

Dopamine (DA) plays a crucial role in the reward component of most substance use disorders, including alcohol, as its release is associated with the positive reinforcing effects of many drugs of abuse (Koob and Volkow 2016). PVN OT neurons are able to modulate the mesolimbic DA system either directly via innervation of dopaminergic terminals in the NAc (Xiao et al. 2018) or indirectly by activating OTRs expressed by dopaminergic VTA neurons which then project to the PFC and NAc (Peris et al. 2017). OT can inhibit reward signaling by reducing DA responses in the NAc and amygdala via the use of a DA reuptake inhibitor (Estes et al. 2019). Additionally, central application of OT decreases methamphetamine-induced increases of DA levels within the NAc (Qi et al. 2008, 2009) and completely blocks DA release in the NAc of alcohol-treated rats (Peters et al. 2017). Local infusion of OT in the NAc has also been shown to block increases in DA activity following the consumption of cocaine (Kovacs et al. 1990). Hence, interactions within the mesolimbic DA pathway seem to be responsible for attenuating effects of OT on drug consumption and relapse.

1.1.2 Serotonin System

The role of serotonin (ST) in drug addiction is more complex than that of DA. ST has several different receptor subtypes (5-HTRs) expressed in various brain areas related to drug addiction such as the VTA, NAc, hippocampus, and amygdala (Hayes and Greenshaw 2011). Some of these receptor subtypes have been shown to exert opposing effects on the reinforcing properties of drugs as types 5-HT1A and 5-HT2C decrease drug reinforcement, while 5-HT1B, 5-HT2A, 5-HT3, and 5-HT4 increase it (De Deurwaerdere and Di Giovanni 2017). Functional links between the OT and ST systems have been shown in several ways. For instance, ST is synthesized primarily by populations of neurons in the raphe nuclei and about half of these populations co-express OTR (Yoshida et al. 2009), which suggests that their activity can be modulated by OT binding. Furthermore, OTRs are likely present in presynaptic axonal terminals of ST neurons and thus OT can stimulate local ST release within the NAc (Dolen et al. 2013). Behaviorally, systemic administration of OT has been shown to reverse autism-like behavioral deficits induced by 5-HT1B agonists in mice (Lawson et al. 2016). Given the fact that 5-HT1B receptor activation is associated with increased cocaine and alcohol consumption (Sari 2013), inhibition of this receptor by OT might be a key mechanism in modulating the positive effects of OT on addiction-related behaviors. In contrast, agonist-induced activation of 5-HT1A and 5-HT2C decreases drug consumption (Higgins et al. 2020; Palacios et al. 2017) and simultaneously upregulates the levels of OT mRNA in the hypothalamus, leading to elevated peripheral OT release (Bagdy and Makara 1995; Jorgensen et al. 2003). This suggests that activation of these receptor subtypes can lead to further reductions in drug consumption via increasing levels of central and peripheral OT.

1.1.3 Noradrenaline System

Noradrenaline (NA) is a molecule that functions both as a hormone and a neurotransmitter with complementary functions. As a peripheral hormone, NA release triggers the activation of sympathetic nervous system responses and when released as a neurotransmitter, NA leads to an increase in general arousal and alertness. NA acts via several types of adrenergic receptors whose activation usually leads to excitation, except for the subtype known as α2 which has an inhibitory effect on neurons and prevents further release of NA when positioned presynaptically. Systemic OT treatment has been previously shown to modulate α2 adrenergic receptors. This modulation may be mediated by the heterodimers of OTR and α2 adrenergic receptors that are present in the amygdala, hypothalamus, and nucleus of the solitary tract, which are regions involved in addiction-related symptoms such as the negative affect experienced during withdrawal (Fuxe et al. 2012). Mechanistically, OT can bind to an OTR-α2 receptor complex and act as an allosteric antagonist such that signal transduction of noradrenaline is reduced (Diaz-Cabiale et al. 2000). In addition, treatment with an α2 adrenergic receptor antagonist has been shown to induce reinstatement of drug-seeking behavior in mice, whereas previous administration of OT attenuates this effect in context of alcohol and methamphetamine (Cox et al. 2017; King and Becker 2019). The administration of OT has been demonstrated to increase noradrenaline release in the SON and noradrenergic neurons in the nucleus of the solitary tract play a role in activating hypothalamic OT neurons upon the occurrence of stressful stimuli (Onaka et al. 2012). Furthermore, OT has been shown to dampen turnover of noradrenaline in the striatum, which is thought to be responsible for the attenuation of morphine-primed reinstatement upon application of an OT analog (Georgiou et al. 2015).

1.1.4 Gamma-Aminobutyric Acid

The majority of OTR neurons release gamma-aminobutyric acid (GABA) and therefore directly contribute to OT-mediated inhibitory processes (Yoshida et al. 2009). Furthermore, OTRs are expressed in GABAergic interneurons located in many addiction-related brain regions, namely the amygdala (Huber et al. 2005), PFC (Li et al. 2016), hippocampus (Zaninetti and Raggenbass 2000), and NAc (Dolen et al. 2013). These interneurons likely play a role in OTs modulatory effects on alcohol-related behaviors. For example, central application of OT in the amygdala of alcohol-dependent rats blocks the enhancing effects of alcohol on local GABA signaling (Tunstall et al. 2019). Another direct mode of interaction between OT and GABA is based on the finding that OT blocks alcohol-mediated positive allosteric modulation of GABA receptors by directly interacting with extra synaptic delta subunit-containing $GABA_A$ receptors (Bowen et al. 2015).

1.1.5 PFC-Glutamatergic System

Cue- and drug-induced reinstatement are regulated by neuronal projections from the PFC to the NAc (Koob and Volkow 2016; Yang et al. 2010). Under normal conditions, OT suppresses glutamatergic neurotransmission in the PFC (Ninan 2011), where glutamatergic signaling has been shown to control firing of DA neurons in the NAc (Geisler and Wise 2008). This reduction of both glutamate overflow and expression of its receptor within the PFC seems to be a major mechanism through which OT can attenuate methamphetamine relapse (Qi et al. 2009). Additionally, one potential mechanism for the attenuating effects of OT on drug-seeking behavior might occur through OTRs located on astrocytes in the NAc, which are known to interact with the local glutamatergic system (Dolen et al. 2013). Indeed, application of a glutamate receptor antagonist blocks OT's attenuating effects on drug-seeking behavior in context of methamphetamine and cocaine addiction (Bernheim et al. 2017; Weber et al. 2018).

1.1.6 Hypothalamic–Pituitary–Adrenocortical Axis

The hypothalamic–pituitary–adrenocortical (HPA) axis is a critical survival mechanism whose activation leads to several physiological changes that facilitate "fight or flight" behaviors in the presence of a perceived threat. HPA activation increases stress and anxiety and it is thought to play a key role in drug relapse as it becomes hyperactive following neural adaptations induced by chronic exposure to several drugs of abuse (Koob and Le Moal 2005). High levels of OT expression are found in different forebrain regions which are key components of the HPA axis due to their role in the regulation of anxiety, stress, and reward (Dabrowska et al. 2011; Gimpl and Fahrenholz 2001; Martinon and Dabrowska 2018). OT inhibits the activity of the HPA axis under basal and stress-related conditions via downregulation of corticotrophin-releasing hormone (CRH), which is a key regulator of the HPA axis (Jurek et al. 2015).

The OT and CRH systems are closely intertwined within the PVN, but the mechanisms of interaction between these two cell types is not fully understood. There is currently no evidence of either direct synaptic or non-synaptic action of OT on CRH neurons. However, OT has been shown to suppress stress-induced synthesis of CRH through specific intracellular signaling pathways within the PVN (Jurek et al. 2015). Additionally, two different CRH receptors (CRHR1 and CRHR2) were found to be expressed on OT neurons (Dabrowska et al. 2011; Winter and Jurek 2019) and CRHR1-expressing cells of the PVN have been shown to directly regulate stress responses (Jiang et al. 2019). Furthermore, treatment with a central CRHR2 agonist or antagonist is able to induce or inhibit OT release in the NAc shell, respectively, which is hypothesized to occur by acting on presynaptic OT neurons projecting from the PVN (Bosch et al. 2016). OT administration can also significantly decrease stress-induced upregulation of CRH mRNA, thus contributing to suppression of anxiety-related behaviors (Jurek et al. 2015). Additionally, drug-induced neural adaptions leading to states of elevated stress and anxiety are believed to be associated with malfunction of the OT system in both animal models and human (King et al. 2020). Therefore, OT may serve as a promising therapeutic option to prevent relapse via its inhibitory effects on the HPA axis and subsequent attenuation of withdrawal-induced stress and anxiety.

1.1.7 Vasopressin

Due to their structural similarity, the hypothalamic magnocellular vasopressin (VP) and OT systems are closely related. Both sister peptides are highly conserved throughout evolution and only differ by two amino acids (Grinevich et al. 2016a, b). Despite their highly conserved structural similarities, activation of these two neuropeptide systems leads to opposing physiological effects. For example, in contrast to the suppression of HPA activity by the OT system, the VP system promotes HPA activity by enhancing CRH synthesis and release (Neumann and Landgraf 2012). However, unlike OT and its sole receptor, OTR, the VP system exerts its effects through a receptor family consisting of three major subtypes: V1a, V1b, and V2 (Devost et al. 2008). All three receptors are expressed in peripheral organs where they regulate several physiological functions such as renal water reuptake and blood pressure. The V1a and V1b receptors are also found in the brain, particularly in areas responsible for governing stress and anxiety responses (Carter 2017a).

Despite the OT and VP peptides being expressed in mostly separated neuronal populations within PVN and SON (3% overlap), there is strong crosstalk between the two systems (Gainer 2012). For example, VP and OT are able to interact with each other's receptors at high concentrations and it has been suggested that OTRs and VP receptors form heterodimers (Bowen and McGregor 2014; Busnelli and Chini 2018; Terrillon et al. 2003).

The VP system has been demonstrated to directly affect alcohol consumption in rodents and humans. For example, administration of a V1b receptor antagonist reduces total alcohol consumption (Katz et al. 2016; Zhou et al. 2011, 2018) and VP is thought to be involved in the development of alcohol dependence (Edwards et al. 2012; Zhou and Kreek 2018). V1b receptor manipulation in particular has also been shown to affect alcohol-related behaviors in other preclinical models of AUD. One group found that repeated exposure to alcohol during adolescence in rats led to a downregulation of OTR but an upregulation of V1b receptors in the hypothalamus during adulthood (Dannenhoffer et al. 2018). Behaviorally, the authors further show that systemic administration of either an OTR or V1b receptor antagonist can reverse the social deficits exhibited by male rats in this model. Another study demonstrated bidirectional control of alcohol withdrawal-induced anxiety-like behavior by infusing either a VP or a V1b receptor antagonist directly into the amygdala (Harper et al. 2019). Overall, the effects of V1b receptor manipulation and OT on alcohol addiction-related behaviors as well as the structural crosstalk between the two systems warrant further investigation of their interactions in the context of addiction.

1.2 Effects of Chronic Alcohol Consumption on the OT System

The chronic relapsing nature of alcohol use disorder (AUD) is thought to be driven in part by the urge to alleviate the anxiety and stress symptoms experienced during withdrawal. These symptoms are believed to be the manifestation of altered circuitry function caused by chronic alcohol-induced neural adaptations, such as dysregulated gene expression and cell signaling (Koob and Volkow 2016). Post-mortem analysis of the brains of AUD patients and a limited number of studies in rodent models have

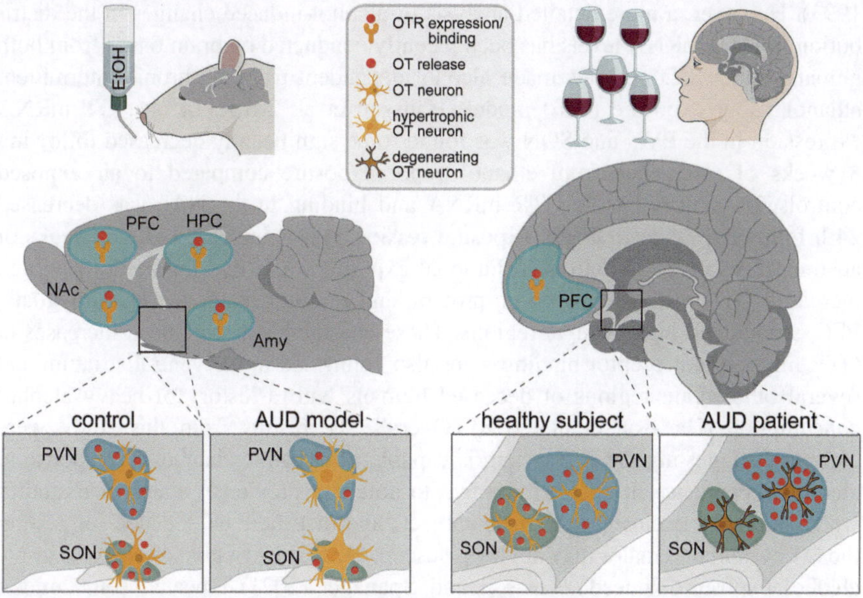

Fig. 2 Known effects of chronic alcohol exposure on the central OT system. Chronic alcohol exposure leads to several dysregulations of the central OT system, both in rodents (left panels) and humans (right panels). These effects occur both in the hypothalamic nuclei and in forebrain structures related to addiction. The effects identified thus far include dysregulated expression of OT and OTR, changes in the distribution of OTR binding sites, altered OT cellular morphology, and degeneration of OT cells

revealed that the central OT system is also susceptible to dysregulations following chronic exposure to alcohol (Fig. 2).

One of the earliest of these post-mortem studies by Sivukhina et al. (2006) reported decreased OT immunoreactivity in the SON, but increased immunoreactivity in the PVN. Interestingly, these differences were accompanied by an altered cellular morphology that was specific to magnOT neurons. This may prove to be an important detail when viewed in the light of more recent work demonstrating that OTs modulation of pain and social behaviors is mediated by a complex microcircuitry between hypothalamic parvOT and magnOT cells (Eliava et al. 2016; Tang et al. 2020). Another study measured mRNA expression in brain areas innervated by hypothalamic OT neurons and found that OT mRNA was significantly increased in the PFC of AUD subjects compared to controls, although no differences in the expression of OTR mRNA were reported in other studied regions (Lee et al. 2017). The latter finding was unexpected as reductions in the expression of a ligand are often accompanied by a compensatory increase in the expression of its receptor. The authors suggest that the lack of a compensatory upregulation of OTR mRNA expression in this case may have been the result of increased levels of OTR protein resulting from cellular hypertrophy of OTR-expressing cells, which has been previously reported in rodent brains following chronic alcohol exposure (Madeira et al.

1993). However, a more detailed analysis of alcohol-induced changes in the distribution of OTR mRNA levels has been recently conducted on brain tissue from both humans with AUD and rats made alcohol dependent using a chronic intermittent ethanol vapor exposure (CIE) model (Hansson et al. 2018). In rats, OT mRNA expression in the PVN and SON was found to be significantly decreased following 3 weeks of abstinence from ethanol vapor exposure compared to air exposed controls. In contrast, while OTR mRNA and binding in the NAc was decreased 24 h following the last ethanol exposure session, it was increased after 3 weeks of abstinence relative to controls. Additional experiments at this time point revealed increased OTR mRNA expression, protein, and receptor binding in the amygdala, PFC, striatal and hippocampal regions. These chronic alcohol-induced increases in OTR mRNA and receptor binding were also confirmed in the ventral striatum and several other brain regions of deceased humans with a history of heavy alcohol consumption (Hansson et al. 2018). Overall, the results from this study were complex, suggesting the OT system is capable of dynamic changes in response to alcohol dependence. It is also important to note that previously mentioned studies were conducted exclusively in male subjects and that there is evidence suggesting the OT system in females may not be as susceptible to the negative effects of chronic alcohol exposure. Indeed, Hansson and Spanagel (2021) repeated some of the experiments from their previous study (Hansson et al. 2018) using post-mortem tissue from female humans with AUD and alcohol-dependent rats, which revealed no significant changes in OTR mRNA expression or receptor binding in any of the brain regions analyzed. The authors further suggest that the lack of effects of alcohol on the OT system in females is justification for future research to be conducted exclusively in males. However, others have argued against this approach by pointing out several studies which have reported at least some neurobiological effects of alcohol on the OT system in females (Ryabinin and Fulenwider 2021). While future work will be needed to address our limited understanding of the precise mechanistic details, it is clear that chronic alcohol exposure negatively impacts the OT system at the neurobiological level in both rodents and humans. Because of these and similar findings from studies of other drugs of abuse, the OT system has become a promising therapeutic target. The results of both preclinical and clinical studies using OT-based interventions in the context of alcohol will be discussed in later sections.

1.3 Considerations for the Clinical Application of OT and OTR Agonists

In order to address OT's efficacy in the treatment of AUD, it has become increasingly important to investigate to what extent peripherally administered OT is able to cross the blood-brain barrier (BBB) and its subsequent effects on brain function. Exogenous OT in humans is administered peripherally either through the intranasal (I.N) or intravenous (i.v) routes. Until recently however, it was commonly accepted that OT was not able to cross the BBB in sufficient amounts to alter behavior directly by central effects, given that only 1–2% of peripherally released OT reaches the brain (Ermisch et al. 1985; Nishimori et al. 2008). Additionally, more recent work from our group and others using a highly sensitive fluorescent OT sensor found no BBB penetration of peripherally administered OT in healthy mice (Ino et al. 2021) or rats (Qian et al. 2022 preprint). Instead of directly crossing the BBB, the main pathway by which peripheral OT modulates behavior was thought to be via stimulating release of endogenous central OT. This pathway has since been confirmed in rats through the observation that peripherally administered OT induces expression of the immediately early gene, cFos, in PVN OT neurons (Leong et al. 2017). However, while peripherally administered OT has been detected in the cerebrospinal fluid (CSF) of rhesus monkeys, OT release into the blood via peripheral activation of the endogenous central OT system was not observed (Lee et al. 2018). Therefore, further investigation is needed to unravel the detailed pathways by which peripherally administered OT activates the endogenous central OT system and how OT crosses the BBB. One proposed mechanism is the transportation of OT across the BBB via carrier proteins (McEwen 2004) and others have suggested that drug-induced (Nishimori et al. 2008) or hypertension-related (Churchland and Winkielman 2012) damage of the BBB may allow leakage of OT from the blood into the CSF.

Beyond the BBB, there are other strong limitations to the therapeutic potential of OT in its native form, namely, its short half-life time and cross reactivity with V1a and V1b receptors at high concentrations. The half-life time of OT is only between 3 and 6 min in the blood and about 20 min in the CSF (Mens et al. 1983; Ryden and Sjoholm 1969; Tanaka et al. 2018). It is therefore unclear if OT in its native form is able to provide long-lasting effects, particularly in deeper brain regions that require a greater diffusion time to reach. As previously described, OT interacts with the VP system, which can induce several side effects, such as hyponatremia and excessive water conservation (Demiselle et al. 2020; Hew-Butler 2010). Recently however, the development of new OT analogs has shown some promise in overcoming these limitations. Two main approaches have been applied to create new molecules with improved pharmacokinetics that do not limit the affinity to bind OTR. For instance, the Chini group demonstrated that deamination of the N-terminal NH_2 group of the OT peptide can reduce its binding efficiency to VP receptors without affecting its binding efficiency to OTR (Chini et al. 2003). Another group engineered an OT analog with a modified disulfide bond that does not affect the binding properties to OTR but can increase the half-life time up to five-fold in the blood (from ~3 to ~14 min) compared to native OT (Beard et al. 2018).

One of the most common peptidergic OTR agonists is the OT analog carbetocin, which has an extended half-life and specifically activates the excitatory $G\alpha_q$ pathway in OTR-expressing cells (Passoni et al. 2016). Carbetocin has further been shown to affect the reinforcing properties of alcohol in mice as measured by the alcohol conditioned place preference assay (CPP). However, the two existing studies reported opposing results such that carbetocin administration was capable of both blocking (Bahi 2015) and facilitating (Rae et al. 2018) ethanol CPP. A likely reason for these mixed results is differences in dosing, schedule and paradigm used, but may also be due to the fact that the currently available peptidergic OT analogs exhibit similar or lower brain penetrance than native OT. Indeed, another OTR agonist, PF-06655075, has been shown to decrease alcohol consumption in rats when administered centrally, but not peripherally (Tunstall et al. 2019). Several other peptidergic OTR agonists that differ in certain properties have been developed such as TGOT (Andres et al. 2004), Atosiban (Akerlund et al. 1999), and DnalOVT (Busnelli et al. 2012). However, none of these have yet to be tested in models of AUD.

Many non-peptidergic OTR agonists have been recently developed and are currently under investigation. For example, TC-OT-39 works as a weak agonist of OTR and an antagonist of VP receptors, but has failed to produce significant effects in a mouse model of social deficits linked to autism-spectrum disorders (Moy et al. 2019). Another OTR agonist, WAY-267464, is mechanistically similar to TC-OT-39 but with better OTR binding efficiency (Jorgensen et al. 2018). This particular OTR agonist is of great interest as it has shown efficacy in the extinction of conditioned fear when centrally administered directly to the amygdala (Lahoud and Maroun 2013), as well as rescuing social deficits in shank3b mice upon systemic application (Resendez et al. 2020). Interestingly, WAY-267464 was also found to enhance otherwise impaired social interactions in adult male rats exposed to ethanol during adolescence while having the opposite effect on water-exposed controls (Dannenhoffer et al. 2018). The authors suggest that the reduction of social interaction found in the control group may be the result of an agonist-induced enhancement of OT system activity that precluded the need for the socially-induced release of endogenous OT. Beyond this particular study however, to our knowledge, there has yet to be any further published reports using these non-peptidergic analogs in preclinical models of AUD. In conclusion, while the OT system has shown great promise as a therapeutic target for many conditions and disease states, there are several mechanistic limitations of the native OT peptide that need to be overcome in order to enhance its efficacy in the clinic. Therefore, the development of alternative OTR agonists that are capable of circumventing these limitations is a promising avenue for current and future research.

2 The Role of OT on Alcohol-Related Behaviors in Animal Models

Numerous studies have demonstrated that OT can affect alcohol-related behaviors in animal models of AUD. In general, the results of selected key studies (Table 1) tend to indicate a positive effect of OT on alcohol-related behaviors despite differences in species, dosing, route of administration, and paradigm used.

Table 1 List of key studies examining the effects of OT system manipulation on alcohol-related phenotypes and behaviors

Administration methodology	Effect on alcohol-related phenotypes and behavior	Reference
10 μg of s.c. OT, single dose, Swiss mice	OT administration attenuated expression of hypothermic tolerance in an i.p. ethanol challenge paradigm	(Rigter et al. 1980)
800 or 2,400 nmol/kg of s.c. OT, three doses, Wistar rats	OT administration inhibited expression of hypothermic tolerance effects in an alcohol i.p. administration paradigm	(Pucilowski et al. 1985)
0.5–2 IU of s.c. OT, three doses, CFLP mice	OT administration blocked expression of hypothermic tolerance in an i.p. ethanol challenge paradigm when applied over 3 days	(Szabo et al. 1985)
0.2–2 IU of s.c. OT, daily, CFLP mice	OT administration reduced withdrawal symptoms induced with picrotoxin in alcohol-dependent mice	(Szabo et al. 1987)
0.3–3 ng of ICV OT, two doses, CFLP mice	OT administration blocked expression of hypothermic tolerance in an i.p. ethanol challenge paradigm	(Szabo et al. 1989)
5 μg of i.p. OT, daily for 5 days, OF1 mice	OT administration attenuates hypothermic, myorelaxant, and akinesic effects in an i.p. ethanol challenge paradigm	(Jodogne et al. 1991; Tirelli et al. 1992)
1 mg/kg of i.p. OT, daily for 10 days, Wistar rats	OT administration reduces alcohol consumption during ad libitum access	(Bowen et al. 2011)
10 mg/kg of i.p. or 0.5 μg of ICV OT, single dose, C57BL/6 mice	i.p OT administration reduced alcohol consumption in a two bottle free choice paradigm. ICV administration did not produce an effect	(Peters et al. 2013)
6.4 mg/kg of i.p. Carbetocin, daily, C57BL/6 mice	Carbetocin administration reduced establishment, reinstatement, and total time spent in an alcohol paired chamber using a CPP paradigm	(Bahi 2015)
6.4 mg/kg of i.p. Carbetocin daily Swiss mice	Carbetocin administration enhanced establishment of alcohol CPP	(Rae et al. 2018)
0.1–0.5 mg/kg of i.p. OT, daily, SD rats	OT administration reduced alcohol intake in a three bottle choice and alcohol gel self-administration paradigm	(MacFadyen et al. 2016)
1 μg of ICV OT, single dose, Wistar rats	ICV administration of OT attenuated self-administration of alcohol in a chronic intermittent access model. OT blocked alcohol-induced dopamine release in the NAc	(Peters et al. 2017)
1–10 mg/kg of i.p. OT, single dose on 2 days, prairie voles	OT administration reduced alcohol intake in a self-administration paradigm in male and female prairie voles	(Stevenson et al. 2017)

(continued)

Table 1 (continued)

Administration methodology	Effect on alcohol-related phenotypes and behavior	Reference
1–10 mg/kg of i.p. OT, single dose daily on 2 days, C57BL/6 mice	OT administration reduced alcohol intake in a self-administration paradigm also in doses not affecting sweet control solution	(King et al. 2017)
10 ng of ICV OT, single dose, Wistar rats	OT reduced cue-reactivity in an alcohol reinstatement paradigm in alcohol-dependent rats	(Hansson et al. 2018)
0.125–1 mg/kg of i.p.; 1 mg/kg of IN OT or 30 μg of ICV PF-06655075, single dose before test, Wistar and SD rats	OT reduced alcohol consumption when administered i.p or IN. In an alcohol dependence model, PF-06655075 reduced alcohol intake only when administered ICV	(Tunstall et al. 2019)
3 mg/kg of i.p. OT, single dose daily, prairie voles	OT application reduced alcohol consumption but not alcohol preference in a social housing paradigm	(Walcott and Ryabinin 2021)
3 mg/kg of i.p. OT, single dose daily, C57BL/6 mice	OT administration significantly reduced ethanol intake on 3 out of 4 test days	(Caruso et al. 2021)

2.1 *Effects of OT in Animal Models of Acute Alcohol Exposure*

Several studies have investigated the effects of exogenous OT and OTR agonists on alcohol consumption. Administration of native OT has been demonstrated to decrease levels of alcohol consumption in rats (Bowen et al. 2011), mice (King et al. 2017; Peters et al. 2013), and voles (Stevenson et al. 2017; Walcott and Ryabinin 2021). The reduction of alcohol consumption caused by OT pretreatment seems to be dose-dependent with effects ranging from as a little as 0.1 mg/kg and as high as 10 mg/kg, although doses above 3 mg/kg could result in confounding locomotor effects (King et al. 2017; Stevenson et al. 2017). Although one study found reductions in consumption lasting at least 6 weeks following a single dose of OT (1 mg/kg i.p.) (McGregor and Bowen 2012), most studies have reported effects lasting only a few hours following exogenous OT administration, which may be related to the limitations of Ots brain penetrance and half-life. Thus, it is essential to test non-peptidergic OT analogs in an effort to achieve long-lasting reductions in alcohol consumption.

2.2 *Effects of OT in Animal Models of Chronic Alcohol Exposure and Withdrawal*

Repeated heavy alcohol drinking leads to neural adaptations resulting in the development of tolerance to both the physical and rewarding effects of alcohol intoxication, as well as increased stress and anxiety during withdrawal. Multiple early studies focused specifically on tolerance to the physiological effects of alcohol intake (Jodogne et al. 1991; Pucilowski et al. 1985; Rigter et al. 1980; Szabo et al. 1985, 1987, 1989; Tirelli et al. 1992). The general consensus of these studies was that systemic OT administration prevents the development of tolerance to the hypnotic, hypothermic, and ataxic effects of alcohol intoxication. Interestingly, this protective effect of OT was no longer present once the tolerance had already developed (Szabo et al. 1985). It was also found that central administration of OT exerts a more profound block of the development of physiological tolerance than peripheral administration (Szabo et al. 1989), which suggests that these effects are centrally mediated and may therefore be involved in the neural adaptions that are thought to drive the negative affective state of withdrawal.

The use of OT as treatment for the symptoms of alcohol dependence was first investigated several decades ago by Szabo and colleagues who reported that OT decreases alcohol-induced withdrawal symptoms in a dose-dependent manner (Szabo et al. 1987). As mentioned previously, the chronic relapsing nature of AUD is thought to be driven by the persistence of a negative affective state when alcohol is no longer in the system, which can last several years into abstinence. The presence of these symptoms leads to a strong craving to consume alcohol for relief, which often triggers a full relapse into the heavy, uncontrolled drinking that is characteristic of AUD (Tabakoff and Hoffman 2013). Preclinical studies model relapse-like behavior using reinstatement paradigms in which an animal with a history of dependence is given the opportunity to resume intake after a period of abstinence. However, while OT has been shown by several studies to decrease the reinstatement of drug-seeking in mice and rats, only a few of those studies focused on alcohol. One such study found that central administration of OT reduced cue-induced reinstatement of alcohol-seeking behavior (Hansson et al. 2018). Interestingly, this reduction was found in alcohol-dependent rats but not in the non-dependent control group. Another study found that OT is able to block the withdrawal-induced escalation of intake commonly observed in alcohol-dependent rats during acute abstinence, and these effects were observed after systemic, central and also IN administration of OT (Tunstall et al. 2019). These findings provide promising evidence that OT system manipulation may be helpful as a therapy to prevent relapse. However, future work examining the effects of OT on relapse-like behavior in rodents is needed to dissect the precise neurobiological underpinnings of these beneficial effects.

3 Clinical Studies on the Effect of OT Administration on Alcohol Consumption

There have been several published reports investigating the effects of IN administered OT on different alcohol-related behaviors in humans. These studies, summarized in Table 2 below, were carried out in three different countries with sample sizes

Table 2 Summary of alcohol-related clinical trials involving IN OT administration in humans

Administration methodology	Effect on alcohol-related phenotypes and behavior	Reference
24 IU of IN OT, twice daily for 3 days, n(OT) = 7; n(placebo) = 4	OT administration reduced withdrawal and craving symptoms	(Pedersen et al. 2013)
40 IU of IN OT, single dose, n (OT) = 32; n(placebo) = 32	OT administration reduced craving in a specific subset of subjects with low attachment anxiety and increased craving in others	(Mitchell et al. 2016)
24 IU of IN OT, single dose, n (OT) = 12; n(placebo) = 15	OT administration reduced alcohol cue-reactivity in brain networks associated with craving	(Hansson et al. 2018)
40 + 20 IU of IN OT, single dose + booster, n(OT) = 35; n(placebo) = 35	OT administration did not alter effects of acute alcohol consumption in healthy, social drinkers	(Vena et al. 2018)
40 IU of IN OT, single dose, n (OT) = 32; n(placebo) = 35	OT administration did not reduce craving in veterans with co-morbid Post-Traumatic Stress Disorder and AUD	(Flanagan et al. 2019)
24 IU of IN OT, single dose, n (OT) = 13; n(placebo) = 13	OT selectively reduces NAc connectivity and subjective craving in male social drinkers during cue-reactivity task	(Bach et al. 2019)
24 IU of IN OT, twice daily, n (OT) = 40; n(placebo) = 40	OT administration did not significantly reduce the dose of oxazepam needed for a 3-day treatment of alcohol intoxication	(Melby et al. 2019)
24 IU of IN OT, twice daily, n (OT) = 20; n(placebo) = 20	OT administration did not significantly alter actigraphy assessment of sleep and motor activity in patients in acute withdrawal	(Melby et al. 2020)
8 IU of IN OT self-administered as needed during periods of craving, up to 6 times daily, n(OT) = 18; n (placebo) = 18	OT administration produced no significant differences in alcohol use amongst alcohol-dependent subjects	(Melby et al. 2021)
24 IU of IN OT, single dose, n (OT) = 13; n(placebo) = 13	OT administration-induced attenuation of amygdala responses to fearful faces in social drinkers was associated with lower ratings of subjective craving and reduced number of heavy drinking days	(Bach et al. 2021)
40 IU of IN OT, single dose, n (OT) = 100, n(placebo) = 100	OT administration did not affect cue-induced alcohol craving or intimate partner aggression	(Flanagan et al. 2022)

ranging between 11 and 84, two-thirds of which were male. The course of treatment varied greatly across the studies, which ranged from a single dose up to 12 weeks of daily administration.

Overall, the results of these studies are quite heterogeneous. For instance, four studies did not report any significant effects of OT administration (Flanagan et al. 2019; Melby et al. 2019, 2020, 2021; Vena et al. 2018). In contrast, five studies showed a decrease in various alcohol-related behaviors (Bach et al. 2019, 2021; Hansson et al. 2018; Pedersen 2017; Pedersen et al. 2013). Finally, one study found bidirectional effects of OT on craving when subjects were grouped by degree of attachment anxiety (Mitchell et al. 2016). The lack of agreement in the direction of the effects observed across these studies precludes a definitive interpretation regarding the therapeutic efficacy of OT as a future treatment for AUD. However, there appears to be an emerging consensus that any potential effects of OT on addiction-related behaviors, particularly with regard to withdrawal and relapse, will likely be mediated through an interaction between the degree of an individual's trait anxiety and central OTs role in the regulation of anxiety. Indeed, Mitchell et al. (2016) suggest that the OT-mediated increase in alcohol craving found in the subjects with low attachment anxiety suggests that there might be an individual variance among the population in terms of response to IN OT in the context of alcohol. On the other hand, while Melby et al. (2021) found a small positive effect of OT on self-reported nervousness during abstinence and Flanagan et al. (2019) found a marginal effect of OT on regulating stress-induced increases in cortisol, neither of these two studies observed a decrease in alcohol craving (Flanagan et al. 2019; Melby et al. 2021). It may be possible that these small effects on stress-/anxiety-related behaviors in the absence of reductions in craving could be due in part to the limitations of native OT's brain penetrance and half-life described above. While there are improved synthetic OTR agonists and antagonists in development, they have yet to be approved for clinical use. Therefore, the utility of targeting the OT system as a treatment for AUD will likely remain unclear for some time yet.

4 Conclusions and Perspectives

OT was initially identified as a key regulator of labor in mammals. It has since garnered great attention for its involvement in an expanding plethora of behaviors, including those relevant to addiction. Several preclinical studies have reported that activating the OT system with either native OT or its optimized analogs leads to reductions in acute alcohol consumption. However, only a few studies have examined the influence of the OT system in the context of chronic alcohol exposure, particularly on the anxiety- and stress-related behaviors that model the relapse-driving negative affective state experienced by humans during withdrawal. Further, mixed results from clinical trials involving the effects of OT administration on alcohol-related behaviors in chronic AUD patients suggest gaps in our understanding of the relationship between OT and AUD. Therefore, more research using

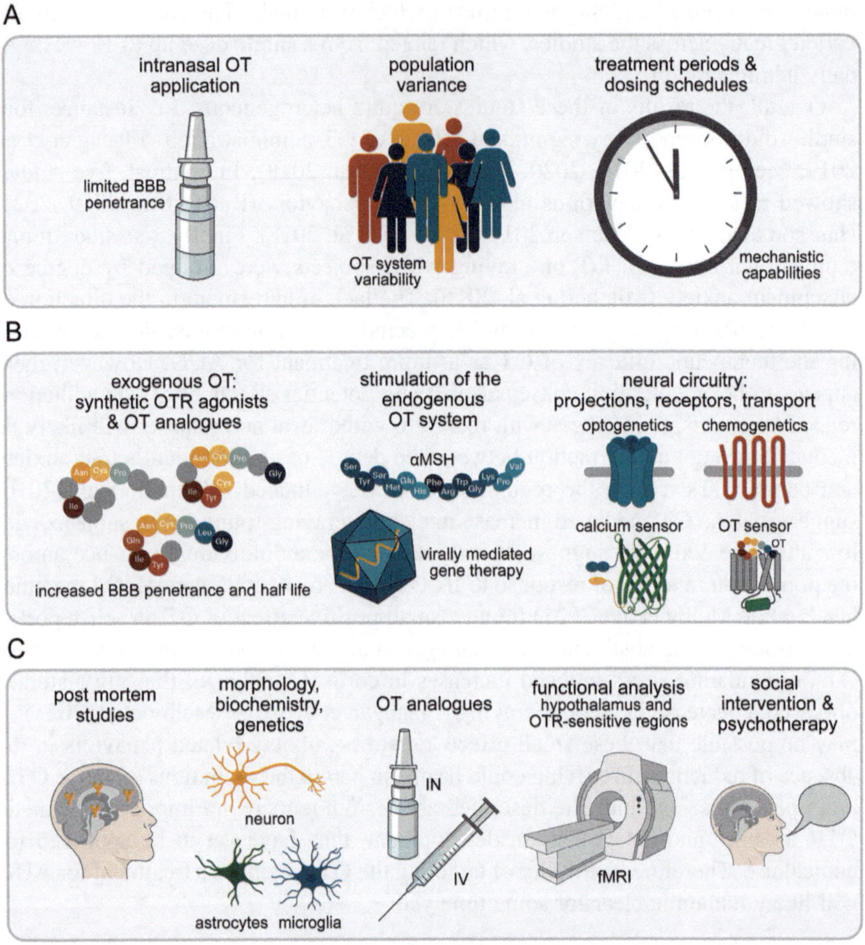

Fig. 3 Overview of the current limitations and future directions of the development of OT-based therapeutics for AUD. (**a**) The intranasal application of native OT in clinical trials for AUD has produced mixed results, likely due to several limitations inherent to OT system biology. (**b**) Future work in rodent models of AUD should focus on the development of novel approaches to modulate the endogenous central OT system in a precise and controlled manner. These efforts can be accelerated by employing modern genetic tools delivered by OT cell-type specific viral vectors. (**c**) The development of novel OT-based therapeutics will require a greater understanding of the anatomical substrates of the OT system in humans, which can be achieved through the effective use of post-mortem brain tissue from human AUD subjects. Future clinical trials should focus on the administration of more potent OT analogs in combination with psychosocial interventions to maximize the potential benefits of an OT-based treatment for AUD

preclinical models of chronic alcohol exposure and withdrawal should be prioritized. To this end, we have provided a graphical summary of the current limitations of OT administration in humans along with suggestions for future directions that may be able to overcome them (Fig. 3).

As depicted in Fig. 3, many clinical trials have opted to administer OT IN due to its ease of use and presumed ability to penetrate the brain more easily. However, as discussed above, the amount of peripherally administered OT that crosses the BBB ranges from incredibly low (Ermisch et al. 1985; Nishimori et al. 2008) to not even detectable (Ino et al. 2021). The mixed results from clinical trials are also likely to be influenced by population variance, insufficient treatment periods, and inconsistent dosing schedules across studies. The common thread among these limitations is the endogenous OT system itself, in terms of both the peptide's mechanistic capabilities and individual variability in OT system physiology across and within populations. We therefore suggest that most, if not all, of these particular limitations could be overcome by conducting more preclinical research using currently available and future synthetic OTR agonists and OT analogs that are more potent and robust in terms of brain penetrance and half-life. Once adequate therapeutic potential is validated in animal models of chronic alcohol exposure, approval should then be sought for use in future clinical trials. Continued research using post-mortem brain tissue from human AUD patients will also be essential to the development of OT-based therapeutics. However, these tissue samples are a rare and highly sought after. To achieve the most efficient use of these precious resources, reverse translational approaches could be considered. For example, spatially resolved brain activation data from clinical trials using fMRI in human AUD patients could serve as a useful anatomical guide for targeted neurobiological research of post-mortem brain samples.

Alternatively, the use of more indirect or unconventional approaches to stimulate the endogenous OT system could overcome the aforementioned limitations while still minimizing off-target effects. Early proof of concept work has demonstrated this possibility with the use of alpha melanocyte-stimulating hormone (αMSH), a hypothalamic peptide hormone which has been shown to specifically trigger the dendritic release of OT from magnocellular neurons in the SON (Sabatier et al. 2003). Additionally, activating the αMSH receptor, MC4R, in voles using highly specific and brain-penetrant agonists can enhance social cognition in an OTR-dependent manner (Modi et al. 2015).

While still in the nascent stages of use in the clinic as an intervention for certain genetic diseases, virally-mediated gene therapy approaches could also be adapted to modulate central OT system activity. The advantage of this approach is derived from the use of specific gene promoters which allow for the cell-type specific expression of custom genetic tools only in cells that also express the targeted gene of interest, such as OT (Grinevich et al. 2016b). However, the use of virally-mediated interventions targeting the OT system in models of AUD is currently limited, but so far promising. For instance, King et al. (2021) used an adeno-associated viral vector (AAV) to express an excitatory chemogenetic receptor that specifically activated PVN OT neurons in mice, which led to significant reductions in binge alcohol consumption (King et al. 2021). Another group employed a lentiviral-based approach to over-express OTR in the NAc of mice, which reduced the reinforcing properties of ethanol in a CPP assay (Bahi 2015). Perhaps the most significant limitation preventing the use of virally-mediated gene-therapy interventions in the clinic is the highly invasive nature of delivering the virus directly into the brain. However, recent advances in AAV technology have led to the development of viral vectors that can cross the BBB and thereby be administered peripherally (Liu et al. 2021). Although there are clearly more limitations to be overcome before gene therapy could be considered for rescuing deficient OT signaling in human AUD patients, it is certainly an avenue of research worth pursuing through further preclinical work.

The continued pursuit of virally-mediated approaches in preclinical research can reveal even more potential therapeutic targets. Indeed, AAVs harboring the OT-promotor in tandem with chemogenetic, optogenetic, and calcium sensor-based tools have already greatly advanced our understanding of the neural circuitry of the central OT system and its role in several behaviors (Eliava et al. 2016; Hasan et al. 2019; Knobloch et al. 2012; Tang et al. 2020; Wahis et al. 2021). One particularly interesting outcome of these studies was the finding that manipulation of OTR-expressing astrocytes in the central amygdala is sufficient to modulate place preference and anxiety-like behavior in rodents (Wahis et al. 2021). In light of the fact that alcohol exposure leads to dysregulated GPCR expression in astrocytes (Adermark and Bowers 2016), future preclinical work that expands the focus beyond neurons may prove highly profitable in terms of finding novel OT-based therapeutic approaches to treat AUD among many other disease conditions.

One final point of consideration is the likelihood that manipulation of the OT system alone is incapable of exerting a degree of influence that is adequate to rescue the behavioral and neurobiological deficits of AUD patients. Given the prominent role of central OT in the regulation of social behaviors and the social deficits observed in AUD patients, it is somewhat surprising that clinical trials have yet to explore the prospect of using OT administration combined with psychosocial interventions, such as cognitive behavioral therapy. Furthermore, this particular approach could be tested immediately with currently available technologies at a relatively low cost.

Acknowledgments RP was a recipient of short-term Deutscher Akademischer Austauschdienst (DAAD) postdoctoral fellowship.

References

Adermark L, Bowers MS (2016) Disentangling the role of astrocytes in alcohol use disorder. Alcohol Clin Exp Res 40(9):1802–1816. https://doi.org/10.1111/acer.13168

Akerlund, M., Bossmar, T., Brouard, R., Kostrzewska, A., Laudanski, T., Lemancewicz, A., . . . Steinwall, M. (1999). Receptor binding of oxytocin and vasopressin antagonists and inhibitory effects on isolated myometrium from preterm and term pregnant women. Br J Obstet Gynaecol, 106(10), 1047–1053. doi:https://doi.org/10.1111/j.1471-0528.1999.tb08112.x

Althammer F, Grinevich V (2017) Diversity of oxytocin neurons: beyond magno- and parvocellular cell types? J Neuroendocrinol. https://doi.org/10.1111/jne.12549

Andres, M., Pena, A., Derick, S., Raufaste, D., Trojnar, J., Wisniewski, K., . . . Guillon, G. (2004). Comparative pharmacology of bovine, human and rat vasopressin receptor isoforms. Eur J Pharmacol, 501(1–3), 59–69. doi:https://doi.org/10.1016/j.ejphar.2004.08.026

Armstrong WE, Warach S, Hatton GI, McNeill TH (1980) Subnuclei in the rat hypothalamic paraventricular nucleus: a cytoarchitectural, horseradish peroxidase and immunocytochemical analysis. Neuroscience 5(11):1931–1958. Retrieved from http://www.ncbi.nlm.nih.gov/pubmed/7432630

Bach P, Reinhard I, Buhler S, Vollstadt-Klein S, Kiefer F, Koopmann A (2019) Oxytocin modulates alcohol-cue induced functional connectivity in the nucleus accumbens of social drinkers. Psychoneuroendocrinology 109:104385. https://doi.org/10.1016/j.psyneuen.2019.104385

Bach, P., Koopmann, A., Bumb, J. M., Zimmermann, S., Buhler, S., Reinhard, I., . . . Kiefer, F. (2021). Oxytocin attenuates neural response to emotional faces in social drinkers: an fMRI study. Eur Arch Psychiatry Clin Neurosci, 271(5), 873–882. doi:https://doi.org/10.1007/s00406-020-01115-0

Bagdy G, Makara GB (1995) Paraventricular nucleus controls 5-HT2C receptor-mediated corticosterone and prolactin but not oxytocin and penile erection responses. Eur J Pharmacol 275(3): 301–305. https://doi.org/10.1016/0014-2999(95)00051-l

Bahi A (2015) The oxytocin receptor impairs ethanol reward in mice. Physiol Behav 139:321–327. https://doi.org/10.1016/j.physbeh.2014.11.046

Beard R, Stucki A, Schmitt M, Py G, Grundschober C, Gee AD, Tate EW (2018) Building bridges for highly selective, potent and stable oxytocin and vasopressin analogs. Bioorg Med Chem 26(11):3039–3045. https://doi.org/10.1016/j.bmc.2018.03.019

Beets I, Temmerman L, Janssen T, Schoofs L (2013) Ancient neuromodulation by vasopressin/oxytocin-related peptides. Worm 2(2):e24246. https://doi.org/10.4161/worm.24246

Bernheim A, Leong KC, Berini C, Reichel CM (2017) Antagonism of mGlu2/3 receptors in the nucleus accumbens prevents oxytocin from reducing cued methamphetamine seeking in male and female rats. Pharmacol Biochem Behav 161:13–21. https://doi.org/10.1016/j.pbb.2017.08.012

Bosch, O. J., Dabrowska, J., Modi, M. E., Johnson, Z. V., Keebaugh, A. C., Barrett, C. E., . . . Young, L. J. (2016). Oxytocin in the nucleus accumbens shell reverses CRFR2-evoked passive stress-coping after partner loss in monogamous male prairie voles. Psychoneuroendocrinology, 64, 66–78. doi:https://doi.org/10.1016/j.psyneuen.2015.11.011

Bowen MT, McGregor IS (2014) Oxytocin and vasopressin modulate the social response to threat: a preclinical study. Int J Neuropsychopharmacol 17(10):1621–1633. https://doi.org/10.1017/S1461145714000388

Bowen MT, Carson DS, Spiro A, Arnold JC, McGregor IS (2011) Adolescent oxytocin exposure causes persistent reductions in anxiety and alcohol consumption and enhances sociability in rats. PloS One 6(11):e27237. https://doi.org/10.1371/journal.pone.0027237

Bowen MT, Peters ST, Absalom N, Chebib M, Neumann ID, McGregor IS (2015) Oxytocin prevents ethanol actions at delta subunit-containing GABAA receptors and attenuates ethanol-induced motor impairment in rats. Proc Natl Acad Sci U S A 112(10):3104–3109. https://doi.org/10.1073/pnas.1416900112

Brown CH, Bains JS, Ludwig M, Stern JE (2013) Physiological regulation of magnocellular neurosecretory cell activity: integration of intrinsic, local and afferent mechanisms. J Neuroendocrinol 25(8):678–710. https://doi.org/10.1111/jne.12051

Burton IW, Martinez Farina CF, Ragupathy S, Arunachalam T, Newmaster S, Berrue F (2020) Quantitative NMR methodology for the authentication of roasted coffee and prediction of blends. J Agric Food Chem 68(49):14643–14651. https://doi.org/10.1021/acs.jafc.0c06239

Busnelli M, Chini B (2018) Molecular basis of oxytocin receptor signalling in the brain: what we know and what we need to know. Curr Top Behav Neurosci 35:3–29. https://doi.org/10.1007/7854_2017_6

Busnelli M, Sauliere A, Manning M, Bouvier M, Gales C, Chini B (2012) Functional selective oxytocin-derived agonists discriminate between individual G protein family subtypes. J Biol Chem 287(6):3617–3629. https://doi.org/10.1074/jbc.M111.277178

Caruso MA, Robins MT, Fulenwider HD, Ryabinin AE (2021) Temporal analysis of individual ethanol consumption in socially housed mice and the effects of oxytocin. Psychopharmacology 238(3):899–911. https://doi.org/10.1007/s00213-020-05741-3

Carter CS (2017a) The oxytocin-vasopressin pathway in the context of love and fear. Front Endocrinol (Lausanne) 8:356. https://doi.org/10.3389/fendo.2017.00356

Carter CS (2017b) The role of oxytocin and vasopressin in attachment. Psychodyn Psychiatry 45(4):499–517. https://doi.org/10.1521/pdps.2017.45.4.499

Chini, B., Chinol, M., Cassoni, P., Papi, S., Reversi, A., Areces, L., . . . Bussolati, G. (2003). Improved radiotracing of oxytocin receptor-expressing tumours using the new [111In]-DOTA-Lys8-deamino-vasotocin analogue. Br J Cancer, 89(5), 930–936. doi:https://doi.org/10.1038/sj.bjc.6601189

Churchland PS, Winkielman P (2012) Modulating social behavior with oxytocin: how does it work? What does it mean? Horm Behav 61(3):392–399. https://doi.org/10.1016/j.yhbeh.2011.12.003

Cox BM, Bentzley BS, Regen-Tuero H, See RE, Reichel CM, Aston-Jones G (2017) Oxytocin acts in nucleus accumbens to attenuate methamphetamine seeking and demand. Biol Psychiatry 81(11):949–958. https://doi.org/10.1016/j.biopsych.2016.11.011

Dabrowska, J., Hazra, R., Ahern, T. H., Guo, J. D., McDonald, A. J., Mascagni, F., . . . Rainnie, D. G. (2011). Neuroanatomical evidence for reciprocal regulation of the corticotrophin-releasing factor and oxytocin systems in the hypothalamus and the bed nucleus of the stria terminalis of the rat: implications for balancing stress and affect. Psychoneuroendocrinology, 36(9), 1312–1326. doi:https://doi.org/10.1016/j.psyneuen.2011.03.003

Dale HH (1906) On some physiological actions of ergot. J Physiol 34(3):163–206. https://doi.org/10.1113/jphysiol.1906.sp001148

Dannenhoffer CA, Kim EU, Saalfield J, Werner DF, Varlinskaya EI, Spear LP (2018) Oxytocin and vasopressin modulation of social anxiety following adolescent intermittent ethanol exposure. Psychopharmacology (Berl) 235(10):3065–3077. https://doi.org/10.1007/s00213-018-5003-8

De Deurwaerdere P, Di Giovanni G (2017) Serotonergic modulation of the activity of mesencephalic dopaminergic systems: therapeutic implications. Prog Neurobiol 151:175–236. https://doi.org/10.1016/j.pneurobio.2016.03.004

Demiselle J, Fage N, Radermacher P, Asfar P (2020) Vasopressin and its analogues in shock states: a review. Ann Intensive Care 10(1):9. https://doi.org/10.1186/s13613-020-0628-2

Devost D, Wrzal P, Zingg HH (2008) Oxytocin receptor signalling. Prog Brain Res 170:167–176. https://doi.org/10.1016/S0079-6123(08)00415-9

Diaz-Cabiale Z, Petersson M, Narvaez JA, Uvnas-Moberg K, Fuxe K (2000) Systemic oxytocin treatment modulates alpha 2-adrenoceptors in telencephalic and diencephalic regions of the rat. Brain Res 887(2):421–425. https://doi.org/10.1016/s0006-8993(00)03017-1

Dolen G, Darvishzadeh A, Huang KW, Malenka RC (2013) Social reward requires coordinated activity of nucleus accumbens oxytocin and serotonin. Nature 501(7466):179–184. https://doi.org/10.1038/nature12518

Dumais KM, Veenema AH (2016) Vasopressin and oxytocin receptor systems in the brain: sex differences and sex-specific regulation of social behavior. Front Neuroendocrinol 40:1–23. https://doi.org/10.1016/j.yfrne.2015.04.003

Duvigneaud V, Ressler C, Swan JM, Roberts CW, Katsoyannis PG (1954) The synthesis of oxytocin. J Am Chem Soc 76(12):3115–3121. https://doi.org/10.1021/ja01641a004

Ebner K, Bosch O, Krömer S et al (2005) Release of oxytocin in the rat central amygdala modulates stress-coping behavior and the release of excitatory amino acids. Neuropsychopharmacology 30:223–230. https://doi.org/10.1038/sj.npp.1300607

Edwards S, Guerrero M, Ghoneim OM, Roberts E, Koob GF (2012) Evidence that vasopressin V1b receptors mediate the transition to excessive drinking in ethanol-dependent rats. Addict Biol 17(1):76–85. https://doi.org/10.1111/j.1369-1600.2010.00291.x

Eliava, M., Melchior, M., Knobloch-Bollmann, H. S., Wahis, J., da Silva Gouveia, M., Tang, Y., . . . Grinevich, V. (2016). A new population of parvocellular oxytocin neurons controlling magnocellular neuron activity and inflammatory pain processing. Neuron, 89(6), 1291–1304. doi:https://doi.org/10.1016/j.neuron.2016.01.041

Ermisch A, Barth T, Ruhle HJ, Skopkova J, Hrbas P, Landgraf R (1985) On the blood-brain barrier to peptides: accumulation of labelled vasopressin, DesGlyNH2-vasopressin and oxytocin by brain regions. Endocrinol Exp 19(1):29–37. Retrieved from https://www.ncbi.nlm.nih.gov/pubmed/3872788

Estes MK, Freels TG, Prater WT, Lester DB (2019) Systemic oxytocin administration alters mesolimbic dopamine release in mice. Neuroscience 408:226–238. https://doi.org/10.1016/j.neuroscience.2019.04.006

Flanagan JC, Allan NP, Calhoun CD, Badour CL, Moran-Santa Maria M, Brady KT, Back SE (2019) Effects of oxytocin on stress reactivity and craving in veterans with co-occurring PTSD and alcohol use disorder. Exp Clin Psychopharmacol 27(1):45–54. https://doi.org/10.1037/pha0000232

Flanagan JC, Nietert PJ, Sippel L, Jarnecke AM, Kirby C, Hogan JN, Massa AA, Brower J, Back SE, Parrott D (2022) A randomized controlled trial examining the effects of intranasal oxytocin on alcohol craving and intimate partner aggression among couples. J Psychiatr Res 152:14–24. https://doi.org/10.1016/j.jpsychires.2022.06.011

Freeman SM, Young LJ (2016) Comparative perspectives on oxytocin and vasopressin receptor research in rodents and primates: translational implications. J Neuroendocrinol 28(4). https://doi.org/10.1111/jne.12382

Fuxe, K., Borroto-Escuela, D. O., Romero-Fernandez, W., Ciruela, F., Manger, P., Leo, G., . . . Agnati, L. F. (2012). On the role of volume transmission and receptor-receptor interactions in social behaviour: focus on central catecholamine and oxytocin neurons. Brain Res, 1476, 119–131. doi:https://doi.org/10.1016/j.brainres.2012.01.062

Gainer H (2012) Cell-type specific expression of oxytocin and vasopressin genes: an experimental odyssey. J Neuroendocrinol 24(4):528–538. https://doi.org/10.1111/j.1365-2826.2011.02236.x

Geisler S, Wise RA (2008) Functional implications of glutamatergic projections to the ventral tegmental area. Rev Neurosci 19(4–5):227–244. https://doi.org/10.1515/revneuro.2008.19.4-5.227

Georgiou, P., Zanos, P., Garcia-Carmona, J. A., Hourani, S., Kitchen, I., Kieffer, B. L., . . . Bailey, A. (2015). The oxytocin analogue carbetocin prevents priming-induced reinstatement of morphine-seeking: involvement of dopaminergic, noradrenergic and MOPr systems. Eur Neuropsychopharmacol, 25(12), 2459–2464. doi:https://doi.org/10.1016/j.euroneuro.2015.09.015

Gimpl G, Fahrenholz F (2001) The oxytocin receptor system: structure, function, and regulation. Physiol Rev 81(2):629–683. http://www.ncbi.nlm.nih.gov/pubmed/11274341

Grinevich V, Neumann ID (2021) Brain oxytocin: how puzzle stones from animal studies translate into psychiatry. Mol Psychiatry 26(1):265–279. https://doi.org/10.1038/s41380-020-0802-9

Grinevich V, Knobloch-Bollmann HS, Eliava M, Busnelli M, Chini B (2016a) Assembling the puzzle: pathways of oxytocin signaling in the brain. Biol Psychiatry 79(3):155–164. https://doi.org/10.1016/j.biopsych.2015.04.013

Grinevich V, Knobloch-Bollmann HS, Roth LC, Althammer F, Domanskyi A, Vinnikov IA, Eliava M, Stanifer M, Boulant S (2016b) Somatic transgenesis (viral vectors). In: Murphy D, Gainer H (eds) Molecular neuroendocrinology. https://doi.org/10.1002/9781118760369.ch12

Hansson AC, Spanagel R (2021) No changes in the oxytocin system in alcohol-dependent female rodents and humans: towards a sex-specific psychopharmacology in alcoholism. Addict Biol 26(2):e12945. https://doi.org/10.1111/adb.12945

Hansson, A. C., Koopmann, A., Uhrig, S., Buhler, S., Domi, E., Kiessling, E., . . . Spanagel, R. (2018). Oxytocin reduces alcohol Cue-reactivity in alcohol-dependent rats and humans. Neuropsychopharmacology, 43(6), 1235–1246. doi:https://doi.org/10.1038/npp.2017.257

Harper KM, Knapp DJ, Butler RK, Cook CA, Criswell HE, Stuber GD, Breese GR (2019) Amygdala arginine vasopressin modulates chronic ethanol withdrawal anxiety-like behavior in the social interaction task. Alcohol Clin Exp Res 43(10):2134–2143. https://doi.org/10.1111/acer.14163

Hasan, M. T., Althammer, F., Silva da Gouveia, M., Goyon, S., Eliava, M., Lefevre, A., . . . Grinevich, V. (2019). A fear memory engram and its plasticity in the hypothalamic oxytocin system. Neuron, 103(1), 133–146 e138. doi:https://doi.org/10.1016/j.neuron.2019.04.029

Hayes DJ, Greenshaw AJ (2011) 5-HT receptors and reward-related behaviour: a review. Neurosci Biobehav Rev 35(6):1419–1449. https://doi.org/10.1016/j.neubiorev.2011.03.005

Hew-Butler T (2010) Arginine vasopressin, fluid balance and exercise: is exercise-associated hyponatraemia a disorder of arginine vasopressin secretion? Sports Med 40(6):459–479. https://doi.org/10.2165/11532070-000000000-00000

Higgins GA, Fletcher PJ, Shanahan WR (2020) Lorcaserin: a review of its preclinical and clinical pharmacology and therapeutic potential. Pharmacol Ther 205:107417. https://doi.org/10.1016/j.pharmthera.2019.107417

Huber D, Veinante P, Stoop R (2005) Vasopressin and oxytocin excite distinct neuronal populations in the central amygdala. Science 308(5719):245–248. https://doi.org/10.1126/science.1105636

Ino D, Hibino H, Nishiyama M (2021) A fluorescent sensor for the real-time measurement of extracellular oxytocin dynamics in the brain. bioRxiv. https://doi.org/10.1101/2021.07.30.454450

Jiang Z, Rajamanickam S, Justice NJ (2019) CRF signaling between neurons in the paraventricular nucleus of the hypothalamus (PVN) coordinates stress responses. Neurobiol Stress 11:100192. https://doi.org/10.1016/j.ynstr.2019.100192

Jodogne C, Tirelli E, Klingbiel P, Legros JJ (1991) Oxytocin attenuates tolerance not only to the hypothermic but also to the myorelaxant and akinesic effects of ethanol in mice. Pharmacol Biochem Behav 40(2):261–265. https://doi.org/10.1016/0091-3057(91)90549-h

Jorgensen H, Kjaer A, Knigge U, Moller M, Warberg J (2003) Serotonin stimulates hypothalamic mRNA expression and local release of neurohypophysial peptides. J Neuroendocrinol 15(6): 564–571. https://doi.org/10.1046/j.1365-2826.2003.01032.x

Jorgensen WT, Gulliver DW, Katte TA, Werry EL, Reekie TA, Connor M, Kassiou M (2018) Conformationally rigid derivatives of WAY-267,464: synthesis and pharmacology at the human oxytocin and vasopressin-1a receptors. Eur J Med Chem 143:1644–1656. https://doi.org/10.1016/j.ejmech.2017.10.059

Jurek B, Neumann ID (2018) The oxytocin receptor: from intracellular signaling to behavior. Physiol Rev 98(3):1805–1908. https://doi.org/10.1152/physrev.00031.2017

Jurek, B., Slattery, D. A., Hiraoka, Y., Liu, Y., Nishimori, K., Aguilera, G., . . . van den Burg, E. H. (2015). Oxytocin regulates stress-induced Crf gene transcription through CREB-regulated transcription coactivator 3. J Neurosci, 35(35), 12248–12260. doi:https://doi.org/10.1523/JNEUROSCI.1345-14.2015

Katz DA, Locke C, Liu W, Zhang J, Achari R, Wesnes KA, Tracy KA (2016) Single-dose interaction study of the arginine vasopressin type 1B receptor antagonist ABT-436 and alcohol in moderate alcohol drinkers. Alcohol Clin Exp Res 40(4):838–845. https://doi.org/10.1111/acer.12996

King CE, Becker HC (2019) Oxytocin attenuates stress-induced reinstatement of alcohol seeking behavior in male and female mice. Psychopharmacology (Berl) 236(9):2613–2622. https://doi.org/10.1007/s00213-019-05233-z

King CE, Griffin WC, Luderman LN, Kates MM, McGinty JF, Becker HC (2017) Oxytocin reduces ethanol self-administration in mice. Alcohol Clin Exp Res 41(5):955–964. https://doi.org/10.1111/acer.13359

King CE, Gano A, Becker HC (2020) The role of oxytocin in alcohol and drug abuse. Brain Res 1736:146761. https://doi.org/10.1016/j.brainres.2020.146761

King CE, Griffin WC, Lopez MF, Becker HC (2021) Activation of hypothalamic oxytocin neurons reduces binge-like alcohol drinking through signaling at central oxytocin receptors. Neuropsychopharmacology 46(11):1950–1957. https://doi.org/10.1038/s41386-021-01046-x

Knobloch HS, Grinevich V (2014) Evolution of oxytocin pathways in the brain of vertebrates. Front Behav Neurosci 8:31. https://doi.org/10.3389/fnbeh.2014.00031

Knobloch, H. S., Charlet, A., Hoffmann, L. C., Eliava, M., Khrulev, S., Cetin, A. H., . . . Grinevich, V. (2012). Evoked axonal oxytocin release in the central amygdala attenuates fear response. Neuron, 73(3), 553–566. doi:https://doi.org/10.1016/j.neuron.2011.11.030

Koob GF, Le Moal M (2005) Plasticity of reward neurocircuitry and the 'dark side' of drug addiction. Nat Neurosci 8(11):1442–1444. https://doi.org/10.1038/nn1105-1442

Koob GF, Volkow ND (2016) Neurobiology of addiction: a neurocircuitry analysis. Lancet Psychiatry 3(8):760–773. https://doi.org/10.1016/S2215-0366(16)00104-8

Kovacs GL, Sarnyai Z, Barbarczi E, Szabo G, Telegdy G (1990) The role of oxytocin-dopamine interactions in cocaine-induced locomotor hyperactivity. Neuropharmacology 29(4):365–368. https://doi.org/10.1016/0028-3908(90)90095-9

Lahoud N, Maroun M (2013) Oxytocinergic manipulations in corticolimbic circuit differentially affect fear acquisition and extinction. Psychoneuroendocrinology 38(10):2184–2195. https://doi.org/10.1016/j.psyneuen.2013.04.006

Lawson SK, Gray AC, Woehrle NS (2016) Effects of oxytocin on serotonin 1B agonist-induced autism-like behavior in mice. Behav Brain Res 314:52–64. https://doi.org/10.1016/j.bbr.2016.07.027

Lee HJ, Macbeth AH, Pagani JH, Young WS 3rd. (2009) Oxytocin: the great facilitator of life. Prog Neurobiol 88(2):127–151. https://doi.org/10.1016/j.pneurobio.2009.04.001

Lee MR, Schwandt ML, Sankar V, Suchankova P, Sun H, Leggio L (2017) Effect of alcohol use disorder on oxytocin peptide and receptor mRNA expression in human brain: a post-mortem case-control study. Psychoneuroendocrinology 85:14–19. https://doi.org/10.1016/j.psyneuen.2017.07.481

Lee, M. R., Scheidweiler, K. B., Diao, X. X., Akhlaghi, F., Cummins, A., Huestis, M. A., . . . Averbeck, B. B. (2018). Oxytocin by intranasal and intravenous routes reaches the cerebrospinal fluid in rhesus macaques: determination using a novel oxytocin assay. Mol Psychiatry, 23(1), 115–122. doi:https://doi.org/10.1038/mp.2017.27

Leong KC, Freeman LR, Berini CR, Ghee SM, See RE, Reichel CM (2017) Oxytocin reduces cocaine cued Fos activation in a regionally specific manner. Int J Neuropsychopharmacol 20(10):844–854. https://doi.org/10.1093/ijnp/pyx058

Li K, Nakajima M, Ibanez-Tallon I, Heintz N (2016) A cortical circuit for sexually dimorphic oxytocin-dependent anxiety behaviors. Cell 167(1):60–72 e11. https://doi.org/10.1016/j.cell.2016.08.067

Liu D, Zhu M, Zhang Y, Diao Y (2021) Crossing the blood-brain barrier with AAV vectors. Metab Brain Dis 36(1):45–52. https://doi.org/10.1007/s11011-020-00630-2

Lukas M, Toth I, Veenema AH, Neumann ID (2013) Oxytocin mediates rodent social memory within the lateral septum and the medial amygdala depending on the relevance of the social stimulus: male juvenile versus female adult conspecifics. Psychoneuroendocrinology 38(6):916–926. https://doi.org/10.1016/j.psyneuen.2012.09.018

MacFadyen K, Loveless R, DeLucca B, Wardley K, Deogan S, Thomas C, Peris J (2016) Peripheral oxytocin administration reduces ethanol consumption in rats. Pharmacol Biochem Behav 140:27–32. https://doi.org/10.1016/j.pbb.2015.10.014

Mack SO, Kc P, Wu M, Coleman BR, Tolentino-Silva FP, Haxhiu MA (2002) Paraventricular oxytocin neurons are involved in neural modulation of breathing. J Appl Physiol 92(2):826–834. https://doi.org/10.1152/japplphysiol.00839.2001

Madeira MD, Sousa N, Lieberman AR, Paula-Barbosa MM (1993) Effects of chronic alcohol consumption and of dehydration on the supraoptic nucleus of adult male and female rats. Neuroscience 56(3):657–672. https://doi.org/10.1016/0306-4522(93)90363-k

Martinon D, Dabrowska J (2018) Corticotropin-releasing factor receptors modulate oxytocin release in the dorsolateral bed nucleus of the stria terminalis (BNST) in male rats. Front Neurosci 12:183. https://doi.org/10.3389/fnins.2018.00183

McEwen BB (2004) Closing remarks: review and commentary on selected aspects of the roles of vasopressin and oxytocin in memory processing. Adv Pharmacol 50(593–654):655–708. https://doi.org/10.1016/S1054-3589(04)50015-7

McGregor IS, Bowen MT (2012) Breaking the loop: oxytocin as a potential treatment for drug addiction. Horm Behav 61(3):331–339. https://doi.org/10.1016/j.yhbeh.2011.12.001

Melby K, Grawe RW, Aamo TO, Salvesen O, Spigset O (2019) Effect of intranasal oxytocin on alcohol withdrawal syndrome: a randomized placebo-controlled double-blind clinical trial. Drug Alcohol Depend 197:95–101. https://doi.org/10.1016/j.drugalcdep.2019.01.003

Melby K, Fasmer OB, Henriksen TE, Grawe RW, Aamo TO, Spigset O (2020) Actigraphy assessment of motor activity and sleep in patients with alcohol withdrawal syndrome and the effects of intranasal oxytocin. PloS One 15(2):e0228700. https://doi.org/10.1371/journal.pone.0228700

Melby K, Grawe RW, Aamo TO, Skovlund E, Spigset O (2021) Efficacy of self-administered intranasal oxytocin on alcohol use and craving after detoxification in patients with alcohol dependence. A double-blind placebo-controlled trial. Alcohol Alcohol 56(5):565–572. https://doi.org/10.1093/alcalc/agaa133

Menon, R., Grund, T., Zoicas, I., Althammer, F., Fiedler, D., Biermeier, V., . . . Neumann, I. D. (2018). Oxytocin signaling in the lateral septum prevents social fear during lactation. Curr Biol, 28(7), 1066–1078 e1066. doi:https://doi.org/10.1016/j.cub.2018.02.044

Mens WB, Laczi F, Tonnaer JA, de Kloet ER, van Wimersma Greidanus TB (1983) Vasopressin and oxytocin content in cerebrospinal fluid and in various brain areas after administration of histamine and pentylenetetrazol. Pharmacol Biochem Behav 19(4):587–591. https://doi.org/10.1016/0091-3057(83)90332-5

Mitchell JM, Arcuni PA, Weinstein D, Woolley JD (2016) Intranasal oxytocin selectively modulates social perception, craving, and approach behavior in subjects with alcohol use disorder. J Addict Med 10(3):182–189. https://doi.org/10.1097/ADM.0000000000000213

Modi ME, Inoue K, Barrett CE, Kittelberger KA, Smith DG, Landgraf R, Young LJ (2015) Melanocortin receptor agonists facilitate oxytocin-dependent partner preference formation in the prairie vole. Neuropsychopharmacology 40(8):1856–1865. https://doi.org/10.1038/npp.2015.35

Moy, S. S., Teng, B. L., Nikolova, V. D., Riddick, N. V., Simpson, C. D., Van Deusen, A., . . . Jarstfer, M. B. (2019). Prosocial effects of an oxytocin metabolite, but not synthetic oxytocin receptor agonists, in a mouse model of autism. Neuropharmacology, 144, 301–311. doi:https://doi.org/10.1016/j.neuropharm.2018.10.036

Nakajima M, Gorlich A, Heintz N (2014) Oxytocin modulates female sociosexual behavior through a specific class of prefrontal cortical interneurons. Cell 159(2):295–305. https://doi.org/10.1016/j.cell.2014.09.020

Neumann ID, Landgraf R (2012) Balance of brain oxytocin and vasopressin: implications for anxiety, depression, and social behaviors. Trends Neurosci 35(11):649–659. https://doi.org/10.1016/j.tins.2012.08.004

Ninan I (2011) Oxytocin suppresses basal glutamatergic transmission but facilitates activity-dependent synaptic potentiation in the medial prefrontal cortex. J Neurochem 119(2):324–331. https://doi.org/10.1111/j.1471-4159.2011.07430.x

Nishimori K, Takayanagi Y, Yoshida M, Kasahara Y, Young LJ, Kawamata M (2008) New aspects of oxytocin receptor function revealed by knockout mice: sociosexual behaviour and control of energy balance. Prog Brain Res 170:79–90. https://doi.org/10.1016/S0079-6123(08)00408-1

Oettl, L. L., Ravi, N., Schneider, M., Scheller, M. F., Schneider, P., Mitre, M., . . . Kelsch, W. (2016). Oxytocin enhances social recognition by modulating cortical control of early olfactory processing. Neuron, 90(3), 609–621. doi:https://doi.org/10.1016/j.neuron.2016.03.033

Onaka T, Takayanagi Y, Yoshida M (2012) Roles of oxytocin neurones in the control of stress, energy metabolism, and social behaviour. J Neuroendocrinol 24(4):587–598. https://doi.org/10.1111/j.1365-2826.2012.02300.x

Palacios JM, Pazos A, Hoyer D (2017) A short history of the 5-HT2C receptor: from the choroid plexus to depression, obesity and addiction treatment. Psychopharmacology (Berl) 234(9–10): 1395–1418. https://doi.org/10.1007/s00213-017-4545-5

Passoni I, Leonzino M, Gigliucci V, Chini B, Busnelli M (2016) Carbetocin is a functional selective Gq agonist that does not promote oxytocin receptor recycling after inducing beta-arrestin-independent internalisation. J Neuroendocrinol 28(4). https://doi.org/10.1111/jne.12363

Pedersen CA (2017) Oxytocin, tolerance, and the dark side of addiction. Int Rev Neurobiol 136: 239–274. https://doi.org/10.1016/bs.irn.2017.08.003

Pedersen, C. A., Smedley, K. L., Leserman, J., Jarskog, L. F., Rau, S. W., Kampov-Polevoi, A., . . . Garbutt, J. C. (2013). Intranasal oxytocin blocks alcohol withdrawal in human subjects. Alcohol Clin Exp Res, 37(3), 484–489. doi:https://doi.org/10.1111/j.1530-0277.2012.01958.x

Peris J, MacFadyen K, Smith JA, de Kloet AD, Wang L, Krause EG (2017) Oxytocin receptors are expressed on dopamine and glutamate neurons in the mouse ventral tegmental area that project to nucleus accumbens and other mesolimbic targets. J Comp Neurol 525(5):1094–1108. https://doi.org/10.1002/cne.24116

Peters S, Slattery DA, Flor PJ, Neumann ID, Reber SO (2013) Differential effects of baclofen and oxytocin on the increased ethanol consumption following chronic psychosocial stress in mice. Addict Biol 18(1):66–77. https://doi.org/10.1111/adb.12001

Peters ST, Bowen MT, Bohrer K, McGregor IS, Neumann ID (2017) Oxytocin inhibits ethanol consumption and ethanol-induced dopamine release in the nucleus accumbens. Addict Biol 22(3):702–711. https://doi.org/10.1111/adb.12362

Pucilowski O, Kostowski W, Trzaskowska E (1985) The effect of oxytocin and fragment (MIF-I) on the development of tolerance to hypothermic and hypnotic action of ethanol in the rat. Peptides 6(1):7–10. https://doi.org/10.1016/0196-9781(85)90067-1

Qi J, Yang JY, Song M, Li Y, Wang F, Wu CF (2008) Inhibition by oxytocin of methamphetamine-induced hyperactivity related to dopamine turnover in the mesolimbic region in mice. Naunyn Schmiedebergs Arch Pharmacol 376(6):441–448. https://doi.org/10.1007/s00210-007-0245-8

Qi J, Yang JY, Wang F, Zhao YN, Song M, Wu CF (2009) Effects of oxytocin on methamphetamine-induced conditioned place preference and the possible role of glutamatergic neurotransmission in the medial prefrontal cortex of mice in reinstatement. Neuropharmacology 56(5):856–865. https://doi.org/10.1016/j.neuropharm.2009.01.010

Qian T, Wang H, Wang P, Geng L, Mei L, Osakada T, Tang Y, Kania A, Grinevich V, Stoop R, Lin D, Luo M, Li Y (2022) Compartmental neuropeptide release measured using a new oxytocin sensor. bioRxiv. https://doi.org/10.1101/2022.02.10.480016

Quintana, D. S., Rokicki, J., van der Meer, D., Alnaes, D., Kaufmann, T., Cordova-Palomera, A., . . . Westlye, L. T. (2019). Oxytocin pathway gene networks in the human brain. Nat Commun, 10(1), 668. doi:https://doi.org/10.1038/s41467-019-08503-8

Rae M, Zanos P, Georgiou P, Chivers P, Bailey A, Camarini R (2018) Environmental enrichment enhances conditioned place preference to ethanol via an oxytocinergic-dependent mechanism in male mice. Neuropharmacology 138:267–274. https://doi.org/10.1016/j.neuropharm.2018.06.013

Resendez, S. L., Namboodiri, V. M. K., Otis, J. M., Eckman, L. E. H., Rodriguez-Romaguera, J., Ung, R. L., . . . Stuber, G. D. (2020). Social stimuli induce activation of oxytocin neurons within the paraventricular nucleus of the hypothalamus to promote social behavior in male mice. J Neurosci, 40(11), 2282–2295. doi:https://doi.org/10.1523/JNEUROSCI.1515-18.2020

Rigter H, Dortmans C, Crabbe JC Jr (1980) Effects of peptides related to neurohypophyseal hormones on ethanol tolerance. Pharmacol Biochem Behav 13(Suppl 1):285–290. https://doi.org/10.1016/s0091-3057(80)80043-8

Ross HE, Freeman SM, Spiegel LL, Ren X, Terwilliger EF, Young LJ (2009) Variation in oxytocin receptor density in the nucleus accumbens has differential effects on affiliative behaviors in monogamous and polygamous voles. J Neurosci 29(5):1312–1318. https://doi.org/10.1523/JNEUROSCI.5039-08.2009

Ryabinin AE, Fulenwider HD (2021) Alcohol and oxytocin: scrutinizing the relationship. Neurosci Biobehav Rev 127:852–864. https://doi.org/10.1016/j.neubiorev.2021.06.009

Ryden G, Sjoholm I (1969) Half-life of oxytocin in blood of pregnant and non-pregnant women. Acta Endocrinol 61(3):425–431. Retrieved from https://www.ncbi.nlm.nih.gov/pubmed/5820054

Sabatier N, Caquineau C, Dayanithi G, Bull P, Douglas AJ, Guan XM, Jiang M, Van der Ploeg L, Leng G (2003) Alpha-melanocyte-stimulating hormone stimulates oxytocin release from the dendrites of hypothalamic neurons while inhibiting oxytocin release from their terminals in the neurohypophysis. J Neurosci 23(32):10351–10358. Retrieved from https://www.ncbi.nlm.nih.gov/pubmed/14614094

Sari Y (2013) Role of 5-hydroxytryptamine 1B (5-HT1B) receptors in the regulation of ethanol intake in rodents. J Psychopharmacol 27(1):3–12. https://doi.org/10.1177/0269881112463126

Sawchenko PE, Swanson LW (1982) Immunohistochemical identification of neurons in the paraventricular nucleus of the hypothalamus that project to the medulla or to the spinal cord in the rat. J Comp Neurol 205(3):260–272. https://doi.org/10.1002/cne.902050306

Sivukhina EV, Dolzhikov AA, Morozov I, E., Jirikowski, G. F., and Grinevich, V. (2006) Effects of chronic alcoholic disease on magnocellular and parvocellular hypothalamic neurons in men. Horm Metab Res 38(6):382–390. https://doi.org/10.1055/s-2006-944522

Smith AS, Williams Avram SK, Cymerblit-Sabba A, Song J, Young WS (2016) Targeted activation of the hippocampal CA2 area strongly enhances social memory. Mol Psychiatry 21(8): 1137–1144. https://doi.org/10.1038/mp.2015.189

Son, S., Manjila, S. B., Newmaster, K. T., Wu, Y. T., Vanselow, D. J., Ciarletta, M., . . . Kim, Y. (2022). Whole-brain wiring diagram of oxytocin system in adult mice. J Neurosci, 42(25), 5021–5033. doi:https://doi.org/10.1523/JNEUROSCI.0307-22.2022

Stevenson, J. R., Wenner, S. M., Freestone, D. M., Romaine, C. C., Parian, M. C., Christian, S. M., . . . O'Kane, C. M. (2017). Oxytocin reduces alcohol consumption in prairie voles. Physiol Behav, 179, 411–421. doi:https://doi.org/10.1016/j.physbeh.2017.07.021

Stoop R (2012) Neuromodulation by oxytocin and vasopressin. Neuron 76(1):142–159. https://doi.org/10.1016/j.neuron.2012.09.025

Swanson LW, Sawchenko PE (1983) Hypothalamic integration: organization of the paraventricular and supraoptic nuclei. Annu Rev Neurosci 6:269–324. https://doi.org/10.1146/annurev.ne.06.030183.001413

Szabo G, Kovacs GL, Szekeli S, Telegdy G (1985) The effects of neurohypophyseal hormones on tolerance to the hypothermic effect of ethanol. Alcohol 2(4):567–574. https://doi.org/10.1016/0741-8329(85)90082-5

Szabo G, Kovacs GL, Telegdy G (1987) Effects of neurohypophyseal peptide hormones on alcohol dependence and withdrawal. Alcohol Alcohol 22(1):71–74. Retrieved from https://www.ncbi.nlm.nih.gov/pubmed/3593487

Szabo G, Kovacs GL, Telegdy G (1989) Intraventricular administration of neurohypophyseal hormones interferes with the development of tolerance to ethanol. Acta Physiol Hung 73(1): 97–103. Retrieved from https://www.ncbi.nlm.nih.gov/pubmed/2711843

Tabakoff B, Hoffman PL (2013) The neurobiology of alcohol consumption and alcoholism: an integrative history. Pharmacol Biochem Behav 113:20–37. https://doi.org/10.1016/j.pbb.2013.10.009

Tanaka, A., Furubayashi, T., Arai, M., Inoue, D., Kimura, S., Kiriyama, A., . . . Yamamoto, A. (2018). Delivery of oxytocin to the brain for the treatment of autism spectrum disorder by nasal application. Mol Pharm, 15(3), 1105–1111. doi:https://doi.org/10.1021/acs.molpharmaceut.7b00991

Tang, Y., Benusiglio, D., Lefevre, A., Hilfiger, L., Althammer, F., Bludau, A., . . . Grinevich, V. (2020). Social touch promotes interfemale communication via activation of parvocellular oxytocin neurons. Nat Neurosci, 23(9), 1125–1137. doi:https://doi.org/10.1038/s41593-020-0674-y

Tang, Y., Benusiglio, D., Lefevre, A., Kuppers, S., Lapies, O., Kerspern, D., . . . Grinevich, V. (2022). Viral vectors for opto-electrode recording and photometry-based imaging of oxytocin neurons in anesthetized and socially interacting rats. STAR Protoc, 3(1), 101032. doi:https://doi.org/10.1016/j.xpro.2021.101032

Terrillon, S., Durroux, T., Mouillac, B., Breit, A., Ayoub, M. A., Taulan, M., . . . Bouvier, M. (2003). Oxytocin and vasopressin V1a and V2 receptors form constitutive homo- and heterodimers during biosynthesis. Mol Endocrinol, 17(4), 677–691. doi:https://doi.org/10.1210/me.2002-0222

Theofanopoulou C, Gedman G, Cahill JA, Boeckx C, Jarvis ED (2021) Universal nomenclature for oxytocin-vasotocin ligand and receptor families. Nature 592(7856):747–755. https://doi.org/10.1038/s41586-020-03040-7

Tirelli E, Jodogne C, Legros JJ (1992) Oxytocin blocks the environmentally conditioned compensatory response present after tolerance to ethanol-induced hypothermia in mice. Pharmacol Biochem Behav 43(4):1263–1267. https://doi.org/10.1016/0091-3057(92)90512-e

Tunstall, B. J., Kirson, D., Zallar, L. J., McConnell, S. A., Vendruscolo, J. C. M., Ho, C. P., . . . Vendruscolo, L. F. (2019). Oxytocin blocks enhanced motivation for alcohol in alcohol dependence and blocks alcohol effects on GABAergic transmission in the central amygdala. PLoS Biol, 17(4), e2006421. doi:https://doi.org/10.1371/journal.pbio.2006421

Vena A, King A, Lee R, de Wit H (2018) Intranasal oxytocin does not modulate responses to alcohol in social drinkers. Alcohol Clin Exp Res 42(9):1725–1734. https://doi.org/10.1111/acer.13814

Wahis, J., Baudon, A., Althammer, F., Kerspern, D., Goyon, S., Hagiwara, D., . . . Charlet, A. (2021). Astrocytes mediate the effect of oxytocin in the central amygdala on neuronal activity and affective states in rodents. Nat Neurosci, 24(4), 529–541. doi:https://doi.org/10.1038/s41593-021-00800-0

Walcott AT, Ryabinin AE (2021) Assessing effects of oxytocin on alcohol consumption in socially housed prairie voles using radio frequency tracking. Addict Biol 26(2):e12893. https://doi.org/10.1111/adb.12893

Weber RA, Logan CN, Leong KC, Peris J, Knackstedt L, Reichel CM (2018) Regionally specific effects of oxytocin on reinstatement of cocaine seeking in male and female rats. Int J Neuropsychopharmacol 21(7):677–686. https://doi.org/10.1093/ijnp/pyy025

Winter J, Jurek B (2019) The interplay between oxytocin and the CRF system: regulation of the stress response. Cell Tissue Res 375(1):85–91. https://doi.org/10.1007/s00441-018-2866-2

Xiao L, Priest MF, Kozorovitskiy Y (2018) Oxytocin functions as a spatiotemporal filter for excitatory synaptic inputs to VTA dopamine neurons. Elife 7:e33892. https://doi.org/10.7554/eLife.33892

Yang JY, Qi J, Han WY, Wang F, Wu CF (2010) Inhibitory role of oxytocin in psychostimulant-induced psychological dependence and its effects on dopaminergic and glutaminergic transmission. Acta Pharmacol Sin 31(9):1071–1074. https://doi.org/10.1038/aps.2010.140

Yoshida M, Takayanagi Y, Inoue K, Kimura T, Young LJ, Onaka T, Nishimori K (2009) Evidence that oxytocin exerts anxiolytic effects via oxytocin receptor expressed in serotonergic neurons in mice. J Neurosci 29(7):2259–2271. https://doi.org/10.1523/JNEUROSCI.5593-08.2009

Young WS, Song J (2020) Characterization of oxytocin receptor expression within various neuronal populations of the mouse dorsal hippocampus. Front Mol Neurosci 13:40. https://doi.org/10.3389/fnmol.2020.00040

Zaninetti M, Raggenbass M (2000) Oxytocin receptor agonists enhance inhibitory synaptic transmission in the rat hippocampus by activating interneurons in stratum pyramidale. Eur J Neurosci 12(11):3975–3984. https://doi.org/10.1046/j.1460-9568.2000.00290.x

Zhang, B., Qiu, L., Xiao, W., Ni, H., Chen, L., Wang, F., . . . Gao, Z. (2021). Reconstruction of the hypothalamo-neurohypophysial system and functional dissection of magnocellular oxytocin neurons in the brain. Neuron, 109(2), 331–346 e337. doi:https://doi.org/10.1016/j.neuron.2020.10.032

Zhou Y, Kreek MJ (2018) Involvement of activated brain stress responsive systems in Excessive and "Relapse" alcohol drinking in rodent models: implications for therapeutics. J Pharmacol Exp Ther 366(1):9–20. https://doi.org/10.1124/jpet.117.245621

Zhou Y, Colombo G, Carai MA, Ho A, Gessa GL, Kreek MJ (2011) Involvement of arginine vasopressin and V1b receptor in alcohol drinking in Sardinian alcohol-preferring rats. Alcohol Clin Exp Res 35(10):1876–1883. https://doi.org/10.1111/j.1530-0277.2011.01532.x

Zhou Y, Rubinstein M, Low MJ, Kreek MJ (2018) V1b receptor antagonist SSR149415 and naltrexone synergistically decrease excessive alcohol drinking in male and female mice. Alcohol Clin Exp Res 42(1):195–205. https://doi.org/10.1111/acer.13544

Neural Circuitries and Alcohol Use Disorder: Cutting Corners in the Cycle

Marie A. Doyle, Anne Taylor, and Danny G. Winder

Contents

Abstract An implicit tenet of the alcohol use disorder (AUD) research field is that knowledge of how alcohol interacts with the brain is critical to the development of an understanding of vulnerability to AUD and treatment approaches. Gaining this understanding requires the mapping of brain function critical to specific components of this heterogeneous disorder. Early approaches in humans and animal models focused on the determination of specific brain regions sensitive to alcohol action and their participation in AUD-relevant behaviors. Broadly speaking, this research has focused on three domains, Binge/Intoxication, Negative Affect/Withdrawal, and

M. A. Doyle, A. Taylor, and D. G. Winder (✉)
Department of Molecular Physiology and Biophysics, Vanderbilt University School of Medicine, Nashville, TN, USA

Vanderbilt Center for Addiction Research, Vanderbilt University School of Medicine, Nashville, TN, USA

Vanderbilt Brain Institute, Vanderbilt University School of Medicine, Nashville, TN, USA
e-mail: danny.winder@vanderbilt.edu

Curr Topics Behav Neurosci (2025) 71: 379–399
https://doi.org/10.1007/7854_2023_454
Published Online: 12 December 2023

Preoccupation/Anticipation, with a number of regions identified as participating in each. With the generational advances in technologies that the field of neuroscience has undergone over the last two decades, this focus has shifted to a circuit-based analysis. A wealth of new data has sharpened the field's focus on the specific roles of the interconnectivity of multiple brain regions in AUD and AUD-relevant behaviors, as well as demonstrating that the three major domains described above have much fuzzier edges than originally thought.

In this chapter, we very briefly review brain regions previously implicated in aspects of AUD-relevant behavior from animal model research. Next, we move to a more in-depth overview of circuit-based approaches, and the utilization of these approaches in current AUD research.

Keywords Neuronal circuits · Alcohol use disorder · Neuroscience · Brain

1 Conceptualization of AUD as a Multi-regional, Multi-component Cycle

A firm foundation of animal model neuroscience over the last 40 years has established several core tenets of alcohol action on the brain: (1) in contrast to drugs of abuse, alcohol acts through interactions with a number of different molecules heterogeneously expressed throughout the brain to alter behavior, (2) acute alcohol is a depressant and anxiolytic, and (3) chronic alcohol produces neural plasticity in a number of brain regions that modulate subsequent alcohol-associated behaviors. These results emphasize a diversity of molecular, cellular, and circuit mechanisms through which alcohol can alter brain function, emphasizing heterogeneous disorder nature of AUD, and the strong likelihood that an umbrella treatment approach is unlikely to have far-reaching effectiveness.

One prominent conceptualization of AUD posits a cycle of behavior moving through periods of *Binge/Intoxication*, *Negative Affect/Withdrawal*, and the development of *Preoccupation/Anticipation* (Koob and Volkow 2010). From animal model work, it is hypothesized that repeated progression through this cycle then produces allostatic alterations in key setpoints that induce a transition from impulsive to compulsive alcohol seeking. Consistent with this idea, brain-wide mapping of neuronal fos expression in rodents that have undergone extensive alcohol exposure reveals patterns of fos activity reminiscent of the regions initially described (and discussed below) for the three sections of the model (Hansson et al. 2008; Vilpoux et al. 2009; Kimbrough et al. 2020; Smith et al. 2020; Roland et al. 2023). Further, recent work has begun to seek to utilize predictions from the cycle model to predict AUD recovery outcomes (Witkiewitz et al. 2022).

1.1 *Binge/Intoxication*

1.1.1 Brief Introduction

The Binge/Intoxication phase refers to periods of heavy alcohol intake resulting in induction of intoxication through the pharmacological actions of alcohol. In animals, aspects of this phase can be modeled with volitional alcohol intake models, as well as alcohol conditioned place preference and vapor exposure. A number of brain regions have been implicated in binge intake and have been heavily reviewed elsewhere (Koob and Volkow 2010; Thiele and Navarro 2014; Jeanblanc et al. 2019). Here we will briefly highlight key identified areas.

1.1.2 Ventral Tegmental Area

The ventral tegmental area (VTA) has long been considered a key region for the reinforcing actions of alcohol, supported by data demonstrating that rats will self-administer intra-VTA infusions of alcohol (Gatto et al. 1994; Rodd et al. 2004). Acute exposure increases the in vivo and ex vivo firing rate of VTA dopamine (DA) neurons (Gessa et al. 1985; Brodie et al. 1990). For additional details of alcohol actions on VTA neurons see (Juarez and Han 2016; Morikawa and Morrisett 2010; You et al. 2018). Early work focused on and further supported VTA DA neurons as a key population of interest, as decreasing VTA DA neuron firing via intra-VTA infusion of the D_2 receptor agonist quinpirole attenuates intra-VTA and oral alcohol self-administration and oral intake (Hodge et al. 1993; Nowak et al. 2000; Rodd et al. 2004). This alcohol-induced increase in DA neuron activity results in an increase in dopamine release in mesocorticolimbic targets, consistent with all drugs of abuse.

1.1.3 Nucleus Accumbens Shell and Core

A major target of VTA DA neurons is the nucleus accumbens (NAc). Here, acute alcohol exposure results in increased extracellular DA levels (Di Chiara and Imperato 1988; Weiss et al. 1993; Robinson et al. 2009). The role of DA signaling in reinforcing alcohol behavior has been studied using NAc microinjections of DA receptor pharmacological agents. Specifically, intra-NAc infusion of D_1R and D_2R antagonists reduces oral alcohol operant responses in rats (Rassnick et al. 1992; Samson et al. 1993; Hodge et al. 1997). However, while alcohol-induced increases in extracellular DA in the NAc play a clear role in the reinforcing properties of alcohol, there is ample support for alcohol's actions through other signaling mechanisms (For review of synaptic effects of alcohol see Lovinger and Roberto (2023)). Alcohol is also known to bind to $GABA_A$ receptors, enhancing receptor function, (Nestoros 1980; Lovinger and Roberto 2023) and can act through NMDAR inhibition (Hoffman et al. 1989; Lovinger et al. 1989; Lovinger and Roberto 2023), as intra-NAc infusion of the glutamate receptor antagonist AP-5 decreases oral alcohol self-administration (Rassnick et al. 1992). There is a large breadth of research defining alcohol's effects on specific cell types and neurotransmitters within key alcohol reward-associated regions, and for the sake of this chapter we seek to focus on higher-level regulation of a region as a whole. We have highlighted key reviews that better dive into details related to specific alcohol mechanisms and cell type-specific actions for additional information.

Focusing on this higher level of regional regulation by alcohol exposure, expression of the immediate early gene cFos is increased in the NAc following an IP injection, intraventricular (ICV), or intragastric administration of alcohol (Hitzemann and Hitzemann 1997; Ryabinin et al. 1997; Herring et al. 2004; Leriche et al. 2008; Segovia et al. 2013), signifying that this region has increased cell activity during exposure. Manipulation of the NAc in the context of alcohol reward is behaviorally relevant, as an electrolytic lesion of the NAc decreases alcohol intake in drinking in the dark paradigm (Cassataro et al. 2014) and prevents the acquisition of alcohol conditioned place preference (CPP) (Gremel and Cunningham 2008). Consistent with NAc function supporting the reinforcing properties of alcohol, pharmacological inhibition via intra-NAc infusion of the $GABA_A$ agonist muscimol decreases alcohol self-administration in rats (Hodge et al. 1995). Chemogenetic inhibition of the NAc using a Designer Receptors Exclusively Activated by Designer Drugs (DREADD) approach similarly decreases alcohol intake in a drinking-in-the-dark paradigm (Cassataro et al. 2014). Together, this work supports the NAc as a region necessary in orchestrating the reinforcing properties of alcohol exposure.

1.1.4 Ventral Pallidum

Downstream from the NAc, the ventral pallidum (VP) is also known to play a role in motivation and hedonic signaling (Kupchik et al. 2015; Lovinger and Alvarez 2017). Intragastric infusion of alcohol decreases VP BOLD signal intensity in conscious rats (Tsurugizawa et al. 2010) and both IP and intraventricular (ICV) administration of alcohol results in increased cfos expression in the VP (Segovia et al. 2013). Inhibition of the VP through direct infusion of muscimol (Kemppainen et al. 2012) or GABA receptor blockade using βCCt (June et al. 2003) both decrease alcohol self-administration.

1.1.5 Dorsal Striatum/Basal Ganglia

While many of the previous brain regions have focused on function in relation to the initial rewarding properties of alcohol, the dorsal striatum (DS, putamen) is highly implicated in habit formation (O'Tousa and Grahame 2014). cFos expression in the dorsal striatum is increased by an IP injection of alcohol (Hitzemann and Hitzemann 1997; Segovia et al. 2013). Research has also shown the behavioral relevance of DS function. Decreased dorsal striatum activity via activation of an inhibitory DREADD decreased alcohol intake (Robins et al. 2018), and, similarly, D_1- and D_2-MSN-specific DREADD manipulations are capable of modulating alcohol intake (Cheng et al. 2017). The DS can also be broken down into medial (DMS) and lateral segments (DLS), with control of different behavioral outcomes. For example, inactivation of the DMS via local infusion of GABA receptor agonists baclofen and muscimol decreases goal-directed responding in early training while local infusion into the DLS decreases habit-like responding after prolonged training (Corbit et al. 2012) (see for review of DS in habituation alcohol intake behaviors (O'Tousa and Grahame 2014)).

The DS projects to the dorsal pallidum, which can be divided into the globus pallidus internal (GPi) and external (GPe) segments (Lovinger and Alvarez 2017). Ex vivo bath application of alcohol to acute brain slices results in a decrease in GPe neuron firing (Criswell et al. 1995), which recently was further characterized as a subpopulation of GPe neurons (Abrahao et al. 2017).

Basal ganglia circuitry is thought to be critically modulated by dopamine input from the substantia nigra. In unanesthetized rats, an IV infusion of alcohol increases SNr DA firing at low doses but results in inhibition at high doses (Mereu et al. 1984). Local injection studies suggest the substantia nigra may contribute to stimulant actions of alcohol (Arizzi-LaFrance et al. 2006).

1.2 Negative Affect/Withdrawal

1.2.1 Brief Introduction

Cessation of alcohol intake following binge exposure leads to a sequelae of withdrawal symptoms dependent upon the prior pattern and history of alcohol intake. It is posited that cycles of binge intoxication and withdrawal lead to the development of what is referred to as allostatic changes in the set point for affective behaviors and hypothalamic setpoint to result in altered stress responses (Koob 2008; Centanni et al. 2019a, b; Koob and Schulkin 2019). Aspects of this phase can be modeled by increased alcohol intake after the development of dependence in animal models, increased conditioned place aversion, and alterations in behavior in tasks that assess affective state. A number of brain regions from animal model studies have been implicated in these withdrawal-associated emergent behaviors, most notably components of the extended amygdala, as described below. These changes are then viewed as contributing stimuli to drive subsequent negative reinforcement-based alcohol seeking in the Preoccupation/Anticipation phase.

1.2.2 The Extended Amygdala (Central Nucleus of the Amygdala and Bed Nucleus of the Stria Terminalis)

Alcohol exposure and withdrawal in rodent models induce negative affective behaviors that appear to heavily involve the central nucleus of the amygdala and the bed nucleus of the stria terminalis (CeA, BNST, for reviews see (Centanni et al. 2019a, b; Roberto et al. 2021). A variety of acute and chronic actions of alcohol on glutamatergic and GABAergic transmission have been identified (Wills et al. 2013; Pati et al. 2020), at least some of which appear to involve the neuropeptide corticotropin-releasing factor (CRF). Consistently, manipulations within the CeA that decrease GABAergic transmission reduced dependence-induced increases in alcohol intake across a range of models.

In a chronic home cage drinking followed by the forced abstinence model (CDFA), female mice show increased anxiety- and depressive-like behaviors after protracted withdrawal from alcohol (Holleran et al. 2016; Vranjkovic et al. 2018). At this time point, BNST cFos is increased in male and female mice in protracted withdrawal from chronic alcohol as well as specifically in BNST CRF cells (Centanni et al. 2019a, b). Further, BNST cells as well as BNST CRF neurons display enhanced frequency of spontaneous excitatory postsynaptic currents (Centanni et al. 2019a, b), suggesting increased glutamatergic input onto these cells during forced abstinence.

1.3 Preoccupation/Anticipation

1.3.1 Brief Introduction

The preoccupation/anticipation phase of AUD represents a particularly relevant potential treatment intervention phase and is associated with increased alcohol craving. Reinstatement of alcohol-seeking behavior is a common means of modeling this component. Studies have revealed a number of involved brain structures, as has been previously reviewed extensively (Mantsch et al. 2016; George and Hope 2017; Domi et al. 2021). Here we will provide brief highlights of this research.

1.3.2 Prefrontal Cortex

Part of the Prefrontal Cortex (PFC), the orbital frontal cortex (OFC) is a region heavily implicated in alcohol-related behaviors. cFos is increased in the OFC of rats during cue- and context-induced reinstatement of alcohol seeking (Jupp et al. 2011; Bianchi et al. 2018). OFC chemogenetic inhibition (Hernandez et al. 2020) and pharmacological inhibition via muscimol and baclofen (Arinze and Moorman 2020) decrease cue-induced alcohol reinstatement in rats, and pharmacological inhibition via muscimol attenuates context-induced reinstatement (Bianchi et al. 2018). The prelimbic cortex, another segment of the PFC, has been implicated in alcohol seeking as well. Rats conditioned to receive alcohol in a distinct context display increased Fos expression in the prelimbic cortex when no alcohol is available (termed “craving” group) compared to naïve controls as well as rats with alcohol access (Topple et al. 1998). Reinstatement of alcohol seeking also increases Fos expression in this region and inactivation via local infusion of muscimol and baclofen significantly decreases lever pressing during this reinstatement task (Palombo et al. 2017). For a general review, see Moorman (2018). Finally, the infralimbic component of the PFC similarly has been implicated in alcohol seeking. Within this region, cFos is increased during cue-induced reinstatement of alcohol seeking (Wedzony et al. 2003; Dayas et al. 2007; Pfarr et al. 2015). Further, this ensemble of cells has been functionally linked to behavior. Pfarr et al. achieved this task using a transgenic approach to label neurons activated by reinstatement in combination with an intra-infralimbic cortex delivery of the prodrug Daun02, resulting in the inactivation of labeled neurons (Pfarr et al. 2015). When Daun02 was administered, active lever presses were significantly increased during a reinstatement session compared to control rats, demonstrating that activity in this infralimbic ensemble is capable of regulating alcohol seeking (Pfarr et al. 2015). Interestingly, Daun02-induced inhibition did not affect reinstatement behavior with either nonspecific infralimbic inactivation or the same activity-targeted approach achieved in the prelimbic cortex of the PFC (Pfarr et al. 2015), highlighting the importance of region and cell specificity in alcohol seeking behaviors. Further work has characterized this infralimbic alcohol seeking ensemble as overlapping with a saccharin seeking ensemble in the same region (Pfarr et al. 2018), prompting exploration of reward seeking ensemble overlap throughout the brain (for further reading see Wandres et al. (2021); Korber and Sommer (2022)).

1.3.3 Hippocampus and Basolateral Amygdala

The dorsal hippocampus has a well-established role in regulating contextual memory. In the frame of other drugs of abuse, the hippocampus is known to be necessary for context-induced reinstatement of drug seeking (Fuchs et al. 2005) (For a review see Feltenstein et al. (2021)). This is no different for ethanol-related behaviors. Rats conditioned to receive ethanol in a designated context have higher Fos expression in the CA3 region of the hippocampus when ethanol is not available when compared to animals with ethanol access (Topple et al. 1998). Context-induced ethanol seeking increases cFos expression in the dorsal hippocampus of rats (Marinelli et al. 2007; Felipe et al. 2021), and inactivation of the dorsal hippocampus via local infusion of cobalt chloride attenuated seeking during the reinstatement task (Felipe et al. 2021).

The basolateral amygdala (BLA) has similarly been implicated in relapse-related behaviors, focusing here on conditioned cues. BLA cFos expression is increased by EtOH cue re-exposure during a reinstatement task (Zhao et al. 2006; Radwanska et al. 2008). $ERK_{1/2}$, a protein associated with learning, memory, and synaptic plasticity, is significantly increased in the BLA of rats during cue-induced ethanol seeking (Schroeder et al. 2008). mGluR5 signaling, upstream of ERK1/2 activation, is implicated in this process, as intra-BLA infusion of the mGluR5 antagonist prior to the reinstatement task blocked ethanol seeking behavior (Sinclair et al. 2012). For review and additional information on both context- and cue-induced seeking in the context of ethanol, see Janak and Chaudhri (2010); Domi et al. (2021).

1.3.4 Insula: Interoceptive Cues

Another cortical region we are beginning to learn more about in relation to its role in alcohol anticipation, preoccupation, and craving is the insular cortex (IC). The insula participates in stress axis circuitry through its projections to the BNST, as well as with limbic, visceral, and somatosensory regions (Craig 2009). These connections position the insula well to be a central hub for regulating interoceptive states. In human imaging studies, this region has been shown to be active when presented with images of drugs and drug-related cues (Naqvi and Bechara 2010; Jasinska et al. 2014; Droutman et al. 2015a, b).

Jaramillo et al. have shown that inhibition of the rodent IC increases sensitivity to interoceptive effects of alcohol in an operant discrimination task (Jaramillo et al. 2015). Furthermore, chemogenetic modulation of insula activity revealed its ability to exert control over how much animals self-administer alcohol as well as their alcohol sensitivity (Jaramillo et al. 2018a, b). For review see Lovelock et al. (2021).

2 Circuit-Based Analysis of AUD

2.1 *Brief Introduction*

The studies discussed above focus on the identification of key brain regions involved in aspects of alcohol exposure-relevant behaviors. As our understanding of the brain has advanced, a clearer vision has emerged that emphasizes the combined actions of circuits of synaptically interconnected neurons running through distinct brain regions. Moreover, it has become apparent that in many cases, circuits routing through the same region may have distinct, or even opposing roles in regulating behavior. Viewed through this lens, it may not be feasible to describe the role of a brain region or nucleus as a whole in AUD, but more appropriate to focus on specific cellular circuits running through distinct brain regions. This type of thinking has precipitated a migration from a brain regional approach to a circuit analysis approach in understanding AUD. Fortunately, over the last two decades a renaissance of neuroscience tool development has ushered in strong new approaches to dissect the roles of specific neural circuits in aspects of physiology and behavior in animal models. These approaches will be briefly summarized below.

2.2 *Animal Model Approaches for Circuit-Based Analysis*

In tandem with our understanding that AUD, like many disorders, reflects a constellation of endophenotypes that need to be broken off and independently examined, the neurosciences have seen unparalleled technical advances in the last two decades that have completely revolutionized the specificity of circuitry-based questions that can be asked and answered. This revolution has impacted every level of experimental design:

1. *Sufficiency experiments*: Optogenetic and chemogenetic strategies utilize signaling proteins co-opted so that, when expressed in a given cell, light or specific chemical ligand, respectively, yields a predictable change in the activity of the cell (Roth 2016; Kim et al. 2017). In combination with genetic and viral delivery vectors, these approaches have provided unparalleled specificity in our abilities to control the activities of precisely defined populations of neurons.
2. *Necessity experiments*: CRISPR, shRNA, and other gene manipulation strategies now allow for reliable and specific control of the expression and activity of individual genes/proteins. These strategies allow generation of resources to perform these experiments in a fraction of the time previously required with homologous recombination approaches and the ability to perform them in a circuit-specific manner (Heidenreich and Zhang 2016).
3. *Massively parallel descriptive experiments*: Omic approaches allow the generation of exhaustive lists of genes, proteins, lipid signaling agents, etc. from individual cells or small tissues, allowing full descriptions of key cellular populations and their alterations by specific pharmacological and/or behavioral stimuli (Hosp and Mann 2017; Armand et al. 2021).
4. *In vivo activity measures*: Novel fluorescent sensor strategies such as the calcium sensor GCaMP as well as all manner of activity reporters from neurotransmitter levels to second messenger signaling approaches through genetic constructs that can be delivered to specific populations of neurons allow for extraordinary insight into the dynamics of signaling in the intact brain (Dong et al. 2022).
5. *Novel imaging approaches*: Fiber photometry, Grin lenses, and head-fixed 2PLSM now provide myriad ways to visualize the above sensors in vivo. Additionally, light sheet imaging provides the incredible ability to image an entire mouse brain in one fell swoop (Ueda et al. 2020).
6. *Machine and vision learning computation approach to analyses*: Artificial intelligence approaches allow for deep, consistent, and precise aspects of specific animal behaviors to be quantified (Mathis et al. 2018). Such advanced computational approaches also allow for the unbiased registration of labeled neurons in a light sheet imaged brain across a mouse brain atlas to do overnight what would take weeks otherwise and to do it in an unbiased manner (Luchsinger et al. 2021).

These approaches have begun to be implemented within the AUD animal model research field. In the following sections, we will focus on how this small but growing number of studies has informed the field's transition from brain regional- to brain circuit-oriented investigations.

2.3 *Circuit-Based Corner Cutting in AUD Models*

Utilizing a variety of the approaches described above in rodent models, a growing number of specific circuits have been investigated within the past 10 years for their modification by and/or roles in alcohol-specific behaviors. An interesting theme that has emerged from these studies is a blurring of the fidelity of specific brain regions within specific components of the AUD cycle (Fig. 1), as will be discussed below.

2.3.1 VTA-Containing Circuits in Binge/Intoxication

Consistent with the brain region data described above, a number of VTA-containing circuits have been implicated in alcohol intake. The VTA-NAc synapse has been repeatedly implicated in positive reinforcement behaviors relevant to feeding and a variety of substances of abuse. Unsurprisingly, manipulation of this specific circuit regulates alcohol intake. Interestingly, however, the relationship between the activity of this pathway and alcohol intake is not 1:1. For example, Juarez et al. (2017) demonstrated that in high alcohol versus low alcohol preferring mice (HAD/LAD), firing in this circuit was higher in the LADs than the HADs. Moreover, specific optogenetic stimulation of this pathway in HAD mice decreased rather than increased alcohol intake. In the context of these described roles, it is interesting to note that chronic alcohol exposure has been demonstrated to increase excitatory input from the ventral hippocampus onto D1 MSNs in the NAc, in part through reversal of long-term depression at the input (Kircher et al. 2019).

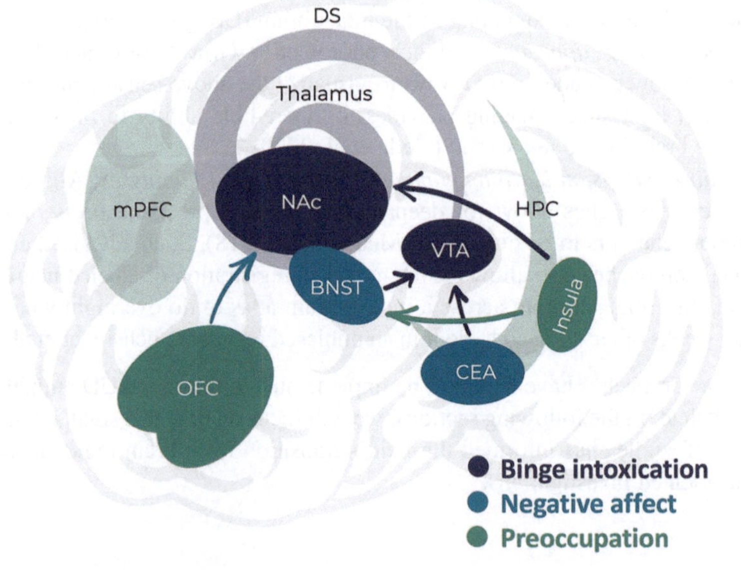

Fig. 1 "Corner Cutting" in the "cycle" of Alcohol Use Disorder. Summary of projections described in this chapter as participating in alcohol-regulated behaviors in rodent models. Note the transition between cycle domains signified by the color changes between regions and arrows. Art rendered by Dr. Kendra Oliver

2.3.2 Extended Amygdala-VTA Circuits in Binge/Intoxication and Negative Affect/Withdrawal

An important source for control of the mesolimbic dopamine system in the context of alcohol bingeing and dependence is the extended amygdala structures, particularly the BNST and CeA. Avegno et al. have demonstrated the VTA-projecting CeA neurons are activated during alcohol exposure and withdrawal (Avegno et al. 2021; Avegno and Gilpin 2022). Further, two studies demonstrate that a projection from the BNST to the VTA plays an important role in alcohol conditioned place preference (Pina and Cunningham 2017) and binge alcohol intake (Rinker et al. 2017). While these data are not unexpected in terms of predicted roles of dopamine signaling in reinforced behaviors, the role of BNST projections here demonstrates an example of "cutting corners," in which a brain region implicated in the negative affect/withdrawal phase of the AUD cycle plays a role in alcohol intake behaviors independent of alcohol dependence.

Circuit-based analysis of extended amygdala structures does provide strong support for the role of circuits involving these structures in the negative affect/ withdrawal phase. For example, a circuit involving CeA CRF neurons projecting into the BNST in rats is activated during chronic alcohol withdrawal, and optogenetic inhibition of this pathway suppresses enhanced alcohol intake as well as physiological signs of withdrawal (de Guglielmo et al. 2016, 2019). Curiously when analogous experiments were performed utilizing chemogenetic approaches in the mouse, significantly distinct outcomes were obtained, suggesting more work needs to be done (Kreifeldt et al. 2022).

BNST CRF neurons projecting to the VTA have been identified as being regulated by chronic alcohol exposure (Silberman et al. 2013), and moreover, direct manipulation of the activity of these cells reveals that activity supports alcohol intake. In mice, these cells are more excitable in females compared to males, and this difference is in part mediated by a stronger glutamatergic projection to these cells from the paraventricular nucleus of the thalamus (PVT) (Levine et al. 2021). BNST CRF neurons also receive input from the insula (Centanni et al. 2019a, b; Fetterly et al. 2019). Excitatory drive in the BNST is upregulated during forced abstinence from chronic alcohol availability (Centanni et al. 2019a, b). Chemogenetic studies indicate that at least a portion of this upregulation occurs at insular afferents. Moreover, alcohol intake is associated with an increase in the excitability of BNST projecting insular neurons (Marino et al. 2021). Driving this circuit produces behavioral phenotypes similar to those observed with alcohol forced abstinence, and chemogenetic inhibition reduces abstinence-induced negative affect-like behavioral disturbances.

2.3.3 Insular Cortex-Driven Circuits Participate in Multiple AUD Cycle Components

Canonically the IC has been posited to be a part of the anticipation/preoccupation stage of the addiction cycle (Lovelock et al. 2021). However, a variety of circuit analysis-oriented studies point to a key role insular cortex in alcohol-dependent behaviors, aligning them with the binge intoxication phase. Seif et al. elegantly demonstrated that insular (and mPFC) inputs to the accumbens were uniquely regulated by GluN2C-containing NMDA receptors, and that reduction of the expression of these receptors in ventral striatum reduced aversion-independent alcohol intake (Seif et al. 2013). In the dorsal striatum, insular afferents are subject to mu opiate receptor-dependent LTD, and the form of synaptic plasticity is disrupted by alcohol exposure (Munoz et al. 2018). Inhibition of insular afferents to the brainstem specifically reduces aversion-resistant alcohol intake (De Oliveira Sergio et al. 2021).

There is also growing evidence supporting the role of the insula in the negative affect/withdrawal phase. Data from the Winder and Sparta labs have shown that the IC is active following abstinence from alcohol in rodent models (Centanni et al. 2019a, b; Marino et al. 2021). Additionally, following a 2-week abstinence period from chronic alcohol intake, chemogenic inactivation of the IC blocked the subsequent increase in BNST activation previously seen during abstinence indicating the IC is a necessary component of the withdrawal phase (Centanni et al. 2019a, b). Moreover, alcohol intake is associated with an increase in the excitability of BNST projecting insular neurons (Marino et al. 2021). Driving this circuit produces behavioral phenotypes similar to those observed with alcohol forced abstinence, and chemogenetic inhibition reduces abstinence-induced negative affect-like behavioral disturbances.

2.3.4 Corticostriatal Circuits in AUD-Like Behaviors During Withdrawal/Negative Affect

Significant circuit-based analyses have also been performed within corticostriatal circuits in the context of alcohol intake and exposure. Dependence following chronic exposure to intermittent alcohol vapor is associated with a reduction in cortical control of habit formation during withdrawal (Renteria et al. 2018). Specifically, orbitofrontal cortex (OFC) neurons that project into the basal ganglia exhibit decreased excitability after chronic intermittent alcohol vapor exposure (Renteria et al. 2018). Further investigation revealed an endocannabinoid-dependent modulation pathway in the OFC-direct pathway circuit that is critical to this process (Renteria et al. 2021). Restoration of this cortical output via optogenetic stimulation produced an increase in goal-directed behavior in dependent mice (Renteria et al. 2018). Additionally, indirect pathway neurons also appear to play an important role in regulating alcohol-related behaviors, optogenetic stimulation of dorsal striatal A2a-expressing neurons reducing alcohol reinforcer approach in a seeking task, aligning this circuit with the anticipation/preoccupation phase of the addiction cycle (Hong et al. 2019).

2.3.5 Novel Connections in the Cycle Identified by Circuit-Based Approaches

In addition to the results described above, new strategies have been utilized to implicate interesting circuits in AUD-relevant behavior, contributing further detail to the addiction cycle and its phases. For example, Dornellas et al., utilizing a pathway-specific chemogenetic strategy, have provided evidence that a synapse from the locus coeruleus to the rostromedial tegmental nucleus (RMTG) selectively regulates alcohol intake (Dornellas et al. 2021). Intriguingly, chemogenetic activation of this pathway reduced alcohol but not sucrose consumption (Dornellas et al. 2021). Similarly, chemogenetic inhibition of LC inputs to the lateral hypothalamus also decreased alcohol intake (Burnham et al. 2021). These data in total suggest that the LC may have a previously underappreciated role in orchestrating alcohol intake via control of lower brain centers.

Increasing emphasis is also beginning to be placed on the role of hypothalamic subnuclei-containing circuits in binge alcohol intake. POMC-expressing neurons are a significant cell population within the arcuate nucleus of the hypothalamus, and they project to multiple regions previously implicated in alcohol behaviors, such as the nucleus accumbens and amygdala. Leyrer-Jackson et al. recently demonstrated that POMC-containing neurons that project to the amygdala are preferentially active during the drinking-in-the-dark binge drinking model, suggesting a potential role for this pathway as well (Leyrer-Jackson et al. 2021).

Finally, one of the most enticing AUD circuit papers to date is Siciliano et al. (2019). In this paper, Siciliano and colleagues measure and modulate in vivo activity of a medial PFC to dorsal periaqueductal gray circuit (Siciliano et al. 2019). They do this within the context of a novel operant alcohol self-administration model that allowed delineation of high and low alcohol drinkers as well as "compulsive" drinkers, defined by their ability to continue high drinking levels in the presence of the bitter tastant quinine (Siciliano et al. 2019). Remarkably, Siciliano and colleagues demonstrated that activity within this pathway was predictive of the classification scheme that the mice would fall into, suggesting the possibility for biomarker-based approaches to assess alcohol use disorder vulnerability based on MRI brain imaging approaches.

3 Conclusions and Future Directions

3.1 Circuit-Based Analysis Evolves Our Models of AUD

The transition from region- to circuit-based AUD animal model research is an important step in the growth of our field toward novel marker and therapeutic target development. One particularly interesting aspect of this approach has been the tendency for circuit-based studies as described above to blur the delineations of specific regions within specific domains of the disorder (Fig. 1). These studies instead suggest that distinct circuits through common nodal areas may play more wide-ranging roles in key behaviors. A challenge moving forward is translating this work into the human brain imaging realm and vice versa. The animal model work has the obvious advantages of an ability to specifically manipulate discrete populations of neurons, and to examine these circuits at very high cellular resolution both pre- and post-mortem. However, these circuits must be studied in the context of models with varying degrees of face and construct validation against aspects of AUD. Thus, human imaging studies in AUD cohorts are a critical convergence point. Although analysis of circuits at cellular resolution remains a challenge, fMRI and related approaches across large cohorts of individuals continue to hold promise for the development of understanding of neuronal networks whose activities are correlated with important AUD behavioral and genetic endophenotypes. The cross-validation of studies such as these in animal models and humans is thus critical for the advancement of translational AUD research.

3.2 *Toward a Circuitry of AUD Vulnerability?*

Great examples of this translation are already beginning to emerge. For example, as described above, PFC inputs to the PAG have been implicated in compulsive alcohol intake in animal model work utilizing sophisticated neuronal circuit control strategies (Siciliano et al. 2019). Analysis of neural networks from AUD cohorts through the IMAGEN consortium revealed a network involving the medial orbitofrontal cortex and dorsal PAG in human alcohol misuse (Jia et al. 2021). This is an exciting example where hypotheses originally tested in animal model work spurred human brain imaging analyses that lead to the potential for considering a circuit's activity as a biomarker of AUD vulnerability in the future.

3.3 *Circuit-Based Therapeutics*

Finally, the potential for circuit-based therapeutics is exciting. First, the "omic" approaches available in modern neuroscience allow the mining of specific circuits as they are identified for potential novel pharmaceutical targets in the treatment of aspects of AUD. Second, a means of delivery of genetic material continues to evolve, consideration may ultimately be reasonable for circuit-based genetic interventions. Finally, current procedures such as transcranial magnetic stimulation may offer ways of modulating the activity of specific pathways in ways that mitigate AUD. These are all exciting near- and long-term possibilities, of which require precise knowledge of key circuits and cell types within those circuits in specific components of AUD.

References

Abrahao KP, Chancey JH, Chan CS, Lovinger DM (2017) Ethanol-sensitive pacemaker neurons in the mouse external Globus pallidus. Neuropsychopharmacology 42(5):1070–1081

Arinze I, Moorman DE (2020) Selective impact of lateral orbitofrontal cortex inactivation on reinstatement of alcohol seeking in male Long-Evans rats. Neuropharmacology 168:108007

Arizzi-LaFrance MN, Correa M, Aragon CM, Salamone JD (2006) Motor stimulant effects of ethanol injected into the substantia nigra pars reticulata: importance of catalase-mediated metabolism and the role of acetaldehyde. Neuropsychopharmacology 31(5):997–1008

Armand EJ, Li J, Xie F, Luo C, Mukamel EA (2021) Single-cell sequencing of brain cell transcriptomes and epigenomes. Neuron 109(1):11–26

Avegno EM, Gilpin NW (2022) Reciprocal midbrain-extended amygdala circuit activity in preclinical models of alcohol use and misuse. Neuropharmacology 202:108856

Avegno EM, Kasten CR, Snyder WB 3rd, Kelley LK, Lobell TD, Templeton TJ, Constans M, Wills TA, Middleton JW, Gilpin NW (2021) Alcohol dependence activates ventral tegmental area projections to central amygdala in male mice and rats. Addict Biol 26(4):e12990

Bianchi PC, Carneiro de Oliveira PE, Palombo P, Leao RM, Cogo-Moreira H, Planeta CDS, Cruz FC (2018) Functional inactivation of the orbitofrontal cortex disrupts context-induced reinstatement of alcohol seeking in rats. Drug Alcohol Depend 186:102–112

Brodie MS, Shefner SA, Dunwiddie TV (1990) Ethanol increases the firing rate of dopamine neurons of the rat ventral tegmental area in vitro. Brain Res 508(1):65–69

Burnham NW, Chaimowitz CN, Vis CC, Segantine Dornellas AP, Navarro M, Thiele TE (2021) Lateral hypothalamus-projecting noradrenergic locus coeruleus pathway modulates binge-like ethanol drinking in male and female TH-ires-cre mice. Neuropharmacology 196:108702

Cassataro D, Bergfeldt D, Malekian C, Van Snellenberg JX, Thanos PK, Fishell G, Sjulson L (2014) Reverse pharmacogenetic modulation of the nucleus accumbens reduces ethanol consumption in a limited access paradigm. Neuropsychopharmacology 39(2):283–290

Centanni SW, Bedse G, Patel S, Winder DG (2019a) Driving the downward spiral: alcohol-induced dysregulation of extended Amygdala circuits and negative affect. Alcohol Clin Exp Res 43(10): 2000–2013

Centanni SW, Morris BD, Luchsinger JR, Bedse G, Fetterly TL, Patel S, Winder DG (2019b) Endocannabinoid control of the insular-bed nucleus of the stria terminalis circuit regulates negative affective behavior associated with alcohol abstinence. Neuropsychopharmacology 44(3):526–537

Cheng Y, Huang CCY, Ma T, Wei X, Wang X, Lu J, Wang J (2017) Distinct synaptic strengthening of the striatal direct and indirect pathways drives alcohol consumption. Biol Psychiatry 81(11): 918–929

Corbit LH, Nie H, Janak PH (2012) Habitual alcohol seeking: time course and the contribution of subregions of the dorsal striatum. Biol Psychiatry 72(5):389–395

Craig AD (2009) How do you feel – now? The anterior insula and human awareness. Nat Rev Neurosci 10(1):59–70

Criswell HE, Simson PE, Knapp DJ, Devaud LL, McCown TJ, Duncan GE, Morrow AL, Breese GR (1995) Effect of zolpidem on gamma-aminobutyric acid (GABA)-induced inhibition predicts the interaction of ethanol with GABA on individual neurons in several rat brain regions. J Pharmacol Exp Ther 273(1):526–536

Dayas CV, Liu X, Simms JA, Weiss F (2007) Distinct patterns of neural activation associated with ethanol seeking: effects of naltrexone. Biol Psychiatry 61(8):979–989

de Guglielmo G, Crawford E, Kim S, Vendruscolo LF, Hope BT, Brennan M, Cole M, Koob GF, George O (2016) Recruitment of a neuronal ensemble in the central nucleus of the Amygdala is required for alcohol dependence. J Neurosci 36(36):9446–9453

de Guglielmo G, Kallupi M, Pomrenze MB, Crawford E, Simpson S, Schweitzer P, Koob GF, Messing RO, George O (2019) Inactivation of a CRF-dependent amygdalofugal pathway reverses addiction-like behaviors in alcohol-dependent rats. Nat Commun 10(1):1238

De Oliveira Sergio T, Lei K, Kwok C, Ghotra S, Wegner SA, Walsh M, Waal J, Darevsky D, Hopf FW (2021) The role of anterior insula-brainstem projections and alpha-1 noradrenergic receptors for compulsion-like and alcohol-only drinking. Neuropsychopharmacology 46(11): 1918–1926

Di Chiara G, Imperato A (1988) Drugs abused by humans preferentially increase synaptic dopamine concentrations in the mesolimbic system of freely moving rats. Proc Natl Acad Sci U S A 85(14):5274–5278

Domi E, Domi A, Adermark L, Heilig M, Augier E (2021) Neurobiology of alcohol seeking behavior. J Neurochem 157(5):1585–1614

Dong C, Zheng Y, Long-Iyer K, Wright EC, Li Y, Tian L (2022) Fluorescence imaging of neural activity, neurochemical dynamics, and drug-specific receptor conformation with genetically encoded sensors. Annu Rev Neurosci 45:273–294

Dornellas APS, Burnham NW, Luhn KL, Petruzzi MV, Thiele TE, Navarro M (2021) Activation of locus coeruleus to rostromedial tegmental nucleus (RMTg) noradrenergic pathway blunts binge-like ethanol drinking and induces aversive responses in mice. Neuropharmacology 199:108797

Droutman V, Bechara A, Read SJ (2015a) Roles of the different sub-regions of the insular cortex in various phases of the decision-making process. Front Behav Neurosci 9:309

Droutman V, Read SJ, Bechara A (2015b) Revisiting the role of the insula in addiction. Trends Cogn Sci 19(7):414–420

Felipe JM, Palombo P, Bianchi PC, Zaniboni CR, Anésio A, Yokoyama TS, Engi SA, Carneiro-de-Oliveira PE, Planeta CDS, Leão RM, Cruz FC (2021) Dorsal hippocampus plays a causal role in context-induced reinstatement of alcohol-seeking in rats. Behav Brain Res 398:112978

Feltenstein MW, See RE, Fuchs RA (2021) Neural substrates and circuits of drug addiction. Cold Spring Harb Perspect Med 11(4)

Fetterly TL, Basu A, Nabit BP, Awad E, Williford KM, Centanni SW, Matthews RT, Silberman Y, Winder DG (2019) alpha2A-adrenergic receptor activation decreases parabrachial nucleus excitatory drive onto BNST CRF neurons and reduces their activity in vivo. J Neurosci 39(3): 472–484

Fuchs RA, Evans KA, Ledford CC, Parker MP, Case JM, Mehta RH, See RE (2005) The role of the dorsomedial prefrontal cortex, basolateral amygdala, and dorsal hippocampus in contextual reinstatement of cocaine seeking in rats. Neuropsychopharmacology 30(2):296–309

Gatto GJ, McBride WJ, Murphy JM, Lumeng L, Li TK (1994) Ethanol self-infusion into the ventral tegmental area by alcohol-preferring rats. Alcohol 11(6):557–564

George O, Hope BT (2017) Cortical and amygdalar neuronal ensembles in alcohol seeking, drinking and withdrawal. Neuropharmacology 122:107–114

Gessa GL, Muntoni F, Collu M, Vargiu L, Mereu G (1985) Low doses of ethanol activate dopaminergic neurons in the ventral tegmental area. Brain Res 348(1):201–203

Gremel CM, Cunningham CL (2008) Roles of the nucleus accumbens and amygdala in the acquisition and expression of ethanol-conditioned behavior in mice. J Neurosci 28(5): 1076–1084

Hansson AC, Rimondini R, Neznanova O, Sommer WH, Heilig M (2008) Neuroplasticity in brain reward circuitry following a history of ethanol dependence. Eur J Neurosci 27(8):1912–1922

Heidenreich M, Zhang F (2016) Applications of CRISPR-Cas systems in neuroscience. Nat Rev Neurosci 17(1):36–44

Hernandez JS, Binette AN, Rahman T, Tarantino JD, Moorman DE (2020) Chemogenetic inactivation of orbitofrontal cortex decreases Cue-induced reinstatement of ethanol and sucrose seeking in male and female Wistar rats. Alcohol Clin Exp Res 44(9):1769–1782

Herring BE, Mayfield RD, Camp MC, Alcantara AA (2004) Ethanol-induced Fos immunoreactivity in the extended amygdala and hypothalamus of the rat brain: focus on cholinergic interneurons of the nucleus accumbens. Alcohol Clin Exp Res 28(4):588–597

Hitzemann B, Hitzemann R (1997) Genetics ethanol and the Fos response: a comparison of the C57BL/6J and DBA/2J inbred mouse strains. Alcohol Clin Exp Res 21(8):1497–1507

Hodge CW, Chappelle AM, Samson HH (1995) GABAergic transmission in the nucleus accumbens is involved in the termination of ethanol self-administration in rats. Alcohol Clin Exp Res 19(6):1486–1493

Hodge CW, Haraguchi M, Erickson H, Samson HH (1993) Ventral tegmental microinjections of quinpirole decrease ethanol and sucrose-reinforced responding. Alcohol Clin Exp Res 17(2): 370–375

Hodge CW, Samson HH, Chappelle AM (1997) Alcohol self-administration: further examination of the role of dopamine receptors in the nucleus accumbens. Alcohol Clin Exp Res 21(6): 1083–1091

Hoffman PL, Rabe CS, Moses F, Tabakoff B (1989) N-methyl-D-aspartate receptors and ethanol: inhibition of calcium flux and cyclic GMP production. J Neurochem 52(6):1937–1940

Holleran KM, Wilson HH, Fetterly TL, Bluett RJ, Centanni SW, Gilfarb RA, Rocco LE, Patel S, Winder DG (2016) Ketamine and MAG lipase inhibitor-dependent reversal of evolving depressive-like behavior during forced abstinence from alcohol drinking. Neuropsychopharmacology 41(8):2062–2071

Hong SI, Kang S, Chen JF, Choi DS (2019) Indirect medium spiny neurons in the dorsomedial striatum regulate ethanol-containing conditioned reward seeking. J Neurosci 39(36):7206–7217

Hosp F, Mann M (2017) A primer on concepts and applications of proteomics in neuroscience. Neuron 96(3):558–571

Janak PH, Chaudhri N (2010) The potent effect of environmental context on relapse to alcohol-seeking after extinction. Open Addict J 3:76–87

Jaramillo AA, Agan VE, Makhijani VH, Pedroza S, McElligott ZA, Besheer J (2018a) Functional role for suppression of the insular-striatal circuit in modulating interoceptive effects of alcohol. Addict Biol 23(5):1020–1031

Jaramillo AA, Randall PA, Frisbee S, Fisher KR, Besheer J (2015) Activation of mGluR2/3 following stress hormone exposure restores sensitivity to alcohol in rats. Alcohol 49(6):525–532

Jaramillo AA, Van Voorhies K, Randall PA, Besheer J (2018b) Silencing the insular-striatal circuit decreases alcohol self-administration and increases sensitivity to alcohol. Behav Brain Res 348: 74–81

Jasinska AJ, Stein EA, Kaiser J, Naumer MJ, Yalachkov Y (2014) Factors modulating neural reactivity to drug cues in addiction: a survey of human neuroimaging studies. Neurosci Biobehav Rev 38:1–16

Jeanblanc J, Rolland B, Gierski F, Martinetti MP, Naassila M (2019) Animal models of binge drinking, current challenges to improve face validity. Neurosci Biobehav Rev 106:112–121

Jia T, Xie C, Banaschewski T, Barker GJ, Bokde ALW, Buchel C, Quinlan EB, Desrivieres S, Flor H, Grigis A, Garavan H, Gowland P, Heinz A, Ittermann B, Martinot JL, Martinot MP, Nees F, Orfanos DP, Poustka L, Frohner JH, Smolka MN, Walter H, Whelan R, Schumann G, Robbins TW, Feng J, IMAGEN Consortium (2021) Neural network involving medial orbitofrontal cortex and dorsal periaqueductal gray regulation in human alcohol abuse. Sci Adv 7(6)

Juarez B, Han MH (2016) Diversity of dopaminergic neural circuits in response to drug exposure. Neuropsychopharmacology 41(10):2424–2446

Juarez B, Morel C, Ku SM, Liu Y, Zhang H, Montgomery S, Gregoire H, Ribeiro E, Crumiller M, Roman-Ortiz C, Walsh JJ, Jackson K, Croote DE, Zhu Y, Zhang S, Vendruscolo LF, Edwards S, Roberts A, Hodes GE, Lu Y, Calipari ES, Chaudhury D, Friedman AK, Han MH (2017) Midbrain circuit regulation of individual alcohol drinking behaviors in mice. Nat Commun 8(1):2220

June HL, Foster KL, McKay PF, Seyoum R, Woods JE, Harvey SC, Eiler WJ, Grey C, Carroll MR, McCane S, Jones CM, Yin W, Mason D, Cummings R, Garcia M, Ma C, Sarma PV, Cook JM, Skolnick P (2003) The reinforcing properties of alcohol are mediated by GABA(A1) receptors in the ventral pallidum. Neuropsychopharmacology 28(12):2124–2137

Jupp B, Krstew E, Dezsi G, Lawrence AJ (2011) Discrete cue-conditioned alcohol-seeking after protracted abstinence: pattern of neural activation and involvement of orexin(1) receptors. Br J Pharmacol 162(4):880–889

Kemppainen H, Raivio N, Kiianmaa K (2012) Role for ventral pallidal GABAergic mechanisms in the regulation of ethanol self-administration. Psychopharmacology (Berl) 223(2):211–221

Kim CK, Adhikari A, Deisseroth K (2017) Integration of optogenetics with complementary methodologies in systems neuroscience. Nat Rev Neurosci 18(4):222–235

Kimbrough A, Lurie DJ, Collazo A, Kreifeldt M, Sidhu H, Macedo GC, D'Esposito M, Contet C, George O (2020) Brain-wide functional architecture remodeling by alcohol dependence and abstinence. Proc Natl Acad Sci U S A 117(4):2149–2159

Kircher DM, Aziz HC, Mangieri RA, Morrisett RA (2019) Ethanol experience enhances glutamatergic ventral hippocampal inputs to D1 receptor-expressing medium spiny neurons in the nucleus Accumbens shell. J Neurosci 39(13):2459–2469

Koob GF (2008) A role for brain stress systems in addiction. Neuron 59(1):11–34

Koob GF, Schulkin J (2019) Addiction and stress: an allostatic view. Neurosci Biobehav Rev 106: 245–262

Koob GF, Volkow ND (2010) Neurocircuitry of addiction. Neuropsychopharmacology 35(1): 217–238

Korber C, Sommer WH (2022) From ensembles to meta-ensembles: specific reward encoding by correlated network activity. Front Behav Neurosci 16:977474

Kreifeldt M, Herman MA, Sidhu H, Okhuarobo A, Macedo GC, Shahryari R, Gandhi PJ, Roberto M, Contet C (2022) Central amygdala corticotropin-releasing factor neurons promote hyponeophagia but do not control alcohol drinking in mice. Mol Psychiatry 27(5):2502–2513

Kupchik YM, Brown RM, Heinsbroek JA, Lobo MK, Schwartz DJ, Kalivas PW (2015) Coding the direct/indirect pathways by D1 and D2 receptors is not valid for accumbens projections. Nat Neurosci 18(9):1230–1232

Leriche M, Mendez M, Zimmer L, Berod A (2008) Acute ethanol induces Fos in GABAergic and non-GABAergic forebrain neurons: a double-labeling study in the medial prefrontal cortex and extended amygdala. Neuroscience 153(1):259–267

Levine OB, Skelly MJ, Miller JD, Rivera-Irizarry JK, Rowson SA, DiBerto JF, Rinker JA, Thiele TE, Kash TL, Pleil KE (2021) The paraventricular thalamus provides a polysynaptic brake on limbic CRF neurons to sex-dependently blunt binge alcohol drinking and avoidance behavior in mice. Nat Commun 12(1):5080

Leyrer-Jackson JM, Hood LE, Olive MF (2021) Alcohol consumption preferentially activates a subset of pro-opiomelanocortin (POMC) producing neurons targeting the amygdala. Neuropharmacology 195:108674

Lovelock DF, Tyler RE, Besheer J (2021) Interoception and alcohol: mechanisms, networks, and implications. Neuropharmacology 200:108807

Lovinger DM, Alvarez VA (2017) Alcohol and basal ganglia circuitry: animal models. Neuropharmacology 122:46–55

Lovinger DM, Roberto M (2023) Synaptic effects induced by alcohol. Curr Top Behav Neurosci

Lovinger DM, White G, Weight FF (1989) Ethanol inhibits NMDA-activated ion current in hippocampal neurons. Science 243(4899):1721–1724

Luchsinger JR, Fetterly TL, Williford KM, Salimando GJ, Doyle MA, Maldonado J, Simerly RB, Winder DG, Centanni SW (2021) Delineation of an insula-BNST circuit engaged by struggling behavior that regulates avoidance in mice. Nat Commun 12(1):3561

Mantsch JR, Baker DA, Funk D, Le AD, Shaham Y (2016) Stress-induced reinstatement of drug seeking: 20 years of progress. Neuropsychopharmacology 41(1):335–356

Marinelli PW, Funk D, Juzytsch W, Li Z, Lê AD (2007) Effects of opioid receptor blockade on the renewal of alcohol seeking induced by context: relationship to c-fos mRNA expression. Eur J Neurosci 26(10):2815–2823

Marino RAM, Girven KS, Figueiredo A, Navarrete J, Doty C, Sparta DR (2021) Binge ethanol drinking associated with sex-dependent plasticity of neurons in the insula that project to the bed nucleus of the stria terminalis. Neuropharmacology 196:108695

Mathis A, Mamidanna P, Cury KM, Abe T, Murthy VN, Mathis MW, Bethge M (2018) DeepLabCut: markerless pose estimation of user-defined body parts with deep learning. Nat Neurosci 21(9):1281–1289

Mereu G, Fadda F, Gessa GL (1984) Ethanol stimulates the firing rate of nigral dopaminergic neurons in unanesthetized rats. Brain Res 292(1):63–69

Moorman DE (2018) The role of the orbitofrontal cortex in alcohol use, abuse, and dependence. Prog Neuropsychopharmacol Biol Psychiatry 87(Pt A):85–107

Morikawa H, Morrisett RA (2010) Ethanol action on dopaminergic neurons in the ventral tegmental area. Functional plasticity and genetic variation: insights into the neurobiology of alcoholism, pp 235–288

Munoz B, Fritz BM, Yin F, Atwood BK (2018) Alcohol exposure disrupts mu opioid receptor-mediated long-term depression at insular cortex inputs to dorsolateral striatum. Nat Commun 9(1):1318

Naqvi NH, Bechara A (2010) The insula and drug addiction: an interoceptive view of pleasure, urges, and decision-making. Brain Struct Funct 214(5–6):435–450

Nestoros JN (1980) Ethanol specifically potentiates GABA-mediated neurotransmission in feline cerebral cortex. Science 209(4457):708–710

Nowak KL, McBride WJ, Lumeng L, Li TK, Murphy JM (2000) Involvement of dopamine D2 autoreceptors in the ventral tegmental area on alcohol and saccharin intake of the alcohol-preferring P rat. Alcohol Clin Exp Res 24(4):476–483

O'Tousa D, Grahame N (2014) Habit formation: implications for alcoholism research. Alcohol 48(4):327–335

Palombo P, Leao RM, Bianchi PC, de Oliveira PEC, Planeta CDS, Cruz FC (2017) Inactivation of the prelimbic cortex impairs the context-induced reinstatement of ethanol seeking. Front Pharmacol 8:725

Pati D, Marcinkiewcz CA, DiBerto JF, Cogan ES, McElligott ZA, Kash TL (2020) Chronic intermittent ethanol exposure dysregulates a GABAergic microcircuit in the bed nucleus of the stria terminalis. Neuropharmacology 168:107759

Pfarr S, Meinhardt MW, Klee ML, Hansson AC, Vengeliene V, Schonig K, Bartsch D, Hope BT, Spanagel R, Sommer WH (2015) Losing control: excessive alcohol seeking after selective inactivation of Cue-responsive neurons in the infralimbic cortex. J Neurosci 35(30): 10750–10761

Pfarr S, Schaaf L, Reinert JK, Paul E, Herrmannsdorfer F, Rossmanith M, Kuner T, Hansson AC, Spanagel R, Korber C, Sommer WH (2018) Choice for drug or natural reward engages largely overlapping neuronal ensembles in the infralimbic prefrontal cortex. J Neurosci 38(14): 3507–3519

Pina MM, Cunningham CL (2017) Ethanol-seeking behavior is expressed directly through an extended amygdala to midbrain neural circuit. Neurobiol Learn Mem 137:83–91

Radwanska K, Wrobel E, Korkosz A, Rogowski A, Kostowski W, Bienkowski P, Kaczmarek L (2008) Alcohol relapse induced by discrete cues activates components of AP-1 transcription factor and ERK pathway in the rat basolateral and central amygdala. Neuropsychopharmacology 33(8):1835–1846

Rassnick S, Pulvirenti L, Koob GF (1992) Oral ethanol self-administration in rats is reduced by the administration of dopamine and glutamate receptor antagonists into the nucleus accumbens. Psychopharmacology (Berl) 109(1–2):92–98

Renteria R, Baltz ET, Gremel CM (2018) Chronic alcohol exposure disrupts top-down control over basal ganglia action selection to produce habits. Nat Commun 9(1):211

Renteria R, Cazares C, Baltz ET, Schreiner DC, Yalcinbas EA, Steinkellner T, Hnasko TS, Gremel CM (2021) Mechanism for differential recruitment of orbitostriatal transmission during actions and outcomes following chronic alcohol exposure. eLife 10

Rinker JA, Marshall SA, Mazzone CM, Lowery-Gionta EG, Gulati V, Pleil KE, Kash TL, Navarro M, Thiele TE (2017) Extended amygdala to ventral tegmental area corticotropin-releasing factor circuit controls binge ethanol intake. Biol Psychiatry 81(11):930–940

Roberto M, Kirson D, Khom S (2021) The role of the central amygdala in alcohol dependence. Cold Spring Harb Perspect Med 11(2)

Robins MT, Chiang T, Mores KL, Alongkronrusmee D, van Rijn RM (2018) Critical role for Gi/o-protein activity in the dorsal striatum in the reduction of voluntary alcohol intake in C57Bl/6 mice. Front Psych 9:112

Robinson DL, Howard EC, McConnell S, Gonzales RA, Wightman RM (2009) Disparity between tonic and phasic ethanol-induced dopamine increases in the nucleus accumbens of rats. Alcohol Clin Exp Res 33(7):1187–1196

Rodd ZA, Melendez RI, Bell RL, Kuc KA, Zhang Y, Murphy JM, McBride WJ (2004) Intracranial self-administration of ethanol within the ventral tegmental area of male Wistar rats: evidence for involvement of dopamine neurons. J Neurosci 24(5):1050–1057

Roland AV, Coelho CAO, Haun HL, Gianessi CA, Lopez MF, D'Ambrosio S, Machinski SN, Kroenke CD, Frankland PW, Becker HC, Kash TL (2023) Alcohol dependence modifies brain networks activated during withdrawal and reaccess: a c-fos-based analysis in mice. Biol Psychiatry

Roth BL (2016) DREADDs for neuroscientists. Neuron 89(4):683–694

Ryabinin AE, Criado JR, Henriksen SJ, Bloom FE, Wilson MC (1997) Differential sensitivity of c-Fos expression in hippocampus and other brain regions to moderate and low doses of alcohol. Mol Psychiatry 2(1):32–43

Samson HH, Hodge CW, Tolliver GA, Haraguchi M (1993) Effect of dopamine agonists and antagonists on ethanol-reinforced behavior: the involvement of the nucleus accumbens. Brain Res Bull 30(1–2):133–141

Schroeder JP, Spanos M, Stevenson JR, Besheer J, Salling M, Hodge CW (2008) Cue-induced reinstatement of alcohol-seeking behavior is associated with increased ERK1/2 phosphorylation in specific limbic brain regions: blockade by the mGluR5 antagonist MPEP. Neuropharmacology 55(4):546–554

Segovia KN, Vontell R, Lopez-Cruz L, Salamone JD, Correa M (2013) c-Fos immunoreactivity in prefrontal, basal ganglia and limbic areas of the rat brain after central and peripheral administration of ethanol and its metabolite acetaldehyde. Front Behav Neurosci 7:48

Seif T, Chang SJ, Simms JA, Gibb SL, Dadgar J, Chen BT, Harvey BK, Ron D, Messing RO, Bonci A, Hopf FW (2013) Cortical activation of accumbens hyperpolarization-active NMDARs mediates aversion-resistant alcohol intake. Nat Neurosci 16(8):1094–1100

Siciliano CA, Noamany H, Chang CJ, Brown AR, Chen X, Leible D, Lee JJ, Wang J, Vernon AN, Vander Weele CM, Kimchi EY, Heiman M, Tye KM (2019) A cortical-brainstem circuit predicts and governs compulsive alcohol drinking. Science 366(6468):1008–1012

Silberman Y, Matthews RT, Winder DG (2013) A corticotropin releasing factor pathway for ethanol regulation of the ventral tegmental area in the bed nucleus of the stria terminalis. J Neurosci 33(3):950–960

Sinclair CM, Cleva RM, Hood LE, Olive MF, Gass JT (2012) mGluR5 receptors in the basolateral amygdala and nucleus accumbens regulate cue-induced reinstatement of ethanol-seeking behavior. Pharmacol Biochem Behav 101(3):329–335

Smith RJ, Anderson RI, Haun HL, Mulholland PJ, Griffin WC 3rd, Lopez MF, Becker HC (2020) Dynamic c-Fos changes in mouse brain during acute and protracted withdrawal from chronic intermittent ethanol exposure and relapse drinking. Addict Biol 25(6):e12804

Thiele TE, Navarro M (2014) "Drinking in the dark" (DID) procedures: a model of binge-like ethanol drinking in non-dependent mice. Alcohol 48(3):235–241

Topple AN, Hunt GE, McGregor IS (1998) Possible neural substrates of beer-craving in rats. Neurosci Lett 252(2):99–102

Tsurugizawa T, Uematsu A, Uneyama H, Torii K (2010) The role of the GABAergic and dopaminergic systems in the brain response to an intragastric load of alcohol in conscious rats. Neuroscience 171(2):451–460

Ueda HR, Dodt HU, Osten P, Economo MN, Chandrashekar J, Keller PJ (2020) Whole-brain profiling of cells and circuits in mammals by tissue clearing and light-sheet microscopy. Neuron 106(3):369–387

Vilpoux C, Warnault V, Pierrefiche O, Daoust M, Naassila M (2009) Ethanol-sensitive brain regions in rat and mouse: a cartographic review, using immediate early gene expression. Alcohol Clin Exp Res 33(6):945–969

Vranjkovic O, Winkler G, Winder DG (2018) Ketamine administration during a critical period after forced ethanol abstinence inhibits the development of time-dependent affective disturbances. Neuropsychopharmacology 43(9):1915–1923

Wandres M, Pfarr S, Molnar B, Schollkopf U, Ercsey-Ravasz M, Sommer WH, Korber C (2021) Alcohol and sweet reward are encoded by distinct meta-ensembles. Neuropharmacology 195: 108496

Wedzony K, Koros E, Czyrak A, Chocyk A, Czepiel K, Fijal K, Mackowiak M, Rogowski A, Kostowski W, Bienkowski P (2003) Different pattern of brain c-Fos expression following re-exposure to ethanol or sucrose self-administration environment. Naunyn Schmiedebergs Arch Pharmacol 368(5):331–341

Weiss F, Lorang MT, Bloom FE, Koob GF (1993) Oral alcohol self-administration stimulates dopamine release in the rat nucleus accumbens: genetic and motivational determinants. J Pharmacol Exp Ther 267(1):250–258

Wills TA, Kash TL, Winder DG (2013) Developmental changes in the acute ethanol sensitivity of glutamatergic and GABAergic transmission in the BNST. Alcohol 47(7):531–537

Witkiewitz K, Stein ER, Votaw VR, Hallgren KA, Gibson BC, Boness CL, Pearson MR, Maisto SA (2022) Constructs derived from the addiction cycle predict alcohol use disorder treatment outcomes and recovery 3 years following treatment. Psychol Addict Behav

You C, Vandegrift B, Brodie MS (2018) Ethanol actions on the ventral tegmental area: novel potential targets on reward pathway neurons. Psychopharmacology (Berl) 235(6):1711–1726

Zhao Y, Dayas CV, Aujla H, Baptista MA, Martin-Fardon R, Weiss F (2006) Activation of group II metabotropic glutamate receptors attenuates both stress and cue-induced ethanol-seeking and modulates c-fos expression in the hippocampus and amygdala. J Neurosci 26(39):9967–9974

Understanding How Acute Alcohol Impacts Neural Encoding in the Rodent Brain

Christopher C. Lapish

Contents

Abstract Alcohol impacts neural circuitry throughout the brain and has wide-ranging effects on the biophysical properties of neurons in these circuits. Articulating how these wide-ranging effects might eventually result in altered computational properties has the potential to provide a tractable working model of how alcohol alters neural encoding. This chapter reviews what is currently known about how acute alcohol influences neural activity in cortical, hippocampal, and dopaminergic circuits as these have been the primary focus of understanding how alcohol alters neural computation. While other neural systems have been the focus of exhaustive work on this topic, these brain regions are the ones where in vivo neural recordings are available, thus optimally suited to make the link between changes in neural activity and behavior. Rodent models have been key in developing an understanding

---

C. C. Lapish (✉)
Department of Anatomy, Cell Biology, and Physiology, Stark Neuroscience Institute, Indiana University School of Medicine, Indianapolis, IN, USA
e-mail: clapish@iu.edu

Curr Topics Behav Neurosci (2025) 71: 401–423
https://doi.org/10.1007/7854_2024_479
Published Online: 11 June 2024

of how alcohol impacts the function of these circuits, and this chapter therefore focuses on work from mice and rats. While progress has been made, it is critical to understand the challenges and caveats associated with experimental procedures, especially when performed in vivo, which are designed to answer this question and if/how to translate these data to humans. The hypothesis is discussed that alcohol impairs the ability of neural circuits to acquire states of neural activity that are transiently elevated and characterized by increased complexity. It is hypothesized that these changes are distinct from the traditional view of alcohol being a depressant of neural activity in the forebrain.

Keywords Alcohol · Behavior · Computation · Mouse · Neuron · Rat

1 Introduction

Humans, and other animals, have consumed alcohol for nearly all of our recorded history seeking out its mind-altering properties. It is argued that alcohol has shaped our evolution and therefore influenced human behavior over thousands of years (Carrigan et al. 2015). Alcohol also has a robust impact on behavior on much shorter timescales. Over the course of a lifetime, repeated alcohol use can lead to changes in brain function that result in changes in behavior (Spear and Swartzwelder 2014). However, the most visible effects of alcohol emerge over the course of minutes to hours where consumption can lead to a wide range of behavioral effects from stress relief to intoxication (Vonghia et al. 2008). Importantly the acute effects of alcohol (e.g., observable from minutes to hours) are typically antecedent to those over longer timescales and typically drive consumption—at least initially. Therefore understanding the acute effects of alcohol has a range of implications from understanding human evolution to treating addiction. This chapter focuses on how alcohol influences neural activity on a timescale of seconds, minutes, and hours. Focusing on this timescale will allow an examination of the known mechanisms that describe how alcohol influences computation in forebrain and midbrain circuits.

Whether examining the transgenerational effects of alcohol or a dose of alcohol delivered over minutes, animal models, especially rodents, have played a critical role in understanding the mechanisms of how alcohol influences behavior. In large part, these advances have been driven by insights gained by directly measuring alcohol concentrations in the brain of behaving animals. In addition, measuring and actuating neural activity *in vivo* from behaving animals has provided insight into how alcohol can alter behavior. While these procedures are accessible to humans and nonhuman primates, they are typically invasive and therefore not ethically justifiable at the scale required to determine how alcohol influences behavior.

The use of rodent models, primarily mice and rats, has facilitated the use of invasive procedures that have led to several important insights into how alcohol effects on brain function result in alterations in behavior. There are clear limitations when making inferences about brain function in human or nonhuman primates from rodent data. However, rodent models play a critical role in narrowing hypotheses to a point where testing them in primates becomes feasible. This is extremely valuable when addressing the intersection of broad, difficult questions such as how alcohol effects on brain function result in changes in behavior. Therefore, the focus of this chapter will be on data acquired from rodent models.

2 Factors that Mediate the Effects of Alcohol on Neural Activity and How They Impact Translating Across Species

The purpose of this chapter is to capture what is known about how alcohol influences neural encoding in the rodent on a timescale from minutes to hours. However, experiments are typically performed in rodents to make inferences about *human* brain function and *human* alcohol drinking behaviors. Therefore, it is reasonable that a rodent model with maximal face validity is one that also drinks alcohol. This brings up important caveats when assessing the effects of alcohol on neural activity, such as working for reinforcement that is delayed following consumption (e.g., intoxication). Therefore, it is important to contextualize rodent data by considering the translational potential and limitations of using rats and mice.

On the surface, determining how alcohol affects brain function and thus leads to changes in behavior seems like a simple question to answer. The most straightforward approach is to have an animal drink alcohol and measure neural activity and behavior following consumption. However, this approach is fraught with pitfalls as there are a number of factors that modify how a given dose of alcohol might affect neural activity. Some of the factors that will be discussed have a clear biological mechanism (e.g., metabolism), whereas the mechanism of others is less clear (e.g., social effects). For example, as discussed in Sect. 2.3, metabolism and tolerance differ between species (Heit et al. 2015) and therefore prevent a 1:1 correspondence between dose consumed and brain concentrations, which presents a major barrier in comparing across species.

This goals of this section are twofold: First, several factors will be discussed that modify an organism's response to alcohol. Second, points of contact and divergence will be identified for each factor that are important to consider when interpreting the effects of alcohol on neural activity and when comparing across species. This section will not provide an in-depth view of each factor but rather selected points to consider when making inferences about alcohol's influence on neural activity and how they might differ across species.

2.1 Dose Dependence

Perhaps the most important feature to account for when interpreting the effects of alcohol on neural activity is that the effects of alcohol on behavior are dose dependent (Grupp 1980; Vonghia et al. 2008). This is true in humans and rodents (Brabant et al. 2014). The effects of alcohol on the biophysical properties of neurons are also clearly dose dependent (Morikawa and Morrisett 2010; Morningstar et al. 2021). Simply put, alcohol tends to stimulate behavior at low doses while it tends to diminish behavior at high doses. Therefore, it is necessary to measure, or infer, brain alcohol levels during behavior in order to make inferences about how it influences neural activity. Measuring brain alcohol concentrations in vivo from awake behaving animals is possible with established techniques such as in vivo microdialysis and implantable biosensors. However, these technologies lack either temporal specificity or precision in measurement. The development of a way to measure brain alcohol concentrations *in vivo* with micromolar precision and a temporal resolution of seconds could be a transformative tool for the field. The explosion of techniques for sampling neurochemicals via genetically encoded sensors expressed in the brain would suggest this is a tractable near-term goal.

2.2 Route of Administration

Both rats and mice are routinely used to measure the effects of alcohol on neural activity and behavior. The fact that rodents will readily consume alcohol for both its taste and post-ingestive properties provides a model to understand how it might alter computations in neural circuits that encode reward, motivation, and cognition. Furthermore, this provides a unique opportunity to model how the neural circuits that govern ingestive behaviors interface with motivational and cognitive circuitry during the progression of addiction in a translation manner with humans.

Like humans, rodents exhibit heterogeneity in their avidity to consume alcohol. In large part, this behavior is line/strain dependent where some populations uniformly either accept or refuse alcohol (Bell et al. 2017; Riley et al. 1977). However, other lines/strains exhibit heterogeneity within a population, where some accept alcohol to varying levels (Carnicella et al. 2014). These innate differences have provided a useful tool to explore how alcohol consumption influences behavior. In addition, these innate differences in alcohol consumption are heritable, thereby providing a tool to examine the genetic basis of avidity for alcohol. This is also an important caveat to consider when mapping consumption to changes in neural activity as strain differences in metabolism are documented (Matson et al. 2013).

Despite the face validity of using a route of administration in rodents that models the human condition, there are caveats to consider with oral self-administration models. For example, rodents usually do not obtain and maintain high blood alcohol levels that can be observed in advanced stages of an AUD (Griffin 2014; Rogers et al. 1979). This is problematic as sustained, high blood alcohol levels are an effective way to evoke a state that models dependence on alcohol in rodents (Griffin 2014; Meinhardt and Sommer 2015; Vendruscolo and Roberts 2014). Alcohol vapor administration models have proven useful in this regard, where rodents can be held at high blood alcohol levels for extended periods of time by passively inhaling alcohol in a closed chamber. Vapor produces several changes in neural activity in mesocorticolimbic circuits that are hypothesized to be important for the expression of behaviors that characterize dependence (Vendruscolo and Roberts 2014). A particular strength of vapor procedures is that they reliably produce lasting impacts on cognitive behaviors in the rodent (Kroener et al. 2012; Trantham-Davidson et al. 2014), which has proven more difficult to produce with oral administration procedures, therefore complicating the link to human studies.

Vapor administration of alcohol is a stressful process for the rodent, which should be considered when interpreting results from this model. Therefore, using the vapor model, it is more difficult to isolate the effects of alcohol on neural circuit function and behavior as distinct from stress (Sommer et al. 2008). However, stress impacts drinking in humans (Sinha et al. 2011), and therefore, this model provides an opportunity to understand stress–alcohol interactions. To date there has not been a study that examines neural activity during vapor exposure. Comparing the effects of vapor to oral exposure on the minutes to hour scale could provide insight into how sustained high brain alcohol levels impact neural activity that would be difficult with oral self-administration models in the rodent.

2.3 Metabolism

Differences between rodents and humans in metabolism of alcohol are important to consider when making inferences across species. The key enzymes involved in the breakdown of alcohol into its metabolic by-products are largely (but not exactly) conserved between rodents in humans (Höög and Ostberg 2011; Thompson et al. 2018). Alcohol is metabolized into acetyl aldehyde and then acetate by alcohol and aldehyde dehydrogenase enzymes that are located primarily in the liver (Edenberg 2007). Rodents metabolize alcohol faster than humans and mice have a faster rate of metabolism than rats (Holmes et al. 1986). As the effects of alcohol on neural activity and behavior are concentration dependent, and brain levels of alcohol are often inferred from oral ingestion, this complicates the comparison of dose consumed to changes in brain function and behavior across species. In addition, metabolism occurs in the brain and the conversion of alcohol to acetate in cerebellar astrocytes is required for the motor-impairing effects of alcohol in mice (Jin et al. 2021). While the enzyme required for ethanol metabolism (ALDH2) was observed in human cerebellar astrocytes, it remains to be determined if this mechanism influences human behavior similarly. These data highlight the potential of using rodent models to identify the mechanisms of alcohol's action on behavior. Identifying how alcohol metabolism in the brain might lead to changes in neural activity that are circuits specific will be critical to understanding the mechanism of how alcohol alters behavior.

2.4 Tolerance

Changes in the way a given dose of alcohol influences behavior is also observed following consumption, typically referred to as tolerance (Kalant 1998). The expression of tolerance is a key feature of the progression of an AUD. Therefore, tolerance is an important phenomenon that must be considered when identifying how alcohol affects brain function and influences behavior. Like humans, tolerance is observed in rodents and provides a useful model to study the factors underlying this phenomenon (Tabakoff and Hoffman 2000).

Tolerance can be mediated by biological factors such as increases in the rate of metabolism and elimination of alcohol (Pietrzykowski and Treistman 2008; Rimondini et al. 2008). It has been argued that the neurobiological effects of tolerance and how they influence behavior are understudied yet critical for understanding AUD (Elvig et al. 2021). When relating a dose of alcohol to changes in neural activity, it is brain concentrations that are most critical. Changes in biological factors that control tolerance therefore will change the relationship between dose consumed and brain concentrations. Since the effects of alcohol on neural activity are concentration dependent, this is important to consider.

Psychological factors, or conditioned effects, can also lead to the expression of tolerance, whereby subjects anticipate the intoxicating effects of alcohol and acquire the ability to behave in a relatively "normal" manner (Vogel-Sprott 1997). Contextual effects have been observed in rodents and humans (Birak et al. 2011; Cunningham 1994; Duncan et al. 2000; McCaul et al. 1989; McCusker and Brown 1990). This is exemplified by the ability of alcohol to elicit changes in behavior (e.g., locomotor activity) or physiological responses (e.g., body temperature) in one environment but not others (Vogel-Sprott 1997). Considering that rodent experiments tend to be very regimented in terms of time and location that alcohol is delivered, the contextual effects on measures of tolerance are critical to consider (White et al. 2002). In contrast, this facet of alcohol consumption is very difficult to recreate in human experiments, which are typically performed in a laboratory setting and thus lack the contextual cues that are associated with alcohol consumption in an individual. Translating data between humans and rats should consider this possible caveat.

2.5 Consummatory Effects

Detecting the direct effects of alcohol on neural activity are also complicated by consummatory or conditioned effects associated with receiving any reinforcing substance. For example, the receipt of a food reward (e.g., sucrose) can elicit robust changes in efflux of neuromodulators (Bassareo and Di Chiara 1999), which will influence neural activity prior to alcohol reaching the brain. Indeed, consumption of alcohol leads to increases in dopamine efflux that precede the rise in brain levels of alcohol (Doyon et al. 2005). These consummatory driven changes in neural activity can co-occur, synergize with, or even mask the effects of alcohol on neural activity and therefore should be considered when making inferences about the effects of alcohol on behavior in both humans and rodents. This caveat should also be considered when using injected alcohol as well as there will be sensory-driven changes in neural activity associated with the stress of injection and the burning sensation possibly elicited with injections of alcohol.

Several of the neural mechanisms that underlie the urge to drink, such as thirst, are conserved across rodents and humans (Leib et al. 2016). However, the lack of control over alcohol consumption is primarily associated with adaptations in motivational and cognitive circuits (Koob and Volkow 2016), while adaptations in thirst-controlling circuits are observed (Jahn et al. 2004). The mechanisms that terminate thirst remain intact in those with an AUD but are not sufficient to stop alcohol consumption. Furthermore, this system is finely tuned to control water absorption and retention; therefore, targeting this system for novel treatments is not historically viewed as a high priority and potentially dangerous (Kiefer et al. 2002; Mutschler et al. 2010; von der Goltz et al. 2014). Notwithstanding, alcohol is a diuretic and therefore can lead to thirst (Inenaga et al. 2017), but this is after the consumption of alcohol and therefore does not play a role in initiating a binge. Chronic alcohol use is associated with increases in measures of baseline measures of thirst (Collins et al. 1992) and leads to alterations in thirst regulation molecules such as vasopressin (Jahn et al. 2004). These adaptations could influence the conditioned and motivational properties of alcohol and therefore are critical to understand the effects of alcohol on neural activity.

2.6 Lifespan

Rodent models have been particularly effective in examining the effects of alcohol on brain function and behavior both in utero and in adolescence. Rodent models have clearly identified that even transient alcohol exposure during adolescence can lead to enduring changes in brain function and behavior that are observable in adult animals (Gass et al. 2014; Gursky et al. 2021; Nentwig et al. 2019). While the condensed lifespan of a rodent makes them excellent models for a number of reasons, this presents a potential limitation when extrapolating to human studies. The effects of alcohol on nervous system function and behavior accumulate throughout the lifespan (Matthews et al. 2017; Spear 2014). Therefore, it is unclear how decades of alcohol consumption may influence brain function or if these changes can be recapitulated in rodent models with exposure to alcohol that lasts, at most, a year.

2.7 *States of Vigilance*

Rodents and humans each display varying states of vigilance. The impact of waxing and waning states of vigilance can be quite pernicious when one considers the time- and concentration-dependent effects of alcohol on behavior. As alcohol accumulates in the brain the behavioral state of the animal is transitioning from an active to inactive one. Some of the most profound changes in neural activity are observed as animals transition across states of vigilance. For example, in the cortex, neural firing becomes synchronized and does not exhibit random, noisy activity patterns that characterize awake behaving states (Bagur et al. 2018; Morningstar et al. 2020; Steriade et al. 1993). These changes are also evident in the rhythms of the brain, which change dramatically across states of vigilance (Bagur et al. 2018; Steriade et al. 1993). Therefore, it is difficult to separate the changes in neural activity that are directly attributable to alcohol effects on neural circuitry from secondary effects associated with reduced states of vigilance. This requires that states of vigilance be measured and considered when inferences are made about the effects of alcohol on neural activity.

A common way to measure states of vigilance is to simultaneously examine the power spectrum of the neural oscillations and the behavior of the animal (Bagur et al. 2018; Morningstar et al. 2020). If an animal remains still for a period of, at least, several seconds and this is accompanied by an increase in low-frequency power (<5 Hz), this would indicate a transition to a reduced state of vigilance (e.g., sleep). This state change dominates the pattern of neural activity and obfuscates measurements of alcohol effects on neural activity. Importantly, changes in neural activity can be detected following more subtle changes in states of vigilance such as (dis)-engaging attentional processes (Klimesch 2012). Therefore, this important factor has to be considered in any experiment where an animal is given alcohol, injected or consumed, and neural activity is measured.

The experimental procedures available to measure neural activity in humans (e.g., EEG, fMRI) typically require that the subject be alert but still and calm. This can contrast dramatically from rodent experiments where neural activity is measured in animals that are freely moving and being reinforced, thus likely in a more activated state of vigilance. Equating the effects of alcohol on neural activity across these experimental preparations is complicated by differences in biological factors (e.g., neurotransmitter efflux) and possibly psychological factors (e.g., motivation).

3 How Does Alcohol Effect Neural Activity?

To understand the mechanisms that underlie the negative and unwanted effects alcohol can have on behavior, it is necessary to first understand the acute pharmacodynamic effects of alcohol on neural function in isolation. This section focuses on what is known about how acute alcohol influences neural activity and computation.

In addition, the reader is directed to an excellent review of the literature of the brain processes that are altered following acute alcohol (Van Skike et al. 2019). Further, the focus of this section is on measures of neural activity with fine temporal and spatial precision (spikes, calcium transients, and local field potentials) in order to make inferences on how ethanol influences computation at the single neuron level.

3.1 Dopamine Neurons

A convergence of evidence from reduced preparations (e.g., ex vivo) as well as in vivo recordings indicates that alcohol increases the firing rate of dopamine neurons at low concentrations (<2 g/kg) and decreases firing at high concentrations (>2 g/kg) (Morikawa and Morrisett 2010). Increases in dopamine neuron firing have been observed in vivo at doses as low as 0.5 g/kg EtOH delivered intravenously, which occur near instantaneously in unanesthetized animals (Gessa et al. 1985; Mereu et al. 1984). Depending on the dose this increase can last anywhere from 10 to 120 min, with larger doses lasting longer (Mereu et al. 1984). Interestingly, an inhibitory effect on neural activity is observed in vivo when animals are anesthetized (Mereu et al. 1984), which seems to be mediated by interactions between alcohol and the type of anesthesia used.

To understand how alcohol influences computations performed by dopamine neurons, it is important to consider what contributes to the two different modes of firing they exhibit and the types of computations they perform. Dopamine neurons fire in a tonic mode that is characterized by consistent, pacemaker-like activity, which is maintained by intrinsic biophysical properties of the dopamine neuron (Morikawa and Morrisett 2010) and contributes to basal levels of dopamine in the forebrain (Grace and Onn 1989). In addition, these neurons display a phasic mode that is characterized by transient increases in firing rate (Grace and Bunney 1984), referred to as "burst" firing. This mode is driven by synaptic inputs (expanded on below) that activate NMDA receptors and is critical for assigning value to stimuli in the environment. Morikawa and Morissett (2010) provide an exquisitely detailed view on the biophysical parameters that control each of these firing modes and conclude that both generally increase in response to alcohol.

Increases in dopamine neuron firing following alcohol administration are observable even when synaptic transmission is blocked (Morikawa and Morrisett 2010), indicating that it can influence tonic firing. Tonic firing plays a critical, but not exclusive, role in maintaining dopamine in afferent brain regions (Venton et al. 2003). The contribution of tonic dopamine to computations that underlie behavior has typically been viewed as distinct from those encoded by phasic release. Tonic dopamine efflux in projection areas has been hypothesized to reflect motivational signals (Cagniard et al. 2006; Mohebi et al. 2019; Niv et al. 2007), which may be independent of DA neuron firing (Mohebi et al. 2019; Threlfell et al. 2012). Therefore, it is not clear if, or how, the well-characterized effects of acute alcohol on the biophysical properties of dopamine neuron tonic firing will translate into changes in tonic dopamine levels in projection regions.

Computational models of the dopamine neuron indicate that the number of spikes in a burst is controlled by a complex interplay of synaptic and intrinsic currents. Phasic firing is initially driven by glutamatergic inputs from outside the VTA including the cortex (Coizet et al. 2003; Floresco et al. 2003; Geisler and Wise 2008). These inputs provide tone to AMPA and NMDA receptors which depolarize the cell and drive it toward threshold. Repolarization of the cell is facilitated by canonical spike producing currents (e.g., voltage-gated sodium and potassium channels). However, other intrinsic and synaptic currents also play a critical role.

Both the substantia nigra and the ventral tegmental area contain GABAergic neurons that provide inhibitory drive to dopamine neurons (Johnson and North 1992; Omelchenko and Sesack 2009; van Zessen et al. 2012). Excitatory inputs to GABAergic neurons in these brain regions tend to overlap with those that project to dopamine neurons setting up a feed-forward inhibitory architecture (Christie et al. 1985; Morozova et al. 2016a; Sesack and Pickel 1992). This is capable of inhibiting firing in the dopamine neuron (Steffensen et al. 1998; van Zessen et al. 2012). However, there may be cases where GABAergic inputs to the dopamine neuron may *increase* firing, which will be expanded upon below.

Hyperpolarization-activated cation channels (HCN) are present in a subpopulation of dopamine neurons (Margolis et al. 2006). These are inward-rectifying currents, typically referred to as Ih, which are activated when the neuron becomes hyperpolarized in order to return it to a more depolarized resting potential. Ih facilitates resonant properties in several types of neurons as well as post-inhibitory rebound spikes (Magee 2000; Shah 2014).

A biophysically detailed microcircuit model of the VTA was used to determine how each of these synaptic and intrinsic properties influences firing and how they might be affected by alcohol. This model contained shared excitatory inputs to the GABA and dopamine neurons and local feed-forward inhibition from the GABA to the dopamine neuron. When glutamatergic inputs were pulsatile (which would be expected in response to salient environmental stimuli) phasic firing of the dopamine neuron was observed. Interestingly, when driven in a pulsatile manner, it was found that the inclusion of GABAergic inputs to the dopamine neuron increased the number of spikes in bursts. This increase was mediated by temporally precise GABAergic synaptic currents that repolarize the neuron, thus inactivating voltage-gated K+ channels, activating Ih, and thus facilitate the dopamine neuron firing again. This GABA-based mechanism for increasing neuronal firing is not unique to dopamine neurons and observed in several brain regions such as the hippocampus, amygdala, and cortex (Alger and Nicoll 1979; Andersen et al. 1980; Gulledge and Stuart 2003; Woodruff et al. 2006).

It is well established that alcohol increases firing in dopamine neurons (see previous). To understand this mechanism and how alcohol might specifically alter the computational properties of dopamine neurons, it is useful to consider a standard measure of neural activity, the frequency–current (F–I) curve (Izhikevich 2006). Two different F–I curves are typically observed that each captures differences in

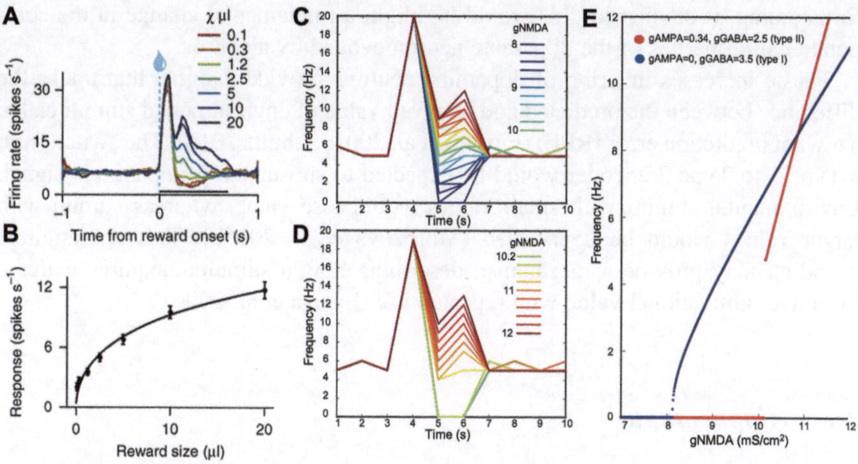

Fig. 1 *Simulations indicate that alcohol changes dopamine neurons from a Type 1 to Type 2 encoder.* In vivo recordings of dopamine neurons to varying amounts of a liquid reinforcer find two components of firing: One at reward onset that is invariant to value and a slower one that reflects value (**a**). Plotting the response of the second component indicates that dopamine neurons encode value in a manner consistent with a Type 1 encoder (**b**). A detailed, conductance-based model of dopamine neuron firing recapitulates both the invariant and value-encoding components under baseline conditions (**c**). Adjusting the conductance of the dopamine neuron to reflect the effects of alcohol selectively alters the value-encoding component of dopamine firing (**d**). Plotting the F–I curves for simulations of the baseline (blue points) and alcohol (red points) conditions finds a change from Type 1 to Type 2 encoding (**e**). *Panels* ***a, b*** *adapted from* Eshel et al. (2016) *Nature Neuroscience; Panels* **c, d, e** *adapted from* Morozova et al. (2016a, b) *PLoS Computational Biology*

how excitatory inputs to the neuron (e.g., electrical current) are translated to an output (e.g., action potential frequency). A linear curve describes a neuron that encodes stimuli in a continuous, or graded, manner. In contrast, a discontinuous curve that more closely resembles a step function describes a neuron that acts more like a binary, or an all or none, encoder. These two different curves capture Type 1 (former) and Type 2 (latter) excitability profiles and are useful to understand how neurons encode information.

Increases in the strength of the synaptic inputs to dopamine neurons typically increase the number of spikes produced in a relatively linear manner, or Type 1 excitation profile (Fig. 1a–c). Simulations indicate that alcohol changes the excitability profile of the dopamine neuron from Type 1 to Type 2 (Fig. 1c–e). This shift is attributable to alcohol enhancing synaptic currents, including Ih, GABA, and the AMPA/NMDA ratio. Alcohol's facilitating effects on Ih and GABA seem to be particularly important as they increase the number of spikes in bursts via the repolarizing mechanisms described above. These changes result in a shift to a Type 2 excitation profile that would be expected to suppress smaller inputs while enhancing the response to larger inputs (Fig. 1e). While experimental corroboration of this shift in the excitability profile of the dopamine neuron is still forthcoming, if confirmed, this would highlight a fundamental change in the computational properties of the dopamine neuron evoked by alcohol.

Phasic increases in firing of dopamine neurons provide a signal that tracks the difference between the predicted and received value of environmental stimuli called a reward prediction error (RPE) (Fiorillo et al. 2003; Schultz 2016). The switch from a Type 1 to Type 2 encoder would be expected to substantially impact this signal: Environmental stimuli with small values would lose value, whereas stimuli with larger values would be overvalued (Morozova et al. 2016b). This alteration in encoding may provide a mechanism describing how a stimulus acquires extreme incentive motivational value with repeated use (Lapish et al. 2006).

3.2 Hippocampus

Several in vivo studies report an inhibitory effect of alcohol on hippocampal neuron firing in rats that are in reduced behavioral states. However, clearly attributing this reduction to alcohol is often complicated by the use of anesthesia and a lack of information about the animals' state of vigilance. Following IP injected alcohol at 1.5 g/kg in animals that were waking from chloral hydrate anesthesia, a transient reduction was observed after ~20 min that was gone by 40 min (Zhang et al. 2016). In a separate study, a biphasic response in neural firing was observed following IV alcohol in restrained but awake rats. Low doses (100, 200 mg/kg) increased firing, whereas high does (400, 800 mg/kg) decreased hippocampal firing (Grupp 1980). These data indicate an inhibiting effect of alcohol at moderate to high doses.

A handful of studies have used freely behaving rats to examine the effects of alcohol on neural activity in the hippocampus and the medial septum, which is a major input to the hippocampus. Givens and colleagues report a biphasic effect on hippocampal theta oscillations in behaving animals where it is increased at doses of 0.25 and 0.5 g/kg and decreased at 1.0 g/kg (Givens 1995). In a subsequent study of neurons from the medial septum, reductions in stimulus-evoked neural activity were observed that were accompanied by a reduction in stimulus-evoked rhythmic firing (Givens 1996) following 0.75 g/kg IP alcohol. Matthews et al. (1996) reported a reduction and smearing of place cell activity in animals performing a radial arm maze task following an IP injection of 2.0 g/kg. Alcohol's disruptive effects on the precision of place fields in hippocampal neurons were replicated (White and Best 2000) following 1.0 and 1.5 g/kg IP alcohol in animals performing a Y-maze task. Interestingly, these effects were limited to firing occurring in place fields and not detected outside them (White and Best 2000). Collectively, these studies highlight the impairing effects of alcohol on attaining modes of neural activity that are transiently elevated and rhythmically structured. However, they do not replicate the broad inhibitory effects observed on firing observed in diminished behavioral states. In these studies, attributing this effect to alcohol alone is complicated by a lack of detailed behavioral information about the state of the animal.

3.3 *Prefrontal Cortex*

A handful of studies have measured in vivo neural activity in the prefrontal cortex of rodents that are in a reduced state of vigilance. Similar to the hippocampus, one of the main findings that come from these data is that alcohol tends to depress neural activity. Dose-dependent reductions in neural activity were observed in anesthetized animals that were injected with increasing doses of alcohol (0.375, 0.75, 1.5, 3.5 g/kg IP) (Tu et al. 2007). These data are consistent with those from slice patch clamp electrophysiology experiments in the medial prefrontal cortex (mPFC) that find reductions in NMDA currents following alcohol administration (Jin and Woodward 2006; Weitlauf and Woodward 2008); however, some important caveats must be considered.

Consistent with data from anesthetized animals, in vivo calcium recordings from the lateral orbitofrontal cortex of freely moving mice found a reduction in the frequency of calcium transients following injection of 2.0 g/kg alcohol (Gioia and Woodward 2021). Electrophysiological recordings from the dorsomedial prefrontal cortex (dmPFC) of freely behaving rats consuming alcohol found no relationship between dose consumed and changes in firing rate (Linsenbardt and Lapish 2015), which is inconsistent with anesthetized data. A subsequent study examined neural activity in the dorsomedial prefrontal cortex following IP injections of identical doses of alcohol and different results were observed in awake, behaving versus anesthetized animals. In awake behaving animals' broad reductions in neural firing were not observed following IP injections of 1.0 or 2.0 g/kg (Morningstar et al. 2020). If anything, a modest increase firing was observed following 1.0 g/kg. With higher doses of injected alcohol (2.0 g/kg IP) decreases in neural activity were observed, but they were concomitant with periods of inactivity and an increase in low-frequency power (<5Hz)—thus indicating the animal was falling asleep. When analysis of neural activity was restricted to periods of vigilance, no decreases in neural activity were observed. These data clearly indicate that the effects of alcohol on neural activity in the awake behaving animal can be confounded by the transition to reduced states of vigilance. In addition, these data indicate that the effects of alcohol in anesthetized animals are likely confounded by interactions with the anesthesia. Therefore, as outlined previously, tracking these variables is critical when relating the effects of alcohol on neural activity to behavior.

There is correspondence on at least one key feature of neural activity following acute alcohol—alcohol clearly inhibits NMDA currents in neurons in cortical and subcortical structures (Jin and Woodward 2006; Li et al. 2002; Lovinger et al. 1990; Nie et al. 1994; Roberto et al. 2004; Weitlauf and Woodward 2008). NMDA receptors are slow to inactivate and, while activated, they hold the neuron in a depolarized state (Iacobucci and Popescu 2017). This can result in plateaus in a depolarized membrane potential accompanied by persistent epochs of burst firing referred to as up-states (Steriade et al. 1993). Alcohol, at doses as low as 25mM, reduces up-states both in vivo and ex vivo (Jin and Woodward 2006; Weitlauf and Woodward 2008). Morningstar et al (2020) showed that 1.0 g/kg alcohol increased

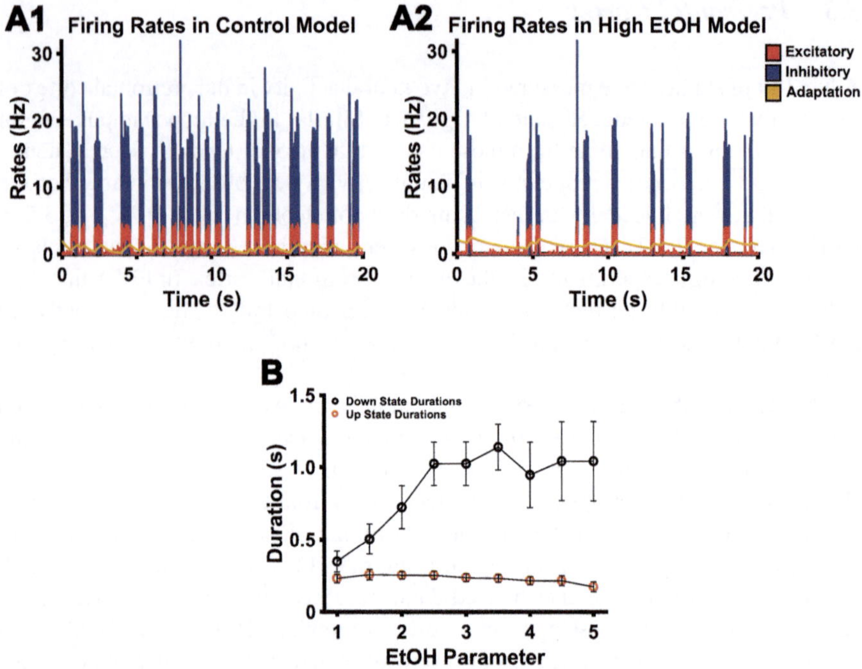

Fig. 2 *Simulations indicate that alcohol suppresses transitions from down-states to up-states.* A reduced model containing excitatory units (red lines), inhibitory units (blue lines), and the addition of an adaptation current (yellow line) recapitulated transitions between cortical down-states and up-states (**a**). The adaptation current was modified in a manner consistent with the effects of alcohol on NMDA adaptation and fewer up-states were observed (**b**). The duration of simulated down-states and up-states is quantified for increasing values of the adaptation current in the model, which models increasing alcohol concentrations (**c**). *Figure is adapted from* Morningstar et al. (2021), *Neuropharmacology*

the durations of down-states and impaired the transition of the down-state to the up-state, with no effect on firing rate during the up-state.

A series of simulations in Morningstar et al. (2021) explored how IP alcohol resulted in neural activity patterns that were less variable in mPFC and impaired transitions to an up-state (Fig. 2). This model used transitions between up- and down-states as a proxy for state remapping across neural ensembles, which is associated with encoding new rules in mPFC (Durstewitz et al. 2010) and context updates in other brain regions such as the hippocampus (Bulkin et al. 2016; Leutgeb et al. 2005). The effects of alcohol on calcium-dependent desensitization (CDD) were assessed. CDD is a slow variable that controls the time constant of NMDA currents—increasing CDD will reduce the open time of the NMDA receptor. Boikov et al. (2020) demonstrated effects of alcohol on NMDA are mediated by enhancing CDD. Further, the effects on NMDA could be seen at a low concentration of alcohol (9mM) in ex vivo slices. This slow variable was included in an established computational model that simulates the dynamics controlling up- and down-state transitions in the cortex (Jercog et al. 2017). Simulating the effects of alcohol on CDD in this model replicated the effects observed from in vivo neural recordings (Fig. 2a, b). The transition from the down-state to the up-state was impaired with no effect on the duration of up-states, therefore (Fig. 2c) suggesting a pivotal role of the NDMA receptor, and more specifically, CDD in networks that exhibit up-states (e.g., reduced state of vigilance). This model also supported the notion that the effect of alcohol is to reduce variability in neural activity in dmPFC neural firing (Morningstar et al. 2021).

Reductions in firing observed in the anesthetized animal are in line with effects on NMDA observed patch clamp electrophysiology experiments performed in the slice. The lack of correspondence between anesthetized and awake preparations in vivo, therefore, would seem to indicate that alcohol impacts other cell types/currents that may offset the effects of NMDA. Indeed, evidence is accumulating that alcohol impacts interneuron activity in PFC. In vivo data indicate that alcohol has a biphasic effect on interneuron activity in mPFC; a 1 g/kg dose increases activity, and at 2–3 g/kg doses, it decreases activity resulting in reductions and increases in pyramidal neuron activity, respectively (Li et al. 2021). However, this effect was not observed in ex vivo slices, thus further complicating the ability to translate the effects of alcohol between in vivo and ex vivo preparations.

4 Conclusions

When assessing the effects of alcohol in vivo it is critical that concomitant changes in behavior/vigilance can be ruled out as responsible for changes in neural activity as distinct from alcohol. This has been a consistent caveat for the field and requires attention in future experiments. With the advent of improved measures to monitor moment to moment changes in behavior (Mathis et al. 2018), and relating them to neural activity, eliminating this caveat is tractable.

Alcohol has an array of effects on the biophysical properties of neurons and understanding how these changes interact to alter computation is critical to understand how it alters behavior. The differences in the effects of alcohol in anesthetized versus awake animals indicate that the effects of alcohol are dependent on the state of cortical and hippocampal networks. When the network is in a high noise state and firing rates are variable, as observed when the animal is vigilant, this may limit the impact of alcohol's effects on NMDA. In this state, the impact of alcohol on interneurons and/or neuromodulators may become more prevalent.

When making inferences about the direct effects of alcohol on neural activity in corticolimbic circuits, it is important to consider that alcohol has direct and rapid effects on the dopamine neuron resulting in increases in firing rate (Morikawa and Morrisett 2010). Dopamine neurons innervate most of the forebrain (Cave and Baker 2009), and increases in dopamine efflux are observed in several forebrain regions following alcohol exposure. Since changes in dopamine efflux can powerfully alter the properties of neural networks (Seamans and Yang 2004) this likely obfuscates the detection of the direct effects of alcohol on neural circuit function and requires ways to parse these effects. Techniques to locally administer alcohol in cortical areas (Engleman et al. 2020) while simultaneously monitoring neural activity in vivo (Panin et al. 2012) will likely be required to determine the local effects of alcohol in the cortex.

The effects of alcohol on reducing the probability of transitioning into an up-state in cortical recordings and simulations are reminiscent of the effects observed in dopamine neurons. The transition from a Type I to a Type II encoder in DA neurons indicates that a stronger input is needed to make the neuron fire. In fact, both results indicate that stronger inputs are required to drive the neuron to fire. However, these results also indicate that alcohol restricts the ability of neural networks to transition into transiently elevated (yet stable) periods of activity, which may be attributable to a different mechanism than just changes in firing at the single neuron level.

To understand how alcohol, even at low doses, can impact behavior, it will be critical to more directly explore how alcohol impacts network level dynamics (e.g., attractors; Durstewitz et al. 2020; Lapish et al. 2015). Network level dynamics are hypothesized to be critical for complex aspects of cognitive function such as attention and decision-making (Dubreuil et al. 2022; Lapish et al. 2015; Mante et al. 2013). Therefore, articulating precisely how alcohol impacts network dynamics and the computations they carry out will provide an explanation for how alcohol negatively impacts these behaviors. Furthermore, articulating how alcohol impacts network dynamics will provide insight into the changes in synaptic and membrane properties at the single neuron level that are critical for changes in behavior.

When using rodent models to infer how alcohol might impact computation in the human brain, it will be critical to understand what aspects of computation are unique to the human brain. If similar computations can be identified in each species, it allows framing the question of how brain function is impaired in terms of how each species implements these computations (Redish et al. 2022). Doing this minimizes the assumption that the pathological function of a human brain region can be modeled as a pathological function of an evolutionary prologue in a homologous rodent brain region. More importantly, if similar computations can be identified across species, this facilitates the construction of predictive models that capture the effects of alcohol and therefore the means to test novel treatment strategies.

References

Alger BE, Nicoll RA (1979) GABA-mediated biphasic inhibitory responses in hippocampus. Nature 281:315–317. https://doi.org/10.1038/281315a0

Andersen P, Dingledine R, Gjerstad L, Langmoen IA, Laursen AM (1980) Two different responses of hippocampal pyramidal cells to application of gamma-amino butyric acid. J Physiol 305:279–296. https://doi.org/10.1113/jphysiol.1980.sp013363

Bagur S, Lacroix MM, de Lavilléon G, Lefort JM, Geoffroy H, Benchenane K (2018) Harnessing olfactory bulb oscillations to perform fully brain-based sleep-scoring and real-time monitoring of anaesthesia depth. PLoS Biol 16:e2005458. https://doi.org/10.1371/journal.pbio.2005458

Bassareo V, Di Chiara G (1999) Modulation of feeding-induced activation of mesolimbic dopamine transmission by appetitive stimuli and its relation to motivational state. Eur J Neurosci 11:4389–4397. https://doi.org/10.1046/j.1460-9568.1999.00843.x

Bell RL, Hauser SR, Liang T, Sari Y, Maldonado-Devincci A, Rodd ZA (2017) Rat animal models for screening medications to treat alcohol use disorders. Neuropharmacology 122:201–243. https://doi.org/10.1016/j.neuropharm.2017.02.004

Birak KS, Higgs S, Terry P (2011) Conditioned tolerance to the effects of alcohol on inhibitory control in humans. Alcohol Alcohol 46:686–693. https://doi.org/10.1093/alcalc/agr084

Boikov SI, Sibarov DA, Antonov SM (2020) Ethanol inhibition of NMDA receptors in calcium-dependent and -independent modes. Biochem Biophys Res Commun 522:1046–1051. https://doi.org/10.1016/j.bbrc.2019.12.007

Brabant C, Guarnieri DJ, Quertemont E (2014) Stimulant and motivational effects of alcohol: lessons from rodent and primate models. Pharmacol Biochem Behav 122:37–52. https://doi.org/10.1016/j.pbb.2014.03.006

Bulkin DA, Law LM, Smith DM (2016) Placing memories in context: hippocampal representations promote retrieval of appropriate memories. Hippocampus 26:958–971. https://doi.org/10.1002/hipo.22579

Cagniard B, Balsam PD, Brunner D, Zhuang X (2006) Mice with chronically elevated dopamine exhibit enhanced motivation, but not learning, for a food reward. Neuropsychopharmacology 31:1362–1370. https://doi.org/10.1038/sj.npp.1300966

Carnicella S, Ron D, Barak S (2014) Intermittent ethanol access schedule in rats as a preclinical model of alcohol abuse. Alcohol 48:243–252. https://doi.org/10.1016/j.alcohol.2014.01.006

Carrigan MA, Uryasev O, Frye CB, Eckman BL, Myers CR, Hurley TD, Benner SA (2015) Hominids adapted to metabolize ethanol long before human-directed fermentation. Proc Natl Acad Sci U S A 112:458–463. https://doi.org/10.1073/pnas.1404167111

Cave JW, Baker H (2009) Dopamine systems in the forebrain. Adv Exp Med Biol 651:15–35. https://doi.org/10.1007/978-1-4419-0322-8_2

Christie MJ, Bridge S, James LB, Beart PM (1985) Excitotoxin lesions suggest an aspartatergic projection from rat medial prefrontal cortex to ventral tegmental area. Brain Res 333:169–172. https://doi.org/10.1016/0006-8993(85)90140-4

Coizet V, Comoli E, Westby GWM, Redgrave P (2003) Phasic activation of substantia nigra and the ventral tegmental area by chemical stimulation of the superior colliculus: an electrophysiological investigation in the rat. Eur J Neurosci 17:28–40. https://doi.org/10.1046/j.1460-9568.2003.02415.x

Collins GB, Brosnihan KB, Zuti RA, Messina M, Gupta MK (1992) Neuroendocrine, fluid balance, and thirst responses to alcohol in alcoholics. Alcohol Clin Exp Res 16:228–233. https://doi.org/10.1111/j.1530-0277.1992.tb01368.x

Cunningham CL (1994) Modulation of ethanol reinforcement by conditioned hyperthermia. Psychopharmacology (Berl) 115:79–85. https://doi.org/10.1007/BF02244755

Doyon WM, Anders SK, Ramachandra VS, Czachowski CL, Gonzales RA (2005) Effect of operant self-administration of 10% ethanol plus 10% sucrose on dopamine and ethanol concentrations in the nucleus accumbens. J Neurochem 93:1469–1481. https://doi.org/10.1111/j.1471-4159.2005.03137.x

Dubreuil A, Valente A, Beiran M, Mastrogiuseppe F, Ostojic S (2022) The role of population structure in computations through neural dynamics. Nat Neurosci 25:783–794. https://doi.org/10.1038/s41593-022-01088-4

Duncan PM, Alici T, Woodward JD (2000) Conditioned compensatory response to ethanol as indicated by locomotor activity in rats. Behav Pharmacol 11:395–402. https://doi.org/10.1097/00008877-200008000-00005

Durstewitz D, Vittoz NM, Floresco SB, Seamans JK (2010) Abrupt transitions between prefrontal neural ensemble states accompany behavioral transitions during rule learning. Neuron 66:438–448. https://doi.org/10.1016/j.neuron.2010.03.029

Durstewitz D, Huys QJM, Koppe G (2020) Psychiatric illnesses as disorders of network dynamics. Biol Psychiatry Cogn Neurosci Neuroimaging. https://doi.org/10.1016/j.bpsc.2020.01.001

Edenberg HJ (2007) The genetics of alcohol metabolism: role of alcohol dehydrogenase and aldehyde dehydrogenase variants. Alcohol Res Health 30:5–13

Elvig SK, McGinn MA, Smith C, Arends MA, Koob GF, Vendruscolo LF (2021) Tolerance to alcohol: a critical yet understudied factor in alcohol addiction. Pharmacol Biochem Behav 204: 173155. https://doi.org/10.1016/j.pbb.2021.173155

Engleman EA, Ingraham CM, Rodd ZA, Murphy JM, McBride WJ, Ding Z-M (2020) The reinforcing effects of ethanol within the prelimbic cortex and ethanol drinking: involvement of local dopamine D2 receptor-mediated neurotransmission. Drug Alcohol Depend 214:108165. https://doi.org/10.1016/j.drugalcdep.2020.108165

Eshel N, Tian J, Bukwich M, Uchida N (2016) Dopamine neurons share common response function for reward prediction error. Nat Neurosci 19(3):479–86. https://doi.org/10.1038/nn.4239. Epub 2016 Feb 8. PMID: 26854803; PMCID: PMC4767554

Fiorillo CD, Tobler PN, Schultz W (2003) Discrete coding of reward probability and uncertainty by dopamine neurons. Science 299:1898–1902. https://doi.org/10.1126/science.1077349

Floresco SB, West AR, Ash B, Moore H, Grace AA (2003) Afferent modulation of dopamine neuron firing differentially regulates tonic and phasic dopamine transmission. Nat Neurosci 6: 968–973. https://doi.org/10.1038/nn1103

Gass JT, Glen WB, McGonigal JT, Trantham-Davidson H, Lopez MF, Randall PK, Yaxley R, Floresco SB, Chandler LJ (2014) Adolescent alcohol exposure reduces behavioral flexibility, promotes disinhibition, and increases resistance to extinction of ethanol self-administration in adulthood. Neuropsychopharmacology 39:2570–2583. https://doi.org/10.1038/npp.2014.109

Geisler S, Wise RA (2008) Functional implications of glutamatergic projections to the ventral tegmental area. Rev Neurosci 19:227–244. https://doi.org/10.1515/revneuro.2008.19.4-5.227

Gessa GL, Muntoni F, Collu M, Vargiu L, Mereu G (1985) Low doses of ethanol activate dopaminergic neurons in the ventral tegmental area. Brain Res 348:201–203. https://doi.org/10.1016/0006-8993(85)90381-6

Gioia DA, Woodward JJ (2021) Altered activity of lateral orbitofrontal cortex neurons in mice following chronic intermittent ethanol exposure. eNeuro 8:ENEURO.0503-20.2021. https://doi.org/10.1523/ENEURO.0503-20.2021

Givens B (1995) Low doses of ethanol impair spatial working memory and reduce hippocampal theta activity. Alcohol Clin Exp Res 19:763–767. https://doi.org/10.1111/j.1530-0277.1995.tb01580.x

Givens B (1996) Behavioral correlates of single units in the medial septal area: the effect of ethanol. Neuroscience 71:417–427. https://doi.org/10.1016/0306-4522(95)00443-2

Grace AA, Bunney BS (1984) The control of firing pattern in nigral dopamine neurons: burst firing. J Neurosci 4:2877–2890

Grace AA, Onn SP (1989) Morphology and electrophysiological properties of immunocytochemically identified rat dopamine neurons recorded in vitro. J Neurosci 9:3463–3481

Griffin WC (2014) Alcohol dependence and free-choice drinking in mice. Alcohol 48:287–293. https://doi.org/10.1016/j.alcohol.2013.11.006

Grupp LA (1980) Biphasic action of ethanol on single units of the dorsal hippocampus and the relationship to the cortical EEG. Psychopharmacology (Berl) 70:95–103. https://doi.org/10.1007/BF00432377

Gulledge AT, Stuart GJ (2003) Excitatory actions of GABA in the cortex. Neuron 37:299–309. https://doi.org/10.1016/s0896-6273(02)01146-7

Gursky ZH, Savage LM, Klintsova AY (2021) Executive functioning-specific behavioral impairments in a rat model of human third trimester binge drinking implicate prefrontal-thalamo-hippocampal circuitry in fetal alcohol spectrum disorders. Behav Brain Res 405:113208. https://doi.org/10.1016/j.bbr.2021.113208

Heit C, Dong H, Chen Y, Shah YM, Thompson DC, Vasiliou V (2015) Transgenic mouse models for alcohol metabolism, toxicity, and cancer. Adv Exp Med Biol 815:375–387. https://doi.org/10.1007/978-3-319-09614-8_22

Holmes RS, Duley JA, Algar EM, Mather PB, Rout UK (1986) Biochemical and genetic studies on enzymes of alcohol metabolism: the mouse as a model organism for human studies. Alcohol Alcohol 21:41–56

Höög J-O, Ostberg LJ (2011) Mammalian alcohol dehydrogenases – a comparative investigation at gene and protein levels. Chem Biol Interact 191:2–7. https://doi.org/10.1016/j.cbi.2011.01.028

Iacobucci GJ, Popescu GK (2017) NMDA receptors: linking physiological output to biophysical operation. Nat Rev Neurosci 18:236–249. https://doi.org/10.1038/nrn.2017.24

Inenaga K, Ono K, Hitomi S, Kuroki A, Ujihara I (2017) Thirst sensation and oral dryness following alcohol intake. Jpn Dent Sci Rev 53:78–85. https://doi.org/10.1016/j.jdsr.2016.12.001

Izhikevich EM (2006) Dynamical systems in neuroscience: the geometry of excitability and bursting, 1st edn. The MIT Press, Cambridge

Jahn H, Döring WKH, Krampe H, Sieg S, Werner C, Poser W, Brunner E, Ehrenreich H (2004) Preserved vasopressin response to osmostimulation despite decreased basal vasopressin levels in long-term abstinent alcoholics. Alcohol Clin Exp Res 28:1925–1930. https://doi.org/10.1097/01.alc.0000148110.34917.c3

Jercog D, Roxin A, Barthó P, Luczak A, Compte A, de la Rocha J (2017) Up-down cortical dynamics reflect state transitions in a bistable network. Elife 6:e22425. https://doi.org/10.7554/eLife.22425

Jin C, Woodward JJ (2006) Effects of 8 different NR1 splice variants on the ethanol inhibition of recombinant NMDA receptors. Alcohol Clin Exp Res 30:673–679. https://doi.org/10.1111/j.1530-0277.2006.00079.x

Jin S, Cao Q, Yang F, Zhu H, Xu S, Chen Q, Wang Z, Lin Y, Cinar R, Pawlosky RJ, Zhang Y, Xiong W, Gao B, Koob GF, Lovinger DM, Zhang L (2021) Brain ethanol metabolism by astrocytic ALDH2 drives the behavioural effects of ethanol intoxication. Nat Metab 3:337–351. https://doi.org/10.1038/s42255-021-00357-z

Johnson SW, North RA (1992) Two types of neurone in the rat ventral tegmental area and their synaptic inputs. J Physiol 450:455–468. https://doi.org/10.1113/jphysiol.1992.sp019136

Kalant H (1998) Research on tolerance: what can we learn from history? Alcohol Clin Exp Res 22: 67–76. https://doi.org/10.1111/j.1530-0277.1998.tb03618.x

Kiefer F, Andersohn F, Jahn H, Wolf K, Raedler TJ, Wiedemann K (2002) Involvement of plasma atrial natriuretic peptide in protracted alcohol withdrawal. Acta Psychiatr Scand 105:65–70. https://doi.org/10.1034/j.1600-0447.2002.0_011.x

Klimesch W (2012) α-band oscillations, attention, and controlled access to stored information. Trends Cogn Sci 16:606–617. https://doi.org/10.1016/j.tics.2012.10.007

Koob GF, Volkow ND (2016) Neurobiology of addiction: a neurocircuitry analysis. Lancet Psychiatry 3:760–773. https://doi.org/10.1016/S2215-0366(16)00104-8

Kroener S, Mulholland PJ, New NN, Gass JT, Becker HC, Chandler LJ (2012) Chronic alcohol exposure alters behavioral and synaptic plasticity of the rodent prefrontal cortex. PLoS One 7: e37541. https://doi.org/10.1371/journal.pone.0037541

Lapish CC, Seamans JK, Chandler LJ (2006) Glutamate-dopamine cotransmission and reward processing in addiction. Alcohol Clin Exp Res 30:1451–1465. https://doi.org/10.1111/j.1530-0277.2006.00176.x

Lapish CC, Balaguer-Ballester E, Seamans JK, Phillips AG, Durstewitz D (2015) Amphetamine exerts dose-dependent changes in prefrontal cortex attractor dynamics during working memory. J Neurosci 35:10172–10187. https://doi.org/10.1523/JNEUROSCI.2421-14.2015

Leib DE, Zimmerman CA, Knight ZA (2016) Thirst. Curr Biol 26:R1260–R1265. https://doi.org/10.1016/j.cub.2016.11.019

Leutgeb JK, Leutgeb S, Treves A, Meyer R, Barnes CA, McNaughton BL, Moser M-B, Moser EI (2005) Progressive transformation of hippocampal neuronal representations in "morphed" environments. Neuron 48:345–358. https://doi.org/10.1016/j.neuron.2005.09.007

Li Q, Wilson WA, Swartzwelder HS (2002) Differential effect of ethanol on NMDA EPSCs in pyramidal cells in the posterior cingulate cortex of juvenile and adult rats. J Neurophysiol 87: 705–711. https://doi.org/10.1152/jn.00433.2001

Li M, Cabrera-Garcia D, Salling MC, Au E, Yang G, Harrison NL (2021) Alcohol reduces the activity of somatostatin interneurons in the mouse prefrontal cortex: a neural basis for its disinhibitory effect? Neuropharmacology 188:108501. https://doi.org/10.1016/j.neuropharm.2021.108501

Linsenbardt DN, Lapish CC (2015) Neural firing in the prefrontal cortex during alcohol intake in alcohol-preferring "P" versus wistar rats. Alcohol Clin Exp Res 39:1642–1653. https://doi.org/10.1111/acer.12804

Lovinger DM, White G, Weight FF (1990) NMDA receptor-mediated synaptic excitation selectively inhibited by ethanol in hippocampal slice from adult rat. J Neurosci 10:1372–1379

Magee JC (2000) Dendritic integration of excitatory synaptic input. Nat Rev Neurosci 1:181–190. https://doi.org/10.1038/35044552

Mante V, Sussillo D, Shenoy KV, Newsome WT (2013) Context-dependent computation by recurrent dynamics in prefrontal cortex. Nature 503:78–84. https://doi.org/10.1038/nature12742

Margolis EB, Lock H, Hjelmstad GO, Fields HL (2006) The ventral tegmental area revisited: is there an electrophysiological marker for dopaminergic neurons? J Physiol 577:907–924. https://doi.org/10.1113/jphysiol.2006.117069

Mathis A, Mamidanna P, Cury KM, Abe T, Murthy VN, Mathis MW, Bethge M (2018) DeepLabCut: markerless pose estimation of user-defined body parts with deep learning. Nat Neurosci 21:1281–1289. https://doi.org/10.1038/s41593-018-0209-y

Matson L, Liangpunsakul S, Crabb D, Buckingham A, Ross RA, Halcomb M, Grahame N (2013) Chronic free-choice drinking in crossed high alcohol preferring mice leads to sustained blood ethanol levels and metabolic tolerance without evidence of liver damage. Alcohol Clin Exp Res 37:194–201. https://doi.org/10.1111/j.1530-0277.2012.01873.x

Matthews DB, Simson PE, Best PJ (1996) Ethanol alters spatial processing of hippocampal place cells: a mechanism for impaired navigation when intoxicated. Alcohol Clin Exp Res 20:404–407. https://doi.org/10.1111/j.1530-0277.1996.tb01660.x

Matthews DB, Novier A, Diaz-Granados JL, Van Skike CE, Ornelas L, Mittleman G (2017) Impact of adolescent alcohol use across the lifespan: long-lasting tolerance to high-dose alcohol coupled with potentiated spatial memory impairments to moderate-dose alcohol. Alcohol 61: 33–42. https://doi.org/10.1016/j.alcohol.2017.01.012

McCaul ME, Turkkan JS, Stitzer ML (1989) Conditioned opponent responses: effects of placebo challenge in alcoholic subjects. Alcohol Clin Exp Res 13:631–635. https://doi.org/10.1111/j.1530-0277.1989.tb00395.x

McCusker CG, Brown K (1990) Alcohol-predictive cues enhance tolerance to and precipitate "craving" for alcohol in social drinkers. J Stud Alcohol 51:494–499. https://doi.org/10.15288/jsa.1990.51.494

Meinhardt MW, Sommer WH (2015) Postdependent state in rats as a model for medication development in alcoholism. Addict Biol 20:1–21. https://doi.org/10.1111/adb.12187

Mereu G, Fadda F, Gessa GL (1984) Ethanol stimulates the firing rate of nigral dopaminergic neurons in unanesthetized rats. Brain Res 292:63–69. https://doi.org/10.1016/0006-8993(84)90890-4

Mohebi A, Pettibone JR, Hamid AA, Wong J-MT, Vinson LT, Patriarchi T, Tian L, Kennedy RT, Berke JD (2019) Dissociable dopamine dynamics for learning and motivation. Nature 570:65–70. https://doi.org/10.1038/s41586-019-1235-y

Morikawa H, Morrisett RA (2010) Ethanol action on dopaminergic neurons in the ventral tegmental area: interaction with intrinsic ion channels and neurotransmitter inputs. Int Rev Neurobiol 91:235–288. https://doi.org/10.1016/S0074-7742(10)91008-8

Morningstar MD, Linsenbardt DN, Lapish CC (2020) Ethanol alters variability, but not rate, of firing in medial prefrontal cortex neurons of awake, behaving rats. Alcohol Clin Exp Res. https://doi.org/10.1111/acer.14463

Morningstar MD, Barnett WH, Goodlett CR, Kuznetsov A, Lapish CC (2021) Understanding ethanol's acute effects on medial prefrontal cortex neural activity using state-space approaches. Neuropharmacology 198:108780. https://doi.org/10.1016/j.neuropharm.2021.108780

Morozova EO, Myroshnychenko M, Zakharov D, di Volo M, Gutkin B, Lapish CC, Kuznetsov A (2016a) Contribution of synchronized GABAergic neurons to dopaminergic neuron firing and bursting. J Neurophysiol 116:1900–1923. https://doi.org/10.1152/jn.00232.2016

Morozova EO, Zakharov D, Gutkin BS, Lapish CC, Kuznetsov A (2016b) Dopamine neurons change the type of excitability in response to stimuli. PLoS Comput Biol 12:e1005233. https://doi.org/10.1371/journal.pcbi.1005233

Mutschler J, Bilbao A, von der Goltz C, Demiralay C, Jahn H, Wiedemann K, Spanagel R, Kiefer F (2010) Augmented stress-induced alcohol drinking and withdrawal in mice lacking functional natriuretic peptide-A receptors. Alcohol Alcohol 45:13–16. https://doi.org/10.1093/alcalc/agp065

Nentwig TB, Starr EM, Chandler LJ, Glover EJ (2019) Absence of compulsive drinking phenotype in adult male rats exposed to ethanol in a binge-like pattern during adolescence. Alcohol 79:93–103. https://doi.org/10.1016/j.alcohol.2019.01.006

Nie Z, Madamba SG, Siggins GR (1994) Ethanol inhibits glutamatergic neurotransmission in nucleus accumbens neurons by multiple mechanisms. J Pharmacol Exp Ther 271:1566–1573

Niv Y, Daw ND, Joel D, Dayan P (2007) Tonic dopamine: opportunity costs and the control of response vigor. Psychopharmacology (Berl) 191:507–520. https://doi.org/10.1007/s00213-006-0502-4

Omelchenko N, Sesack SR (2009) Ultrastructural analysis of local collaterals of rat ventral tegmental area neurons: GABA phenotype and synapses onto dopamine and GABA cells. Synapse 63:895–906. https://doi.org/10.1002/syn.20668

Panin F, Cathala A, Piazza PV, Spampinato U (2012) Coupled intracerebral microdialysis and electrophysiology for the assessment of dopamine neuron function in vivo. J Pharmacol Toxicol Methods 65:83–92. https://doi.org/10.1016/j.vascn.2012.01.003

Pietrzykowski AZ, Treistman SN (2008) The molecular basis of tolerance. Alcohol Res Health 31: 298–309

Redish AD, Kepecs A, Anderson LM, Calvin OL, Grissom NM, Haynos AF, Heilbronner SR, Herman AB, Jacob S, Ma S, Vilares I, Vinogradov S, Walters CJ, Widge AS, Zick JL, Zilverstand A (2022) Computational validity: using computation to translate behaviours across species. Philos Trans R Soc Lond B Biol Sci 377:20200525. https://doi.org/10.1098/rstb.2020.0525

Riley EP, Worsham ED, Lester D, Freed EX (1977) Selective breeding of rats for differences in reactivity to alcohol. An approach to an animal model of alcoholism. II. Behavioral measures. J Stud Alcohol 38:1705–1717. https://doi.org/10.15288/jsa.1977.38.1705

Rimondini R, Sommer WH, Dall'Olio R, Heilig M (2008) Long-lasting tolerance to alcohol following a history of dependence. Addict Biol 13:26–30. https://doi.org/10.1111/j.1369-1600.2007.00079.x

Roberto M, Schweitzer P, Madamba SG, Stouffer DG, Parsons LH, Siggins GR (2004) Acute and chronic ethanol alter glutamatergic transmission in rat central amygdala: an in vitro and in vivo analysis. J Neurosci 24:1594–1603. https://doi.org/10.1523/JNEUROSCI.5077-03.2004

Rogers J, Wiener SG, Bloom FE (1979) Long-term ethanol administration methods for rats: advantages of inhalation over intubation or liquid diets. Behav Neural Biol 27:466–486. https://doi.org/10.1016/s0163-1047(79)92061-2

Schultz W (2016) Dopamine reward prediction error coding. Dialogues Clin Neurosci 18:23–32

Seamans JK, Yang CR (2004) The principal features and mechanisms of dopamine modulation in the prefrontal cortex. Prog Neurobiol 74:1–58. https://doi.org/10.1016/j.pneurobio.2004.05.006

Sesack SR, Pickel VM (1992) Prefrontal cortical efferents in the rat synapse on unlabeled neuronal targets of catecholamine terminals in the nucleus accumbens septi and on dopamine neurons in the ventral tegmental area. J Comp Neurol 320:145–160. https://doi.org/10.1002/cne.903200202

Shah MM (2014) Cortical HCN channels: function, trafficking and plasticity. J Physiol 592:2711–2719. https://doi.org/10.1113/jphysiol.2013.270058

Sinha R, Fox HC, Hong K-IA, Hansen J, Tuit K, Kreek MJ (2011) Effects of adrenal sensitivity, stress- and cue-induced craving, and anxiety on subsequent alcohol relapse and treatment outcomes. Arch Gen Psychiatry 68:942–952. https://doi.org/10.1001/archgenpsychiatry.2011.49

Sommer WH, Rimondini R, Hansson AC, Hipskind PA, Gehlert DR, Barr CS, Heilig MA (2008) Upregulation of voluntary alcohol intake, behavioral sensitivity to stress, and amygdala crhr1 expression following a history of dependence. Biol Psychiatry 63:139–145. https://doi.org/10.1016/j.biopsych.2007.01.010

Spear LP (2014) Adolescents and alcohol: acute sensitivities, enhanced intake, and later consequences. Neurotoxicol Teratol 41:51–59. https://doi.org/10.1016/j.ntt.2013.11.006

Spear LP, Swartzwelder HS (2014) Adolescent alcohol exposure and persistence of adolescent-typical phenotypes into adulthood: a mini-review. Neurosci Biobehav Rev 45:1–8. https://doi.org/10.1016/j.neubiorev.2014.04.012

Steffensen SC, Svingos AL, Pickel VM, Henriksen SJ (1998) Electrophysiological characterization of GABAergic neurons in the ventral tegmental area. J Neurosci 18:8003–8015

Steriade M, Nuñez A, Amzica F (1993) A novel slow (<1 Hz) oscillation of neocortical neurons in vivo: depolarizing and hyperpolarizing components. J Neurosci 13:3252–3265

Tabakoff B, Hoffman PL (2000) Animal models in alcohol research. Alcohol Res Health 24:77–84

Thompson CE, Freitas LB, Salzano FM (2018) Molecular evolution and functional divergence of alcohol dehydrogenases in animals, fungi and plants. Genet Mol Biol 41:341–354. https://doi.org/10.1590/1678-4685-GMB-2017-0047

Threlfell S, Lalic T, Platt NJ, Jennings KA, Deisseroth K, Cragg SJ (2012) Striatal dopamine release is triggered by synchronized activity in cholinergic interneurons. Neuron 75:58–64. https://doi.org/10.1016/j.neuron.2012.04.038

Trantham-Davidson H, Burnett EJ, Gass JT, Lopez MF, Mulholland PJ, Centanni SW, Floresco SB, Chandler LJ (2014) Chronic alcohol disrupts dopamine receptor activity and the cognitive function of the medial prefrontal cortex. J Neurosci 34:3706–3718. https://doi.org/10.1523/JNEUROSCI.0623-13.2014

Tu Y, Kroener S, Abernathy K, Lapish C, Seamans J, Chandler LJ, Woodward JJ (2007) Ethanol inhibits persistent activity in prefrontal cortical neurons. J Neurosci 27:4765–4775. https://doi.org/10.1523/JNEUROSCI.5378-06.2007

Van Skike CE, Goodlett C, Matthews DB (2019) Acute alcohol and cognition: remembering what it causes us to forget. Alcohol 79:105–125. https://doi.org/10.1016/j.alcohol.2019.03.006

van Zessen R, Phillips JL, Budygin EA, Stuber GD (2012) Activation of VTA GABA neurons disrupts reward consumption. Neuron 73:1184–1194. https://doi.org/10.1016/j.neuron.2012.02.016

Vendruscolo LF, Roberts AJ (2014) Operant alcohol self-administration in dependent rats: focus on the vapor model. Alcohol 48:277–286. https://doi.org/10.1016/j.alcohol.2013.08.006

Venton BJ, Zhang H, Garris PA, Phillips PEM, Sulzer D, Wightman RM (2003) Real-time decoding of dopamine concentration changes in the caudate-putamen during tonic and phasic firing. J Neurochem 87:1284–1295. https://doi.org/10.1046/j.1471-4159.2003.02109.x

Vogel-Sprott M (1997) Is behavioral tolerance learned? Alcohol Health Res World 21:161–168

von der Goltz C, Jahn H, Mutschler J, Wiedemann K, Kiefer F (2014) Intraperitoneal atrial natriuretic peptide attenuates anxiety-related behaviour during alcohol withdrawal in mice. Pharmacopsychiatry 47:97–100. https://doi.org/10.1055/s-0034-1372645

Vonghia L, Leggio L, Ferrulli A, Bertini M, Gasbarrini G, Addolorato G, Alcoholism Treatment Study Group (2008) Acute alcohol intoxication. Eur J Intern Med 19:561–567. https://doi.org/10.1016/j.ejim.2007.06.033

Weitlauf C, Woodward JJ (2008) Ethanol selectively attenuates NMDAR-mediated synaptic transmission in the prefrontal cortex. Alcohol Clin Exp Res 32:690–698. https://doi.org/10.1111/j.1530-0277.2008.00625.x

White AM, Best PJ (2000) Effects of ethanol on hippocampal place-cell and interneuron activity. Brain Res 876:154–165. https://doi.org/10.1016/s0006-8993(00)02629-9

White AM, Roberts DC, Best PJ (2002) Context-specific tolerance to the ataxic effects of alcohol. Pharmacol Biochem Behav 72:107–110. https://doi.org/10.1016/s0091-3057(01)00731-6

Woodruff AR, Monyer H, Sah P (2006) GABAergic excitation in the basolateral amygdala. J Neurosci 26:11881–11887. https://doi.org/10.1523/JNEUROSCI.3389-06.2006

Zhang Y, Yu H, Li W, Yang Y, Wang X, Qian Z (2016) Effect of acute ethanol administration on the hippocampal region neural activity using a microelectrode array. Alcohol Clin Exp Res 40:1857–1864. https://doi.org/10.1111/acer.13144

GPSR Compliance

The European Union's (EU) General Product Safety Regulation (GPSR) is a set of rules that requires consumer products to be safe and our obligations to ensure this.

If you have any concerns about our products, you can contact us on ProductSafety@springernature.com

In case Publisher is established outside the EU, the EU authorized representative is:

Springer Nature Customer Service Center GmbH
Europaplatz 3
69115 Heidelberg, Germany

Batch number: 09003054

Printed by Printforce, the Netherlands